T0291818

CAMBRIDGE LIBRARY COLLECTION

Books of enduring scholarly value

Earth Sciences

In the nineteenth century, geology emerged as a distinct academic discipline. It pointed the way towards the theory of evolution, as scientists including Gideon Mantell, Adam Sedgwick, Charles Lyell and Roderick Murchison began to use the evidence of minerals, rock formations and fossils to demonstrate that the earth was older by millions of years than the conventional, Bible-based wisdom had supposed. They argued convincingly that the climate, flora and fauna of the distant past could be deduced from geological evidence. Volcanic activity, the formation of mountains, and the action of glaciers and rivers, tides and ocean currents also became better understood. This series includes landmark publications by pioneers of the modern earth sciences, who advanced the scientific understanding of our planet and the processes by which it is constantly re-shaped.

Fossil Plants

A.C. Seward (1863–1941) was an eminent English geologist and botanist who pioneered the study of palaeobotany. After graduating from St John's College, Cambridge, in 1886 Seward was appointed a University Lecturer in Botany in 1890. In 1898 he was elected a Fellow of the Royal Society, and was appointed Professor of Botany in 1906. These volumes, published to great acclaim between 1898 and 1919, provide a detailed discussion and study of an emerging science. In the early nineteenth century, research and critical literature concerning palaeobotany was scattered across disciplines. In these volumes Seward synthesised and revised this research and also included a substantial amount of new material. Furnished with concise descriptions of fossil plants, detailed figures and extensive bibliographies these volumes became the standard reference for palaeobotany well into the twentieth century. Volume 4, first published in 1919, contains systematic descriptions of fossil ginkgoes and conifers.

CAMBRIDGE UNIVERSITY PRESS

Cambridge, New York, Melbourne, Madrid, Cape Town, Singapore,
São Paolo, Delhi, Dubai, Tokyo, Mexico City

Published in the United States of America by Cambridge University Press, New York

www.cambridge.org
Information on this title: www.cambridge.org/9781108015981

© in this compilation Cambridge University Press 2010

This edition first published 1919
This digitally printed version 2010

ISBN 978-1-108-01598-1 Paperback

CAMBRIDGE BIOLOGICAL SERIES

FOSSIL PLANTS

CAMBRIDGE UNIVERSITY PRESS

C. F. CLAY, Manager

London: FETTER LANE, E.C. 4

London: H. K. LEWIS AND CO, Ltd., 136, GOWER STREET, W.C. 1
London: WILLIAM WESLEY AND SON, 28, ESSEX STREET, STRAND, W.C. 2
New York: G. P. PUTNAM'S SONS
Bombay, Calcutta and Madras: MACMILLAN AND CO., Ltd.
Toronto: J. M. DENT AND SONS, Ltd.
Tokyo: THE MARUZEN-KABUSHIKI-KAISHA

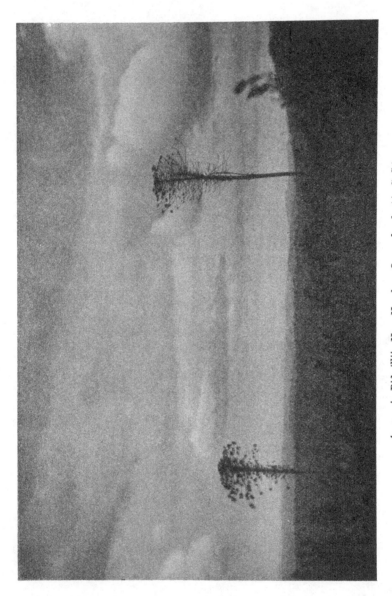

Araucaria Bidwillii. Near Nambur, Queensland. (A. C. S.)

FOSSIL PLANTS

A TEXT-BOOK FOR STUDENTS
OF BOTANY AND GEOLOGY

BY

A. C. SEWARD
M.A., F.R.S., Hon. Sc.D. Dublin

PROFESSOR OF BOTANY IN THE UNIVERSITY;
MASTER OF DOWNING COLLEGE AND HONORARY
FELLOW OF EMMANUEL COLLEGE, CAMBRIDGE

WITH 190 ILLUSTRATIONS

VOLUME IV

GINKGOALES, CONIFERALES, GNETALES

CAMBRIDGE:
AT THE UNIVERSITY PRESS
1919

PREFACE

IN the Preface to Volume III I expressed my thanks for help given to me by many friends in the course of the preparation of the subject-matter of Volumes III and IV, but Dr Scott has again earned my gratitude by very willingly and to very good purpose continuing the tedious task of reading the proofs. It is also a pleasure to acknowledge the help received from the Staff of the University Press.

Since the publication of Volume III Palaeobotany has been deprived of the services of three senior investigators, Professor C. E. Bertrand of Lille, Monsieur Grand'Eury, and Mr Clement Reid, men whose researches along different lines of inquiry have played a prominent part in the progress of the science during the last few decades. By the death of Miss Ruth Holden, a graduate of Harvard University and a Fellow of Newnham College, Cambridge, Palaeobotany has lost an unusually gifted and promising worker: though a citizen of a country which was then neutral her strong sense of duty led her to lay aside, temporarily as we hoped, botanical research for work with a British Medical Unit in Russia where she died in April of last year. Miss Holden's last contribution to Palaeobotany ('On the Anatomy of two Palaeozoic stems from India'; *Annals of Botany*, vol. XXXI. p. 315, 1917) was published too late to be considered in Volume III.

If it is possible to carry out my intention of supplementing the descriptive treatment of plants, which forms the basis of Volumes I–IV, by a general review of the Floras of the Past the results will be published as an independent work more intelligible, I hope, to the general reader than the text-book which, with a certain sense of relief, is now brought to a conclusion.

A. C. SEWARD.

BOTANY SCHOOL, CAMBRIDGE,
May, 1918.

POSTSCRIPT

THROUGH the death of Dr Newell Arber in June of last year at the comparatively early age of forty-seven Botany and Geology have lost an able and indefatigable investigator. Since 1901, the date of publication of his first paper, he laboured incessantly and with success to advance palaeobotanical science. To his activity the Sedgwick Museum of Geology at Cambridge is deeply indebted, and through his personal influence several younger men acquired something of their teacher's enthusiasm.

The proofs of this Volume were passed for press in June, 1918, but owing to the exigencies of war conditions publication has been unexpectedly and inevitably delayed.

A. C. S.

June 18, 1919.

TABLE OF CONTENTS

CHAPTER XL

GINKGOALES. Pp. 1—60.

CHAPTER XLI

GENERA BELIEVED TO BELONG TO THE GINKGOALES BUT WHICH ON THE AVAILABLE EVIDENCE CANNOT BE REFERRED WITHOUT HESITATION TO THAT GROUP. Pp. 61—75.

CHAPTER XLII

GENERA OF UNCERTAIN POSITION. Pp. 76—105.

CHAPTER XLIII

CONIFERALES (RECENT). Pp. 106—164.

CHAPTER XLIV

CONIFERALES (FOSSIL). Pp. 165—244.

LIST OF ILLUSTRATIONS

CHAPTER XL.

GINKGOALES.

A. RECENT.

IN the account of the Coniferae contributed to *Die Natür-lichen Pflanzenfamilien*[1] the genus *Ginkgo*, in accordance with the prevailing custom, was included in the Taxeae with *Taxus, Cephalotaxus,* and *Torreya.* Eichler had previously referred *Ginkgo,* or *Salisburia,* to a separate family, the Salisburyeae[2]. Hirase's discovery of motile antherozoids in the pollen-tube of *Ginkgo biloba* in 1896, 'the most remarkable event in plant morphology during the last decade of the 19th century,' confirmed the suspicion that the association of this 'unicum de la création actuelle' with *Taxus* and other Conifers was inconsistent with a natural scheme of classification. At a later date Engler adopted the family-name Ginkgoaceae, and in his survey of the Embryophyta Siphonogama the isolation of *Ginkgo* is emphasised by the reference of the Ginkgoaceae to a special class, the Ginkgoales[3].

Ginkgo biloba L. (*Salisburia adiantifolia* Smith) has a preeminent claim to be described in Darwin's words as a living fossil.' It is sometimes said to occur in China as a wild plant, but there appears to be no sufficient reason to believe that it would have escaped extinction had it not been carefully tended as a sacred tree in the gardens of temples[4]. Since its introduction into Europe in 1730, *Ginkgo* has become familiar in cultivation in the northern hemisphere and thus through man's agency this monotypic genus has been restored to regions where it survived as late as the Tertiary epoch. In habit *Ginkgo*[5] resembles many Conifers and its long and short shoots recall those of *Cedrus* and *Larix*: the short shoots may also be compared with the main trunk of a

---

[1] Eichler (89) p. 108.   [2] Seward and Gowan (00) B. p. 113.
[3] Engler (97) pp. 19, 341.   [4] Elwes and Henry (06) Vol. I. p. 58.
[5] For a full account of the genus, see Sprecher (07).

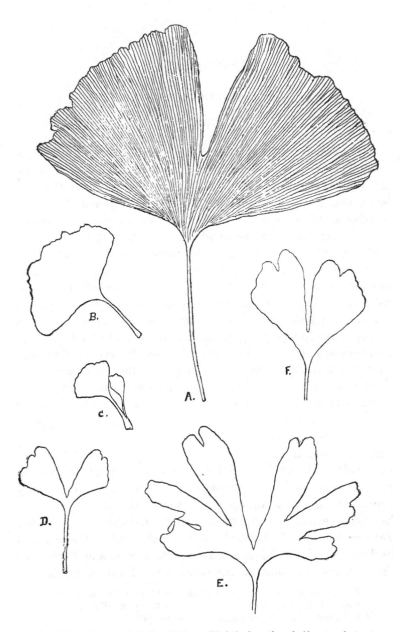

FIG. 630. Leaves of *Ginkgo biloba*. (Slightly less than half natural size.)

Cycad. The large thin leaves with long and slender petioles are scattered on long shoots or crowded on slow-growing branches covered with leaf-scars. These short shoots are occasionally branched[1] and, as Tupper[2] states, they may branch within the wood of the axis out of which they grow, a feature exhibited by the Triassic Conifer *Woodworthia*. The deciduous leaves are usually more or less deeply bilobed (fig. 630, A, D, F) but those on short shoots are often smaller, and the margin may be entire or uneven (fig. 630, C). On young and vigorous shoots or on seedlings the lamina is often deeply divided into several cuneate segments (fig. 630, E). In exceptional cases the lamina may reach a breadth of 20 cm. (fig. 630, A) though as a rule it seldom exceeds half that size. The leaf-scars show two small cicatrices. The considerable range in size and form of the foliage-leaves is an important consideration in connexion with the determination of fossil specimens. Two vascular bundles pass up the petiole: at the summit each divides and the two outer branches follow the outer edge of the lamina, giving off a succession of forked veins. Objection is taken by Prof. Johnson[3] to the statement that there are two marginal veins on the lower edge of the lamina; he regards the 'marginal' vein as the product of the successive fusions of the forked veins of the lamina as they pass towards the leaf-base. Whatever interpretation is adopted, the presence of two broadly diverging marginal veins is a noteworthy feature, and the correct explanation is probably that each is derived from one of the two strands in the petiole and gives off a succession of dichotomously branched veins as it passes along the margin of the leaf-blade. The presence of short secretory tracts at intervals between the veins is a characteristic feature sometimes recognisable in fossil examples. Throughout the greater part of their course in the lamina the veins are accompanied by a small number of reticulate transfusion-tracheids (fig. 631, G, t): these increase in amount near the distal end of each vein and the water-conducting elements may be eventually replaced by a group of short, pitted, tracheids[4]. A group of large cells with brown contents occurs above and below each collateral endarch bundle. The

[1] Seward and Gowan (00) B. Pl. IX. fig. 42.					[2] Tupper (11) p. 376.
[3] Johnson (14) p. 171.					[4] Sprecher (07) pp. 68—71; Bertrand, C. E. (74).

stomata, irregularly scattered over the lower epidermis, consist
of two guard-cells surrounded by 4—6 accessory cells which project towards the centre of the stoma as blunt cuticularised papillae[1]
(fig. 636, C). The epidermal cell-walls are slightly undulate[2].
The distinctive form of *Ginkgo* leaves renders almost negligible
the danger of confusion with those of other Gymnosperms; but
impressions of certain Fern fronds, *e.g. Lindsaya reniformis* Dry.,
Pterozonium (Gymnogramme) reniforme Mart., *Trichomanes reniforme* Forst., and *Scolopendrium nigripes* Hook. might be mistaken
for imperfectly preserved specimens of *Ginkgo*.

Ginkgo is dioecious. The male flowers occur in loose catkins
(fig. 631, B) borne on short shoots in the axil of a scale-leaf : each
microsporophyll consists of a short, slender, filament with a small
terminal scale or knob bearing as a rule 2 but not infrequently
3 or 4 elliptical microsporangia (fig. 631, A, A′) The microspores
recall those of Cycads. Jeffrey[3] has recently called attention to
the occurrence of 'wings' on the microspores of *Ginkgo*: these are
very slightly developed and hardly warrant the use of the term
wing; they present the appearance of very small shoulders giving
the spores a form similar to that of a brachiopod shell. The same
author expresses the view that *Ginkgo* presents striking resemblances to the Abietineae. It has recently been pointed out that
the microsporangia have a hypodermal layer of cells with thickening bands comparable with the fibrous layer in the anthers of
Angiosperms[4]. Jeffrey and Torrey[5] claim that certain anatomical
features in the microsporangia of *Ginkgo* indicate a closer affinity
to the Abietineae than to any other section of the Gymnosperms.
The vascular bundles of the microsporophylls end in transfusion-
tissue which passes almost imperceptibly into the mechanical
elements of the sporangial wall : a similar distribution of mechanical
tissue occurs in Abietineous microsporangia and there is the same
intimate relationship as in *Ginkgo* between the tracheary and
mechanical tissues. The female flower consists of a comparatively long peduncle borne in the axil of a foliage-leaf, with two

---

[1] Strasburger (66) figs. 139—142; Sprecher (07) figs. 79—81.
[2] Bertrand, C. E. (74) Pl. IV. figs. 9, 10.
[3] Jeffrey (14) Pl. XXIII. figs. 7, 8.   [4] Starr (10).
[5] Jeffrey and Torrey (16).

ovules at the summit, one on each side of the actual apex. Frequently one of these is larger than the other. The occasional occurrence of abnormal female flowers is interesting from the point of view of palaeobotanical comparison. In extreme cases

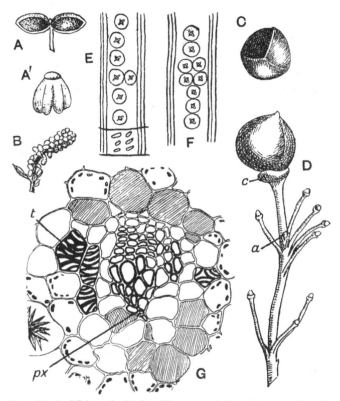

FIG. 631. *Gingko biloba.* A, A′, B. Microsporophylls and sporangia. C. Seed with the outer flesh removed showing an exceptional, tricarinate (radiospermic), form. D. Abnormal megasporophyll; *a*, vegetative bud; *c*, collar. E, F. Tracheids from a stem. G. Transverse section of a leaf-vein; *px*, protoxylem, *t*, transfusion-tracheids. (A, B, after Seward and Gowan; D, after Fujii.)

the partially modified lamina of a foliage-leaf may bear marginal ovules, the lamina being continuous with the collar (fig. 631, D, *c*) at the base of the ovule. In other cases the peduncle may give off several stalked ovules, as in the specimen described by Fujii[1]

[1] Fujii (96); Seward and Gowan (00) B. Pl. IX. figs. 1—5; Sprecher (07) p. 138.

and reproduced in fig. 631, *D*: the apex of the shoot is seen at *a*.
Fertilisation is said to occur after the fall of the ovule, but Hirase
states that some seeds contain an embryo while still attached to
the tree. The seeds are comparable in size with large cherries;
the broad integument consists externally of a thick sarcotesta
rich in secretory tissue but without a vascular supply, and an
inner sclerotesta which is usually two-angled (platyspermic) but
occasionally three-angled and radiospermic (fig. 631, C). An
account has recently been published[1] of some remarkable examples
of *Ginkgo* seeds gathered from one tree: the stony coat showed 2,
3 and 4 ribs and many transitional forms. The sclerotesta is lined
by a few layers of loose cells which form a papery membrane in
ripe seeds. The absence of vascular tissue in the sarcotesta is a
character in which the seeds differ from those of Cycads and
Taxads. At the base of the ovule is a single concentric strand
which splits into two branches and these pass through the shell
and divide into several bundles on the inner face of the integument
forming a continuous mantle[2] of short reticulate tracheids as in
the Palaeozoic seed *Stephanospermum*[3]. The base of the ovule
is enclosed in a shallow cup or collar, a structure that is probably
homologous with the lamina of a foliage-leaf but which has received various interpretations. A suggestion has been made that
the collar may be homologous with the cupule of *Lagenostoma*[4].
The nucellus is joined to the integument except at the apex where
it forms a prominent cone in which a pollen-chamber is developed:
this chamber becomes roofed over by nucellar tissue and at a
later stage a blunt outgrowth is produced from the summit of the
prothallus, serving as a 'tent-pole' to support the roof of the
pollen-chamber. There are two or more archegonia at the apex
of the prothallus differing from those of Conifers and Cycads in
the shorter and more spherical form of   egg-cell and similar to
those in some Palaeozoic seeds. Fujii[5] draws attention to the
remarkable capacity for pollination exhibited by *Ginkgo* and
speaks of the conveyance of microspores over a distance of 500
to 1000 metres. Another fact worthy of remark in view of the
wide distribution of the Ginkgoales in the Mesozoic era is the

[1] Affourtit and La Riviere (15).   Sprecher (07) figs. 120, 147, 148.
[3] See Vol. III. p. 326.   [4] Shaw, F. J. F. (08).   [5] Fujii (10) p. 216.

germination of *Ginkgo* seeds after 45 days' immersion in sea-water[1]. The embryo has normally two hypogean cotyledons though three are not uncommon. Velenovsky[2] mentions a peculiarity, another indication of the isolated position of the genus, in which seedlings of *Ginkgo* differ from those of other Phanerogams; the cotyledons are succeeded by two elongated scales with a forked apex; the next higher leaves, in which a small bilobed lamina is a characteristic feature, show at the base two divergent prongs representing the fork of the lower scales. The lamina of the foliage-leaf thus arises in the angle of the V-shaped distal end of the earlier scale-leaf.

A microspore on germination developes 2—3 prothallus-cells- and the generative cell forms two large ($110\mu \times 80\mu$) spirally ciliated antherozoids. After fertilisation the egg-nucleus divides, as in some Cycads, until 256 free nuclei are formed[3], but in *Ginkgo* the subsequent production of walls results in a tissue, called by Lyon[4] the protocorm, which completely fills the egg; whereas in *Cycas* this tissue is massed at the base and in *Zamia* wall-formation is also restricted. In Conifers the number of nuclei is much less and the proembryo still further reduced. It is probably legitimate to deduce from these facts that *Ginkgo* is in respect of its embryogeny the most primitive of the Gymnosperms: in this and other characters it is allied more closely to the Cycads than to the Conifers. Saxton[5] who has described the later stages in the embryogeny of *Encephalartos* draws attention to certain features shared by that genus and *Ginkge*.

The leaf-traces arise from the stele as a pair of collateral bundles, as in the Palaeozoic genus *Mesoxylon*, which pass up the petiole. Annual rings are fairly distinct though, as Nicol[6] recognised, less obvious than in Conifers. The walls of the late-summer tracheids are hardly thicker than those of the spring-elements and the difference between the early and late wood is often slight[7]. Circular bordered pits occur either in a single or double row on the radial walls of the tracheids and are fairly common on the tangential walls. The pits may be separate or in contact, occasionally

[1] Ewart (08) p. 78.   [2] Velenovský (07) p. 457, fig. 291a.
[3] Coulter and Chamberlain (03).   [4] Lyon (04).   [5] Saxton (10⁴).
[6] Nicol (34) A. p. 147.   [7] Nakamura (83) p. 25; Fujioka (13) Pl. xviii.

slightly flattened and alternate, but usually opposite (fig. 631, E, F). The pores of the pits are often crossed. Rims of Sanio form a well-marked feature on the tracheal walls, and Jeffrey[1] points out that they are clearly shown in the mature wood but not in close proximity to the primary xylem or in the wood of the reproductive shoots and leaves. True bars of Sanio frequently occur on the tracheids[2]. The secondary phloem consists of discontinuous rows of fibres in addition to sieve-tubes and parenchyma. Characteristic features are presented by the uniseriate medullary rays: these are often 1—2 or 1—5 cells deep and do not appear to exceed 11 cells in depth; they are comparatively large and in tangential sections of the wood present an inflated appearance. There are 2—7 elliptical pits in the field of the ray-cells. Xylem-parenchyma though not abundant occurs here and there among the tracheids; the cells have thin walls and are larger than the medullary-ray cells and characterised by the occurrence of stellate calcium oxalate crystals[3].

B. FOSSIL.

i. PETRIFIED WOOD REFERRED TO THE GINKGOALES.

The characters of the wood of *Ginkgo biloba* summarised above are in general agreement with those of many Conifers, and such anatomical features as have been described by authors as more or less distinctive of the genus do not afford very trustworthy guides to the identification of fossil wood. The comparatively large size and rounded contour of the medullary-ray cells, as seen in tangential section, though worthy of note as characteristic features, are hardly satisfactory criteria when applied to wood that may have undergone partial decay and been exposed to influences affecting the original form of the more delicate tissues before petrification. The untrustworthy evidence afforded by the size of the medullary rays has been emphasised by Essner[4] who states that the ray-cells of *Ginkgo* are larger than those in any genus of Conifers. It has been claimed by Felix[5] that *Ginkgo* is

[1] Jeffrey (12) p. 548.   [2] Müller (90) Pl. XIV.
[3] For anatomical details, see also Kleeberg (85); Essner (86); Strasburger (91); Seward and Gowan (00) B.; Penhallow (07); Sprecher (07); Burgerstein (08); Tupper (11).   [4] Essner (86).   [5] Felix (94).

an exception to the general truth of Essner's conclusions and that the large dimensions and rounded form of the ray-cells are features of diagnostic value, though in the Tertiary specimens compared by him with the wood of *Ginkgo* the ray-cells do not appear to differ appreciably in size or form from those of true Conifers. Given well-preserved material, it is not improbable that in favourable cases the characters of fossil wood might furnish adequate grounds for referring it to *Ginkgo*: the numerous obliquely elliptical pits in each 'field,' the swollen medullary-ray cells, and the frequent crossing of the pores of the tracheal pits are the features mentioned by Gothan[1] who considers that the wood of *Ginkgo*— though difficult to define precisely in an analytical key—may be distinguished from that of Conifers.

Among the specimens of wood assigned to *Ginkgo* there are none, so far as I am aware, that can safely be accepted as entirely above suspicion. In 1850 Goeppert[2] proposed the generic name *Physematopitys*[3] for some Tertiary wood that he believed to possess the anatomical characters of *Ginkgo biloba*. Kraus[4] subsequently recognised resin-cells in the wood of Goeppert's type-species, *Physematopitys salisburioides*, and identified the specimens as the root-wood of a *Cupressinoxylon*: he did not, therefore, include *Physematopitys* in the list of woods contributed by him to Schimper's *Traité de Paléontologie*, but mentioned it as a synonym of *Cupressinoxylon*. Beust[5] and Barber[6] among other authors adopt the same course. It has more recently been stated by Kräusel[7] that Goeppert's genus *Physematopitys* has the characters of *Protopiceoxylon*. Goeppert[8] afterwards described a second species, *Physematopitys succinea*, founded on a tangential section of a piece of Oligocene wood from the Baltic amber, but the data are clearly insufficient to justify its identification as *Ginkgo*: Conwentz[9] includes the specimen in *Pinus succinifera*.

Schroeter[10] described some wood from beds on the Mackenzie river in North Canada, referred to the Miocene period, as *Ginkgo* sp.

---

[1] Gothan (05) p. 103.   [2] Goeppert (50) p. 242, Pl. xlix. figs. 1—5.
[3] φύσημα, that which is blown out.
[4] Kraus in Schimper (72) A. p. 370; Kraus (83); Schenk in Zittel (90) A. p. 871.
[5] Beust (85).   [6] Barber (98).   [7] Kräusel (13).
[8] Goeppert and Menge (83) A. p. 32, Pl. x. fig. 74.
[9] Conwentz (90) A. p. 26.   [10] Schroeter (80) p. 32, Pl. iii. fig. 27—29.

on the ground of the large size of the medullary cells: no pits are described either on the tracheids or on the medullary-ray cells and the unusual size of the ray-cells may well be a pathological or post-mortem phenomenon. The species *Physematopitys excellens* described by Felix[1] from beds, probably Eocene in age, in the Caucasus agrees with *Cupressinoxylon* in the presence of rows of resin-parenchyma in the wood, and the depth of the rays greatly exceeds that in *Ginkgo biloba*. Penhallow[2] described some calcified wood from Upper Cretaceous beds in the Queen Charlotte Islands as *Ginkgo pusilla*, but the reasons for assigning it to that genus are not convincing. A fuller description of another specimen regarded as the wood of a *Ginkgo* has been published by Dr Platen[3] under the name *Physematopitys Goepperti* from material collected in Miocene beds in Milam County, Texas. The relatively large size of the medullary-ray cells is mentioned as the chief character on which the determination was based.

It may be said that such fossil specimens as have been referred to *Physematopitys* or *Ginkgo* have very little value as records of the occurrence of the genus *Ginkgo*: in view of the abundance of leaves in Mesozoic and Tertiary strata that are hardly distinguishable from those of the surviving type it is remarkable—if the anatomical characters of the genus afford in themselves a trustworthy basis of identity--that more satisfactory specimens have not been found.

ii. LEAVES.

GINKGOITES. Gen. nov.

It has been customary to use the generic name *Ginkgo* both for the recent species and for fossil leaves from Mesozoic and Tertiary strata, and in a few cases for Palaeozoic leaves. In certain instances, for example such leaves as those from the Island of Mull and other Tertiary localities referred to *Ginkgo adiantoides* (fig. 644) there can be no doubt as to generic identity with the recent species and indeed, so far as concerns form and venation, the Eocene leaves might well belong to *Ginkgo biloba*. On the other hand even in the case of *Ginkgo adiantoides* we lack

---

[1] Felix (94) p. 107, Pl. IX. fig. 4.
[2] Penhallow (02) p. 43, Pls. XII., XIII.   [3] Platen (08) p. 143.

the confirmatory evidence of flowers and seeds. From Wealden and Jurassic rocks numerous leaves have been described that in some cases are practically identical with those of the living species, but for the most part they are characterised by certain features denoting at least a specific difference. For these and for other Ginkgo-like leaves it would seem desirable to follow the usual custom and adopt a designation that does not necessarily imply even generic identity. A few examples of seeds and male flowers are known from Jurassic strata bearing a close resemblance to those of *Ginkgo biloba,* but such specimens are not common and some of the few that have been found, though probably belonging to the Ginkgoales, may not be correctly included in *Ginkgo.* I therefore propose to employ the name *Ginkgoites* for leaves that it is believed belong either to plants generically identical with *Ginkgo* or to very closely allied types.

It is impossible in some cases to draw a sharp line between the genera *Ginkgo* and *Baiera*: typical examples of the latter genus are easily recognised by their narrow, relatively longer, and more numerous segments, but it is obvious that characters based on the degree of division of a lamina and on the breadth of the segments are at best unsatisfactory, and the inclusion of certain specimens in one or other genus is purely arbitrary.

A difficulty is presented by several types of Palaeozoic leaves assigned by many authors to the Ginkgoales and referred to *Ginkgophyllum, Psygmophyllum,* and other genera which, while bearing a general resemblance to the leaves of *Ginkgo,* cannot be regarded as evidence of the occurrence of the class that is now represented by *Ginkgo biloba.* It has been suggested that *Psygmophyllum, Ginkgophyllum, Rhipidopsis,* and certain other genera should be included in a distinct group, the Palaeophyllales[1], a group of which the affinities are unknown. Though the adoption of a distinctive group-name has the advantage of indicating the absence of any trustworthy evidence of relationship to the Ginkgoales, it is open to question whether anything substantial is gained by the use of a term suggestive of relationship between different leaves that in themselves afford no clue as to the position of the parent-plants.

[1] Arber, E. A. N. (12) p. 405.

The name *Ginkgoites* as used in this chapter is restricted to leaves that are regarded as records of the Ginkgoales, while the genera referred by Dr Arber to the Palaeophyllales are briefly described as fossils that may or may not be closely related to one another but which cannot as yet be assigned to any place in a natural system of classification.

The leaves discovered by Grand'Eury in Permian Uralian beds and described by Saporta as *Salisburia primigenia*[1] should probably be referred to the genus *Psygmophyllum*: like many other supposed Palaeozoic species assigned to the Ginkgoales or to *Ginkgo* they afford no satisfactory evidence of affinity to the surviving genus. Other examples of leaves from Palaeozoic rocks described as species of *Ginkgo* or *Salisburia* on inadequate grounds are described in the latter part of this chapter. The Rhaetic leaves described by Brauns as *Cyclopteris crenata* and afterwards referred by Nathorst, with some doubt, to *Ginkgo* are described in the account of *Psygmophyllum*[2].

Ginkgoites obovata Nathorst.

Fig. 632 A shows the form of the specimen from the Rhaetic beds of Scania on which Nathorst[3] founded the species *Ginkgo*

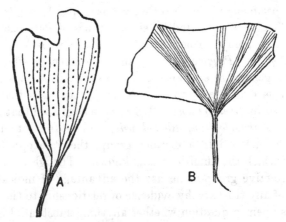

Fig. 632. A. *Ginkgoites obovata*. B. *Ginkgoites antarctica*. (Nat. size; A, after Nathorst; B, drawn from a specimen in the University Museum, Brisbane.)

[1] Saporta and Marion (85), p. 145, fig. 74.   [2] See page 88.
[3] Nathorst (86) p. 93, Pl. xx. fig. 5.

obovata: the obovate lamina is 5·6 cm. long with a maximum breadth of 2·5 cm.; the upper edge is partially torn and the forked veins are about 3 mm. apart. Some dark patches between the veins are probably, as Nathorst suggests, secretory sacs similar to those in the leaves of *Ginkgo*. In shape but not in venation this type resembles *Ginkgodium Nathorsti* Yok.[1] (fig. 659, p. 62); except in the absence of a deep median sinus it is, however, nearer to a species from the Jurassic of Dzungaria described originally as *Ginkgo Obrutschewi*[2] (fig. 642, p. 26).

Ginkgoites Geinitzi Nathorst.

The leaf from the Rhaetic beds of Scania on which this species was founded was originally referred by Nathorst[3] to *Ginkgo* but subsequently transferred by him to *Baiera*; it consists of a slender stalk and a sub-triangular lamina deeply divided into 4—6 linear truncate segments with 2—4 veins dichotomously branched near the base (fig. 645, B, p. 38). An examination of the original specimen leads me to prefer the designation *Ginkgoites* to *Baiera*. This species like many others from Rhaetic rocks is hardly distinguishable from some Jurassic types.

Ginkgoites antarctica Saporta.

Under the name *Salisburia antarctica* Saporta[4] described a single leaf from Australia believed to be of Lower Lias age, but no precise information is given with regard to the locality. Shirley[5] has also figured a specimen as *Ginkgo antarctica* from rocks that are probably of Rhaetic age at Denmark Hill, Ipswich (Queensland). The lamina of Saporta's specimen is broadly obcuneate and 3·5 cm. broad, characterised by the presence of two marginal veins like those in *Ginkgo biloba* from which forked branches are given off. This leaf is practically identical with some of the smaller, entire, examples on the short shoots of the recent type. The rather larger specimen figured by Shirley does not present so striking a similarity to those of the existing species. The lack of definite information as to the provenance of the type-specimen

[1] See page 61.   [2] Seward (11) p. 46, Pls. III.—VI.
[3] Nathorst (78) B. p. 26, Pl. XIII. fig. 17.
[4] Saporta and Marion (85) p. 142, fig. 71, A.; Ratte (88) Pl. III. fig. 1; Renault (85) A. Pl. II. fig. 19.
[5] Shirley (98) Pl. I. fig. 1.

is unfortunate, but whether or not Shirley's fossil is identical
with Saporta's specimen there would seem to be no reasonable
doubt that it should be included in the genus *Ginkgoites*.

Fig. 632 B is drawn from a photograph of a specimen in the
Brisbane Museum which I recently had an opportunity of examining: it is from the Ipswich beds and is undoubtedly specifically
identical with Saporta's type.

Ginkgoites digitata (Brongniart).

This widely spread Jurassic species founded on leaves from
the Yorkshire coast, was first figured by Phillips in 1829[1] as
Sphenopteris latifolia, but under the same name Brongniart[2] had
a year previously recorded a Carboniferous Fern.   In 1830 Brongniart[3] figured and described another Yorkshire specimen as *Cyclopteris digitata*: the generic name *Cyclopteris* was adopted by Dunker
and other authors until Heer[4] drew attention to the very close
agreement between the Jurassic leaves and those of the Maidenhair
tree, a similarity that led him to adopt the generic designation
Ginkgo.   Leaves hardly distinguishable from the Jurassic impressions had previously been recorded from Tertiary rocks as species
of *Salisburia* or *Ginkgo*.

It is impossible to define precisely the several species of *Ginkgoites* founded on leaves: in the account of the recent species
attention is called to the range in leaf-form and its bearing on
the determination of fossils.   All that can be done is to adopt
certain specific names as a matter of convenience, recognising
that the differences on which the classification is based are not
either sufficiently sharply defined or morphologically important
to be regarded as criteria of true specific distinctions.   Many
authors have employed the specific name *Huttoni,* first used by
Sternberg[5], for leaves identical in size and outline with *G. digitata*
but characterised by a deeply-lobed lamina; this difference is,
however, not greater than or even as great as differences met
with within the species *Ginkgo biloba*.   To facilitate description
the designation *Huttoni* is retained as a form-designation for the
more deeply lobed examples included in the species *G. digitata* (*e.g.*

[1] Phillips (29) A. Pl. vii. fig. 18.   See Fontaine in Ward (05) B. p. 121.
[2] Brongniart (28) A. p. 51.   [3] Brongniart (28²) A. p. 219, Pl. lxi *bis*, figs. 2, 3.
[4] Heer (81²); (77) i. p. 40.   [5] Seward (00) B. p. 256.

fig. 633). The number of *Ginkgoites* leaves from Jurassic strata is considerable and the student who attempts to classify specimens in a large collection under specific heads soon finds himself confronted in an acute form with the constantly recurring difficulty of fixing boundaries. As Knowlton[1] says, 'In dealing with such an abundance of specimens and multiplicity of forms one must needs make either many " species" to accommodate this diversity, or only one or two, and in view of the known variation exhibited by the single living species, the latter plan seems preferable.' In advocating this use of specific names in a liberal sense I admit the probability or indeed the certainty that forms specifically distinct will be grouped

Fig. 633. *Ginkgoites digitata* var. *Huttoni*. A leaf from the Upper Jurassic of Helmsdale, Scotland. (Stockholm Museum; nat. size.)

under one designation. It is, however, clearly impossible in the case of impressions of leaves of *Ginkgoites* to impose limitations based on the form of the lamina, the degree of dissection, and similar variable features that cannot be accepted as trustworthy criteria of true specific distinctions. As material accumulates data may be furnished that will enable us to recognise characters of morphological significance : in carbonised impressions from which cuticular preparations can be made the form of the epidermal cells and the structure of the stomata may supply a valuable aid to more accurate diagnosis. The spacing of the veins is a feature worthy of attention in the description of well-preserved specimens.

[1] Knowlton (14) p. 55.

There is also a further difficulty in regard to terminology: the employment of the two generic names *Ginkgo* or *Ginkgoites* and *Baiera* reveals a striking lack of uniformity among authors, and the artificial nature of the characters determining the use of one or other generic name necessarily lead to diversity in practice. As with the definition of species within the genus *Ginkgoites*, so also the adoption of *Ginkgoites* or *Baiera* is to a large extent the

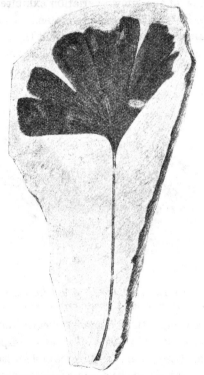

FIG. 634. *Ginkgoites digitata.* Leaf from Kap Boheman, Spitzbergen.
(Stockholm Museum; nat. size.)

result of individual preference and merely expresses an attempt to classify in an arbitrary fashion the numerous types of leaves that in themselves afford no sure guide as to precise affinity. The South African, Rhaetic, specimen shown in fig. 635, L was originally described as *Baiera moltenensis*[1] but it might equally well be referred to *Ginkgoites*.

[1] Seward (08) B. Pl. II. fig. 4.

The leaves of *Ginkgo digitata* have a long slender petiole (fig. 634); the lamina is semiorbicular or obcuneate, entire, or more or less deeply divided into equal lobes, or irregularly divided into several

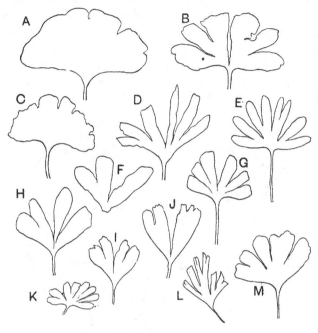

FIG. 635. *Ginkgoites.* (⅓ nat. size.)

A. *Ginkgoites adiantoides*, Tertiary, Island of Mull.
B. *G. pluripartita*, Wealden, North Germany (after Schenk).
C. *G. digitata*, Jurassic, Japan (after Yokoyama).
D. *G. digitata* var. *Huttoni*, Jurassic, Australia (after Stirling).
E. *G. sibirica*, Jurassic, Siberia (after Heer).
F. *G. digitata*, Jurassic, Turkestan.
G. *G. multinervis*, Lower Cretaceous, Greenland (after Heer).
H. *G. digitata*, Jurassic, Oregon (after Fontaine).
I. *G. digitata*, Jurassic, Yorkshire.
J. *G. digitata*, Jurassic, Scotland (after Stopes).
K. *G. digitata*, Jurassic (or Wealden), Franz Josef Land (after Nathorst).
L. *G. moltenensis*, Rhaetic, South Africa.
M. *G. digitata*, Jurassic (or Wealden), Spitzbergen (after Heer).

segments; the number and size of the segments and the form of their distal ends, truncate or obtuse, vary within wide limits (figs. 635, 637, 639, etc.). Numerous dichotomously branched

veins spread from the base of the lamina, the veins in the middle
of the leaf being generally about 0·8—1 mm. apart.

The stomata are practically confined to the lower surface of
the lamina. The epidermal cells are polygonal and the walls
slightly sinuous as in the recent species, and over the veins the
cells are longer and narrower (fig. 636, A). Many of the larger
epidermal cells have a cuticular ridge in the middle of the outer
wall, represented in the figure by a black line. The stomata agree
closely with those of *Ginkgo biloba*; the two guard-cells are surrounded by a group of subsidiary cells characterised by their

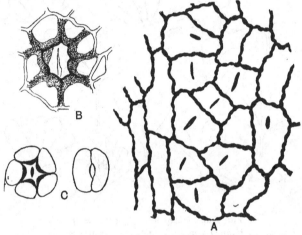

FIG. 636. *Ginkgoites digitata*; epidermal cells (A) and stoma (B).
C. Stoma of *Ginkgo biloba* in two planes. (A, B, drawn by Miss N. Bancroft.)

papillose heavily cuticularised walls overarching the stoma (fig.
636, B). The features shown in fig. 636 are not brought out in
drawings from cuticles of the same specimen reproduced by Dr
Stopes[1] in her account of fossil plants from Brora: this may be
due in part to a difference in the level at which the stomata were
drawn. The stoma of *Ginkgo biloba* represented in fig. 636, C
illustrates the considerable difference produced by viewing a stoma
in slightly different planes[2].

No seeds have been found attached to stems bearing leaves of
G. digitata, but seeds closely resembling those of the recent species

¹ Stopes (07) p. 380.   ² Seward (11) p. 47, Pl. v. fig. 62.

occasionally occur in association with the foliage of this and other Jurassic species. Male flowers[1] similar in habit to those of *Ginkgo biloba* are also found in beds containing impressions of *Ginkgoites*.

The abundance and wide geographical range of *Ginkgoites digitata* precludes anything more than a brief reference to some representative types selected in illustration of the range in form and the widespread occurrence of the species in Jurassic floras.

The leaf represented in fig. 637 is an unusually complete example from the Middle Jurassic beds of Scarborough; the lamina is 3·8 cm. deep and 6 cm. broad, the venation agrees with that of *Ginkgo biloba*. A very similar type of leaf is figured by Heer from Upper Jurassic (or Wealden) strata of Spitzbergen as *G. integriuscula*[2], but with the proviso that it may be merely a variety of *G. digitata*, a view that Nathorst[3] has wisely adopted. The latter author in speaking of the occurrence of *G. digitata* in Spitzbergen states that

FIG. 637. *Ginkgoites digitata.* (⅔ nat. size.) M. S.

'sometimes the surface of the schists [shales] is as completely covered with the leaves of *Ginkgo* as the soil beneath a living *Ginkgo* tree may be in autumn[4].' In some specimens from the Yorkshire coast the lamina is practically entire as in a leaf from Scarborough in the York Museum figured in 1900[5]. An exceptionally large form is shown in fig. 638; the lamina, 8 cm. broad, is divided into several short and comparatively broad obtuse or truncate lobes[6]. Fig. 639 shows a leaf from the Stonesfield Slate, now in the Cirencester Museum; the lamina is deeply divided into two broad cuneate lobes as in some forms of the recent species. The Stonesfield Slate specimens were originally named by Buckman *Noeggerathia* (?) and later *Stricklandia*

---

[1] See page 51.   [2] Heer (77) i. p. 44, Pl. x. figs. 7—9.

[3] Nathorst (97) p. 15; for a discussion of the age of the Spitzbergen beds, see Nathorst (13²).

[4] Nathorst (11³) p. 221.

[5] Seward and Gowan (00) B. Pl. x. fig. 54.

[6] Seward (00) Pl. II. fig. 5.

FIG. 638. *Ginkgoites digitata.* (Manchester Museum; nat. size.)

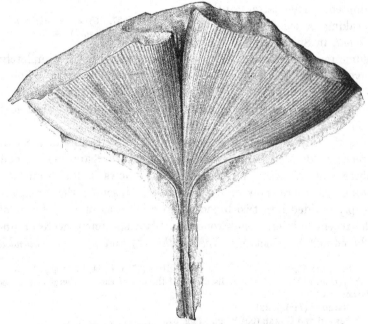

FIG. 639. *Ginkgoites digitata.* (Cirencester Museum; nat. size.)

acuminata[1]. A deeply divided obcuneate leaf, only 2—4 cm. broad, is figured by Dr Stopes[2] from Jurassic strata (Lower Oolite) on the Sutherland coast at Brora (fig. 635, J). The specimen shown in fig. 633, also from the Sutherland coast, a few miles north of Brora and of Kimeridgian age, illustrates the type of leaf that may be conveniently referred to *Ginkgoites digitata* var. *Huttoni*[3]. Heer's Jurassic species, *Ginkgo Jaccardi*[4], from Switzerland is almost certainly *Ginkgoites digitata*.

Arctic regions and northern Europe and Asia.

The leaf reproduced in fig. 634 from a drawing made for me in the Stockholm Museum through the kindness of Prof. Nathorst was originally figured by Heer[5] from Cape Boheman, Spitzbergen, from beds assigned by Nathorst to the Upper Jurassic series. Nathorst includes in this series strata agreeing in their fossil plants with the Wealden of England and North Germany; the 'Ginkgo beds' may be Portlandian or even younger. The veins in this specimen are approximately 1 mm. apart.

The leaf represented in fig. 640 is one of several specimens collected by Dr Koettlitz in Franz Josef Land from beds that are probably Jurassic. The lamina is 2·5 cm. in depth and the veins are about 0·8 mm. apart, the lamina between the veins shows a fine transverse striation, not shown in the drawing, a feature occasionally seen in impressions of *Ginkgoites* and due in all probability to the shrinkage of the mesophyll tissue into transverse bands[6]. This form of leaf has been described as

Fig. 640. *Ginkgoites digitata* var. *polaris*. Franz Josef Land. (Museum of the Geological Survey, London; nat. size.) M. S.

Gingko polaris Nath.[7]; it is smaller than most examples of *Ginkgoites*

---

[1] Seward (04) B. p. 98.

[2] Stopes (07) Pl. xxvii. fig. 4. For a map of the Sutherland Coast, see Seward (11²).

[3] Seward and Bancroft (13) p. 886.   [4] Heer (76) Pl. lviii. fig. 20.

[5] Heer (77) i. Pl. x. fig. 2.

[6] For similar transverse wrinkling, see Schenk (71) B. Pl. xxiv. figs. 7, 8; Seward (11) Pl. xi. fig. 74; Heer (77) ii. Pl. xi. fig. 1 *b*.

[7] Newton and Teall (97) Pl. xxxviii. figs. 1, 2; (98) Pl. xxix. fig. 3; Nathorst (99) Pl. i. fig. 8 and Thomas (11) Pl. iv. fig. 8.

digitata and the veins are more crowded. For this type I suggest the designation *G. digitata* var. *polaris.* Solms-Laubach[1] figures an incomplete leaf from Franz Josef Land with a broader lamina divided into several broadly rounded segments as in many British specimens but with rather closer venation. A similar leaf from the same locality is figured by Newton and Teall[2]. Fig. 635, K shows a lobed leaf similar to that represented in fig. 640, described by Nathorst[3] from Franz Josef Land and compared by him with *G. sibirica* Heer and *G. flabellata* Heer from the Jurassic of East Siberia. A very small specimen similar in form to the larger example shown in fig. 635, K was figured by Nathorst[4], also from Franz Josef Land, as *Ginkgo polaris* var. *pygmaea.* From the west coast of Greenland Hartz described a leaf very similar to some of the Yorkshire examples as *Ginkgo (Baiera) Hermelini*[5]: Hartz regarded the beds as Liassic or Rhaetic, and the occurrence of shells pointing to a Kellaways horizon immediately above the plant-beds suggests that the latter may belong to the Middle Jurassic series[6]. *Ginkgo digitata* is represented also in Jurassic strata in the New Siberian Islands by a leaf figured by Nathorst as *Ginkgo* sp. which agrees with the type *G. digitata* var. *Huttoni*[7], and Krystofovič[8] has described *G. digitata* from Jurassic beds in Ussuriland at the northern end of the Muravjev-Amurskyj peninsula. Some good specimens are recorded from Bornholm[9], of Middle or Lower Jurassic age, as *G. Huttoni* which are identical in form and size with British specimens.

North America.

Several examples of leaves of the *G. digitata* type and some of the form *Huttoni* have been figured from Middle Jurassic rocks in Oregon[10] (fig. 635, H). Some particularly large examples are named by Fontaine *G. Huttoni* var. *magnifolia* but these are not

---

[1] Solms-Laubach (04) Pl. i.. fig. 10.
[2] Newton and Teall (97) Pl. xli. fig. 10.
[3] Nathorst (99) p. 11, Pl. i. figs. 8—19.   [4] *Ibid.* Pl. i. figs. 20, 21.
[5] Hartz (96) Pl. xix. fig. 1.   [6] Johnstrup (83).
[7] Nathorst (07) Pl. i. fig. 20.
[8] Krystofovič (10) Pl. iii. fig. 1.
[9] Bartholin (94) p. 96, Pl. xii. figs. 1—3; (10) Pl. iii. figs. 9, 10.
[10] Fontaine in Ward (05) B. Pls. xxx —xxxii., xlvi.

specifically distinct from *G. digitata*. *G. digitata* is represented in Upper Jurassic or Wealden beds in Alaska (Cape Lisburne)[1]

Other Localities.

From Southern Russia Thomas[2] has recently described good specimens of *Ginkgo digitata* of Middle Jurassic age some of which agree closely with the large leaves figured by Fontaine from Oregon as *G. Huttoni* var. *magnifolia*; the specimens previously figured by Eichwald[3] as *Cyclopteris incisa* from the same district are examples of *G. digitata*. The species is recorded also from Jurassic rocks in Turkestan[4] (fig. 635, F), Chinese Dzungaria[5] on the west border of Mongolia, from the region to the east of Lake Baikal[6], and from Afghanistan[7]. The incomplete specimen figured by Feistmantel[8] from the Jabalpur group of India as *Ginkgo lobata*, part of which is shown in fig. 643, A, agrees in the form of the lamina and in venation with *G. digitata*: the veins in the middle of the lamina are from ·8 to 1 mm. apart. Feistmantel compares his species with *G. digitata*, and an examination of one of his figured specimens leads me to assign it to that type; it is indistinguishable from the Afghan specimen already quoted. The piece of a leaf figured by Feistmantel as *Ginkgo* sp. belongs to a similar leaf, but the venation is finer and it may be identical with *Ginkgo crassipes* Feist.[9]

Leaves of the *G. digitata* type are recorded from Jurassic beds in Victoria[10] (fig. 635, D) and from Japan[11] (fig. 635, C). Tuzson has figured a bilobed petiolate leaf from Jurassic rocks in Hungary as *Ginkgo parvifolia*[12]; it is similar in form to most of the specimens referred to *G. digitata*, but has relatively broader segments: it is interesting as being the first recorded example of *Ginkgoites* from Hungary.

[1] Knowlton (14) Pl. XLIV.
[2] Thomas, H. H. (11) p. 73, Pl. IV. fig. 7; Pl. VIII. fig. 2.
[3] Eichwald (65) Pl. IV. fig. 6.   [4] Seward (07²) Pl. VII.
[5] Seward (11) Pl. III. fig. 40.   [6] Krasser (05) Pl. II. fig. 3.
[7] Seward (12) Pl. IV. fig. 51.
[8] Feistmantel (77⁴) Pl. I. fig. 1.
[9] See page 28.   [10] Seward (04²) B. fig. 35.
[11] Yokoyama (88) B. Pl. XIII, fig. 2.   [12] Tuzson (14) Pl. XIV. fig. 1.

Ginkgoites sibirica Heer.

The specific name *sibirica* was given by Heer[1] to one of the most abundant forms in the rich plant-beds at Ust Balei near Irkutsk in Siberia; the specimens figured from Siberia as *Ginkgo sibirica* (fig. 635, E), *G. Schmidtiana*, and *G. lepida* cannot be regarded as well-defined species; they agree in the deep division of the lamina into several linear segments with obtuse or in some cases more pointed apices. Heer draws attention to the resemblance between *G. sibirica* and *G. pluripartita* from Wealden rocks but, as he says, the venation is rather coarser in the Siberian leaves and the segments are generally narrower in *G. sibirica*. Fontaine in his description of Jurassic leaves from Oregon apparently identical with Heer's *G. sibirica* states that *G. Schmidtiana* is a smaller form of the same species[2], but Ward[3] points out that as *G. Schmidtiana* is described on one page and *G. sibirica* is defined on the following page the former-designation must be preserved. It may be urged that as the name *sibirica* is the more widely used and familiar term, considerations of convenience should override this meticulously strict interpretation of the rule of priority. A revision of Heer's Siberian material would, I have no doubt, result in the reduction of his specific terms; on comparing several specimens in the Museums of Copenhagen and Stockholm with the illustrations in the *Flora Fossilis Arctica* I found that several of the published figures are far from accurate. For the present the most convenient course would seem to be the retention of *Ginkgoites sibirica* for leaves similar to some of the more deeply divided forms of *G. digitata* and to *G. pluripartita,* but normally characterised by a lamina divided almost or quite to the base into oblong, obtuse or more or less acute segments. Leaves of the *G. sibirica* type, using the term in the wider sense and including Heer's other species *Ginkgo Schmidtiana* and *G. lepida*, are fairly common in Jurassic rocks and occur also in Cretaceous floras; they are recorded from Kimeridgian beds in Scotland[4] (fig. 641, A), Jurassic strata in Siberia, China[5] (described by Yokoyama as

---

[1] Heer (77) ii p. 61. Pl. VII. fig. 6; Pl. IX. fig. 5 b; Pl. XI.; Heer (82) ii A. p. 16 Pls. IV., V.

[2] Fontaine in Ward (05) B. p. 125, Pl. XXXIII.   [3] *Ibid.* p. 126 (footnote).

[4] Seward (11²) p. 679, fig. 9, A.

[5] Yokoyama (06) B. Pl. VII.; Krasser (05) Pl. II. fig. 5.

G. flabellata, also the similar leaves figured by Krasser as *G. Schmidt-iana* var. *parvifolia*), Turkestan[1], and Western America[2]; also from Upper Jurassic beds in Japan[3] and Franz Josef Land[4] and from

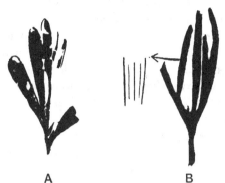

A   B

Fig. 641. A. *Ginkgoites sibirica*. B. *Baiera Brauniana*; from Upper Jurassic (Kimeridgian) beds on the coast of Sutherland.

Lower Cretaceous beds in Canada[5]. A similar form of *Ginkgoites* is recorded from Jurassic rocks in Victoria[6].

It should be added that some Jurassic specimens described as species of *Baiera*, *e.g. Baiera Phillipsi*[7] Nath. are very near to *G. sibirica* and in such cases the choice of *Baiera* or *Ginkgoites* is not determined by any satisfactory standard.

[*Ginkgo digitata* (Brongn.), Schmalhausen (79) A. p. 33, Pl. v. fig. 4 *b*. *Gingko sibirica* Heer?, *Ibid.* p. 34, Pl. IV. fig. 2 *b*.]

The incomplete leaf-fragments from the Altai mountains re-ferred by Schmalhausen to these species are too incomplete to be determined with any degree of certainty. The precise age of the beds is uncertain but, as Zeiller[8] has shown, they are probably Permian. There is a certain resemblance between the specimen referred to *Ginkgo digitata* and some leaves from Permo-Carboni-ferous strata in Kashmir described as *Psygmophyllum Hollandi*[9].

[1] Seward (07²) Pl. VII.   [2] Fontaine in Ward (05) B. Pl. XXXIII.
[3] Geyler (77) B. Pl. XXXI. fig. 6; Yokoyama (89) B. Pl. XIV.
[4] Newton and Teall (97) Pl. XXXVIII.; Nathorst (99).
[5] Dawson (85) Pl. II.   [6] Seward (04²) B. p. 177.
[7] Seward (00) B. p. 270.
[8] Zeiller (96) A.   [9] Seward (07) p. 59, Pl. XIII. figs. 3—6.

Ginkgoites whitbiensis (Nathorst).

This name was proposed by Nathorst[1] for a leaf from the Jurassic rocks of the Yorkshire coast in the British Museum similar to some of the smaller forms referred to *G. digitata* but characterised by the deltoid form of the lamina, its deep dissection into six more or less pointed segments, three on each side of a broad median V-shaped sinus, and by the small size (1·5 cm. broad and 2·5 cm. deep) of the lamina. It is hardly possible to decide whether this and similar small leaves should be regarded as varieties, *e.g. G. digitata* var. *polaris,* or assigned to a distinct species. Fontaine[2] compares some leaves figured by him from Oregon as *Ginkgo* sp. with Nathorst's species, but they are probably nearer to the examples described by Nathorst and others as *G. polaris.* A small bilobed leaf figured by Raciborski[3] from Rhaetic beds near Cracow as *Ginkgo* aff. *whitbiensis* is more likely to be a young leaf of the Fern *Hausmannia.*

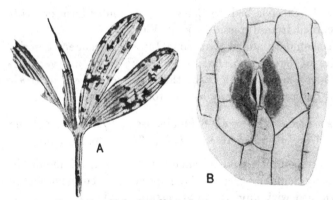

FIG. 642. *Ginkgoites Obrutschewi.* (A, nat. size.)

Ginkgoites Obrutschewi Seward.

This species, named after Prof. Obrutschew who made a collection of plants from Jurassic rocks in Chinese Dzungaria[4] on the western border of Mongolia (lat. 85° N. long. 45° E.), agrees fairly closely with some forms of *G. digitata.* The lamina is deeply

[1] Nathorst (80) A. p. 74; Seward (00) B. p. 261, Pl. IX. fig. 8
[2] Fontaine in Ward (05) B. Pl. XXXIV.
[3] Raciborski (91) Pl. IV. fig. 13.   [4] Seward (11) p. 46, Pls. III.—VII.

bilobed and each half may be similarly divided: the segments are obtuse and not truncate (fig. 642, A); the veins are approximately 1 mm. apart. The walls of the epidermal cells are straight and not sinuous; the stomata (fig. 642, B) are practically confined to the lower surface and are less numerous than in *Ginkgo biloba*. The guard-cells are surrounded by a group of broadly triangular cells with papillose, thickly cuticularised, inner walls, but the overarching papillae are rather less prominent than in *G. digitata*. It is interesting to find short secretory tracts at intervals in the intercostal regions of this species agreeing with those in the leaves of the recent species. A leaf described from Jurassic rocks in Amurland as *Ginkgo* sp., cf. *G. Obrutschewi*[1] agrees closely with the type-specimens.

Ginkgoites crassipes (Feistmantel).

This Upper Gondwana species (fig. 643, B) was described from the Madras coast as *Ginkgo crassipes*[2]. An examination of the type-specimens enables me to confirm generally the accuracy of Feistmantel's figures except in one point, namely the supposed presence of a median ridge extending from the petiole through the lower third of the lamina, which gives the impression of a midrib: this is merely a shallow groove that is clearly accidental. The leaves appear to be entire; the lamina is obcuneate and passes into a fairly stout petiole; the veins are occasionally forked and approximately 0·5 mm. apart. The piece of lamina described by Feistmantel as *Ginkgo* sp.[3] may be specifically identical with this species.

Ginkgoites pluripartita (Schimper).

This Wealden species was first described by Dunker[4] as *Cyclopteris digitata* Brongn. and Ettingshausen[5] also regarded the German specimens as identical with the English Jurassic type. Schimper[6] proposed the name *Baiera pluripartita* because of the deeply divided lamina and the comparatively narrow segments:

[1] Seward (12³) Pl. I. fig. 9.
[2] Feistmantel (77) p. 197, figs. 6, 7; (79) p. 31, Pls. xv., xvi.
[3] Feistmantel (79) Pl. xvi. fig. 12.
[4] Dunker (46) A. p. 9, Pl. I. figs. 8, 10; Pl. v. figs. 5, 6; Pl. vi. fig. 11.
[5] Ettingshausen (52) p. 12, Pl. iv. fig. 2.
[6] Schimper (69) A. p. 423.

while substituting *Ginkgoites* for *Baiera,* it is advisable to retain
Schimper's specific designation on the ground that the deep division
of the lamina appears to be the rule in the Wealden leaves whereas
in *G. digitata* the leaves are usually much less deeply dissected and

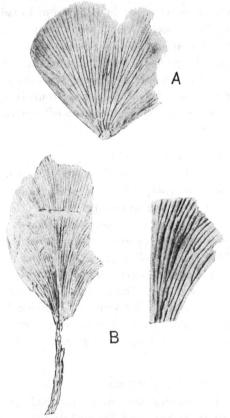

FIG. 643.   *Ginkgoites lobata* (A) and *Ginkgoites crassipes* (B).
(Indian Geological Survey, Calcutta.)

have broader segments.   The largest specimen is one figured by
Ettingshausen with a lamina 7·5 cm. broad and 4·5 cm. deep;
there is generally a deep median sinus and each half of the lamina
is subdivided into relatively narrow obtuse or truncate obcuneate
segments (fig. 635, B).   Schenk[1] describes the epidermal cells as

[1] Schenk (71) B. p. 212, Pls. xxiv., xxv. fig. 7.

polygonal with straight walls; the stomata are surrounded by 5—6 accessory cells as in *Ginkgo biloba* and they are confined to the lower surface. Although there is practically no difference, as regards form and venation, between this Wealden species and some of the Jurassic leaves referred to *G. digitata* var. *Huttoni* the distinctive specific name is retained for the reason already mentioned.

Leaves of a very similar form are figured by Heer as *Ginkgo multinervis*[1] (fig. 635, G), from Upper Cretaceous (Cenomanian) beds of West Greenland, and *Baiera arctica* from the Kome beds (Urgonian) of West Greenland[2].

The specimen from the Atane (Cenomanian) beds of Greenland figured by Heer as *Ginkgo primordialis*[3] appears to be an entire leaf with a long petiole 2·5 mm. broad: the original impression in the Stockholm Museum, too incomplete to serve as the type of a species, shows a very imperfect lamina and a long axis that has probably no connexion with the leaf.

Ginkgoites adiantoides (Unger).

The Tertiary leaves on which this species was founded were in the first instance described as *Ginkgo biloba*[4] and, as several writers have pointed out, so far as regards form and venation there is no good reason for drawing a distinction between the fossils and the leaves of the recent species. In a note published in 1913 Depape[5] definitely adopts the name *Ginkgo biloba* for Tertiary leaves which he regards as specifically identical with Unger's species. In the absence of any satisfactory evidence as to the nature of the reproductive organs and in view of the considerable interval that separates the Tertiary and recent plants, it is clearly inadvisable to assume specific identity. In adopting the generic designation *Ginkgoites* instead of *Ginkgo* I am following the custom generally recognised of distinguishing fossil forms by a special termination, though there is no implication that all species so named are generically distinct from the surviving type. In proposing the

---

[1] Heer (82) ii. B. p. 46, Pl. VIII. figs. 2—4; Pl. IX. fig. 3 *b*. For an account of the stratigraphy and maps of these plant-beds on the coasts of the Noursoak peninsula, see Johnstrup (83) and White and Schuchert (98).

[2] Heer (75) ii. B. p. 37, Pl. III. fig. 3.

[3] Heer (75) ii. B. p. 100, Pl. XXVII. fig. 1.

[4] See Gardner (86) p. 99.   Depape (13).

name *Salisburia adiantoides* for the Miocene leaves from Sengallia
in North Italy Unger[1] indicated their probable identity with
Ginkgo biloba L., the generally adopted name for the existing
species which Smith in 1797 proposed should be called *Salisburia
adiantifolia*[2]. The Miocene leaves from Sengallia figured by
Massalongo and Scarabelli[3] have an entire, irregularly crenulate
or a more or less deeply bilobed lamina very like that of *Ginkgo
biloba*: the specimen named by them *S. Procaccini*[4] should also·
be included in *G. adiantoides*. The resemblance to the recent
leaves extends to the presence of short secretory tracts between
the veins, but these were referred by Massalongo to a fungus
which he named *Sclerotites Salisburiae*[5].

The two specimens reproduced in fig. 644 from the Eocene beds
in the Isle of Mull illustrate the broad fan-like lamina that varies
from 5 to 10 cm. in breadth and may be entire, unevenly lobed or
symmetrically bilobed. The venation is identical with that of the
living species: the characteristic marginal veins on the lower edge
of the lamina are clearly seen in fig. 644, A. The preservation
of these British leaves described by Mr Starkie Gardner[6] is
exceptionally good; they occur as purple impressions in white
clay interbedded with basaltic sheets in the cliffs of Ardtun Head
in the Island of Mull. Though perhaps on the average these
Eocene leaves from Mull are larger than those of the Maidenhair
tree some examples of the latter exceed in size any of the fossils.

Leaves identical with or very similar to the Italian and Scottish
specimens are recorded from both Tertiary and Upper Cretaceous
rocks in many parts of the world. Specimens collected by Dr Lyall
from Tertiary (Miocene or Eocene) beds on Disco Island off the
West coast of Greenland, lat. N. 70°, were described by Heer[7] as
Salisburia borealis and he also speaks of them as *Salisburia adiantoides* var. *borealis*[8]. An examination of the original specimens
in the Kew Museum and an impression in the Dublin Museum

---

[1] Unger (45) p. 211; (50) A. p. 392.   [2] Smith, J E. (1797).
[3] Massalongo and Scarabelli (58) p. 163, Pl. I. fig. 1; Pl. VI. fig. 18; Pl. VII.
fig. 2; Pl. XXXIX. fig. 12.
[4] *Ibid.* p. 165, Pl. XXXIX. fig. 1.
[5] *Ibid.* Pl. I. fig. 1 *a*.   [6] Gardner (86) p. 99, Pl. XXI.; (87) A.
[7] Heer (68) i. p. 95, Pl. II. fig. 1; Pl. XLVII. fig. 4 *a*.
[8] *Ibid.* p. 183.

collected by Sir Leopold McClintock enables me to confirm Gardner's view as to the identity of ′S. *borealis* and *S. adiantoides.*
Examples of *G. adiantoides* were also obtained from Atanekerdluk[1]
on the Noursoak peninsula to the north of Disco Island. Heer

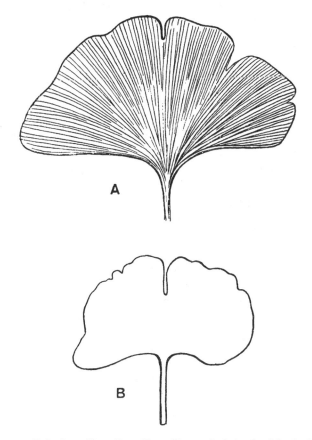

FIG. 644.   *Ginkgoites adiantoides.*   From Eocene beds in the Island of Mull.
(British Museum, V. 1060.)

figures a very well preserved leaf with a long petiole from Atanekerdluk which he refers to *S. adiantoides.*   In a later account of
this flora he refers the species to *Ginkgo.*   Heer regarded these

[1] Heer (68) i. p. 57, Pl. LXXVII. figs. 9—12; Pl. XVII. fig. 14.

Greenland beds as Miocene, but it has been suggested that they may belong to the Eocene period[1].

Heer[2] also described specimens of *Ginkgoites adiantoides* from beds, assigned to the Miocene period, in Sachalin Island. This species has been obtained from the Laramie series, a thick succession of brackish water strata deposited on both sides of the Rocky Mountains 'extending from Mexico far into British North American Territory' and including both Upper Cretaceous and Lower Tertiary strata[3]. Leaves described by Lester Ward as *Ginkgo laramiensis*[4] and regarded by him as intermediate between *G. adiantoides* and *G. biloba* are indistinguishable by any definite character from *G. adiantoides*. Ward's species is also recorded by Knowlton[5] from the Montana formation, a series of beds formerly included in the Laramie group, in Wyoming. *Ginkgoites adiantoides* occurs in Upper Cretaceous beds in British Columbia and specimens described as *Salisburia pusilla*[6] by Dawson, which I believe to belong to this species, were found in the Upper Cretaceous of Vancouver Island. From beds of the same age Penhallow[7] has described some wood as *Ginkgo pusilla* though it is not clear on what grounds it is assigned to the genus *Ginkgo*.

It is interesting to find leaves of this type recorded from Lower Pliocene beds at Saint Marcel-d'Ardèche in France[8] and from Upper Pliocene beds in the Lower Main valley in Germany[9].

It is therefore abundantly clear that trees, apparently indistinguishable as regards the form of the leaves from *Ginkgo biloba*, flourished as recently as the Pliocene period in western Europe and in the Eocene period grew as far north as latitude 70°.

Records of seeds referred to *Ginkgo* are very meagre and add nothing of importance to our knowledge of Tertiary species. Some pyritised seeds were described by Ettingshausen and Gardner[10]

---

[1] White, D. and Schuchert (98) p. 367.
[2] Heer (78) v. p. 21, Pl. II. figs. 7—10.
[3] Ward (87) p. 15, Pl. xxx. figs. 5, 6; (85) Pl. xxxi. pp. 4—6.
[4] *Ibid.* Pl. I. fig. 4; Pl. xxxi. fig. 4.
[5] Knowlton (00) p. 31, Pl. iv. figs. 7—10; Pl. v. fig. 5.
[6] Dawson (93) p. 56, Pl. vi. figs. 11—14. Heer's species *Ginkgo pusilla* is founded on Jurassic leaves; Heer (77) ii. p. 61.
[7] Penhallow (02) p. 43, Pls. xii., xiii.   [8] Depape (13).
[9] Engelhardt and Kinkelin (08) p. 196, Pl. xxiii. fig. 18.
[10] Ettingshausen (79) p. 392.

from the London clay of Sheppey as *Salisburia eocenica* and afterwards figured by Gardner as *Ginkgo*? *eocenica*[1]. The specimens (11 × 9 mm.) are smaller than the seeds of the recent species but in shape and in the keeled shell there is a fairly close resemblance. The hard sclerotesta forms the surface of the fossils.

Similar seeds have been assigned to *Ginkgoites adiantoides* from the Upper Pliocene of the Frankfurt district[2], but neither the German nor English specimens possess any interest as records of *Ginkgoites* seeds.

Ginkgocladus. Ettingshausen.

An imperfect leaf-like impression described by Ettingshausen[3] from Eocene beds in New Zealand was made by him the type of a new genus *Ginkgocladus* and interpreted as a stalked phylloclade similar to those of the recent Conifer *Phyllocladus,* but because of the presence of a slender stalk and the resemblance of the lateral veins to the venation of *Ginkgo* Ettingshausen suggested an affinity to that genus. The existence of a midrib is, however, an important difference. Neither the New Zealand species nor similar fragments from Eocene strata in New South Wales (*Ginkgocladus australiensis*[4]) are of value as botanical records.

BAIERA. Braun.

This generic name was first used by Braun[5] for some Triassic and Jurassic leaves agreeing in shape with those of *Ginkgoites* but distinguished by the greater number and less breadth of the linear segments. Braun's definition states that the primary veins are dichotomously branched while between them secondary veins form irregular hexagonal meshes. Schenk[6] examined Braun's Rhaetic specimens and failed to discover any indication of the presence of secondary veins. In 1877 Heer[7] emended the original definition of the genus: he refers to the presence of finer veins between the main vascular strands and this feature is shown in

---

[1] Gardner (86) p. 46, Pl. IX. figs. 31—34.
[2] Engelhardt and Kinkelin (08) p. 196, Pl. XXIII. figs. 16—18.
[3] Ettingshausen (87) p. 39, Pl. VII. fig. 19.
[4] *Ibid.* (88) p. 103, Pl. VIII. fig. 32.   [5] Braun, C. F. W. (43) p. 20.
[6] Schenk (67) A. p. 42. Schenk includes some of Braun's species of *Baiera* in Unger's genus *Jeanpaulia* which has since been discarded.
[7] Heer (77) ii. p. 51.

some of his figures : these interstitial 'veins' probably mark the
position of hypodermal strands of stronger cells, a feature that is
not represented in recent or fossil *Ginkgo* leaves and is by no means
generally characteristic of *Baiera*.   Both Braun and Heer describe
male and female reproductive organs : Braun interpreted some
small specimens as sporocarps but these were recognised by
Schenk as young foliage leaves.   The male organs are described
by Heer as 'amenta staminifera pedunculata, nuda, filamenta
filiformia, antherae loculis 5—12, verticillatis.   Semen drupae-
forme, basi cupula carnosa cinctum' and compared with Schenk's
Stachyopitys Preslii which that author afterwards regarded as
microstrobili of *Baiera Muensteriana*.   Reference is made to
these and similar fossils in the account of examples of male
flowers[1].   Seeds have been referred to *Baiera* on evidence
furnished by their occasional association with leaves and by their
resemblance to those of *Ginkgo*.   It has been suggested that
specimens described from the Potomac group of Virginia and
Maryland as *Carpolithus ternatus*[2] and other species may be seed-
bearing organs of *Baiera,* but there is no satisfactory evidence in
support of this view.   In all probability some of the associated
seeds belong to *Baiera,* also some of the microstrobili, *e.g.* Leuthardt's
Swiss specimens described on a later page, but in the present
state. of our knowledge it is preferable to regard these specimens
as reproductive organs that cannot be assigned with certainty to
any particular species of *Baiera* or *Ginkgoites*.

Leaves assigned to *Baiera* vary within wide limits as regards
size, the number of linear segments and their angle of divergence.
In many cases the leaves are petiolate though in several instances
the petiole is represented by a narrow basal region of the lamina
as in *Psygmophyllum*.   It is stated by some authors that the veins
are undivided, but though dichotomy is less frequent in *Baiera*
and may be absent in narrow parallel-sided segments it is by no
means rare.   The difference in venation between such leaves as
Ginkgoites digitata and typical species of *Baiera, e.g. Baiera gracilis,*
is mainly the result of the different form and degree of dissection
of the lamina.   The choice between *Baiera* or *Ginkgoites* as the

---

[1] Page 51.
[2] Fontaine (89) B. pp. 265 etc., Pls. 134 etc.;   Berry (11) p. 372.

more suitable name for certain forms of leaf is not governed by
any definite criterion: specimens described as *Ginkgo sibirica,
G. lepida, G. concinna*[1], etc. are indistinguishable from leaves
referred by authors to *Baiera*. Similarly such a species as
B. Lindleyana differs very slightly from some forms usually included in the genus *Czekanowskia*. Although leaves of *Ginkgoites*
and *Baiera* are abundantly represented in plant-bearing beds we
know very little of the habit of the foliage-shoots; in a few cases
there is evidence of the occurrence of several leaves on a single
short shoot (fig. 646); in *Baiera paucipartita*, for example[2], the
habit is the same as that of *Czekanowskia* and *Phoenicopsis*, but
in view of the frequent preservation of *Czekanowskia* leaves still
attached to an axis it is surprising that the leaves of *Baiera* almost
always occur as detached specimens. The explanation may be
that in *Ginkgoites* and *Baiera* the foliage-leaves were borne on
long and dwarf-shoots as in the recent species *Ginkgo biloba*,
whereas in *Czekanowskia* the leaves were confined to shoots of
limited growth as in *Pinus*. Some specimens described by Salfeld[3]
from the Solenhofen beds of Bavaria as *Baiera*? *longifolia* Heer
are interesting in this connexion; they consist of fairly stout
branches bearing alternate leaf-like organs having the habit of
Baiera longifolia but subtended by a short and thick recurved
spinous process. There is no means of deciding from the available material whether the resemblance of the leaves to those of
Baiera is an expression of relationship or merely a case of parallel
development, nor have we any means of determining the morphology of the leaves and the subtending spines. The Solenhofen
plant agrees in habit with *Sewardia latifolia*[4] from the Wealden
of England and is included under that genus. While relationship
between *Sewardia* and *Baiera* is by no means excluded, it is clear
that the species of the former genus differ considerably from
typical representatives of *Baiera* and *Ginkgoites*. Certain species
of *Baiera* exceed in size the leaves of any example of *Ginkgoites*,
notably *B. Simmondsi, B. spectabilis,* and others.

The specimens described by Fontaine and Berry from the

---

[1] Heer (77) ii. Pl. XIII. figs. 6—8.
[2] Nathorst (78²) B. Pl. XXI.
[3] Salfeld (07) B. p. 195 Pl. xx. fig. 3; Pl. XXI. fig. 1.
[4] Page 105.

Potomac group in Virginia as *Baiera foliosa*[1], consisting of an axis bearing crowded leaves with a deeply and rather irregularly divided lamina and a comparatively broad and flat basal region, are not typical examples of the genus but agree more closely in habit with the older genus *Dicranophyllum*: their precise position cannot be definitely determined.

The structure of the cuticles is known in a few species from accounts published by Schenk[2], Nathorst[3] and Thomas[4]; the epidermal cells are sometimes characterised by fairly prominent papillae on the outer walls (fig. 647) and the stomata, more abundant on the lower surface but present also on the upper surface, closely resemble those of *Ginkgo* and *Ginkgoites*; the guard-cells are slightly depressed and are surrounded by 5—6 subsidiary cells with strongly cuticularised and projecting walls (fig. 647). Nathorst[5] has drawn attention to the presence in *B. spectabilis* of traces of some secreted substance in the mesophyll recalling the secretory tracts in the leaves of *Ginkgo*.

Braun and other authors have included *Baiera* in the Filicales, and attention has been called to the danger of confusing true Fern-fronds with leaves of *Baiera*. Berry's discovery of sporangia on the linear segments of *Baiera*-like leaves from the Potomac beds, originally referred by Fontaine to his genus *Baieropsis* and regarded by him as Ginkgoaceous, illustrates the possibilities of error in determinations founded on leaves alone. The fertile examples of *Baieropsis* have been made the type of a new genus *Schizaeopsis*[6], other species of Fontaine's genus being transferred to the genus *Acrostichopteris*; they differ from *Baiera* in their attachment to slender axes and are no doubt portions of compound Fern-fronds. It is impossible to define with confidence the precise geological range of the genus; leaves from Permian and Upper Carboniferous strata agreeing with *Baiera* in the deep dissection of the lamina have been assigned to the genus *Ginkgophyllum* (*Psygmophyllum*) and compared with Saporta's *Ginkgophyllum Grasserti* (fig. 669, p. 87). In imperfect specimens it is not always possible to draw a sharp line between *Baiera* and species of *Psygmo-*

---

[1] Fontaine (89) B. p. 213, Pl. xciv. fig. 13; Berry (11) p. 372, Pl. lix.
[2] Schenk (67) A. Pl. vi. figs. 1, 2; Pl. ix. figs. 11—13.
[3] Nathorst (06).   [4] Thomas (13²) p. 244, fig. 5.
[5] Nathorst (06) p. 9, fig. 9.   [6] Berry (11) p. 214; (11²).

phyllum. There are, however, certain Permian leaves that are legitimately included in *Baiera*; the genus appears to have been widespread in Triassic floras, though more especially in those of the Rhaetic and Jurassic age. *Baiera* shares with *Ginkgoites* an important position in the Jurassic vegetation of both hemispheres, but in the Cretaceous period *Baiera* appears to have been a comparatively rare genus and in the Tertiary floras it was entirely replaced by members of the Ginkgoales with leaves of the type that still survives. *Baiera* is clearly an older form than *Ginkgoites*; it is not recorded from India and it has not been found in the Permo-Carboniferous rocks of Gondwana Land.

Baiera though unknown in a petrified condition may confidently be included in the Ginkgoales; the habit of the leaves, the structure of the epidermal cells and such evidence as there is with regard to the fertile shoots favour this conclusion. It must, however, be added that the position of the Palaeozoic examples is less firmly established.

Baiera virginiana Fontaine and White.

This species, from Permian beds in Virginia[1], is based on imperfect portions of laminae deeply divided into bifurcate segments with truncate apices and several parallel veins; it agrees in the form of the lamina and in the linear divisions with the type-specimen of Brongniart's *Fucoides digitatus*[2] from Permian beds of Mansfeld, a species which Geinitz[3] also recorded, but under the generic name *Zonarites*. Potonié[4] and other authors, following Heer, transferred the species to *Baiera*. Heer's combination, *B. digitata*, had, however, already been used by Schimper[5] for the Jurassic species usually called *Ginkgoites digitata*.

Leaves of the form represented by *B. virginiana* may be closely allied to Saporta's Permian species *Ginkgophyllum* (*Psygmophyllum*) *Grasserti*[6]. In the absence of the basal part of the lamina a complete diagnosis or accurate identification is impossible. Some authors have referred fragments of similar leaves to the genus

---

[1] Fontaine and White (80) B. p. 103, Pl. xxxvii. figs. 11, 12.
[2] Brongniart (28²) A. p. 69, Pl. ix. fig. 1.
[3] Geinitz (62) Pl. xxxi. figs. 1 2.
[4] Potonié (93) A. p. 237, Pl. xxxii. fig. 2; Pl. xxxiii. fig. 6.
[5] Schimper (69) A. p. 423. [6] Page 87.

Schizopteris, e.g. S. Guembeli from Upper Carboniferous and Permian strata[1], but other Permian specimens assigned to *Schizopteris, e.g. S. dichotoma* and *S. trichomanoides*[2] are generically distinct and probably belong to Pteridosperms. *B. virginiana* and the leaves referred to *Baiera digitata*—possibly specifically with Fontaine's type—represent Permian forms that agree closely with the larger species *B. Simmondsi,* Heer's Keuper species *B. furcata,* and with the larger examples of *B. Muensteriana* (Presl) from Rhaetic beds.

Baiera Raymondi Renault.

This French, Permian (Autunien), species[3] (fig. 645, A) differs but little from *B. virginiana*; the narrow cuneate leaf is divided

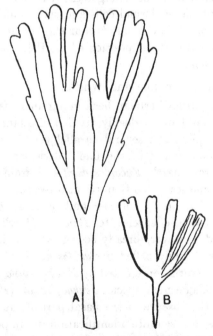

FIG. 645. A, *Baiera Raymondi.* B, *Ginkgoites Geinitzi.*
(Nat. size; A, after Renault; B, after Nathorst.)

into very regularly bifurcate narrow linear segments diverging at a small angle. The largest specimen is 10·8 cm. long, the basal

[1] Weiss (69) B. Pl. XII. fig. 7; Goeppert (64) A. Pl. IX. figs. 6, 7.
[2] Zeiller (06) B. Pl. I. figs. 7, 8.   [3] Renault (88) p. 324, fig. 48.

portion of the lamina is 5 mm. broad and the ultimate segments with obtuse apices have a breadth of 2·5 mm. The venation is imperfectly shown in the specimens figured by Renault and Zeiller[1] *Baiera furcata* Heer.

A type similar to *B. multifida* Font., but characterised by the more uniformly narrow segments (2—2·5 mm.), is described by Heer[2] and Leuthardt[3] from Keuper beds of Switzerland. Leuthardt's figures show a single vein in the segments a feature which may be a peculiarity of the species. It was in association with this species that Leuthardt found the male flowers referred to on another page (p. 53).

Baiera paucipartita Nathorst.

The leaves of this Swedish, Rhaetic, species[4] may reach a length of 10 cm.; the lamina is narrow and cuneate, deeply divided into bifurcate linear segments with obtuse apices. It differs from

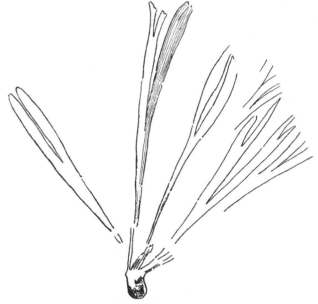

Fɪɢ. 646.   *Baiera paucipartita.*   (After Nathorst; ¾ nat. size)

[1] Zeiller (06) B. p. 202, Pl. xlviii. figs. 1, 2.
[2] Heer (76) A. Pls. xxix., xxx., xxxvi.
[3] Leuthardt (03) p. 7, Pls. ii.—iv.
[4] Nathorst (86) p. 94, Pls. xx.—xxii.

B. longifolia Heer and other similar species in the smaller number of segments. There is no petiole but the leaves are attached by a narrow basal portion to a short axis (fig. 646). The occurrence of several leaves on a very short scale-covered dwarf-shoot is an interesting feature which affords evidence of relationship with *Czekanowskia* and *Phoenicopsis*. *Baiera paucipartita* is described by Yokoyama[1] from Rhaetic beds in Japan and an imperfect specimen from Rhaetic strata in New Zealand has been assigned by Arber[2] to this species. In the case of imperfect leaves it is impossible to distinguish specifically between many 'species' of *Baiera* characterised by the division of the lamina into bilobed linear segments.

Baiera spectabilis Nathorst.

The leaves of this Rhaetic species from the south of Sweden[3] reach a length of 25 cm.; the coriaceous lamina is obcuneate and fan-like but narrower than the leaves of such a species as *B. Simmondsi*. There is no well-defined petiole; the lamina is deeply divided into two symmetrical halves each of which is further subdivided into bifurcate linear segments, and the strongly contracted ultimate segments are a characteristic feature (fig. 647). The basal region forms a stalk-like portion varying in length and, in the specimens so far obtained, not exceeding 3 cm. The veins are parallel to the sides of the segments and on the average 1 mm. apart; they are occasionally dichotomously branched. Stomata occur on both sides of the lamina but are more numerous on one surface, presumably the lower; the guard-cells are slightly depressed and surrounded by 5—6 subsidiary cells with thickly cuticularised papillose walls (fig. 647, B) as in *Ginkgo*. The epidermal cells on the lower surface are also papillose like those described by Thomas in *B. longifolia*[4]. Several dark spherical and spindle-shaped patches found by Nathorst between the cuticularised layers are believed to be the remains of some resinous or other substance comparable with that formed in the secretory tracts in the leaves of *Ginkgo* and some species of *Ginkgoites*.

Baiera spectabilis is one of the larger forms of the species; it

[1] Yokoyama (05) Pl. II. fig. 5.   [2] Arber, E. A. N. (13) Pl. VII. figs. 2, 3.
[3] Nathorst (06).   [4] Thomas (13) Pl. xxv. figs. 3, 4.

FIG. 647. A, *Baiera spectabilis*; B, stoma　C, D, epidermal cells and stoma of *Baiera longifolia*. (A, B, after Nathorst A, ¾ nat. size　C, D, after Thomas.)

resembles *B. pulchella,* a Jurassic species described by Heer[1] from East Siberia and by Bartholin[2] from Bornholm, but in typical examples of *B. pulchella* the lamina is divided into two segments only. Comparison may be made also with *B. longifolia* a Jurassic species distinguished by its narrower segments. A specimen from Bornholm referred by Möller[3] to *B. pulchella* is probably, as Nathorst suggests, a piece of a *B. spectabilis* leaf.

Baiera Simmondsi (Shirley).

The leaves described by Shirley[4] as *Ginkgo Simmondsi* from Denmark Hill near Ipswich in Queensland, from rocks that are probably of Rhaetic age, are of the same type as the leaf on which Ratte[5] founded his species *Jeanpaulia* (?) *palmata,* which he afterwards transferred to *Salisburia,* from the Wianammata beds (Trias) near Sydney. The precise age of the rich flora from Ipswich is difficult to determine: a recent examination of several specimens in the Brisbane collections led me to regard the plants as Rhaetic, but further light on this question will be afforded by Mr Walkom who is engaged in an investigation of the material. The Australian leaves agree closely with Fontaine's Triassic species, *Baiera multifida*[6], from Virginia: the plant-beds of the Richmond coalfield are correlated with the Lunz plant-beds in Austria[7], the flora of which has never been adequately illustrated. The specimens from Virginia on which Fontaine founded his species do not afford any evidence of a true petiole and the basal portion of the cuneate lamina is narrower than in the Australian leaves: it is, therefore, not improbable that *B. multifida* is a distinct though very similar species. Ratte's name *B. palmata* cannot be retained as Heer had previously employed the same name for a Jurassic Siberian form[8]: I have adopted Shirley's designation in the belief that there are no differences of specific value between the Sydney and Ipswich specimens.

The leaf reproduced in fig. 648 is Ratte's type-specimen in the Australian Museum, Sydney: the whole leaf is nearly 30 cm.

---

[1] Heer (77) ii. p. 114, Pl. xx. fig. 3 *c*; Pl. xxii. fig. 1 *a*; Pl. xxviii. fig. 3.
[2] Bartholin (94) Pl. xi. fig. 5.   [3] Möller (03) Pl. iv. fig. 19.
[4] Shirley (98) p. 12, Pl. ii.   [5] Ratte (87) Pl. xvii; (88).
[6] Fontaine (83) B. p. 87, Pls. xlv.—xlvii.
[7] Berry (12).   [8] Heer (77) ii. p. 115, Pl. xxviii. fig. 2.

long and 23 cm. in breadth; there are nearly 60 ultimate linear segments with obtuse apices and, in the smaller subdivisions,

Fig. 648. *Baiera Simmondsi*. (Australian Museum, Sydney; ⅔ nat. size.)

3—5 veins. There is a well-defined petiole and in outline the whole leaf is identical with typical examples of *Ginkgo biloba*.

Fontaine speaks of the lamina of his species as reaching a length of 25 cm.; both in the method of division and in the form of the segments, *B. multifida* agrees closely with the specimen shown in fig. 648. A similar form of leaf is figured by Solms-Laubach[1] from Rhaetic beds in Chile as *Baiera? Steinmanni*, but the lamina only is preserved. Schenk's *B. taeniata*[2] from the Rhaetic flora of Franconia is another similar type.

Baiera stormbergensis Seward.

The specimens described from the Stormberg series (Rhaetic) of South Africa[3] are portions of leaves that must have reached a length of 12 cm. or more and a breadth of 10 cm. The lamina is deeply divided into broad linear segments which are further subdivided into narrower distal segments. In the lower part of the lamina the venation is comparatively coarse, but as the result of repeated dichotomy the veins are much more numerous in the upper portion. This species may be merely a larger form of Feistmantel's *B. Schenki*[4] from the same beds, in which the lobes are narrower as in *B. longifolia* Heer. *B. stormbergensis* resembles Nathorst's *B. spectabilis* from the Rhaetic of Scania, but the segments of the South African leaves have a coarser venation.

Baiera Muensteriana (Presl).

This Rhaetic species, originally figured by Presl as *Sphaerococcites Muensterianus* and subsequently described by Braun as *Baiera dichotoma*, was named by Schenk *Jeanpaulia Muensteriana*. Schenk[5] examined Braun's specimens from Franconia and identified the supposed sporocarps as partially expanded segments of foliage-leaves. The leaves are petiolate and the fan-like lamina is deeply dissected into bifurcate linear segments; the veins are numerous and dichotomously branched. The epidermal cells are elongate over the veins and elsewhere polygonal; their walls are straight or slightly sinuous. The stomata are of the usual type met with in Ginkgoaceous plants.

Baiera Muensteriana cannot be distinguished by any definite character from leaves that are referred to *B. gracilis*: in the

[1] Solms-Laubach (99) Pl. xiv. fig. 1.
[2] Schenk (67) A. p. 26, Pl. v. figs. 1—4; Pl. vi. figs. 1, 2.
[3] Seward (03) B. p. 64, Pl. viii. fig. 3.
[4] Feistmantel (89) p. 72, Pl. iii.   [5] Schenk (67) A. p. 39, Pl. ix.

type-specimen of the latter species the segments are fewer than
in *B. Muensteriana,* but in some Jurassic forms (*e.g.* fig. 651)
this difference no longer holds good. This is only one among
several instances where Rhaetic and Jurassic 'species' cannot be
separated by any constant differentiating feature. *B. Muensteriana* is recorded also from Persia[1] and several European localities,
but it is impossible to determine its geographical range apart from
that of *Baiera gracilis.*

Baiera gracilis Bunbury ex Bean MS.

The type-specimen of this species from the Middle Jurassic
rocks of Yorkshire, as shown in fig. 649, is an impression of an

FIG. 649. *Baiera gracilis.* Type-   FIG. 650. *Baiera gracilis.* (British Museum, 39208.)
specimen of Bunbury (Bun-
bury   Collection,   Botany
School, Cambridge).

imperfect leaf with a fan-shaped lamina deeply divided into forked
linear segments[2]: a better example is reproduced in fig. 650. Bean

---

[1] Schenk (87) B. Pl. VIII. fig. 44; Zeiller (05) p. 194.
[2] Bunbury (51) A. p. 182, Pl. XII. fig. 3.

referred this type to *Schizopteris* but Bunbury, while adopting Bean's MS. specific name, substituted the generic designation *Baiera*. The leaves are petiolate and the lamina is divided almost or quite to the base into a varying number of linear segments with obtuse apices. The veins, frequently indistinct, run parallel to the edges of the lamina and there are several in each segment.

Leaves identical with or very similar to *Baiera gracilis* are very widely distributed among Jurassic floras in both hemispheres. Some of the specimens described by authors as *Ginkgo lepida* are hardly distinguishable from Bunbury's species; *G. concinna*[1] Heer from the Siberian Jurassic flora is another very similar form; also *Baiera incurvata* Heer[2] from the Lower Cretaceous of Greenland, *B. angustiloba* Heer[3], as figured from Siberia and China, *B. bidens* (Ten.-Woods)[4] from Queensland, *B. australis* McCoy and *B. delicatula* Sew. from Jurassic rocks in Victoria[5], also leaves recently referred by Halle[6] to *B. australis* from the Lower Cretaceous plant-beds of Patagonia. *Baiera gracilis* is recorded from Upper Jurassic (or Lower Cretaceous) beds in Alaska[7], but leaves of this

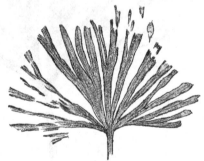

FIG. 651.	*Baiera gracilis* forma *Mucnsteriana.*
(British Museum, ¾ nat. size.)	M. S.

type are rare in the Jurassic strata of North America. The Rhaetic species *B. Muensteriana*[8] (Presl) described by Schenk from Franconia and by other authors is a closely allied type which

[1] Heer (77) ii. Pl. XIII. figs. 6—8.	[2] *Ib d.* (82) B. Pl. XIII. fig. 6.

[3] *Ib d.* (78) ii. Pl. VII. fig. 2; Krasser (05) Pl. II. fig. 10; Schenk (83) A. Pl. LIII. fig. 1.

[4] Tenison-Woods (83) A. Pl. IV. fig. 3.

[5] Seward (04²) B. figs. 36—38.			[6] Halle (13) Pls. IV., V.

[7] Fontaine in Ward (05) B. Pl. XLIV. fig. 2.		[8] Schenk (67) A. Pl. IX.

cannot always be distinguished from *B. gracilis*. The example shown in fig. 651 from the Yorkshire coast has been named *B. gracilis* forma *Muensteriana* to denote its close resemblance to the Rhaetic species[1]. *B. Guilhaumali*[2] described by Zeiller from Rhaetic rocks in Tonkin is another similar form but the leaves are narrower and the apices of the segments more obtuse. On the one hand *Baiera gracilis* approaches close to *B. Lindleyana*, a species characterised by still narrower segments, and on the other it shades into leaves agreeing with *Ginkgoites sibirica*.

Baiera longifolia (Pomel).

Pomel[3] described this Jurassic species as *Dicropteris longifolia* and Heer substituted the generic name *Baiera*[4]. The leaves resemble those of *B. Simmondsi* in the division of the lamina into narrow linear segments 2—9 mm. in breadth, but the leaf is narrower and cuneate; the segments have obtuse apices. Heer describes the veins as parallel and simple, 3—7 in each segment. With this species Heer associates some male flowers similar to those shown in fig. 654, also some detached seeds, but in neither case is there any convincing evidence of connexion. The Siberian species *B. Czekanowskiana*[5], recorded also by Möller from Bornholm, is probably not a distinct type. Thomas[6] records *B. longifolia* from the Middle Jurassic series of Yorkshire and gives new facts with regard to the structure of the epidermal cells: one of his specimens of an incomplete lamina is 12 cm. long, the whole leaf being at least 18 cm. in length. The epidermal cells have a very thick cuticle; those on the lower surface are arranged in longitudinal rows and most of them have a prominent papilla; on the lower face the cells are more rounded or hexagonal and the stomata are much more numerous; each pair of guard-cells is surrounded by a group of 5—6 subsidiary cells (fig. 647, C, D) as in *Ginkgo*. Krasser[7] records this species from Jurassic rocks

[1] Seward (00) B. p. 264.   [2] Zeiller (03) B. Pl. L. figs. 16—19.
[3] Pomel (49) p. 9.
[4] Heer (77) ii. p. 52, Pl. VII figs. 2, 3; Pl. VIII.; Pl. IX. figs. 1—11; Pl. X. figs. 6, 7; Pl. XV. fig. 11 *b*.
[5] *Ibid.* p. 56, Pl. X. figs. 1—5; Pl. VII. fig. 1; (82) B. Pl. III. figs. 4—8; Möller (03) Pl. V. fig. 3.
[6] Thomas (13) p. 243, Pl. XXV. figs. 3, 4.
[7] Krasser (05) p. 18, Pl. I. fig. 16.

in China but the photographic reproduction is unfortunately too obscure to afford any indication as to the nature of the specimen.

Baiera Phillipsi Nathorst.

This Jurassic species[1] (fig. 652) illustrates the absence of any definite dividing line between *Baeira* and *Ginkgoites*; it agrees very closely with *G. sibirica* and with leaves assigned to *G. lepida* and other 'species.' Fig. 652 is drawn from Phillips' type-specimen[2] which he named *Sphenopteris longifolia* and afterwards transferred to *Cyclopteris*; his specific name is not retained because Pomel adopted it for a type subsequently called by Heer *Baiera longifolia*[3]. Krasser records *B. Phillipsi* from Jurassic strata in Sardinia[4]. This species shades into *B. gracilis* and the very similar *B. australis* McCoy, especially resembling some leaves included by Halle[5] in the latter species.

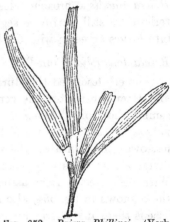

FIG 652. *Baiera Phillipsi*. (York Museum; ⅔ nat. size.) M. S.

Baiera Lindleyana (Schimper).

Leaves of this type were first figured by Lindley and Hutton[6] as *Solenites*? *furcata* and transferred by Braun to *Baiera*. Schimper[7] subsequently substituted *Jeanpaulia* and proposed the specific name *Lindleyana* on the ground that Heer had employed the designation *furcata* for a Rhaetic species of *Baiera*. Saporta included this species in *Trichopitys*. *Baiera Lindleyana* is characterised by the deep dissection of the lamina into very narrow, filiform, segments and by the presence of a long and slender petiole (fig. 653). Some forms of this type with rather broader segments are hardly distinguishable from *Baiera gracilis*.

[1] Nathorst (80) A. p. 76.   [2] Phillips (75) A. Pl. VII. fig. 17.
[3] Seward (00) B. p. 270.   [4] Krasser (13) p. 5.
[5] Halle (13) Pls IV. V.
[6] Lindley and Hutton (37) A Pl. 209.   [7] Schimper (69) A. p. 683.

In a former account of this species[1] I included the specimen reproduced in fig. 661 (p. 66), also a similar specimen figured by Phillips[2]
as a distinct type, *Baiera microphylla*. The examination of additional material collected from the Yorkshire coast by Mr Hamshaw Thomas leads me to substitute *Czekanowskia* for *Baiera* as the more appropriate name for the bunch of leaves represented in fig. 661 which is in all probability identical with *B. microphylla* as figured by Phillips. In the case of incomplete leaves it is by no means easy to distinguish *B. Lindleyana* from *Czekanowskia microphylla*; but in the latter the branches of the lamina are separated by a smaller angle and if cuticular preparations are available the stomata afford a means of differentiation: in *Baiera* the guard-cells are surrounded by a circular group of cells, while in *Czekanowskia* the subsidiary cells

FIG. 653. *Baiera Lindleyana.* (British Museum; A, 39208; B, V. 3682.)

are longer and narrower, forming a more oblong group.

Baiera Lindleyana is recorded also from Middle Jurassic rocks in Chinese[3] Dzungaria and from Upper Jurassic rocks in Scotland[4]. Some specimens described by Fontaine[5] from the Black Hills (Lower Cretaceous) as *Czekanowskia nervosa* Heer are, as Berry[6] points out, probably leaves of a *Baiera*, and I am disposed to refer them to *B. Lindleyana*.

Baiera Brauniana (Dunker).

This species[7], represented by leaves from Wealden and Upper Jurassic rocks, agrees in the form and dissection of the lamina

---

[1] Seward (00) B. p. 266, fig. 46 (p. 268).
[2] Phillips (75) A. p. 200, fig. 9.   [3] Seward (11) Pl. IV. fig. 44.
[4] Seward (11²) Pl. V. fig. 105.
[5] Fontaine in Ward (99) B. p. 685, Pl. 169, figs. 1, 2.
[6] Berry (11) p. 374.   [7] Dunker (46) A. p. 11, Pl. V. fig. 4.

with *B. gracilis* but is distinguished by the smaller dimensions. The imperfect example shown in fig. 641, B from the Kimeridge beds of Sutherland (Scotland)[1] illustrates the unsatisfactory characters on which specific distinctions are drawn in the case of *Baiera* leaves agreeing in habit with *B. gracilis*. Better examples are figured by Schenk[2] from the Wealden of North Germany.

Baiera spetsbergensis Nathorst.

This species, one of the smallest representatives of the genus, is described by Nathorst[3] from Upper Jurassic rocks of Spitzbergen; it is characterised by the very narrow but apparently cylindrical segments and, except in its smaller size, resembles *B. Lindleyana*.

iii. FLOWERS AND SEEDS.

Our knowledge of seeds assigned to Mesozoic and Tertiary representatives of *Ginkgo* or to *Baiera* is limited to casts and impressions of detached examples: no reproductive organs have been discovered either in a petrified state or in connexion with a foliar shoot. Reference has already been made to some small *Ginkgo*-like seeds from the Eocene beds of Sheppey described by Gardner as *Ginkgo ? eocenica*. Many similar seeds are figured by Heer from Jurassic strata in Siberia and elsewhere, in most cases as detached seeds but in a few instances borne singly or in pairs on an axis resembling the peduncle of *Ginkgo biloba*[4]. Heer's seeds are correlated with *G. digitata, G. sibirica* and other species but only on evidence afforded by association with leaves; they are preserved as oval nuts, sometimes enclosed in a carbonaceous envelope possibly representing an outer flesh, and resemble *Ginkgo* seeds in shape though they differ from them in their smaller size (8—9 mm. long and 6—8 mm. in diameter): in some of the Jurassic specimens the nuts have an apical beak. All that can be said is that seeds similar except in their smaller size to those of the recent species are not infrequently found in association with different species of *Ginkgoites*.

[1] Seward (11²) p. 680.
[2] Schenk (71) B. p. 224, Pl. III. figs. 9—14.
[3] Nathorst (97) p. 53, Pl. III. figs. 6—12.
[4] Heer (82) ii. A. p. 16, Pls. IV, v; (77) ii. p. 57, Pl. XI.

It has been suggested that the seed-bearing shoots, which Carruthers named *Beania,* from Jurassic beds on the Yorkshire coast may have belonged to a member of the Ginkgoales, but it is at least equally probable that *Beania* is Cycadean and possibly the seed-bearing axis of *Nilssonia.* The genus is described in Ch. xxxviii[1]. It is possible that specimens from Cretaceous and Jurassic rocks regarded by Heer as male flowers of *Ginkgoites sibirica* and other species, also specimens described by him as *Antholithus Schmidtianus*[2], may be fertile shoots, which bore seeds and not microsporangia, belonging to *Ginkgoites* or some other member of the Ginkgoales: the nature of these fossils is, however, uncertain and they are described under the generic name *Stenorachis.*

a.   Male Flowers.

As with seeds so also with regard to the microsporophylls our information is scanty and indecisive. Nathorst[3] first suggested that some small carbonised bodies from Yorkshire Jurassic beds figured by Phillips[4] as 'unknown leaves' are probably fragments of male flowers of some species of *Ginkgoites.* The specimen of which Phillips figured a small portion is shown in fig. 654, B; it consists of a slender axis with several short and partially broken lateral branches bearing terminal groups of oblong bodies 4 mm. long and 1 mm. wide, 2—4 in each group: these suggest comparison with pollen-sacs with longitudinal dehiscence, and the habit of the whole fertile shoot agrees with that of a male flower of *Ginkgo biloba.* In the recent species the microsporangia are only about 2 mm. long, but in the occurrence of two to four microsporangia on a single microsporophyll the resemblance between the fossil and recent form is fairly close[5]. Unfortunately it has not been possible to make any preparations of the cuticularised remains showing microspores, and while the probability is that the oblong bodies are microsporangia it is not impossible that they are small seeds. A collection of identical bodies showing what appears to be a median line of dehiscence is illustrated in Part I of *The Jurassic Flora of Yorkshire*[6]. A larger specimen is shown in fig. 654, A;

---

[1] Vol. iii. p. 502.   [2] Heer (82) A. p. 21, Pl. ix.
[3] Nathorst (80) A. p. 75.   [4] Phillips (29) A.; (75) A. Pl. vii. fig. 23.
[5] See page 5.   [6] Seward (00) B. p. 260, fig. 45.

the axis is 2·3 cm. long and some microsporangia are seen in their
original position, while others are detached[1]. It is by no means
unlikely that these specimens are portions of male flowers of
Ginkgoites digitata or of some other species, but this cannot be
definitely settled until better material is available. Some Rhaetic
fossils described by Nathorst[2] as *Antholithus Zeilleri* present a
certain resemblance to these supposed male flowers. One of

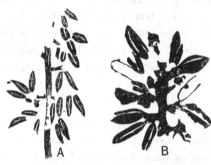

FIG. 654. *Antholithus* sp. (Sedgwick Museum,   FIG. 655. *Antholithus Zeilleri.*
   Cambridge; A,*ca.* × 1½; B,*ca.* × 2.) A, drawn   (After Nathorst; × 2⅔.)
   by M. Seward; B, drawn by L. D. Sayers.

Nathorst's specimens from Scania is reproduced in fig. 655 twice
natural size; the photograph, for which I am indebted to Prof.
Nathorst who published it in 1908, shows a cuticular preparation
of the axis and microsporangia. The axis of *Antholithus* is
dichotomously branched and bears terminal clusters of microsporangia about 3·5 mm. long, usually eight in a cluster; several of
them have dehisced longitudinally and the apices show a slight
separation of the two halves. In some of the sporangia Nathorst
found microspores with an average length of 40—43 μ agreeing
closely with the spores of *Ginkgo* and recent Cycads. Nathorst
considered that *Antholithus Zeilleri* may be a male flower of some
Ginkgoaceous plant though a correlation with a Cycadean type is
by no means excluded. There is, however, a general resemblance
between the English Jurassic specimens shown in fig. 654 and the
Rhaetic species; the latter is distinguished by a greater tendency

---

[1] Seward and Gowan (00) B. Pl. IX. fig. 28.
[2] Nathorst (08) p. 20, Pl. IV.

towards a dichotomous habit of the axis, and in the Jurassic specimen we have no proof as to the nature of the 'microsporangia.'

Arguments have recently been brought forward[1] in favour of regarding *Antholithus Zeilleri* as the male organ of the plant which bore the fronds known as *Lepidopteris Ottonis* (Gopp.), originally described by Goeppert as *Alethopteris Ottonis* and made by Schimper the type of a new genus *Lepidopteris*. Various statements have been made by authors with regard to the occurrence of sori on this interesting Rhaetic species, but Antevs believes that the evidence hitherto adduced in favour of a fern-like type of fructification is untrustworthy. It is now suggested that these Rhaetic compound fronds with thick linear pinnules belong to some seed-bearing plant and that *Antholithus Zeilleri* represents the microspore-bearing organ: there is no proof of connexion, but there is a very close resemblance in the epidermal characters of *Lepidopteris* and *Antholithus* and the latter is only found in beds containing the fronds. Nathorst called attention to resemblances between the cuticle of *Antholithus* and that of *Baiera* leaves, but according to the later investigations of Antevs, with which it would appear that Nathorst is in sympathy, there is a closer correspondence as regards cuticular structure with *Lepidopteris*.

Fossils regarded by Nathorst as closely allied to his species are described by Leuthardt[2] from the Keuper of Basel as male flowers of *Baiera furcata* Heer; these appear to be almost identical with the English Jurassic specimens; the specimens reproduced by Leuthardt consist of long axes—in one of those shown on his plate the axis is 4 cm. long—with short lateral branches bearing terminal groups of three or four microsporangia 4—5 mm. long and 1·5—2 mm. broad very similar to those shown in figs. 654, 655.

A comparison may also be made with Schenk's *Stachyopitys Preslii*[3] from the Rhaetic of Franconia which he subsequently regarded as the male flower of *Baiera Muensteriana* Heer, a comparison previously made by Heer[4]. This type consists of an axis bearing short lateral appendages terminating in oval bodies opening

[1] Antevs (14).   [2] Leuthardt (03) p. 9, Pl. III.
[3] Schenk (67) A. p. 185, Pl. XLIV. figs. 9—12.
[4] Heer (77) ii. p. 52; Schenk (90) A. p. 261.

at maturity into 10—12 spreading lobes each of which resembles in appearance a microsporangium of *Antholithus Zeilleri*. No spores have been isolated and, as Nathorst points out, the agreement with the Scanian specimens is to a large extent superficial.

The Australian specimens, probably of Rhaetic age, described by Shirley[1] from Ipswich, Queensland, as *Stachyopitys annularioides* and *S. Simmondsi* require further investigation; they may be allied to *Stachyopitys Preslii* Sch., though neither their morphological nature nor systematic position can be settled without fresh data. Halle[2] describes some examples of a similar kind from the Jurassic beds of Graham Land as *Stachyopitys*, *cf. annularioides* Shir. and thinks it probable that they are portions of some Gymnospermous male strobilus, but, as he points out, the absence of any member of the Ginkgoales in these southern beds is noteworthy. Specimens similar to those described by Schenk, Shirley, and Halle are also figured from Rhaetic beds in South America as *Sphenolepis rhaetica*[3] and from the Stormberg (Rhaetic) series of South Africa as *Stachyopitys* sp.[4]

The generic name *Ginkgoanthus* has been adopted by Nathorst[5] for a fragmentary specimen from the Upper Jurassic of Franz Josef Land which he considers may be a male flower of a *Ginkgoites*; but the preservation is too imperfect to admit of satisfactory determination. As regards terminology, in the present state of our knowledge it is preferable to use the non-committal designation *Antholithus*[6] for the English, Scanian, Swiss, and Franz Josef Land fossils, leaving Schenk's *Stachyopitys Preslii* as a type apart. As regards the English and Swiss specimens, the probability would seem to be that they are the microstrobili of some members of the Ginkgoales.

STENORACHIS.

This generic name[7] was first used by Saporta[8] for Nathorst's *Zamiostrobus scanicus* from Rhaetic and Liassic rocks in Scania[9]

---

[1] Shirley (98) p. 13, Pl. xviii.   [2] Halle (13) p. 88, Pl. vi. fig. 13.
[3] Geinitz (76) B. p. 12, Pl. ii.   [4] Seward (03) B. p. 66, Pl. ix. fig. 2.
[5] Nathorst (99) p. 13 Pl. i. figs. 33, 49.
[6] Used by Nathorst in Linnaeus's and not Brongniart's sense; Nathorst (08 p. 23.   [7] στενός, narrow; ῥάχις, the backbone.
[8] Saporta (75) A. Pl. cxvii.; (79) A. p. 193.
[9] Nathorst (75); (97) p. 20; (02) Pl. i. pp. 16, 17.

and for a Liassic species from Belgium, *S. Ponceleti*. Nathorst
subsequently adopted Saporta's genus. I have elsewhere suggested[1]
the application of *Stenorachis* to various species described by Heer
from Jurassic and Cretaceous beds as male flowers of Ginkgoaceae.
Although there is no proof as to the morphological nature of the
specimens included in this genus some of them, *e.g. S. scanicus,*
present the appearance of seed-bearing shoots though, as Nathorst
is careful to point out, the seed-like bodies may not be true seeds.
I am inclined to regard Heer's supposed male flowers[2] (fig. 657) as
possibly fertile shoots of some members of the Ginkgoales which
originally bore seeds, but this view is merely tentative. *Stenorachis*
is employed as a designation for specimens consisting of a central
axis, generally fairly stout, bearing lateral appendages, whether
axial or foliar cannot be definitely determined, either simple or
forked and in some cases with terminal seed-like bodies but usually
with a small distal swelling or a few spreading lobes as in *S.
Schmidtianus* (Heer)[3]. Some at least of the specimens included
in this genus probably belong to Ginkgoaceous plants, though
in regard to others, *e.g. S. scanicus*, it should be remembered that
Nathorst inclines towards a Cycadean affinity. The genus *Beania*
was founded on specimens similar in general habit to species of
Stenorachis but in *Beania* the appendages have a comparatively
large terminal shield bearing on its adaxial side two seeds.

Stenorachis scanicus (Nathorst).

The type-specimen, first described in 1875 as *Zamiostrobus
scanicus*[4] and afterwards transferred to *Stenorachis*, is represented
by a comparatively slender axis 10 cm. long bearing, at a wide
angle, several lateral appendages, spoken of by Nathorst as
sporophylls; these are split into two divergent arms each of which
bears on the side away from the fork an oval, longitudinally
striated, body described as thick and woody (fig. 656). The
nature of these bodies is uncertain and Nathorst is inclined to
think they are not seeds; he suggests as an alternative interpretation that they are laminar structures in which microsporangia
are embedded. The morphology of this Rhaetic and Liassic

---

[1] Seward (12) p. 23.
[2] Heer (77) ii. Pl. XI.; (78) ii. Pl. VI. fig. 8; (82) ii. A. pp. 18, 21, Pls. VI., IX.
[3] Heer (82) ii. A. p. 21.   [4] Nathorst (75).

species referred by Nathorst to the Cycadophyta[1] is, therefore, uncertain. It is by no means certain that it has not an equal claim to inclusion in the Ginkgoales; there are no substantial

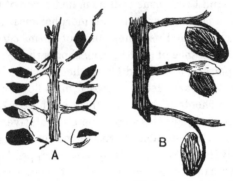

FIG. 656. *Stenorachis scanicus.* (After Nathorst; A, nat. size; B, ×2.)

grounds for such relationship, but the resemblance of this and other species of *Stenorachis* to abnormal seed-bearing shoots of *Ginkgo biloba* may be significant (cf. fig. 631, D, p. 5).

A similar but rather smaller type was described by Heer[2] from Upper Jurassic rocks in Spitzbergen as *Carpolithes striolatus.* Nathorst[3] examined Heer's figured specimens and recognised one of them as an example of *Stenorachis,* agreeing in the possession of forked appendages with *S. scanicus* and bearing seed-like bodies.

The fossils described by Shirley[4] from Rhaetic (?) beds in Queensland as *Beania geminata* are similar in habit to *Stenorachis scanicus* and differ from *Beania gracilis* Carr. in the absence of distally expanded sporophylls.

Another Rhaetic species is described by Nathorst[5] as *Stenorachis Solmsi* in which the 'sporophylls' have a different form and are characterised by a distal, erect, laminar expansion deeply divided into two segments: no seeds or microsporangia have been found.

Stenorachis lepida (Heer).

The species for which this name has been suggested was originally regarded by Heer as the male flower of the Jurassic species *Ginkgo*

[1] Nathorst (02) p. 16.
[2] Heer (77) i. p. 47, Pl. IX. fig. 17.   [3] Nathorst (97) p. 20, Pl. I. fig. 15.
[4] Shirley (98) p. 16, Pl. XX.   [5] Nathorst (02) p. 17, Pl. I. figs. 18—21.

lepida (= *Ginkgoites sibirica*) and described by him under that
name[1]. Similar specimens are correlated by Heer with other
species of *Ginkgo* leaves, and *Ginkgo grandiflora*[2] Heer is represented by supposed male flowers only. Similar though rather
larger examples are described by Heer from Jurassic beds in
Siberia as *Antholithus Schmidtianus*[3] and regarded as male flowers
of some member of the Ginkgoales, possibly *Phoenicopsis*; whatever the parent-plant may have been it is clearly a type closely
allied to those he refers to different species of *Ginkgo*. Fig. 657, B

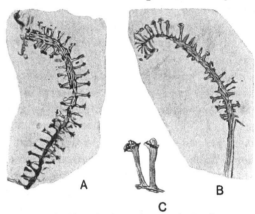

Fig. 657. *Stenorachis lepida*. A, from Amurland; B, C, from Afghanistan.
(A, B, nat. size.)

shows a specimen of *Stenorachis lepida* from Jurassic beds in
Afghanistan[4] which is undoubtedly of the same type as Heer's
European examples. One of Heer's specimens from the Jurassic
beds of Amurland[5] is shown in fig. 657, A: a curved and fairly
stout axis bears numerous spirally disposed, appendages with
slightly expanded ends which in a few cases are more or less
definitely bilobed. No remains of seeds or microsporangia are
preserved, but the swollen ends of the appendages suggest the
former presence of some reproductive organs: some of the appendages are bilobed as in the Afghan example.

Heer states that some of his specimens bear 2—3 pollen-sacs
at the tips of the appendages, but the published figures afford no

[1] Heer (77) ii. Pl. xi.   [2] Heer (82) A. p. 18, Pl. vi. figs. 1—6.
[3] *Ibid.* p. 21, Pl. ix.   [4] Seward (12) p. 23, Pl. iv. fig. 52.
[5] Seward (12³) p. 28, Pl. i. fig. 8.

confirmation of this and an examination of some of Heer's material lent to me through the kind offices of Dr Zalessky failed to reveal any indication of spores or sporangia. In Heer's *Antholithus Schmidtianus* the lateral appendages are said to bear 3—4 pollen-sacs in a terminal whorl, but Heer also suggested the possibility that these bodies are the segments of a calyx-like envelope, a more probable interpretation. It may be that the terminal bodies are homologous with the slightly expanded distal ends of the appendages in *S. lepida* and possibly with the collar at the base of the ovules of *Ginkgo biloba,* which in the case of *S. Schmidtianus* has the form of a lobed cupular organ which enclosed a seed. It is noteworthy that Heer's figures show a central scar surrounded by the spreading lobes.

The incomplete Jurassic specimens from Victoria (Australia)[1] described as possibly parts of a female shoot of a Ginkgoaceous plant resemble *Stenorachis lepida* and should be referred to the same genus.

A specimen like those represented in fig. 657 has been figured by Krystofovič[2] from Jurassic rocks in Ussuriland as *Ginkgo* sp. An imperfect fossil described from Jurassic beds in Australia as possibly a seed-bearing shoot of a Ginkgoaceous plant[3] should be included in *Stenorachis,* as also Raciborski's *Ixostrobus Siemiradzkii*[4] from Rhaetic beds of Poland.

In no case have we any decisive evidence with regard to the parentage or morphological nature of the specimens referred to *Stenorachis,* but any material that may represent fertile shoots belonging to Ginkgoales or Cycadophyta should be described in the hope that additional facts may be obtained.

ERETMOPHYLLUM. Thomas.

A genus founded[5] on some well-preserved leaves from the Middle Estuarine (Middle Jurassic) beds of Gristhorpe Bay on the Yorkshire coast, and named *Eretmophyllum* from the paddle-like form of the lamina[6] Leaves oblanceolate to linear reaching a length of 12 cm. and a breadth of 2 cm.; in the type-species,

[1] Seward (04²) B. p. 179 figs. 39, 40.
[2] Krystofovič (10) Pl. III. fig. 5.   [3] Seward (04²) B. Pl. XIX.
[4] Raciborski (92) Pl. II. figs. 5—8.
[5] Thomas (13).   [6] ἐρετμόν, a paddle.

E. pubescens Thom., the leaf is from 7 to 10 cm. long and 1—3 cm. broad; the apex is rounded or retuse (fig. 658, B), the base tapering gradually towards the petiole. Veins 1—1·5 mm. apart, dichotomously branched in the proximal part of the lamina and usually parallel and simple except where they converge at the apex. The epidermal cells (preserved in the Yorkshire species) are polygonal, with or without papillae (fig. 658, D); the stomata are characterised by an enclosing group of subsidiary cells as in *Ginkgo.*

Eretmophyllum pubescens Thomas.

Secretory tracts occur between the veins of the smooth lamina (fig. 658, C) like those in the leaves of *Ginkgo.* The polygonal cells with straight or slightly undulate walls are characterised by papillae (fig. 658, D), one on each cell: these are particularly conspicuous on the lower epidermis to which the stomata are confined; the slightly depressed stomata are in regular rows and the guard-cells are surrounded by 4—7 subsidiary cells. In another species, *E. whitbiense* from Whitby, the surface of the lamina, which may be 7 cm. long and 1·2 cm. broad and slightly falcate, is rough, and between the veins are strands of elongated cells, possibly denoting the presence of hypodermal stereome. Stomata occur on both surfaces and the papillae are confined to the subsidiary cells.

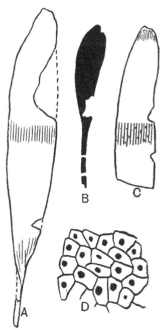

Fig. 658. A, *Eretmophyllum saighanense.* B—D, *E. pubescens.* D, epidermal cells. (A, after Seward; B—D, after Thomas.)

A leaf figured by Ettingshausen[1] from Wealden rocks as *Cyclopteris squamata,* which Schenk[2] suggests may be a segment of a *Ginkgoites,* should probably be included in *Eretmophyllum.*

---

[1] Ettingshausen (52) Pl. IV. fig. 1.

[2] Schenk (71) B. p. 213.

Eretmophyllum saighanense (Seward).

This species (fig. 658, A), originally referred with some misgiving to *Podozamites*[1] and compared with Yokoyama's *Ginkgodium*, is from Jurassic beds in Afghanistan. There can be little or no doubt as to its generic identity with the Yorkshire leaves. The broadly linear lamina tapers gradually to a slender petiole and the veins, 1 mm. apart, are simple except at the proximal end.

Mr Thomas is certainly justified in his opinion that *Eretmophyllum* is a member of the Ginkgoales. In shape the leaves resemble *Ginkgodium* and differ but little from some Jurassic specimens referred to *Ginkgoites*. They agree in venation, in the presence of short secretory tracts, in the structure of the epidermal cells and stomata with *Ginkgo* and species of *Ginkgoites*. *Eretmophyllum* is distinguished from *Feildenia* by its larger leaves, a coarser venation, and a more definite petiole. Some leaves figured by Fontaine[2] from Jurassic-Cretaceous rocks of Alaska as *Nageiopsis longifolia?* Font. have little claim to be included in that genus[3] and may perhaps be allied to *Eretmophyllum*.

[1] Seward (12) p. 35, Pl. IV. fig. 53.
[2] Fontaine in Ward (05) B. Pl. XLV. figs. 1—5.
[3] Berry (10) p. 190.

CHAPTER XLI.

GINKGODIUM. Yokoyama.

Yokoyama[1] defined the genus as follows: 'Leaf coriaceous, entire or lobed, gradually narrowed towards the base which is thickened at its margin and gradually passes into a short petiole; veins numerous, simple, parallel; interstitial veins very fine.' He draws attention to the thickening of the lower margin of the lamina, a feature reminiscent of *Ginkgo* and to the course of the veins which run parallel to the median axis of the lamina instead of spreading from the base as in *Ginkgoites* and *Baiera*. *Gink-godium* resembles the Palaeozoic genus *Whittleseya* in the position of the veins but the genera are unlikely to be confused; a com- parison may also be made with the Jurassic genus *Eretmophyllum* (fig. 658) which has longer and narrower leaves with a coarser venation. We have no information with regard to the cuticular structure, the nature of the supporting axes or reproductive organs. The supposed affinity to *Ginkgo* rests therefore on leaf-form alone.

Ginkgodium Nathorsti Yokoyama.

The type-species was founded on specimens from strata in Japan assigned by Yokoyama[2] to the Middle Jurassic series, but the flora suggests a somewhat higher horizon in the Jurassic system. Some of the leaves are entire, obovate, and have a truncate distal end; others are cuneate and broader at the apex which may be lobed, while in some forms the leaf is divided by a

---

[1] Yokoyama (89) B. p. 56.
[2] *Ibid.* p. 57, Pls. II., III., VIII., IX., XII.

deep median sinus into divergent obtuse segments (fig. 659).   One
leaf is described as 6·6 cm. long, 2·1 cm. broad with thirty veins
and an interstitial 'vein' between
each pair: the interstitial 'vein'
is due to the presence of an intercostal stereome strand.   Thomas[1]
records this species from the
Bathonian series of Kamenka
in the south of Russia (fig. 659,
B). The specimens from Alaska
named by Fontaine[2] *Ginkgodium?
alaskense* agree more closely with
Ginkgoites.

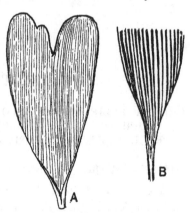

FIG. 659. *Ginkgodium Nathorsti*. (Nat.
size; A, after Yokoyama; B, after
Thomas.)

CZEKANOWSKIA. Heer.

Heer[3] gave this name (after
Czekanowski who discovered the
specimens) to fascicules of long
and narrow, filiform, leaves with a simple or occasionally
forked lamina borne on a short supporting axis covered with
broader and shorter scale-leaves.   The deciduous fascicules or
dwarf-shoots are similar to those of *Phoenicopsis*.   Bunches
of *Czekanowskia* leaves with their short scale-covered supporting
axes resemble the dwarf-shoots of Pines[4].   Heer assigned to this
genus some seeds associated with the leaves, also what he believed
to be a male flower[5], an example of a reproductive shoot of the
type described on page 57 as *Stenorachis*.   There is, however,
no conclusive evidence as to the nature of the reproductive organs.
The venation is seldom shown on the carbonised laminae; some
leaves are finely striated while on others there may be one or two
narrow ridges that represent veins, but as a rule the impressions
afford no indication of the venation.   *Czekanowskia* was placed
by the author of the genus in the Ginkgoales, the short shoots
being compared with those of *Ginkgo* though, except in the larger
number of the leaves, they closely resemble the foliar spurs of

---

[1] Thomas (11) p. 75, Pl. IV. figs. 9—11; Pl. VIII. fig. 3.
[2] Fontaine in Ward (05) B. p. 168, Pl. XLIV. figs. 3, 4.
[3] Heer (77) ii. p. 65.   [4] Cf. *Pinus flexilis*; Bot. Mag. Tab. 8467.
[5] Heer (82) ii. Pl. VI. fig. 7.

Pines. The dichotomous branching of the lamina in some forms
is another feature in which *Czekanowskia* resembles *Baiera* and
Ginkgo, a resemblance which derives a certain significance from
the occurrence of stomata of the Ginkgoaceous type. Nathorst[1]
has described the cuticular membranes of the superficial layers:
the epidermal cells have straight walls and the stomata, more
numerous on the lower surface, are accompanied by four or five
subsidiary cells: these do not form a circular group as in *Baiera*
and *Ginkgoites*, but, as the result of elongation in the direction of
the long axis of the leaf, the group is relatively long and narrow.
Cuticular preparations can often be made from the well-preserved
leaves that occur in great abundance in the shales of Gristhorpe
Bay and elsewhere on the Yorkshire coast[2], and some particularly
good examples were collected by Prof. Obrutschew from Jurassic
rocks in the Djair Mountains in Chinese Dzungaria[3]; these occur
with carbonised remains of *Ginkgoites* leaves in papery masses
similar in the manner of preservation to the Palaeozoic paper
coal from Russia[4]. The epidermal cells of the Dzungaria *Czekanowskia*, possibly identical with *C. rigida* but too incomplete to be
determined with certainty, have straight walls and are relatively
long; the stomata are scattered and appear as dark patches, their
darker colour being due to the thick cuticles of the two or three
flanking cells on the sides of the stoma; the epidermal features are
similar to those described by Nathorst[5], but in the Rhaetic specimens
from Scania the heavily cuticularised accessory cells are generally
more numerous.

It has been suggested by Jeffrey[6] that *Czekanowskia* may be
Araucarian in its affinities, but this opinion rests on the slender
evidence of association of *Czekanowskia*-like leaves in Middle
Cretaceous rocks with a stem described as *Araucariopitys* so
named because of the association of Araucarian and Abietineous
features. Such evidence of affinity as we have would seem to be
in favour of relationship with *Baiera* and *Ginkgo* though decisive
data are not as yet available. The genus is very widely spread in

---

[1] Nathorst (06); see also Seward (00) B. p. 278.
[2] Seward (00) B. p. 278, fig. 48.   [3] Seward (11) p. 49, Pls. IV., v., VI., VII.
[4] Vol. I. p. 68.   [5] Nathorst (06).
[6] Jeffrey (07); Hollick and Jeffrey (09) B. p. 63, Pl. VI. figs. 1—3.

Jurassic floras and a few examples are recorded from Cretaceous strata.

Czekanowskia Murrayana (Lindley and Hutton).

On the specimen shown in fig. 660, A Lindley and Hutton[1] founded the species *Solenites Murrayana* which they compared

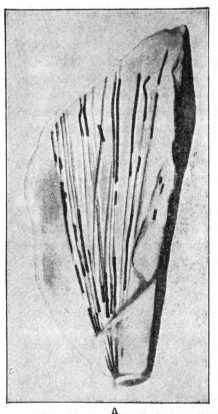
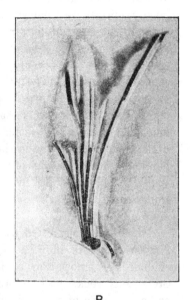

A   B

FIG. 660. A, B, *Czekanowskia Murrayana*. A, The type-specimen of *Solenites Murrayana*, Lind. and Hutt., from the Middle Jurassic plant-beds of Yorkshire. (British Museum, no. 3685, no. V. 3684.)

with *Isoetes* and *Pilularia*. The type-specimen is from the Middle Jurassic plant-bed at Gristhorpe Bay on the Yorkshire coast. They describe the narrow leaves as converging to a common

[1] Lindley and Hutton (34) A. Pl. cxxi.

point, but the actual axis is not preserved; the lamina is longitudinally striated but no veins are shown either on the carbonised lamina or in the magnified cuticle figured in the original description. In the specimen reproduced in fig. 660, B the leaves are seen to be attached to a short and relatively broad axis covered with scale-leaves, one of which is shown bent over on one side of the dwarf-shoot. The comparatively large size and the pendulous position of the scales are characteristic features of the genus which are well seen in figures of *Czekanowskia* published by Nathorst[1]. This species was included by Saporta in *Jeanpaulia* and later transferred to *Trichopitys*, while Zigno[2] and some other authors regarded *Solenites Murrayana* as a species of *Isoetes*. The comparison with *Isoetes* suggested by the form of the leaves is not borne out by the structure of the epidermal cells. Phillips[3] figured a specimen in 1829 as *Flabellaria? viminea*: this specific name though employed before the publication of *Murrayana* has not been adopted by authors. Some of the specimens included by Heer in his account of the species *C. rigida* should be referred to *C. Murrayana*, but in a previous description[4] of the species I went too far in uniting *C. Murrayana* and *C. rigida*. In *Czekanowskia Murrayana* the leaves, usually about 1 mm. broad but sometimes narrower, reach a length of more than 17 cm.; they are unbranched and in this respect and in their slightly greater breadth differ from *C. rigida*.

The species is characteristic of Middle Jurassic floras.

Czekanowskia microphylla (Phillips).

The specimen figured by Phillips[5] from the Yorkshire coast as *Baiera microphylla* is undoubtedly identical specifically with that reproduced in fig. 661, and both were formerly included in *Baiera Lindleyana*[6]. The chief reasons for transferring them to *Czekanowskia* are the more acute angle of divergence of the filiform segments, the difference in the shape of the leaves, the absence of a petiole, and the occurrence of the leaves in a fascicle, a habit not shown by any typical examples of *B. Lindleyana* though not

[1] Nathorst (06) Pl. II.
[2] Zigno (56) A. p. 216. For other references, see Seward (00) B. p. 280.
[3] Phillips (29) A. Pl. x. fig. 12.   [4] Seward (00) B. p. 279.
[5] Phillips (75) A. p. 200, fig. 9.   [6] Seward (00) B. p. 266.

unknown in the genus (*e.g. B. paucipartita*). Some specimens which may be identical with this type were obtained several years ago by Dr Nathorst from Yorkshire but never fully described: an examination of his unpublished drawings and of specimens collected by Mr Hamshaw Thomas convinced me that some forms of *Czekanowskia* are more freely branched and exhibit more variation in the breadth of the lamina than I had formerly supposed. In specimens of the type first noticed by Nathorst some of the segments are comparatively broad and fern-like, a feature that is not seen in the leaves shown in fig. 661. This species affords a striking contrast to *Czekanowskia Murrayana* in which the long leaves are unbranched, and as Nathorst[1] suggests the name

FIG. 661. *Czekanowskia microphylla* (Phillips). (British Museum No. 39, 283; nat. size.)

Solenites might be revived for the unbranched type; but in the absence of any difference in the epidermal characters, it would seem undesirable to raise to generic rank a feature depending on the simple or branched habit of the leaves of otherwise similar leaf-fascicles.

Czekanowskia rigida Heer.

This species, founded on specimens from Siberian Jurassic rocks[2], is characterised by its branched filiform leaves borne on short shoots enclosed by scale-leaves, triangular or lanceolate in form and in some specimens pendulous on slender stalks. Nathorst[3] states that the epidermal structure of the scale-leaves is similar to that of the scale-leaves on short shoots of *Ginkgo*. The characters of the stomata are mentioned in the account of the

[1] Nathorst (06) p. 11.
[2] Heer (77) ii. p. 70; (78) ii p. 7; (82) A. p. 19.
[3] Nathorst (06).

genus. The leaves are generally slightly narrower than the unbranched needles of *C. Murrayana,* but the habit of the dwarf-shoots is the same. The leaves often show fine striations; in most specimens there is no indication of clearly marked veins though two or three vascular strands are sometimes visible. Heer on very slender evidence refers to this species some seeds and a 'male flower.'

It is not always easy to distinguish between imperfect examples of *C. rigida* and *Baiera Lindleyana*: the leaves of the latter type are petiolate and the segments diverge at a wider angle. Two leaves with spreading bifurcate segments figured by· Fontaine[1] from Lower Cretaceous rocks in the Black Hills as *Czekanowskia nervosa* Heer afford no indication that they were borne in clusters on dwarf-shoots but resemble the petiolate leaves of *Baiera Lindleyana.* Berry[2] points out a similarity between Fontaine's fossils and *Baiera foliosa* Font. Heer's type-specimens of *C. nervosa* from the Wealden of Portugal[3] are more like *Czekanowskia* leaves. The leaves described from Siberia as *C. setacea* Heer[4], though narrower than some forms of *C. rigida,* are probably not specifically distinct.

Czekanowskia rigida is characteristic of Jurassic strata, and occurs in Europe, including Greenland, also in Siberia, China, and Japan.

Czekanowskia dichotoma Heer and *C. capillaris* Newberry.

The branched leaves described under these names[5] from Cretaceous rocks in Greenland and North America are in most cases not sufficiently complete to be assigned with certainty to the genus *Czekanowskia*; the examples figured by Hollick and Jeffrey[6] from Middle Cretaceous beds as *C. capillaris* occur in closely packed groups, but no specimens have been discovered showing any scale-covered supporting axis. While admitting the probability that these species and *C. nervosa*[7] from Wealden strata

[1] Fontaine in Ward (99) B. p. 685, Pl. CLXIX. figs. 1, 2.
[2] Berry (11) p. 374.   [3] Heer (81) Pl. XVII.
[4] Heer (77) ii. p. 68; (78) ii. p. 26; (82) A. p. 18.
[5] Heer (82) A. p. 8.   Newberry and Hollick (95) p. 61.
[6] Hollick and Jeffrey (09), B. p. 63, Pl. VI. figs. 1—3.
[7] Heer (81) Pl. XVII.

may be allied to *Czekanowskia rigida*, such evidence as is available points to a maximum development of *Czekanowskia* in the Jurassic period.

FEILDENIA. Heer.

In 1870 Heer[1] described some small linear leaves from Tertiary strata in Spitzbergen for which he proposed the generic name *Torellia*, defining it as follows: 'Folia rigida coriacea, basin versus angustata, articulata, tenuiter costata, costis interstitiisque subtilissime striatis.' On the discovery by Capt. Feilden of additional specimens in Miocene beds in Grinnell Land (81° 46′ N.) Heer published a further account of the genus and substituted *Feildenia* for *Torellia* because of the previous use of the latter name by Zoologists. Heer compared *Feildenia* with *Podocarpus*, *Araucaria*, and other Conifers but, mainly because of the occurrence of a leaf with a lobed lamina, he provisionally included the genus in the Taxineae[2]. The leaves are usually found as detached specimens but in one case several are spirally disposed on a stout axis and one imperfect example shows at the base what appears to be a scale-leaf, suggesting that leaves were also borne on short shoots like those of *Phoenicopsis* and *Czekanowskia*. Heer lays stress on the ribbing and striation of the surface of the lamina as distinguishing features between *Feildenia* and *Phoenicopsis*, but Nathorst[3], in his revision of the genus, expresses the opinion that it is only in the tendency to a sickle-like form and a feeble expansion of the slightly curved base that *Feildenia*, at best an ill-defined genus, can be distinguished from *Phoenicopsis*.

Feildenia rigida Heer.

This species, from Miocene beds of Spitzbergen[4] and Grinnell Land[5], is represented by linear leaves 6—8 cm. long and 5—8 mm. broad at the widest part, usually rather nearer the apex than the base; the lamina is often slightly falcate and tapers gradually to a narrow base. There are 8—-11 veins for the most part parallel but occasionally feebly convergent at the bluntly rounded apex.

---

[1] Heer (71) iii. p. 44.   [2] Heer (78) i. p. 20.
[3] Nathorst (97), p. 55.
[4] Heer (71) i.i. p. 44, Pl. VI. figs. 3—12; Pl. XVI. fig. 1 *b*.
[5] Heer (78) i. p. 20, Pls. I., II., VIII.

Feildenia Nordensköldi Nathorst.

A species from Upper Jurassic rocks in Spitzbergen founded by Nathorst[1] on leaves similar to those of *F. rigida* but smaller; the lamina is generally 3—4 mm. broad and may reach a length of 4·5 cm. There are usually six veins and as in other species finer longitudinal lines occur between the true veins. A few small leaves very similar to *F. Nordensköldi* are described by Nathorst as *Feildenia* sp.[2] from Franz Josef Land, probably of Wealden age.

Until further evidence is available it is impossible to fix precisely the position of the genus. Though often distinguished by the sickle-shaped lamina and the broad apical region from leaves of *Podozamites* it is not always possible to separate the leaves of the two genera.

PHOENICOPSIS. Heer.
DESMIOPHYLLUM. Lesquereux.

Phoenicopsis was founded by Heer[3] on linear leaves from Middle Jurassic strata in Siberia; the leaves, in extreme cases 20 cm. long and varying in breadth from 2 mm. to 2 cm., occur in clusters of six or more and even as many as twenty on very short and relatively broad axes covered with small scale-leaves. These dwarf-shoots were deciduous: the lamina is fairly uniform in breadth but tapers gradually towards the slender base and is usually obtusely rounded at the apex; the veins are parallel and very rarely dichotomously branched. The features on which species are founded are often of little systematic value: they are the breadth of the lamina, the arrangement of the veins, the presence or absence of interstitial 'veins.' It is very doubtful whether much confidence can be placed on the occurrence of the so-called interstitial veins: in some species of *Phoenicopsis* the parallel veins show no trace of a smaller 'vein' between them, but occasionally in a leaf of the same species there are indications of interstitial 'veins.' The breadth of the lamina is also an uncertain guide: well preserved specimens show that the leaves may reach a considerable length and that the lamina gradually decreases in breadth towards the narrow base. Species have been needlessly multiplied particularly

---

[1] Nathorst (97) p. 56, Pl. III. figs. 16—27.
[2] Nathorst (99) p. 15, Pl. I. figs. 25—30, 32.   [3] Heer (77) ii. p. 49.

in the case of detached leaves which it is often impossible to
determine even generically. The characters usually employed for
the separation of different forms are conveniently shown in a table
published by Krasser[1]. There is no information available as to
the epidermal structure of the various types of *Phoenicopsis* leaves,
nor have we any data with regard to the reproductive organs. The
genus is generally included in the Ginkgoales: the dwarf-shoots
agree closely with those of *Czekanowskia* which, in the structure
of the epidermis and in the bifurcation of the leaves, resembles
Ginkgo and *Baiera*. The precise position of *Phoenicopsis* cannot
be regarded as settled. The only evidence with regard to anatomical structure is that furnished by Solms-Laubach[2] who
described petrified leaves from Jurassic rocks in Franz Josef
Land which are probably examples of *Phoenicopsis*; but, assuming
that they belong to this genus, the anatomical data are insufficient
to determine the position of the genus within the Gymnosperms.
Incomplete and detached leaves agreeing in their venation and in
the form of the lamina with those of *Phoenicopsis* cannot as a
rule be distinguished from leaves of *Podozamites*, *Feildenia*, or
even from narrow forms of *Cordaites*. The Jurassic specimens
from North Germany on which Salfeld[3] founded the genus *Phyl-
lotenia* should probably be assigned to *Phoenicopsis*. Solms-
Laubach refrained from assigning the imperfect Franz Josef
Land leaves to *Phoenicopsis* because no dwarf-shoots were found;
he employed the non-committal generic name *Desmiophyllum*, a
designation that might with advantage be more frequently used
for specimens which cannot be proved to belong to *Phoenicopsis*,
Podozamites or other genera with similar leaves.

Desmiophyllum. Lesquereux established the genus *Desmiophyllum*[4] for some narrow sublinear leaves from the Coal Measures
of Pennsylvania similar to those of *Poacordaites* and attached to
an imperfectly preserved axis either singly or in small groups.
The type-species *D. gracile* is probably a species of *Cordaites*: the
name *Desmiophyllum* never came into general use until its revival
by Solms-Laubach in 1904 as a convenient term to apply to linear

[1] Krasser (05) p. 612.
[2] Solms-Laubach (04) Pls. I., II.   [3] Salfeld (09) B. p. 26, Pl. IV. fig. 3.
[4] Lesquereux (78) p. 322; (80) A. p. 556, Pl. 82, fig. 1.

leaves that in the absence of evidence as to the habit of the shoots cannot be assigned to more precisely defined genera such as *Phoenicopsis* or *Podozamites*. Nathorst[1] employs *Desmiophyllum* for some *Phoenicopsis*-like leaves from Jurassic rocks in the New Siberian Islands which may be specifically identical with those described by Graf Solms from Franz Josef Land, which I propose to call *Desmiophyllum Solmsi*. In view of the probability that these leaves belong to a species of *Phoenicopsis*, a brief description of their structure may be conveniently inserted here.

Desmiophyllum Solmsi sp. nov.

The collection of plants obtained by the Jackson-Harmsworth Expedition (1894—96) to Franz Josef Land includes several specimens of matted linear leaves some of which are figured by Newton and Teall[2]. Similar leaves collected by Nansen are described by Nathorst[3]. Subsequently Solms investigated the structure of the leaves figured by Newton and Teall and the photographs reproduced in fig. 662 were taken from sections of the silicified material in the Museum of the Geological Survey. The largest specimens reach a length of 10 cm. and are 5—10 mm. broad; the veins are unbranched and there are six in a breadth of 3 mm. of lamina. In transverse section the lamina is of fairly uniform breadth or, owing to the partial collapse of the intercostal mesophyll, it is characterised by prominent ribs (fig. 662, A). The vascular bundles are collateral, enclosed in a sheath of rather thick-walled cells: the xylem elements are spiral and scalariform and, as Solms states, the occasional preservation of single rows of circular bordered pits and the occurrence of lateral sieve-plates point to a Gymnospermous affinity. The mesophyll is fairly homogeneous and lacunar as shown in fig. 662, A and B: in the tangential section (B) the lacunar mesophyll is seen between the veins and their associated rows of rectangular cells. The epidermal cells are thickly cuticularised and papillose. In one section a few stomata were found showing two dark guard-cells, 40 μ long, surrounded by six faintly outlined cells (fig. 662, C) agreeing with those of *Ginkgo* and *Ginkgoites*, except in the absence of any overarching papillate subsidiary cells. This may, however, be

[1] Nathorst (07) p. 4.
[2] Newton and Teall (97); (98).   [3] Nathorst (99).

due to the section having passed slightly below the level of the
epidermal surface.   The cells of the epidermis are short and have
straight walls.

Though we should not be justified in asserting that the leaves
named *Desmiophyllum Solmsi* are examples of *Phoenicopsis*, the
probability is that they belong to that genus.   *Phoenicopsis* is

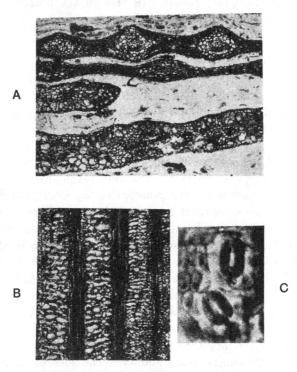

Fig. 662.   *Desmiophyllum Solmsi*.   A, transverse sections of leaves; B, tangentia
section of the lamina showing veins and mesophyll; C, stomata.   (From
sections in the Museum of the Geological Survey, London.)

especially characteristic of Jurassic rocks and is best represented
in the Middle Jurassic series of Siberia.   The genus is recorded also
from Spitzbergen[1], Franz Josef Land, and Bornholm[2]: a species,
P. Gunni[3], has been described from Upper Jurassic rocks in Scotland, the only example of the genus in Britain.   Feistmantel[4]

[1] Nathorst (97), p. 16.   [2] Möller (03) p. 30.
[3] Seward (11²) p. 681, Pl. IX. fig. 35   [4] Feistmantel (77) fig. 9

has figured an imperfect specimen from Upper Gondwana rocks in India that may be correctly referred to this genus. Incomplete leaves from the Rhaetic beds of Sweden described by Nathorst[1] as *Phoenicopsis cf. speciosa* Heer may be examples of the genus, but the name *Desmiophyllum* would be more appropriate.

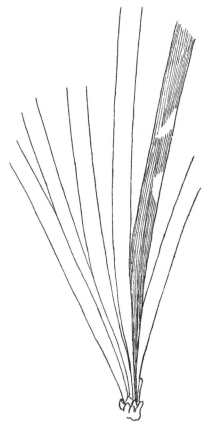

FIG. 663. *Phoenicopsis speciosa.* (After Heer; ¾ nat. size.)

Phoenicopsis.

Phoenicopsis speciosa Heer.

In this Siberian Jurassic species[2] the leaves reach a length of 20 cm. and a breadth of 7—8 mm.; there are from 15 to 23 veins

---

[1] Nathorst (86) p. 96, Pl. xxv. figs. 25, 26.
[2] Heer (77) ii. p. 112, Pls. xxix., xxx.

in the lamina and a faintly marked interstitial 'vein,' probably the impression of a stereome strand, between each pair of veins. In one specimen Heer found 21 leaves in a single cluster but usually the number on a single dwarf-shoot is smaller. It was the superficial resemblance of a cluster of these leaves (fig. 663) to the leaf of some Palms that suggested the name *Phoenicopsis*. The leaves described by Heer as *P. latior*[1] are not distinguished from *P. speciosa* by any very definite character. Examples of detached leaves from Lower Jurassic rocks in Bornholm described by Möller[2] as *cf. Phoenicopsis latior* may equally well be referred to *Podozamites*.

Phoenicopsis angustifolia Heer.

The leaves are 4 mm. broad or less and have 6—8 veins without interstitial veins[3]. This species is recorded from Russia[4], Siberia, China, and the Arctic regions, and leaves of similar type are represented by *Phoenicopsis media* Krasser[5], which is probably merely a form of *P. angustifolia*[6]; *P. taschkessiensis* Krass. from China; also Chinese specimens first described by Potonié[7] without a specific name and afterwards named by Krasser[8] *P. Potoniei*.

The species *Phoenicopsis Gunni* from Scottish Kimeridge beds is a similar type with leaves 3—4 mm. broad and 12 cm. long with eight veins and indications of interstitial 'veins.'

? *Phoenicopsis elongatus* (Morris).

Morris[9] founded this species on a linear leaf, now in the British Museum, from the Jerusalem Basin (Triassic), Tasmania, which he referred to *Zeugophyllites*, a genus founded by Brongniart on a specimen from the Lower Gondwana rocks of India but never figured. To the same species McCoy[10] referred some broader leaves from Permo-Carboniferous strata in New South Wales: these were shown by Etheridge[11] and Arber[12] to be distinct from Morris's type and the latter author identified them with *Noeg-*

---

[1] Heer (77) ii. p. 113, Pls. XXIX., XXXI.   [2] Möller (03) p. 31.
[3] Heer (77) ii. pp. 51, 113, Pls. I., II., XXX.
[4] Seward (07²), Pl. VIII. fig. 69.   [5] Krasser (00) B. p. 147, Pl. III. fig. 4.
[6] Nathorst (07) p. 7. Seward (11) p. 50.
[7] Potonié (03).   [8] Krasser (05) p. 23.
[9] Morris in Strzelecki (45) B. p. 250, Pl. VI. figs. 5, 5 a.
[10] McCoy (47) B.   [11] Etheridge (93) p. 75.   [12] Arber (02²) B. p. 17; (03²

gerathiopsis Goepperti (Schmal.) a species that is now recognised as identical with *Cordaites* (*Noeggerathiopsis*) *Hislopi* (Bunb.). Feistmantel[1], who reproduces Morris's original figure of *Zeugophyllites elongatus*, assigns the Tasmanian plant to *Podozamites*. In 1903[2] I expressed the opinion that this species is a *Phoenicopsis* and figured some specimens from the Stormberg (Rhaetic) series of South Africa as *Phoenicopsis elongatus*. The leaves reach a length of more than 16 cm.; the lamina is gradually tapered to an acuminate termination which may be the base, the distal end having a bluntly rounded apex. The veins are parallel and simple. Similar leaves have been described by Szajnocha and by Kurtz[3] from Rhaetic rocks in South America. In the absence of specimens attached to an axis it is impossible to speak with confidence as to the systematic position of the detached leaves, but they bear a very close resemblance to some of the European examples of *Phoenicopsis*. The occurrence of these leaves in Australia, South Africa, and South America and the *Phoenicopsis*-like leaves recorded from India[4], though not proving the existence of *Phoenicopsis* in the later vegetation of Gondwana Land, afford some evidence of its occurrence in the southern floras.

---

[1] Feistmantel (90) A. p. 150, Pl. xxi. fig. 6.
[2] Seward (03) B. p. 67, Pl. ix. figs. 1, 9, 10.
[3] Szajnocha (88) B. p. 19, Pl. ii. fig. 4; Kurtz (03).
[4] Feistmantel (77) fig. 9.

CHAPTER XLII.

GENERA OF UNCERTAIN POSITION.

In this section are included several Palaeozoic genera most of which have been assigned to the Ginkgoales on evidence which in most cases is wholly inadequate.

Probably the oldest specimens referred to the Ginkgoales are some imperfect leaves from Middle Devonian rocks in Bohemia described by Potonié and Bernard as *Barrandeina Dusliana* (Krejci)[1], but there is no evidence of affinity to this or to any other class of plants.

GLOTTOPHYLLUM. Zalessky.

This designation has recently been proposed by Zalessky[2] for Schmalhausen's Permian species *Ginkgo cuneata* which has no substantial claim to be regarded as generically identical with *Ginkgo biloba*. Zalessky considers that it may belong to the Ginkgoales though the available data hardly justify more than a suggestion of possible relationship.

Glottophyllum cuneatum (Schmalhausen).

A Permian species from the Altai mountains[3] represented by an obovate spathulate leaf 11 cm. long and 4 cm. broad with a comparatively long 'stalk' consisting of a narrow portion of the lamina: the lamina is traversed by slightly spreading veins about 1 mm. apart. In form this leaf resembles the genus *Eretmophyllum*[4] (cf. fig. 658) and may be a species of that genus, but in the absence of any information with regard to the epidermal features it is inadvisable to adopt a name implying affinity to the Ginkgoales.

---

[1] Potonié and Bernard (04) p. 45.
[2] Zalessky (12[2]) p. 28 (footnote).
[3] Schmalhausen (79) A. p. 34, Pl. IV. fig. 5
[4] Thomas, H. H. (13).

GINKGOPSIS. Zalessky.

This generic term proposed by Zalessky[1] for Schmalhausen's *Ginkgo Czekanowskii*, but not defined by him, may be adopted for certain leaves resembling those of *Ginkgo* but which there is no adequate reason for regarding as the leaves of a member of the Ginkgoales.

Ginkgopsis Czekanowskii (Schmalhausen) and *Ginkgopsis minuta* (Nathorst).

The small leaves, or leaflets, described by Schmalhausen[2] from the Permian of the Lower Tunguska river as *Ginkgo Czekanowskii* are more or less orbicular and the lamina, approximately 1·5 cm. broad, is divided into several bilobed segments (fig. 664, C). It is by no means certain that all the fragments included in this species are specifically identical; some bear a close resemblance to *Ginkgo minuta* Nath.[3] from the Rhaetic of Scania, a type which he assigns with hesitation to *Ginkgo* and compares with the leaves of *Acrostichum peltatum*. It is impossible without additional data to determine either the systematic position of the specimens included in these two species or their morphological nature. Through the kindness of Prof. Zalessky I have been able to examine photographs of some of the original material and, as shown in fig. 664, C, the supposed leaves are borne on a rachis-like axis, possibly of some Pteridosperm, a circumstance which though not proving that they are leaflets of a compound frond, favours that interpretation.

NEPHROPSIS Zalessky.

This name is proposed by Zalessky[4] for the leaves described by Schmalhausen and Renault respectively as *Ginkgo integerrima* and *Ginkgo martensis* on the ground that these Permian specimens do not afford satisfactory evidence of close affinity to the genus *Ginkgo*. The two sets of leaves are probably specifically identical.

Nephropsis integerrima (Schmalhausen).

In fig. 664, A, B, are reproduced two of the leaves from the

---

[1] Zalessky (12²) p. 28 (footnote).
[2] Schmalhausen (79) A. p. 84, Pl. xvi. figs. 8—10.
[3] Nathorst (86) p. 93, Pl. xiii. fig. 112; Pl. xx. figs. 14—16.
[4] Zalessky (12²) p. 28 (footnote).

Permian of the Lower Tunguska on which Schmalhausen founded
his species *Ginkgo integerrima*[1]; the characteristic features are:
the transversely elongate form of the entire lamina, the spreading,
dichotomously branched, veins and the short and relatively broad
stalk-like basal portion of the lamina. As Zeiller[2] says, this
species is probably identical with *Ginkgo martensis*[3] from Permian
beds near Toulon-sur-Arroux (Saône-et-Loire) founded on a single
leaf 3 cm. broad and 1·6 cm. deep. Zeiller reproduces Renault's
figure in his volume on the flora of Blanzy and Creusot[4] and states
that the specimen could not be found: he retains the name *Ginkgo*

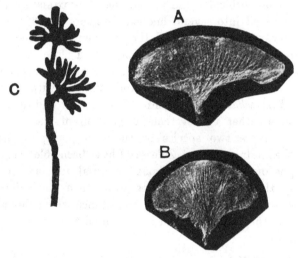

Fig. 664.   C, *Ginkgopsis Czekanowskii*.   A, B, *Nephropsis integerrima*.   (From the
originals of Schmalhausen's figures, supplied by Prof. Zalessky; nat. size.)

but with the addition of a question-mark. The Permian leaves
differ from those of *Ginkgo* in the absence of marginal veins—though
this is of secondary importance—in the absence of a true petiole,
in the form of the lamina and, from the majority of recent *Ginkgo*
leaves, in the entire margin. The supposed leaves may be leaflets
of a compound frond or possibly bracts from some fertile shoot.

[1] Schmalhausen (79) A. p. 85, Pl. xvi. figs. 12—15.
[2] Zeiller (96) A. p. 475.
[3] Renault (88) p. 232, fig. 47 c.
[4] Zeiller (06) B. Pl. xlviii. fig. 3.

PSYGMOPHYLLUM. Schimper.

This generic name[1] was instituted by Schimper[2] for large cuneate leaves (figs. 665—667) from Upper Carboniferous and Permian rocks that had previously been included in Sternberg's genus *Noeggerathia*: the type-species is *Noeggerathia flabellata* Lind. and Hutt.[3] *Psygmophyllum* is thus defined by Schimper: 'Folia pinnatisecta, pinnis erecto-patentibus, e basi valde angustata flabelliformis, longitudinaliter flabellatim plicatis, plus minus profunde pinnatisectis vel margine lobatis seu crenatis; nervis pluries dichotomis, erecto-radiantibus.' Among other species referred by Schimper to *Psygmophyllum* are *Noeggerathia expansa* and *N. cuneifolia* from Permian beds in the Ural mountains figured by Kutorga[4], Brongniart and later authors. In 1878 Saporta[5] published a note giving the results of an examination of Brongniart's specimens: he expressed the opinion that these species are portions of compound fronds comparable with *Eremopteris*, differing morphologically from *Psygmophyllum flabellatum* which he regarded as a shoot bearing simple leaves. He also pointed out that in the Ural fossils the leaflets have a more or less well defined midrib in contrast to the regular flabelliform dichotomous venation in the leaves of the English species, *P. flabelliforme*: the latter he assigns to the genus *Ginkgophyllum* previously established for a Permian species, *G. Grasserti*[6], consisting of an axis bearing spirally disposed cuneate and deeply divided leaves (fig. 669) very similar to some of the older *Baiera* leaves; while Schimper's genus *Psygmophyllum* is applied to the two Russian species *P. expansum* and *P. cuneifolium*. Saporta included a third Ural species in *Psygmophyllum*, *P. santagoulourensis*. An examination of some specimens of Brongniart's *Noeggerathia expansa* in the British Museum leads me so far to agree with Saporta in the opinion that some of the specimens referred to that species are generically distinct from *P. flabellatum*. Schmalhausen's figures of the Ural species, which he refers to

---

[1] ψῦγμα, a fan.   [2] Schimper (70) A. Vol. II. p. 192.
[3] Lindley and Hutton (32) A. Vol. I. Pls. XXVIII.—XXIX.
[4] Kutorga (44); Brongniart (45) B. Pl. E; Schmalhausen (87) Pls. III., IV. See also Arber (12) p. 401.
[5] Saporta (78); (78²); (78³).   [6] Saporta (75).

Psygmophyllum, show a considerable variation in the venation of the leaves or leaflets: those represented in his Plate III. figs. 8—10[1] agree with *P. flabellatum* while others differ in the presence of a midrib (*e.g.* Schmalhausen's Pl. IV. fig. 3). There has been considerable confusion in regard to the determination of these Russian specimens: as Arber says, Kutorga's *Cyclopteris gigantea*[2] is probably a true *Psygmophyllum* though other specimens subsequently referred to *P. expansum* should not be included in that genus. Zalessky[3] in 1912 proposed a new generic name *Palamophyllum* for the Russian species but retained *Psygmophyllum* for a Mongolian Permian specimen which he named *P. mongolicum*: this fossil is clearly a portion of a compound frond with leaflets like those of some forms of *Palaeopteris*. In a later paper[4] this author assigns *Psygmophyllum mongolicum* to *Palamophyllum*, but on his attention being called by Zeiller to Saporta's note of 1878, Zalessky[5] decided to abandon his proposed genus *Palamophyllum* in favour of *Psygmophyllum*.

Confusion has also been caused by lack of uniformity in the use of the two generic names *Psygmophyllum* and *Ginkgophyllum*. The type-species of these genera I believe to be generically identical; they agree in the general form of the leaves, the lamina being much more deeply divided in the type-species of *Ginkgophyllum*, also in the decurrence of the narrow basal portion of the lamina, and both are probably shoots, though the morphological nature of these and other types included in *Psygmophyllum* is by no means clear.

Arber retains both names: as a matter of convenience he restricts *Psygmophyllum* to leaves that are entire or only slightly lobed, *e.g. P. flabellatum* (fig. 665), and the more deeply dissected leaves such as those of *Psygmophyllum Grasserti* (fig. 669) he refers to *Ginkgophyllum*. This distinction is, however, purely arbitrary and on the analogy of the leaves of *Ginkgo biloba* it would seem preferable to include both deeply divided and more or less entire leaves in the same genus. Cambier[6] and Renier prefer the name *Psygmophyllum* to *Ginkgophyllum* on the ground that in the leaves

---

[1] Schmalhausen (87).   [2] Kutorga (44) Pl. II. fig. 7.
[3] Zalessky (12) p. 38, Pl. VII. fig. 5.   [4] Zalessky (12²) p. 27.
[5] This decision was communicated in a letter (October, 1913). A. C. S.
[6] Cambier and Renier (10).

assigned by some authors to *Ginkgophyllum* there are no marginal veins like those in the lamina of a *Ginkgo* leaf. This objection, though not in itself fundamental, is based on a sound principle, namely an objection to the assumption of affinity to *Ginkgo* implied by *Ginkgophyllum*, an assumption that rests on a superficial resemblance unsupported by any evidence of real value.

The name *Psygmophyllum* is adopted both for entire and deeply divided leaves of larger dimensions than the similar leaflets of fronds included in such genera as *Palaeopteris* and *Adiantites*. Specimens usually occur as detached leaves, but when the leaves are attached to an axis the lamina is usually contracted into a fairly long, decurrent, basal portion; there is no true petiole.

The veins spread from the base of the lamina and are repeatedly forked; they may be very numerous and in some forms occasionally anastomose, as in *P. flabellatum,* or much farther apart, as in *P. majus* Arb. and *P. Brownii* (Daws.). The genus is at best a purely artificial one; we know nothing as to the reproductive organs or the anatomical structure, nor is it possible to determine in many instances whether the specimens are portions of compound fronds or shoots bearing simple leaves.

Psygmophyllum ranges from Devonian to Permian strata, and if the Rhaetic leaves named *Psygmophyllum ? crenatum* (Nath.) are accepted as evidence, the vertical distribution must be extended. The genus occurs in England, Ireland, the continent of Europe, Spitzbergen, and North America; it is also a member of the Permo-Carboniferous floras of South Africa and India.

Psygmophyllum flabellatum (Lindley and Hutton).

The name *Noeggerathia* was given by Lindley and Hutton[1] to some specimens from the Newcastle Coal Measures; of the two examples figured one consists of a slender supporting axis bearing several torn cuneate leaves, which they speak of as part of a compound frond, and the other is a single leaf or leaflet. It is stated by Prof. Lebour and Dr Arber[2] that the original specimens cannot be found in the Hutton collection in the Newcastle Museum. In Mr Howse's *Catalogue*[3] a specimen is described as possibly the original of the larger example, and some years ago I examined a

[1] Lindley and Hutton (32) A. Pls. xxviii., xxix.
[2] Arber (12) p. 394.   [3] Howse (88) A. p. 109.

leaf in the Hutton collection which was believed to be the specimen reproduced on Plate 29 of the *Fossil Flora*. Fig. 665 represents an impression in the British Museum[1] from the Newcastle Coal Measures; the cuneate and partially torn lamina, 15 cm. long, is characterised by the very large number of forked veins, approximately three to a breadth of 1 mm. In this and another specimen[2]

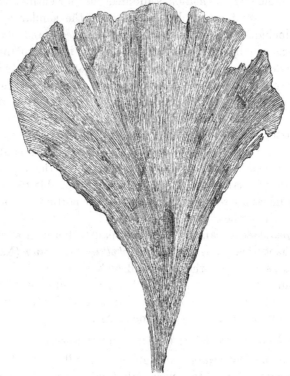

FIG. 665. *Psygmophyllum flabellatum*. (Brit. Mus. No. 40578; ca. ⅔ nat. size.) M. S.

it can be seen that the veins occasionally anastomose, but this feature is more clearly shown in some better specimens in the Sunderland Museum[3]. One of these is represented half natural size in fig. 666. The leaves have no true petiole and are attached to the axis by narrow decurrent bases. A careful examination of the specimen did not enable me to detect any satisfactory evidence

[1] No. 40578.   [2] No. 38927.   [3] For other figures, see Arber (12).

of the sheathing nature of the leaf-bases described by Arber, but
it was on this example that I noticed the anastomosing of the
veins (fig. 666, A—C), a character not hitherto recorded in the
genus[1]. The lamina is torn and it is difficult to determine the
degree of original lobing. Dr Arber believes that such specimens
as that reproduced in fig. 666 are portions of a herbaceous plant

Fig. 666. *Psygmophyllum flabellatum.* A—C, anastomosing veins.
(From a specimen in the Sunderland Museum; ½ nat. size.) M. S.

and quite distinct from *Ginkgo*. While agreeing with the latter
statement I do not regard the shoot-nature as definitely demonstrated, though that is the interpretation usually accepted
and not improbably correct. The habit appears to be identical

[1] A good specimen in the Manchester Museum also shows anastomosing veins.

with that of the Permian species named by Saporta *Ginkgophyllum Grasserti* (fig. 669) and his figure[1] of that type suggests a shoot rather than part of a compound frond. Some species referred by authors to *Psygmophyllum* are certainly pinnae, while others bear a closer resemblance to shoots with simple leaves. Until better material is available we cannot determine either the morphological nature or the systematic position of the various examples assigned to this provisional genus.

Psygmophyllum Kolderupi Nathorst.

Nathorst[2] has recently founded this species, naming it after Dr Kolderup, on specimens from Devonian strata in West Norway consisting of pieces of shoots, or possibly compound fronds, bearing spirally disposed fan-like leaves or leaflets on long stalks and reaching a breadth of 15 to 30 mm.; the veins are fine and repeatedly forked. The habit appears to be similar to that of *Psygmophyllum flabellatum*, but it is hardly possible to say whether we are dealing with fragments of a frond or branches bearing simple leaves.

Psygmophyllum Kidstoni Seward.

This species, very similar to *P. flabellatum*, is founded on specimens discovered by Mr Leslie in the Permo-Carboniferous rocks at Vereeniging, South Africa[3]. The cuneate leaves reach a length of 13 cm. and in some cases are deeply divided into two truncate lobes (fig. 667). The veins appear to be identical with those of *P. flabellatum* though no definite anastomosing has been detected. A photograph recently received of a new specimen shows some indication of a few cross veins, but the occasional anastomosing of veins should not be regarded as a feature of great importance. The axis of this species is broader than any so far found in the case of the English species, and the leaves are attached by a similar decurrent base. Incomplete leaves similar to *P. Kidstoni* though probably not specifically identical are described by Dun[4] from Permo-Carboniferous strata at Sydney as *Rhipi-*

[1] Saporta (84) Pl. 152, fig. 2.
[2] Nathorst (15) p. 25, Pl. I. figs. 6—11; Pl. II. figs. 2—5.
[3] Seward (03) B. p. 93, Pl. XII.; Arber (05) B. p. 213, fig. 47.
[4] Dun (10) Pl. 51.

dopsis ginkgoides var. *Suessmilchi*; they are probably referable
to *Psygmophyllum*.

Psygmophyllum Williamsoni Nathorst.

A species[1] founded on imperfect leaves from the Upper Devonian
of Spitzbergen agreeing closely in shape and venation with *P.*
flabellatum. This is the oldest European example of the genus.
In answer to an enquiry with regard to the venation Prof. Nathorst
kindly informed me that he was unable to detect any definite
traces of anastomosis and that the veins agree in their spacing
with those of *P. flabellatum*.

FIG. 667. *Psygmophyllum Kidstoni*. From Vereeniging, S. Africa.
(⅓ nat. size.)

Psygmophyllum majus Arber.

The large flabellate leaves[2], often more than 16 cm. long and
15 cm. broad, representing this Lower Carboniferous or Upper
Devonian species from Newfoundland, are distinguished from
P. flabellatum by their broader and less kite-like lamina and by
the coarser venation. The distal margin is almost entire or
characterised by broad and shallow lobes: as in *P. Williamsoni*
no axis occurs with the leaves. As Arber points out, this species

---

[1] Nathorst (94) A. Pl. II. figs. 1, 2; Arber (12) Pl. XLII. fig. 4.
[2] Arber (12) p. 392. Pls. XLII.—XLIV.

bears some resemblance to *Psygmophyllum Brownii* originally described by Dawson[1] as *Cyclopteris Brownii* from Upper Devonian strata in Maine.

Psygmophyllum Haydeni Seward.

In 1905 some incomplete specimens were described from Permo-Carboniferous rocks in Kashmir as *Psygmophyllum* sp.[2] for which, on the discovery of better material, a specific name was proposed[3]. The leaves reach a length of 13 cm. and are characterised by the division of the lamina into six or more obcuneate

FIG. 668.   *Psygmophyllum Haydeni.*   (Nat. size.)

segments, the divisions sometimes extending to the base of the broad part of the leaf (fig. 668). In the upper part of the lamina there are three to four veins per millimetre but lower in the lamina the veins are 1 mm. apart. Dr Arber[4] suggests that this species should rather be referred to *Ginkgophyllum* or *Rhipidopsis*.

Psygmophyllum Hollandi Seward.

This less satisfactory species is represented by some imperfect leaves from Carboniferous rocks of Kashmir[5]. The lamina is

[1] Dawson (63) Pl. XVII. fig. 6.   [2] Seward and Woodward (05) B.
[3] Seward (07⁵) B.   [4] Arber (12) p. 400 (footnote).   [5] Seward (07⁵) B.

divided by a deep median sinus into two bilobed segments and agrees closely with some species of *Baiera*. By some authors this species would be included in *Ginkgophyllum* but, as already stated in the account of the genus, the degree of dissection of the leaves is too variable and unimportant a character to be made the basis of a generic differentiation.

Psygmophyllum Grasserti (Saporta).

This Permian species from Lodève[1], France, agrees closely with *P. flabellatum* in the size and outline of the leaves as also in their

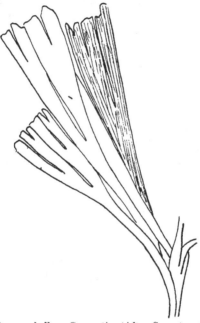

Fɪɢ. 669. *Psygmophyllum Grasserti*. (After Saporta; ⅔ nat. size.)

method of attachment to the axis, but it is distinguished by the division of the lamina into linear segments (fig. 669). A species with similar leaves, from the Permian beds of the Ural mountains, is figured by Saporta as *Ginkgophyllum kamenskianum*[2]. Johnston[3] describes a species, *Ginkgophyllum australe*, from 'Mesozoic' rocks

[1] Saporta (75) p. 1018; (84) Pl. 152, fig. 2.
[2] Saporta and Marion (85) p. 144, fig.73, B; Saporta (82).
[3] Johnston (86) p. 178.

in Tasmania which he compares with *G. Grasserti* Sap., but as he gives no illustration of the fossil no opinion can be formed of its true nature.

Psygmophyllum kiltorkense Johnson (= *Kiltorkensia devonica* John.).

A species[1] recently described from the Upper Devonian grits of Kiltorcan, Ireland, characterised by fan-shaped leaves 7 cm. long and 5 cm. broad, deeply divided into two symmetrical halves each of which is again divided into two ribbon-like segments with a *Ginkgo*-like venation. The leaves agree closely with those of *P. Grasserti* in their general form and in the lobing of the lamina. Johnson believes this type to be an ancestral form of *Ginkgo* though there are no adequate grounds for such a view.

Psygmophyllum ? crenatum (Brauns).

The close resemblance presented by the Rhaetic leaves from near Braunschweig, originally described by Brauns[2] as *Cyclopteris crenata* and subsequently figured by Nathorst[3] as *Ginkgo ? crenata*, to some Permo-Carboniferous leaves included in *Psygmophyllum* suggests affinity with that genus. The obovate lamina, approximately 12 cm. long, is slightly lobed on the upper margin and contracted below into a stalk-like base; the forked veins are nearly 3 mm. apart. Nathorst compares the species with *Psygmophyllum flabellatum* Lind. and Hutt., but the Rhaetic specimens differ from the English type in their much coarser venation: in the lobing of the lamina and in the coarse venation there is a much greater similarity to the broader leaves described by Arber[4] as *Psygmophyllum majus*. In view of the incomplete nature of the material it is inadvisable to adopt the name *Psygmophyllum* without reservation. An examination of Nathorst's specimen in the Stockholm Museum led me to regard it as more probably an example of *Psygmophyllum* than of *Ginkgo*.

---

[1] Johnson (14). Since this account was written Prof. Johnson has described additional material including stems and foliage [Johnson (17)] demonstrating the occurrence of repeatedly forked filamentous leaves [or leaflets] attached to slender axes bearing also the broader form of lamina. The plant, which he now refers to a new genus *Kiltorkensia*, may well be a Pteridosperm with compound fronds and dimorphic pinnules.

[2] Brauns (66) p. 52, Pl. XIII. fig. 8.   [3] Nathorst (78).
[4] Arber (12) p. 392, Pls. XLII.—XLIV.

Other records of Psygmophyllum.

Psygmophyllum primigenium (Saporta). Some leaves discovered by Grand'Eury in Permian rocks of the Urals were described by Saporta[1] as *Salisburia primigenia* and regarded as the prototype of the surviving species. The original specimens are unfortunately not available, but from the published figures it would seem that the species is of the same general type as *P. flabellatum.*

Reference has already been made to an American Devonian species referred by Dawson to *Cyclopteris* and recently transferred to *Psygmophyllum.* A leaf or leaflet described by Dawson from Gilboa, New York, as *Noeggerathia gilboensis*[2] affords a good example of a specimen that may be a *Psygmophyllum* leaf or a leaflet of a frond of the *Noeggerathia* type. Lesquereux[3] considers Dawson's specimen to be a pinnule of *Palaeopteris.* The same remark applies to Lesquereux's species *Noeggerathia obtusa*[4] from the Coal Measures of Pennsylvania included by Arber in *Psygmophyllum*; it is probably a pinnate frond.

The Russian species *P. expansum* and *P. cuneifolium* are discussed in the account of the genus. The species *Psygmophyllum Delvali*[5] Camb. and Ren. from the Westphalian of Belgium is now admitted to be a leaf of *Cordaites.* A species described by Sandberger[6] from the Permian of the Black Forest as *Ginkgophyllum minus* has been assigned by Sterzel[7] to *Dicranophyllum.* A leaf figured by Schmalhausen[8] from the Permian of East Russia as *Baiera gigas* is no doubt a *Psygmophyllum* allied to *P. Kidstoni.* A fragment figured by Schenk from China as *Ginkgophyllum* sp. is too imperfect to determine, and the specimens from the same locality described as *Psygmophyllum angustilobum*[9] are, as Zeiller points out, pinnules of a frond of the *Eremopteris* type.

A sufficient number of examples have been described to illustrate the range of the genus and the unsatisfactory nature of

[1] Saporta (82); Saporta and Marion (85) p. 145, fig. 74.
[2] Dawson (63) p. 463, Pl. XVII. fig. 6; (71) A. p. 46, Pl. XVI. fig. 172.
[3] Lesquereux (80) A. p. 305.   [4] *Ibid.* Pl. XLIX. figs. 6, 7.
[5] Cambier and Renier (10) Pl. VI. fig. 1.
[6] Sandberger (90) p. 101.   [7] Sterzel (07) p. 820.
[8] Schmalhausen (87) Pl. V. fig. 10.
[9] Schenk (83) A. p. 221, figs. 7, 8; Pl. XLIII. figs. 22—24.

the material from a botanical point of view. Failing reproductive organs or petrified specimens some useful evidence might be afforded by an examination of the cuticular structure of well preserved leaves.

RHIPIDOPSIS. Schmalhausen

Schmalhausen[1] instituted this genus for large petiolate oval leaves from the Permian rocks of the Petschora district, characterised by the division of the lamina into several obcuneate or obovate segments closely resembling in their form and venation some forms of *Psygmophyllum* especially *P. Haydeni*[2]. We have no definite information as to the systematic position of the parent-plant; the genus has usually been regarded as a representative of the Ginkgoales on the ground of similarity in the leaves, but while admitting that a relationship between *Rhipidopsis* and *Ginkgo* is not improbable it is the safer course to regard *Rhipidopsis* as a genus of Gymnosperms of uncertain affinity. Schmalhausen attributes to *Rhipidopsis* some *Samaropsis* seeds[3] found in association with the leaves, and Kurtz[4] states that he has found leaves and 'fruits' in the Argentine. No proof of any connexion between leaves and seeds has so far been discovered. The genus is recorded from Russia, South America, and India from strata that are Permian or approximately Permian in age.

Rhipidopsis (fig. 670) is distinguished from *Psygmophyllum* by the presence of a petiole and from most forms of that genus by the deeper dissection of the lamina, as also by a more pronounced difference in form and size between the several segments of the lamina. Zeiller[5] has drawn attention to a close resemblance between *Rhipidopsis* and a specimen figured by Schmalhausen from the Artinsk Permian beds as *Psygmophyllum expansum*[6].

Rhipidopsis ginkgoides Schmalhausen.

The type-species (fig. 670) is characterised by the large size of the leaves which, according to Schmalhausen[7], may reach a

---

[1] Schmalhausen (79) A. p. 50, Pls. vi., viii.
[2] See p. 86, fig. 668.
[3] Schmalhausen (79) A. Pl. viii. figs. 9—11.
[4] Zeiller (96) A. p. 467.   [5] *Ibid.* p. 471.
[6] Schmalhausen (87) Pl. iii. fig. 10.
[7] Schmalhausen (79) A. p. 50, Pls. vi., viii.

length including the petiole of 30 cm. and a breadth of 11 cm.
The segments, 6—10 in number, are often free to the summit of
the petiole; they vary considerably in shape and size, the median
segments are obcuneate with a broad rounded truncate margin,

FIG. 670. *Rhipidopsis ginkgoides.* (From a photograph of the original of one
of Schmalhausen's figures supplied by Prof. Zalessky.)

while the lateral lobes are obovate asymmetrical.   The repeatedly
forked veins are 1—1·5 mm. apart in the lower part of the lamina
but much more crowded in the apical region.   The slender petiole

reaches a length of 10 cm. This species has been recorded from the Argentine but no figures have been published. It is not improbable that the seeds of the *Samaropsis* type associated with the leaves in the Russian and Argentine localities may belong to this genus, but proof is lacking.

Rhipidopsis densinervis Feistmantel.

This Indian species from the Raniganj group of the Damuda series[1] (Lower Gondwana) is founded on some leaf-impressions very similar in size and form to *Rhipidopsis ginkgoides*. The presence of a petiole is shown on one of the figured specimens: the lamina is deeply divided into about six obcuneate segments that appear to be irregularly lobed on the truncate margin. *Rhipidopsis densinervis* is distinguished by its dense venation and by a difference in size between the lateral and median segments less than in the leaves of *R. ginkgoides*. Kurtz[2] states that some specimens found by him in Permo-Carboniferous beds in the Argentine may belong to this species.

Rhipidopsis gondwanensis sp. nov.

The specimens for which this name is proposed were described by Feistmantel as *Rhipidopsis ginkgoides*[3] from the Barakar group of the Damuda series. My examination of the type-specimens confirms Feistmantel's statement that they agree closely with Schmalhausen's Russian leaves except in their much smaller size: the Indian leaves reach a length of 3 cm. while in Schmalhausen's species the lamina may be 14 cm. in length. In view of this difference and the wide geographical separation of the two forms it would seem preferable to adopt a distinctive name. The lamina is divided, almost to the base, into 6—10 segments; the larger are cuneate and the smaller obovate and obtuse.

SAPORTAEA. Fontaine and White.

Fontaine and White[4] instituted this generic term for some incomplete impressions of large leaves from Permian rocks in Virginia having a broadly cuneate or suborbicular lamina characterised by

[1] Feistmantel (80) B. p. 121, Pl. xlvi. A.
[2] Zeiller (96) A. p. 467.
[3] Feistmantel (81) p. 257, Pl. ii. fig. 1; (86) Pl. iii. A. figs. 1, 2.
[4] Fontaine and White (80) B. p. 99, Pl. xxxviii.

a thickened lower margin extending horizontally a short distance
on either side of the petiole and presenting the appearance of
being formed by the bifurcation of the summit of the leaf-stalk
at right-angles to its long axis.   The lamina is irregularly dissected,
but from the published figures it is difficult to distinguish between
original lobing and divisions due to tearing.   The dichotomously
branched veins spread through the lamina from the centre of the
base and are given off at a wide angle from the thick lower edge of
the lamina.   In *Saportaea grandifolia* the petiole has a length of
10 cm. and the rest of the incomplete leaf is 8 cm. long and 9·5 cm.
broad : the second species *S. salisburioides* is represented by portions of similar but smaller leaves with a slender petiole.   While
comparing these fossils with *Ginkgo* the authors of the genus call
attention to the peculiar features of the lower edge of the lamina
and of the venation.   The general resemblance in leaf-form between
Saportaea and *Ginkgo* is hardly sufficient to warrant any definite
statement as to relationship and this Permian genus must for the
present be relegated to the class of *Plantae incertae sedis*.

DICRANOPHYLLUM.   Grand'Eury.

This genus was first described by Grand'Eury[1] who, before the
publication of the full description of the type-species, suggested
the substitution of *Eotaxites* for *Dicranophyllum*[2] the name finally
adopted[3].   The genus is fairly abundant in the Upper Carboniferous
rocks of France and occurs also in Portugal, Belgium, and Germany ;
it has recently been recorded from England and is represented in
the Coal Measures of the United States and Canada.   It occurs in
Permian strata in Germany but with a few exceptions the genus
is characteristic of Stephanian beds.

The systematic position of *Dicranophyllum* is far from settled ;
by many authors it is considered to be a member of the Ginkgoales
and is compared also with the Taxeae.   In all probability the
genus is allied to the Cordaitales, though, as stated in the case of
Trichopitys, it cannot be assigned to a definite position in the
Gymnosperms until we possess fuller information with regard to
the reproductive organs or the anatomical structure.

[1] Compt. Rend. Vol. LXXX. p. 1021, 1875.   [2] δίκρανος, two-pointed.
[3] Grand'Eury (77) A. p. 272, Pls. XIV., XXX.

In habit *Dicranophyllum* resembles *Lepidodendron*; it is an
arborescent plant sparsely and irregularly though sometimes
dichotomously branched; the leaves are crowded and spirally
disposed, in some species persistent—in the sense in which the
leaves of *Araucaria* are persistent—while in others they probably
fell at an earlier stage. The leaves (fig. 671) exhibit a wide range
in size, in the amount of lobing and the angle of divergence of the
segments; there is no differentiation into a lamina and petiole
nor are there any short foliage-shoots as in *Ginkgo*; the whole
leaf is represented by a narrow lamina, in some species almost
spinous, which consists in the basal portion of a simple linear
'stalk' reaching in extreme cases a breadth of about 7 mm.,
attached by a decurrent base which persists as an elongated cushion
closely resembling the leaf-base on some Lycopodiums or the
projecting cushions of *Picea* (*cf.* fig. 140, Vol. II. p. 94). The
cushions are contiguous and cover the surface of a branch as in
Lepidodendron, but they are distinguished by the occurrence of
the leaf-scar at the apex of the cushion in contrast to its sub-
apical position in *Lepidodendron*. The typical form of the leaf-
base is shown in fig. 671, A, but in *Dicranophyllum Beneckianum*
Sterz. the transversely elongated leaf-scars are almost contiguous
as in some species of *Sigillaria*. At a·distance from the base
varying in different species the lamina is divided into two, generally
equal, branches that diverge at an acute or wide angle, and in
most species each arm undergoes one or more bifurcations in a
single plane. The whole leaf may reach a length of over 20 cm.
In the basal portion of the lamina there are two or more parallel
veins, but in branches in which the leaf-scars are well preserved
there is only a single vascular-bundle scar indicating a single
leaf-trace up to the base of the lamina. Each segment of the
leaf has usually two veins and the acutely pointed ultimate
segments have a median vein. The so-called secondary or interstitial veins are no doubt due to the presence of hypodermal
stereome strands. The narrower *Dicranophyllum* leaves are very
similar to the deeply divided pinnae of *Macrozamia heteromera*
(fig. 671; *cf.* fig. 396, F, Vol. III. p. 26). The branching of the lamina
is generally regular but in several instances the subdivision is
irregular (fig. 671, D). On young shoots the leaves may be

nearly vertical but in most species they become widely extended and on older branches may be reflexed as in some Lycopods (fig. 121, B, Vol. II. p. 35). There is some evidence that the pith was discoid as in *Cordaites*[1]. The microsporophylls are said to be borne in small ovoid strobili in the axils of foliage leaves, but

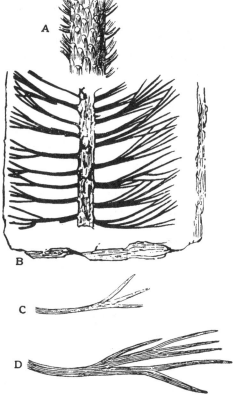

FIG. 671. *Dicranophyllum gallicum.* A, piece of a stem showing leaf-cushions. B—D, foliage-shoot and leaves. (After Grand'Eury.)

the only evidence as to their structure so far adduced is furnished by an imperfectly preserved specimen described by Zeiller[2] associated with a shoot of *Dicranophyllum robustum* but not actually attached; this consists of a small axis expanded into

---

[1] Renault and Zeiller (88) A. Pl. LXXI. figs. 3, 4.
[2] Zeiller (78) Pl. X. figs. 1 *a* and 3.

a radially segmented distal portion bearing some imperfectly
preserved ovoid bodies on its lower face which are probably
microsporangia. No spores are recorded. Zeiller compares this
sporangiophore with the microsporophyll of a *Taxus*. On some
stems small scale-covered buds occur immediately above the
attachment of a leaf; these are probably fertile shoots but we
have no definite information with regard to their structure. Some
specimens from Commentry[1] demonstrate the occurrence of small
oval ovules or seeds, 4×3 mm., along the length of ordinary
leaves (fig. 672), and seeds are sometimes found associated with the
basal portions of foliage leaves though not in organic connexion
with them, except in an example described by Renault from
Autun as *Dicranophyllum gallicum* var. *Parchemineyi*[2]. Some
leaves of *D. striatum*[3] are described as enlarged at the base and
slightly concave as if to hold a seed, but if this supposition is
correct it involves the admission of two types of seed-bearing
organs within the genus. The more probable conclusion is that
the seeds were borne along the length of the lamina of the sporophylls and on the expanded bases.

Some specimens from Lower Cretaceous beds in Virginia
described as *Baiera foliosa*[4] resemble *Dicranophyllum*, but in view
of the vast chronological gap between these beds and those in
which *Dicranophyllum* occurs it is unlikely that the similarity
has any significance.

Dicranophyllum gallicum Grand'Eury.

This species, one of the two described by Grand'Eury in 1877[5],
is the commonest representative of the genus; it is characterised
by persistent leaves with a base that is unbranched for a distance
of 15—20 mm. and then bifurcates into two equal or approximately
equal segments at an angle of about 30°; these reach a length of
10—15 mm. and divide into two acute segments 8—10 mm. long.
There are three veins in the basal portion of the lamina, one of
which branches below the dichotomy, and each of the divergent
arms has two veins. The leaf-cushions are 2—3 times as long as

[1] Renault and Zeiller (88) A. Pl. LXXI. fig. 5.
[2] Renault (96) A. p. 375.   [3] Renault and Zeiller (88) A. p. 632.
[4] Fontaine (89) B. p. 213, Pl. XCIV. fig. 3; Berry (11) Pl. LIX.
[5] Grand'Eury (77) A. p. 272, Pls. XIV., XXX. Zeiller (80) A. Pl. LXXVI. figs. 1, 2.

broad and the median vein of the lamina is continued as a keel in the middle of the persistent base. The sporophylls have the form of foliage leaves and bear numerous ovules (fig. 672). This species is recorded from the coal-fields of the Loire, Commentry, Gard, Brive[1] and elsewhere; it occurs also in the Coal Measures of Portugal[2].

The specimens described by Grand'-Eury from Gard as *D. tripartitum*[3], which I had an opportunity of examining in the Ecole des Mines, Paris, are not specifically distinguishable from *D. gallicum*. A large decorticated stem of *D. gallicum* in the Paris Collection recalls a decorticated stem of *Lepidodendron*. Some

FIG. 672. *Dicranophyllum gallicum*. (After Zeiller from Renault; ⅔ nat. size.)

imperfect specimens described by White[4] from the Coal Measures of Missouri as *Dicranophyllum* sp. are compared by him with *D. gallicum*; one of them consists of a forked foliage-shoot with short and repeatedly bifurcate leaves illustrating the superficial resemblance between *Dicranophyllum* and *Lepidodendron*. Another specimen shows an irregularly branched leaf which might equally well be referred to *Trichopitys*.

Dicranophyllum lusitanicum (Heer).

This species was first figured by Gomes[5] as *Cyperites?* sp. on the ground of the similarity of the lamina to some fragments, probably of Lepidodendroid leaves, described by Lindley and Hutton[6] as *Cyperites bicarinata* and subsequently included by Heer[7] in his genus *Distrigophyllum*. In a note to his account of Mesozoic plants from Portugal Heer[8] renames the plant *Distrigophyllum lusitanicum* and compares it with *Dicranophyllum gallicum* Grand'Eury. De Lima[9] recognised the true nature of the specimens from the Stephanian of Portugal and published a full

[1] Zeiller (92²) A. p. 96.   [2] Lima, de (88) Pls. I., III.
[3] Grand'Eury (90) A. p. 335, Pl. VI. figs. 12, 13.
[4] White (99) B. p. 272, Pl. XLI. fig. 10; Pl. LXXIII. fig. 1.
[5] Gomes (65) p. 32, Pls. I., V.   [6] Lindley and Hutton (32) A. Pl. XLIII.
[7] Heer (76) A. p. 39.   [8] *Ibid.* (81) p. 11, footnote 1.
[9] Lima, de (88).

description of the species. The leaves are 14—16·5 cm. long and
2—4 mm. broad at the base; the lamina is once forked and the
forks diverge at a very small angle as in *D. longifolium* Ren.
Exceptionally good specimens figured by Gomes and de Lima
show numerous long leaves spreading radially in the matrix from
a comparatively slender axis. In *D. longifolium* the leaves are
given off at a much more acute angle.

Dicranophyllum robustum Zeiller.

This type[1] is similar to *D. gallicum* but the leaves are only
preserved in their basal portions; the lamina is 5—6 mm. broad
and bifurcates at a distance of about 15 mm. from the base at an
angle of 20°—30°. Ovoid buds occur in the axils of some of the
leaves. It was in association with this specimen from the Gard
coal-field that Zeiller found the microsporophylls already described.
The surface shows particularly well-preserved large and slightly
depressed cushions 3—4 cm. long and 4—5 mm. broad.

Dicranophyllum Beneckianum Sterzel[2].

In the form of the leaves this Permian species from Baden
closely resembles *D. gallicum*; it is chiefly of interest because of
the almost complete absence of leaf-cushion; the leaf-scars,
characterised by their acute lateral angles, are almost contiguous
as in some species of *Sigillaria*.

Dicranophyllum Richiri Renier[3].

In this Belgian Westphalian species the leaves are dichotomously
branched into two linear segments at an angle of about 60°; it
differs from *D. gallicum* in the single bifurcation of the lamina,
the wider angle of divergence, and in the feebler relief of the
leaf-cushions.

Dicranophyllum anglicum Kidston.

This, the only British species, has recently been described from
the Westphalian beds of Staffordshire[4] The crowded leaves,
3·50 cm. long, are dichotomously branched three or four times into
slightly spreading linear rigid segments with a maximum breadth

---

[1] Zeiller (78).   [2] Sterzel (07) p. 381, Pls. XIV., XV.
[3] Renier (07) p. 186. Pl. XVII. figs. 3—7: (10²) Pl. CXVII.
[4] Kidston (14) p. 170, Pl. XIV. figs. 3, 3 *a*.

of 1·25 mm. The undivided portion of the lamina is about 7 mm.
long. Kidston speaks of the rhomboidal outline of the leaf and
the repeated dichotomy of the lamina as distinguishing features.

Some fragments of forked leaves are figured by Schenk[1] from
the Coal Measures of China as *Dicranophyllum latum*, but the
material is too meagre for accurate determination. It is noteworthy that the broader type of *Dicranophyllum* leaf may easily
be confused with an impression of a branched Stigmarian rootlet.
The narrower specimens described by Schenk from China as
D. angustifolium[2] are also too fragmentary to be accepted as
trustworthy evidence of the occurrence of the genus in the southern
flora.

Dicranophyllum striatum Grand'Eury.

This species like several others is founded on detached leaves[3],
a circumstance that has led some authors to draw a distinction
between species with persistent leaves and those with caducous
leaves. There is, however, no good reason for assuming that all
species were not of the evergreen type. The leaves of this species
are characterised by their great length which may be 24 cm.;
the lamina is once or twice forked and is 5—6 mm. broad at the
base, which contains 4—7 veins.

Dicranophyllum longifolium Renault.

In this Commentry species[4] the leaves, which reach 14 cm. in
length, are characterised by the very small angle of the divergence
of the segments, 3° as contrasted with a divergence of 30° in
D. gallicum. The leaves are almost erect and twice bifurcate.

In addition to *Dicranophyllum Beneckianum* Sterzel has described a second species, *D. latifolium*[5], from the Lower Permian
of Baden characterised by leaves similar to those of *D. striatum*
but generally longer. The species is founded on leaves and is not
a well-defined type.

Two species are recorded by Lesquereux[6] from the Coal Measures

---

[1] Schenk (83) A. p. 222, Pl. XLII. figs. 11, 12.
[2] *Ibid.* Pl. XLII. figs. 17, 18.
[3] Grand'Eury (77) A. p. 275. Renault and Zeiller (88) A. Pl. LXXI. fig. 2.
[4] Renault and Zeiller (88) A. p. 631, Pl. LXXI. fig. 1.
[5] Sterzel (07) p. 391, Pl. xv. figs. 9—11.
[6] Lesquereux (80) A. pp. 553, 554, Pls. LXXXIII., LXXXVII.

of Pennsylvania, but neither is represented by very satisfactory specimens: *Dicranophyllum dichotomum* Lesq. is founded on a dichotomously branched shoot bearing long and narrow leaves in the apical region only and very similar in appearance to *Lepidodendron* except in the branched lamina. The second species *D. dimorphum* Lesq. is represented by leaves and branches, which however are not well preserved. The peculiar subdivision of the apical portion of the laminae suggests a simple leaf with a frayed termination.

Dicranophyllum glabrum (Dawson).

Under this name Dr Stopes[1] has recently described a well-preserved leaf from the Westphalian series of New Brunswick. The specific name was first applied by Dawson[2] to specimens which he referred doubtfully to *Psilophyton*. The type-specimen is 9 cm. long and 3 mm. broad at the base and the lamina is repeatedly branched. This specimen bears a close resemblance to the leaf from Autun described by Renault as *Trichopitys Milleryensis*[3].

The imperfect specimens described by Dawson[4] from Devonian rocks in Queensland as *Dicranophyllum australicum* and subsequently figured by Jack and Etheridge[5] consist of a slender axis, 3 mm. wide, with elongate leaf-bases bearing leaves 3 mm. long with two widely divergent apical segments like those characteristic of the sporophylls of *Gomphostrobus*. The fragments have no claim to be included in *Dicranophyllum*.

There has been confusion between *Dicranophyllum* and *Gomphostrobus*[6]: as shown by drawings reproduced by Potonié[7] of specimens of *Gomphostrobus* from the Permian of Thuringia, there is a close resemblance in habit between the two genera, but in *Gomphostrobus* the foliage-leaves are falcate and entire, while the bifurcate sporophylls differ from the leaves of *Dicranophyllum* in their widely divergent and small apical fork.

[1] Stopes (14) p. 79, Pl. XVIII. fig. 47.
[2] Dawson (62) p. 315: for other references, see Stopes *loc. cit.*
[3] Renault (96) A. p. 378; (93) A. Pl. LXXXII. fig. 2.
[4] Dawson (81) A. p. 306, Pl. XIII. figs. 15, 16.
[5] Jack and Etheridge (92) B. p. 49.
[6] Schenk (90) A. erroneously includes *Sigillariostrobus bifidus* Geinitz, (73) Pl. III. figs. 5—7, in *Dicranophyllum*. See also Sterzel (93) A. p. 111.
[7] Potonié (93) A. Pl. XXVIII. figs. 1, 2.

A foliage-shoot described by Renault[1] from Autun as *Pinites permiensis*, though too imperfect to be identified, is worthy of notice as possibly an example of *Dicranophyllum* or *Trichopitys*; it consists of an axis 3 mm. in diameter bearing numerous spirally disposed leaves 3 cm. long, barely 1 mm. broad and triangular in section, at an angle of 45°. The leaf-cushions are elongate and slightly prominent. It is, however, impossible to decide whether this fossil should be referred to the Lycopodiales or to the Gymnosperms. There is no evidence that the leaves are attached to short shoots and the use of the generic name *Pinites* cannot be justified by any trustworthy test.

TRICHOPITYS. Saporta.

Saporta[2] proposed the generic name *Trichopitys* in 1875 for some shoots from the Permian beds of Lodève bearing long, narrow, and deeply divided leaves; he defined the genus as follows: 'Folia verosimiliter rigida cartilagineaque, dichotome partita etiamque pedato-partita, petiolo plus minusve elongato, sursum in lacinias 4—6, anguste lineares, uninerviasque dissecta[3].' Many palaeobotanists have followed Saporta in regarding *Trichopitys* as a member of the Ginkgoales, but the evidence in support of this view is by no means conclusive. The only species so far described that affords any information with regard to the habit or fertile shoots of the plant is the type-species *T. heteromorpha* (fig. 673). A fairly stout branched axis bears leaves varying considerably in size and form; they may be long and filiform, apparently rigid, simple or deeply divided, or short and entire, and in some cases resembling the leaves of certain smaller species of *Baiera* except in the less regular forking of the lamina. In the axil of some foliage leaves are short, simple or branched, axes bearing seed-like bodies originally described as buds and afterwards regarded as seeds. A specimen figured by Zeiller[4] from Lodève (fig. 673) shows a branched axillary shoot bearing several small ovules comparable with an abnormal ovuliferous shoot of *Ginkgo* (cf. fig. 631, D).

---

[1] Renault (96) A. p. 377; (93) A. Pl. LXXXII. fig. 1.
[2] Saporta (75) p. 1020.
[3] *Ibid.* (84) p. 263, Pl. CLII. fig. 1.
[4] Zeiller (00²) B. p. 254, fig. 182.

Renault[1] has figured a portion of a large leaf from Autun as *Trichopitys Milleryensis* which may belong to the closely allied genus *Dicranophyllum*; it is 12 cm. long and 3 mm. broad at the base; the narrow basal part of the lamina has three parallel veins and forks into two arms, each of which again branches into divergent linear segments. The leaf is larger and broader than the leaves of *T. heteromorpha* and agrees very closely with those of some species referred to *Dicranophyllum*: the fact that the branching of the lamina is not absolutely regular cannot be accepted as a constant difference between *Dicranophyllum* and *Trichopitys*: the leaves shown in fig. 671, which were found attached to undoubted *Dicranophyllum* branches, are no more regular in the

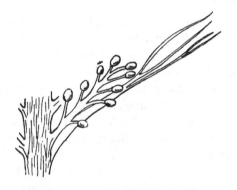

Fig. 673.   *Trichopitys heteromorpha.*   (After Zeiller; ⅔ nat. size.)

forking of the segments than *T. Milleryensis,* and a leaf recently figured by Dr Stopes[2] from New Brunswick as *Dicranophyllum glabrum* (Daws.) bears a close resemblance to Renault's figure. In some cases the more regular dichotomy of the leaves is a characteristic of *Dicranophyllum*, but it is not a constant feature. Some imperfectly preserved specimens figured by White[3] from the Coal Measures of Missouri as *Dicranophyllum* sp. bear irregularly branched leaves which are hardly distinguishable from some of those on Saporta's type-specimen of *Trichopitys heteromorpha.*

[1] Renault (93) A. Pl. LXXXII. fig. 2; (96) A. p. 378.
[2] Stopes (14) p. 79, Pl. XVIII. fig. 47.
[3] White (99) B. p. 272, Pl. XLI. fig. 10; Pl. LXXIII. fig. 1.

In the present state of our knowledge it is impossible to give a
satisfactory definition of the genus or to state precisely on what
grounds it is separated from *Dicranophyllum*. In *Trichopitys*,
as represented by *T. heteromorpha*, the leaves are more variable
in form than in *Dicranophyllum* and less regular in the subdivision
of the lamina; there are no persistent leaf-bases like those of
Dicranophyllum, but this is a character that could not be seen in
imperfectly preserved or partially decorticated specimens. A more
important difference would seem to be that in *Trichopitys* the
seeds are borne on special axillary shoots, while in *Dicranophyllum*
they occur on ordinary leaves. Such evidence as we have suggests
that *Trichopitys* is a Gymnosperm possibly allied to the Cordaitales
and Ginkgoales, but the facts hardly justify its inclusion in either
group. Its affinity to *Dicranophyllum* cannot be definitely determined though in all probability the two genera are closely related
if not indeed generically identical.

Saporta included in *Trichopitys* two Jurassic species, *T. lacini-
ata*, originally referred to the genus *Jeanpaulia*, and *T. Lindleyana*[1];
in the latter species he included the specimens doubtfully assigned
by Lindley and Hutton to *Solenites* as *Solenites ? furcata*[2]. These
and other Jurassic leaves that are referred by some authors to
Trichopitys are usually regarded as examples of *Baiera*[3]; there are
no adequate grounds for believing them to be closely related to
the Permian *Trichopitys*.

Zeiller[4] records a fossil from Triassic beds in Madagascar that
he thinks may be an example of *Trichopitys*.

SEWARDIA. Zeiller.

This generic name was proposed by Zeiller in place of *Withamia*
which, in ignorance of its previous use, I employed for some
specimens from the Wealden rocks of Sussex. The inclusion of
a second species, *Sewardia longifolia*, necessitates an extension of
the definition of the genus to include spinous branches bearing
spirally disposed leaves or leaf-like organs, either orbicular and
entire or fan-shaped and deeply divided, in the axil of recurved
spinous processes.

[1] Saporta (84) Pl. CLV. figs. 1—9.   [2] Lindley and Hutton (37) A. Pl. 209.
[3] Seward (00) B. p. 266.   [4] Zeiller (11²) p. 234.

Sewardia latifolia (Saporta).

1849. *Otozamites latifolia* Brongniart, Tableau, p. 106.
1872. *Sphenozamites latifolius* Schimper[1], Traité, Vol. II. p. 163.
1875. *Cycadorachis armata* Saporta, Plant. Jurass. p. 196, Pl. 117, fig. 1.
1895. *Withamia armata* Seward, Wealden Flora, Vol. II. p. 174, Pl. II.
    figs. 1, 2: Pl. v. fig. 1.
1900. *Sewardia latifolia* Zeiller, Éléments Paléobot. p. 233.

This species is represented by woody axes, about 1 cm. in
breadth reaching a length of 50 cm., from the Wealden beds of
Sussex, bearing more or less orbicular, entire, leaves or leaf-like

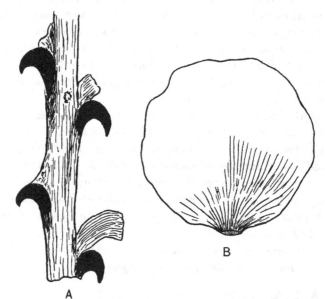

FIG. 674. *Sewardia latifolia.* Axis (A) and single leaf (B). (British
Museum; ¾ nat. size.)

organs, 6 cm. or more long, in the axil of stout recurved spinous
processes (fig. 674, A, B). The leaves are sessile and the venation
is of the *Cyclopteris* type. Spinous axes of the same form had
previously been described by Saporta from Kimeridgian rocks in
France as *Cycadorachis armata,* but these show no indication of
leaves and were regarded as Cycadean. In a letter written to me
in 1895, shortly before his death, the Marquis of Saporta[2] suggested

---

[1] See also Saporta (75) A. p. 188, Pls. 112, 113.   [2] Seward (95) A. p. 175.

the generic name *Acanthozamites* as a substitute for *Cycadorachis* in view of the new data afforded by the English material, but it seemed preferable to adopt some provisional name which did not imply affinity to the Cycads. The leaflets described by Saporta as *Sphenozamites latifolius* are apparently identical with those found in the English beds, but none of the French specimens were attached to a supporting axis. The relation of spines and 'leaves' suggests that the latter may be phylloclades borne in the axil of modified spinous leaves, but their morphological nature cannot be determined. In this connexion attention may be called to *Dioncophyllum Tholloni* Baill. a West African shrub which bears on the long shoots leaves 2—3½ inches long each of which has a pair of strong revolute hooks at the apex: in the axils of these leaves are short shoots with larger leaves without hooks. It is suggested[1] that the apparent lamina of the hooked leaves is a winged petiole, the hooks representing lateral leaflets.

Sewardia longifolia (Salfeld).

This species was described by Salfeld[2] from the Solenhofen beds (Upper Jurassic) of Bavaria as *Baiera? longifolia* Heer: it is founded on branches nearly 30 cm. long bearing large fan-shaped deeply divided leaves, or leaf-like organs, in the axils of recurved spines similar to those in *S. latifolia.* The 'leaves' are identical in habit with those of some species of *Baiera*, but we have no information with regard to the structure of the epidermis. In view of the uncertainty as to the morphological nature of the leaves or their relationship to leaves of *Baiera*, it is inadvisable to adopt a generic title that implies affinity to the Ginkgoales.

[1] Sprague (16).   [2] Salfeld (07) B. p. 195, Pls. xx., xxi.

Fig. 674*. *Sequoia sempervirens.* Near Crescent City, California. (From a photograph by Professor A. Henry.)

CHAPTER XLIII.

CONIFERALES (RECENT).

THE Coniferales, by far the largest section of the Gymnosperms, present considerable difficulty to the student of fossil plants. There is great divergence of opinion with regard to the relative antiquity of the several families, and their position in an evolutionary series. The Abietineae are by some botanists regarded as the most primitive; on the other hand, and this is the view that in my opinion receives most support from the available evidence, it is held that the Araucarineae are both the most primitive and the oldest representatives of the Coniferales. Until recent years the study of fossil Conifers has suffered neglect and little help has been afforded by palaeobotanists to the solution of the morphology of the ovulate shoots of the different genera, a

problem that has long exercised the ingenuity of investigators.
The view expressed by Jeffrey[1] that recent work on fossil Conifers
corroborates the interpretation of the seed-bearing scales as
metamorphosed shoots is based on facts furnished by a study of
vegetative organs, which in themselves do not afford any decisive

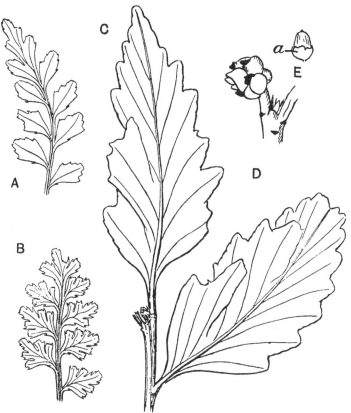

Fig. 675.   A, B, *Phyllocladus trichomanoides*.   C, *P. hypophylla*.   D, E, Megastro-
    bilus and seed of *Phyllocladus alpina*; *a*, arillus.   [A—C from specimens in
    the British Museum; D, E, after Miss Robertson (Mrs Arber.)]

evidence as to the morphology of reproductive shoots.   In view
of these considerations it is important that an attempt should be
made, even at the risk of disproportionate treatment, to give a
general account of recent genera which, though necessarily far

[1] Jeffrey (10) p. 331.

from complete, may afford assistance to students prepared to undertake a critical study of the fragmentary records of the rocks.

Conifers are trees or shrubs exhibiting a fairly wide range in habit; the 'great ones of the forest' such as the Sequoias (fig. 674*), the sugar Pines (*Pinus Lambertiana*) and Douglas Firs (*Pseudotsuga Douglasii*) of the Rocky Mountains, *Taxodium mucronatum*[1] of Mexico, remarkable for its enormous bulk, the tall and slender Cypresses, the less formal Podocarps of the southern hemisphere, the shrubby Junipers, the dwarf *Dacrydium laxifolium*[2] of New Zealand afford examples of recent types. In most species the leaves are small and crowded, not infrequently dimorphic, and in *Phyllocladus* (fig. 675) reduced to inconspicuous and caducous scales subtending phylloclades. *Agathis* is exceptional in having narrow ovate leaves reaching a length of nearly 20 cm. (fig. 695) and a similar but smaller leaf is characteristic of some species of *Podocarpus* (fig. 676). The presence of long and short shoots is a striking feature of *Pinus, Larix, Pseudolarix, Cedrus,* and *Sciadopitys*: the short shoot, as Goebel says 'takes no part in the construction of the permanent skeleton of the tree[3].' The whorled arrangement of leaves characteristic of several Cupressineae and the Callitrineae is not a constant feature and, as in *Lycopodium*, both whorled and spiral foliage may occur on the same shoot.

Conifers are monoecious or dioecious; the microsporophylls and megasporophylls are borne spirally or in whorls on separate shoots, and in some genera on separate trees, except in the case of abnormal bisporangiate strobili[4]. Proliferous cones are not uncommon in some genera: the prolonged axis of the cone of *Cryptomeria japonica* shown in fig. 677 bears microstrobili in the axils of the small leaves. The microstrobili are for the most part constructed on a uniform plan; they are usually short-lived, small shoots, and each microsporophyll often consists of a slender axis bearing two microsporangia on its lower surface and prolonged as a small upturned distal expansion. In *Pinus* the sporangia dehisce longitudinally, while *Abies* (fig. 684, E) affords an example of

[1] *Gard. Chron.* Nov. 26, 1892, p. 648.
[2] Hooker, J. D. (52) Pl. 815.   [3] Goebel (05) p. 444.
[4] For examples see Sterzel (76); Eichler (82); Worsdell (04); Bartlett (13); Shaw, W. R. (96); Robertson (06); Renner (04); Bayer (08); etc.

transverse dehiscence. The microstrobili of *Cedrus* are similar but longer. In *Torreya* (fig. 684, D), *Taxodium, Widdringtonia,*

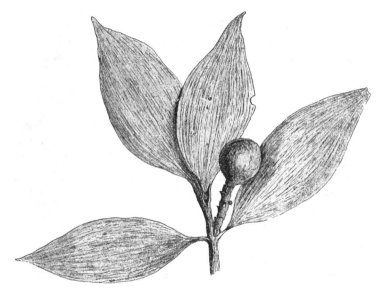

FIG. 676. *Podocarpus latifolia.* (⅔ nat. size.) M. S.

FIG. 677. *Cryptomeria japonica.* Proliferous cone. (Nat. size.)

and some other genera there are 4—6 microsporangia on each sporophyll. In *Araucaria* and *Agathis* there may be as many as

10—20 sporangia, longer and relatively narrower than in other
genera and attached by one end, in contrast to the more complete
union of sporophyll and sporangium
in *Pinus*. In *Araucaria Muelleri* and
A. Rulei (fig. 678) the microstrobili
reach a length of 25 cm.: in *A. excelsa*
and *A. Cookii* they are much smaller
(fig. 679, A, B). In *Taxus* 4—7 sporangia are radially disposed on the
inner face of a flat distal expansion.
The microstrobili of *Cunninghamia*[1],
Pseudolarix[2], and *Keteleeria* are borne
in umbels, while in *Cryptomeria* and
Taxodium[3] they occur in spikes. The
microspores may be winged or wingless: in the Abietineae there are as a
rule two conspicuous wings or bladders
(fig. 684, B), but the spores of *Pseudo-*
tsuga are wingless and in *Tsuga* both
types occur. In *Microcachrys*[4] the
wings vary from 2 to 6 (fig. 684, C)
and in *Dacrydium*[5] and *Podocarpus*[6]
(fig. 684, A) there are 2 or 3 small
bladders. In *Taxus*, *Cephalotaxus*,
Torreya, *Sciadopitys*, the Cupressineae,
and some other Conifers there are no

FIG. 678. Microstrobilus of
Araucaria excelsa (A) and
Araucaria Rulei (B). (After
Seward and Ford; A, nat.
size; B, ½ nat. size.)

prothallus cells: the microspores of the Abietineae are characterised by the occurrence of 2, or occasionally 3 or 4[7], evanescent
prothallus cells (fig. 684, B); in *Dacrydium* there are 4—6 prothallus
cells; in *Microcachrys* 3 or 4; in *Podocarpus* (fig. 684, A) as many
as 8, while in *Araucaria* 15 cells have been recorded and as many
as 30 nuclei. The two male gametes are non-motile.

The term Conifer though appropriate as regards the majority
of the plants so styled is misleading in the case of several genera

[1] Siebold (70) Pl. CIII.   [2] *Bot. Mag.* Jan. 1908.
[3] *Gard. Chron.* Nov. 25, 1893, p. 659.
[4] Thomson (09).   [5] Young (07).
[6] Jeffrey and Chrysler (07); see also Thibout (96); Burlingame (08); (13);
(15); Sinnott (13).   [7] Hutchinson (14).

which possess ovulate shoots differing widely from cones as the
term is generally understood.   The cones of *Araucaria* (figs. 680,
681) and *Agathis* reach a considerable size; those of *Araucaria*
Bidwillii[1], similar to some cones of *Encephalartos*, may be 28 cm.
in diameter and in some species of *Agathis*[2] they exceed 11 cm.

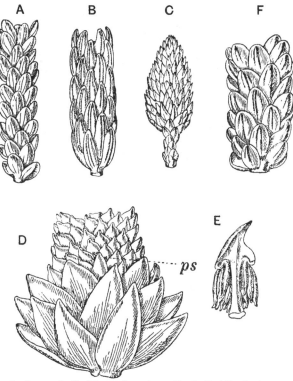

FIG. 679.   A, *Araucaria Cookii* var. *luxurians*. B, *A. Cookii*. C, *Araucaria Cookii*,
   microstrobilus. D, E, *Araucaria Muelleri*, part of a microstrobilus and a single
   sporophyll; *ps*, microsporangia. F, *Araucaria Montana*, branch. (After
   Seward and Ford.)

and are 14 cm. long.   The cone-scales of *Agathis* are flat, woody
structures bearing a single ovule (fig. 682): in *Araucaria* the
single seed is embedded in the scale, and a more or less prominent
appendage, the ligule, forms a characteristic feature (fig. 683, *l*).

---

[1] *Gard. Chron.* April 14, 1894, p. 465.
[2] Seward and Ford (06) B.

FIG. 680. *Araucaria brasiliensis*, cone. (⅜ nat. size; from a specimen in the Royal Gardens, Kew.)

FIG. 681. *Araucaria Cunninghamii*, cone. (After Seward and Ford; ½⅔ nat. size.)

The cone-scales of some species, *e.g.* *A. Cookii*, *A. excelsa* (fig. 683, A, D) are flat and laterally winged, while in *A. brasiliensis* the thick distal ends closely resemble those of the seed-scales of some Pines: the cone-scale of *A. imbricata* is larger and deeper, and that of *A. Bidwillii* broad and woody (fig. 683, B, C). In *Pinus*, with cones reaching a length of 2 feet, the mature scales are apparently simple like those of the Araucarineae: the distal end is broad and rounded (*P. silvestris*) with a central umbo or,

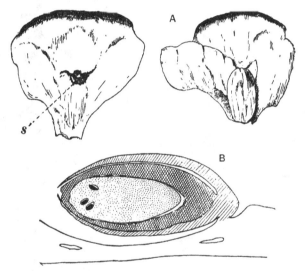

FIG. 682. A, *Agathis Moorei*, cone-scales; *s*, scar of seed. B, *Agathis loranthifolia*, section of ovule showing integument, nucellus, and megaspore with three archegonia, also part of the cone-scale with a projection close to the base of the ovule. (After Seward and Ford.)

as in *P. Coulteri*, the umbo is prolonged as a strong recurved spine, while in *P. excelsa* (fig. 704) and *P. Cembra* the scales are flatter like those of *Picea*. In the young Pine cone each scale is clearly a double structure consisting of a lower portion, the bract or carpellary scale, and an upper portion, the ovuliferous scale, bearing two ovules. In the course of development the seminiferous scale alone increases in size, and the bract-scale is hardly visible in the ripe cone or is represented by a small remnant. In *Abies*, *Larix*, *Pseudotsuga* the dual nature of the scales is obvious at

maturity, the bract-scale usually extending beyond the edge of
the seminiferous scale (fig. 705), in *Abies bracteata* reaching a
length of 5 cm. In *Sequoia* (fig. 702 B) the cone-scales show no
outward sign of a double structure, but each scale contains two

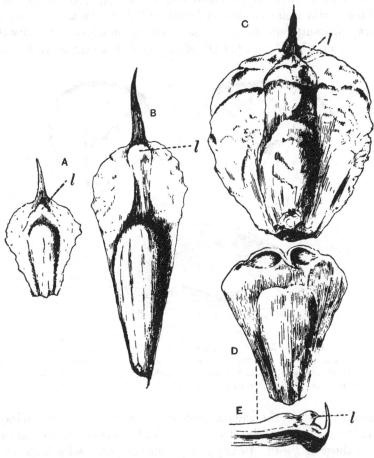

FIG. 683.   Cone-scales of *Araucaria Cookii* (A), *A. imbricata* (B), *A. Bidwillii* (C),
and *A. excelsa* (D, E); *l*, ligule.   (After Seward and Ford.)

sets of bundles, the lower being normally orientated as in a leaf
and the upper vascular strands inversely orientated (*cf.* fig. 684, R).
The occurrence of these two sets of bundles is often quoted in
support of the view that the double cone-scale of the Abietineae

represents the lowest term of a series, the upper end of which is represented by the scales of *Agathis* which have lost all external signs of their supposed dual nature and retain only the inversely orientated bundles as evidence of their descent from an ancestral type in which the ovuliferous and bract-scales were separate organs[1]. The scales of such genera as *Sciadopitys, Athrotaxis, Cryptomeria* (fig. 684, S, N, M), on this hypothesis, occupy an intermediate position. The seminiferous scale of the Abietineae is considered by many botanists to be a leaf or leaf-like organ borne on an axillary shoot subtended by a bract, and it is believed that the simple scale of *Agathis* has been produced by the gradual fusion of two originally distinct organs. The ligule of *Araucaria* is held to be the outward and visible sign of the seminiferous scale that has almost lost its individuality, and with this ligular relic are homologised the upper half of the scale of *Sequoia*, the deeply toothed upper part of the scale of *Cryptomeria* (fig. 684, M), the rounded ridge on the abaxial side of the seeds in *Athrotaxis* (fig. 684, N), the membranous outgrowth on the scales of *Cunninghamia* (fig. 684, K, *m*), and the seminiferous scale of the Abietineae. It has been pointed out in support of this hypothesis that two vascular bundles are given off from the axis of a Pine cone, one of which forms the bract-scale bundles and the other the vascular supply of the seminiferous scale[2].

In a recently published paper on the vascular anatomy of the megasporophylls of Conifers by Miss Aase[3] additional facts are given with regard to the origin and behaviour of the vascular bundles of the cone-scales. In the upper part of a cone of *Pinus maritima* the bract-supply arises as a single bundle at the base of a gap in the stele, and the bundles of the seminiferous scale are given off from the sides of the gap above the point of origin of the bract-bundle: in the lowest sporophylls, on the other hand, the bract and scale-bundles have a common origin. A separate origin for bract and seminiferous scale-bundles is recorded in several Abietineae and in some other Conifers. The origin of the vascular

---

[1] For references to literature on the morphology of cones, see Coulter and Chamberlain (10); Worsdell (04); Rendle (04); Lotsy (11); also Čelakovský (82); Kramer (85); Bayer (08); Aase (15).

[2] Worsdell (99).   [3] Aase (15).

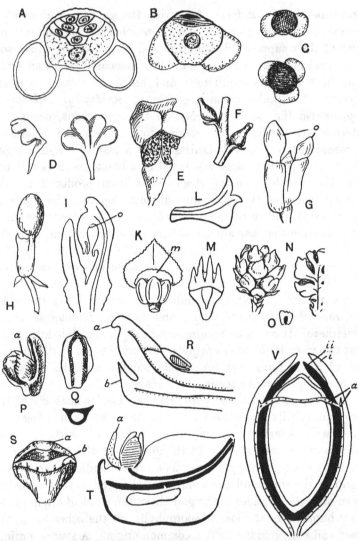

FIG. 684. A, B, C, microspores of *Podocarpus Totara* (A) *Pinus Laricio* (B), *Microcachrys tetragona* (C). D, E, microsporophylls of *Torreya californica* (D) and *Abies alba* (E). Megasporophylls etc. of *Podocarpus spicata* (F); *P. Totara* (G). *o*, ovule; *P. neriifolia* (H); *P. imbricata* (I), *o*, ovule; *Cunninghamia sinensis* (K), *m*, membrane; *Cryptomeria* (L, M). N, O, megastrobilus and seed of *Athrotaxis laxifolia*. P, megasporophyll and seed of *Dacrydium Balansae*; *a*, epimatium. Q, seed of *Cupressus sempervirens*. R, S, megasporophyll of *Sciadopitys*; *a*, ovuliferous scale; *b*, bract-scale. T, megasporophyll of *Microcachrys*; *a*, epimatium. V, seed of *Torreya*; *i*, *ii*, integuments; *a*, vascular strands. [After Burlingame (A), Coulter and Chamberlain (B), Thomson (C), *Gard. Chron.* (D, N, O), Kirchner, Loew and Schröter (E, Q), Pilger (F, H, P), Gibbs (G, I), Eichler (K—M, S), Worsdell (R, T), Oliver (V).]

supply of the double cone-scale is, however, not constant even in
the same cone, and in *Araucaria Bidwillii*[1] each cone-scale is
supplied by two separate strands from the vascular axis though
in other species a single bundle enters the cone-scale and divides
later. It would appear, therefore, that the single or double origin
of the lower normally orientated bundles and of the upper set of
inversely orientated bundles is far from constant, and the data
derived from anatomical study do not afford a satisfactory means
of determining the morphological nature of the cone-scales. It is
held by Jeffrey and his school that the Abietineae represent the
oldest members of the Coniferales and that the Araucarineae are
a more recent development, the apparently single cone-scale of
Araucaria and *Agathis* being derived from the double cone-scales
of the Abietineae. Some botanists, *e.g.* Vierhapper[2], while
believing that the seminiferous scale of the Abietineae is an organ
belonging to an axillary shoot subtended by a bract-scale and that
the cone-scales of other Conifers are also double structures, whether
or not externally divided, regard the Araucarineae as earlier in
origin than the Abietineae.   If the cone-scales of the Araucarineae,
to take the extreme type, are in origin double and homologous
with the obviously double cone-scales of the Abietineae it is more
logical to regard the Abietineae as the precursors of the Araucarineae.   The evidence afforded by fossils in my opinion lends strong
support to the greater antiquity of the Araucarineae, and I venture
to believe that no adequate reasons have been given for regarding
the cone-scales of the Araucarineae as other than simple leaves
bearing ovules.   If, as seems probable, the Coniferales are mono-
phyletic in origin the cone-scales of the different families are in
all probability variants of a common type and, in opposition to
the view which is most in favour, I regard the double cone-scales
of the Abietineae and the corresponding organs of other Conifers
which afford evidence of a double structure as derivatives of a
simple form of sporophyll strictly comparable with the sporophyll
of a *Lycopodium*, the placental outgrowth assuming an increasing
degree of individuality in the different lines of evolution illustrated
by various types of strobilus.   It is noteworthy that the transition
from foliage leaves to megasporophylls in the Araucarineae is

----

[1] Worsdell (99).   [2] Vierhapper (10)

often very gradual in contrast to the much more sharply defined
break between leaves and scales in most of the Abietineae and
many other Conifers. The cone-scales of the Abietineae are
recognised as more complex and more recent developments, the
seminiferous scale being an excessively enlarged placental outgrowth from a megasporophyll, while in the intermediate types
such as *Sequoia, Cryptomeria*, and others the separation between
the two parts of the cone-scale is much less complete[1]. This
morphological question is too complex to discuss fully in a
general summary: students should, however, be warned that
several botanists do not agree with the opinion that is here
expressed. It is at least fair to add that the views expressed
by Prof. Jeffrey and his pupils with regard to the relative positions
of the Araucarineae and the Abietineae in an evolutionary series
are stated with an assurance which is misleading to those unfamiliar
with the nature of the evidence[2].

In *Saxegothaea* (fig. 685), *Dacrydium* (fig. 684, P), and some other
genera each ovule is surrounded by a cup-like integument (fig. 684,
T *a*), formerly called the arillus but recently styled the epimatium[3];
this is by some authors considered to be the equivalent of the
seminiferous scale.

This inadequate account may serve to call attention to a complex
morphological problem which has an important bearing on questions
connected with the relative positions of the several genera. It would
be out of place to enter fully into this difficult subject, but it is one
that demands careful attention by students of extinct types.

The number of seeds borne on each scale is an important
feature in the recognition of genera. In the Abietineae each
scale usually bears two seeds though it is not uncommon to find
single-seeded seminiferous scales such as those of *Pinus monophylla*
(fig. 686, A—C) in which there is a deep cavity showing that the
seed was partially embedded in the supporting organ. Such a
scale might, as a fossil, be easily mistaken for an Araucarian cone-
scale. In *Cunninghamia* there are three seeds to each scale (fig.
684, K); in *Athrotaxis* and *Cryptomeria* 3—6; in *Sequoia* 5; and

---

[1] Eichler (81).

[2] For a recent discussion on the origin and relationship of the Araucarineae see
Burlingame's paper (15²) which appeared after this chapter was written.

[3] Pilger (03) p. 16.

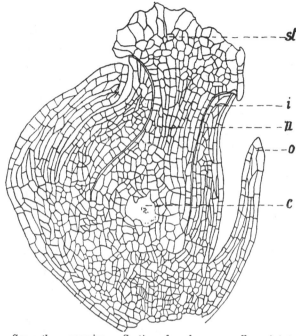

FIG. 685. *Saxegothaea conspicua.* Section of ovule; *n*, nucellus; *i*, integument; *st*, stigma-like apex of nucellus; *o*, epimatium; *c*, young megaspore. (After Stiles.)

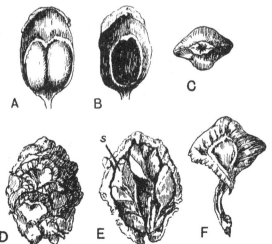

FIG. 686. A—C, *Pinus monophylla,* cone-scales with two seeds (A), one seed (B), and in distal view (C). D, E, *Taxodium mucronatum,* cone in surface-view (D) and section (E); *s*, seed. F, *Taxodium distichum,* scale. (From specimens in the British Museum.) M. S.

in *Sciadopitys* 7—9.   Fig. 686, E represents half a cone of *Taxodium* in which the distally expanded woody ends of the scales are tightly joined by their edges and form a hard case enclosing as in an ovary several angular seeds, the slender stalks being shrivelled and inconspicuous.   The cones of the Cupressineae and Calli- trineae are characterised by a whorled arrangement and a com- paratively small number of the scales.   In *Cupressus* the cones are oval or spherical and each scale bears 6—20 seeds: in the Callitrineae the cones are valvular (figs. 703, 762, B—D) and the scales vary from 2 to 3 in *Callitris* and from 7 to 8 in *Widdringtonia*.   The small cones of *Saxegothaea* consist of one-seeded mega- sporophylls which become fleshy and par- tially concrescent (fig. 687); in *Juniperus* the strobilus has the appearance of a berry; in *Microcachrys* the leaves pass gradually into the single-seeded verticillate megaspo- rophylls, each with two vascular strands[1] (fig. 684, T) and an epimatium, *a*, on one side of the ovule; in the ripe cone the mega- sporophylls are fleshy but not connate as in *Saxegothaea*[2].   In *Dacrydium* the megaspo- rophylls differ but slightly from the foliage leaves in some species, *e.g.* in *D. Balansae* (fig. 684, P) a single leaf at the apex of a branch bears an ovule partially covered by a hood-like epimatium.   In *Torreya*[3] a very short shoot in the axil of a leaf bears two bracts and each subtends an ovule and two pairs of bracteoles.   The seeds of *T. cali-*

FIG. 687. *Saxegothaea con-spicua.* (After Stiles.)

fornica, which may be 4 cm. long, are enclosed by a thick integument differentiated into a sarcotesta and sclerotesta surrounding a ruminated endosperm (fig. 688): there is a ring of vascular bundles at the limit of the free part of the integument and this is regarded by Oliver[4] as homologous with the tracheal plate at the base of

[1] Worsdell (99).
[2] Stiles (08); Norén (08); Tison (09); *Bot. Mag.* Tab. 8664 (1916).
[3] Robertson (04); (07).   [4] Oliver (02); (03).

the nucellus of Cycadean seeds. The portion of the *Torreya* seed (fig. 684, V) below the free part of the nucellus has, according to Oliver, been produced by the intercalation of a new basal region that has pushed up the chalaza. *Cephalotaxus*[1] has plum like seeds similar to those of *Torreya*. In *Phyllocladus*[2] (fig. 675, E) an ovule enclosed in a papery epimatium occurs in the axil of a succulent bract, and in *Taxus* a terminal ovule is borne on a short shoot

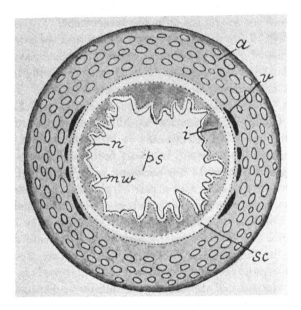

FIG. 688. *Torreya nucifera*, transverse section of seed; *a*, arillus; *v*, vascular tissue; *sc*, outer zone of integument in which the hard shell will be formed; *i*, inner part of integument; *mw*, wall of megaspore; *n*, nucellus; *ps*, prothallus. (After Oliver.)

without any megasporophylls. In *Podocarpus*[3] the megasporophylls are few and a part or whole of the reproductive apparatus is fleshy: the strobilus of *P. Totara* (fig. 684, G) consists of 3—4 bracts two of which are usually fertile. Similarly the strobilus of *P. imbricata* is formed of 2—4 bracts with swollen bases forming the receptacle; in the section shown in fig. 684, I, one bract is

[1] *Gard. Chron.* Oct. 14, 1895, p. 717.   [2] Robertson (06).
[3] Pilger (03); Brooks and Stiles (10); Stiles (12); Gibbs (12); Sinnott (13).

fertile and its apex overtops the ovule, while a second bract is sterile. Fig. 684, H shows the strobilus of *P. neriifolia* in which the lowest bracts are leaf-like and the swollen bases of the upper bracts are fused with the axis and each other to form a receptacle analogous to the flower-axis of *Anacardium*. *P. spicata* has a long loose strobilus bearing several ovules (fig. 684, F); and in *P. Nagi* a single seed occurs on an axillary branch bearing small decussate scales: the flesh of the seed is formed from the epimatium, and the sclerotesta from the integument.

The seeds of Conifers vary greatly in size and shape: those of some species of *Pinus* and several other Abietineae have a large wing while others are wingless. The absence of a wing in a fossil seed does not necessarily denote an original feature. The seeds of *Cedrus, Abies, Picea, Pseudotsuga, Keteleeria* and *Tsuga* are winged. Glück[1] calls attention to differences in the relation of seed to wing in certain Abietineae: in *Picea* the base of the wing covers the seed like a spoon; in *Abies, Larix,* and *Cedrus* the seed lies in a pocket formed by the enclosing wing, while in *Pinus* the wing embraces the seed like a pair of pincers. In *Agathis* (fig. 682, A) there are two very unequal wings. The much smaller seeds of many Conifers have 2—3 wings: in *Cupressus* they are more or less equal (fig. 684, Q), in *Libocedrus* and *Fokienia* markedly unequal; in *Fitzroya* and *Cunninghamia* there are 2—3 wings. Our knowledge of the relative vitality of Conifer seeds is meagre[2], and from the point of view of the possibilities of dispersal further research is desirable. The viviparous habit of *Rhizophora* and other Mangrove Dicotyledons, believed by Guppy[3] to be a primitive feature, is recorded in *Podocarpus Makoyi*[4].

The relation between nucellus and integument is less uniform in Conifers than in Cycads. In some genera, *e.g. Agathis* (fig. 682, B), *Dacrydium, Phyllocladus, Fitzroya, Callitris* and a few others the nucellus is free from the integument to the base; in *Podocarpus* the relation is variable; in *Pinus* and other Abietineae, in *Torreya* (fig. 684, V) and some other genera the nucellar apex alone is free. The free summit often has the form of a steep cone: in *Araucaria*[5] and to a greater degree in *Saxegothaea* (fig. 685, *st*)

[1] Glück (02) p. 402.   [2] Coker (09).   [3] Guppy (06).
[4] Lloyd (02).   [5] Seward and Ford (06) B.

this protrudes through the micropyle: in *Fitzroya patagonica* the prolongation of the integument as a micropylar tube with a stigma-like terminal expansion is particularly striking. There is no regular pollen-chamber as in Cycads and Ginkgo, but in *Pseudo-tsuga*[1] a two-storied chamber, analogous to the pollen-chamber, is formed by a knee-like bend in the integument. A peculiar type of pollination characterises *Araucaria*: the microspores germinate on the ligule or on the megasporophyll and their tubes grow over or into the scale-tissues on their way to the ovule[2]. The archegonia of Conifers usually occur at the apex of the prothallus and are few in number, they are separated by a few layers of cells (Abietineae) or form a compact group (Cupressineae). In some Podocarps there may be as many as 14, in *Taxodium* 34, in *Agathis* 60, irregularly distributed on the sides of the prothallus. In *Widdring-tonia*[3] 100 archegonia are recorded occupying a lateral position; in *Actinostrobus*[4] Saxton has discovered groups of laterally placed archegonia. In *Sequoia*[5] the archegonia are also numerous and not confined to the apex. It is an open question whether or not the greater number and irregular disposition of the archegonia are primitive features. The occasional occurrence of lateral archegonia in *Pinus* may be a revival of an older habit.

Classification.

The result of recent research into the morphology and life-histories of genera demand certain changes in the generally adopted grouping. The following classification is an attempt to give clearer expression to the inter-relationships of existing genera[6]. Arnoldi[7] proposed to withdraw *Sciadopitys* from *Sequoia*, *Taxodium*, and other members of the Taxodineae as the type of a separate family; he also suggested the isolation of *Sequoia*. The more recent work of Coker[8] and Lawson[9] favours the removal of *Taxodium* and *Cryptomeria* to the Cupressineae. Miyake's re-searches[10] point to a similar affinity in the case of *Cunninghamia*. The genera *Athrotaxis*, *Fokienia*, and *Taiwania* are placed tenta-tively in the Cupressineae. The family-name Callitrineae, first

[1] Lawson (09).   [2] Thomson (07); Eames (13); Burlingame (13); (15).
[3] Saxton (10).   [4] *Ibid.* (13).   [5] Shaw, W. R. (96); Arnoldi (01).
[6] See also Saxton (13²).   [7] Arnoldi (01); Lawson (10); Radais (94).
[8] Coker (03).   [9] Lawson (04).   [10] Miyake (10).

used by Masters, has been revived by Saxton[1] to give expression
to the distinctive characters of *Callitris, Widdringtonia,* and
Actinostrobus. Saxton's work on *Tetraclinis* leads him to assign
it to the Cupressineae. Pilger[2] makes *Phyllocladus* the sole genus
of Phyllocladoideae, while Miss Robertson[3] includes it in the
Podocarpeae though recognising leanings towards the Taxineae.
It should be stated that the changes in classification suggested
are based mainly on characters of the gametophyte though
anatomical and vegetative features have not been entirely
neglected[4].

   I.   ARAUCARINEAE. *Agathis, Araucaria.*

   II.   CUPRESSINEAE. *Cupressus, Chamaecyparis, Libo-
cedrus, Thuya, Juniperus, Fitzroya, Diselma, Thujopsis, Taxodium,
Glyptostrobus, Cryptomeria, Cunninghamia, Taiwania, Fokienia,
Athrotaxis, Tetraclinis.*

   III.   CALLITRINEAE. *Callitris, Actinostrobus, Widdringtonia.*

   IV.   SEQUOIINEAE. *Sequoia.*

   V.   SCIADOPITINEAE. *Sciadopitys.*

   VI.   ABIETINEAE. *Pinus, Cedrus, Larix, Pseudolarix,
Picea, Tsuga, Abies, Pseudotsuga, Keteleeria.*

   VII.   PODOCARPINEAE. *Podocarpus, Dacrydium, Microcachrys, Acmopyle, Pherosphaera, Saxegothaea.*

   VIII.   PHYLLOCLADINEAE. *Phyllocladus.*

   IX.   TAXINEAE. *Taxus, Torreya, Cephalotaxus.*

The order of the families is not intended to indicate their
natural sequence in an evolutionary series, though the Araucarineae are considered to be the most primitive. As certain
authors have suggested, *Saxegothaea* is probably closely allied to
Araucaria, but this is not indicated in the order adopted.

Geographical · Distribution.

The distribution of the Conifers[5], though too wide a subject
for more than a brief notice, is of great interest from a palaeontological point of view. The ABIETINEAE, comparable in their
present dominant rôle with the Polypodiaceae among the Ferns,

[1] Saxton (10²); (13²).   [2] Pilger (03).   [3] Robertson (06).
[4] For other views on classification. see Vierhapper (10).
[5] Drude (90); Engler (89); Graner (94); Hildebrand (61); Vierhapper (10).

are the most widely spread; for the most part restricted to the northern hemisphere, they are not unrepresented south of the equator. *Pinus* reaches the tree-limit in the north and extends as far south as Formosa[1], Siam, the Malay region, the Philippines[2], S. Africa, and the West Indies. *Picea* has a similar distribution in the north and reaches to the temperate regions of the southern hemisphere. *Abies* ranges from Europe and Algeria to Siberia, the Himalayas, Japan, and Formosa. *Larix* flourishes in northern Europe and Siberia, Canada and the northern United States, the Himalayas, and Japan. *Tsuga* is more especially a North American

Fig. 689. *Araucaria imbricata* on the Andes, Argentina. (From a photograph by Dr Wieland.)

genus, but it occurs in the Himalayas and in Japan. *Pseudotsuga* is characteristic of N.W. America and is recorded from Formosa. *Pseudolarix* is a native of N.E. China and Formosa. *Cedrus*[3] occurs in Algeria, Morocco, Syria, Cyprus, and the western Himalayas. The distribution of the ARAUCARINEAE affords a striking example of the contrast between the present and past range of a family. *Araucaria* occurs in Brazil, Chile and Argentina, in Australia, New Caledonia, New Guinea, the Pacific islands. *Agathis* is confined to the Australian and Malay region, New Zea-

[1] Hayata (10).   [2] Foxworthy (11).   [3] Hooker, J. D. (62).

land, New Caledonia, and the Queen Charlotte Islands. The two trees of *Araucaria Bidwillii* shown in the Frontispiece are survivors of a forest on the hills of Queensland. The photograph reproduced in fig. 689, for which I am indebted to Dr Wieland[1], illustrates the habit of *Araucaria imbricata* on the eastern slopes of the Andes in South-West Argentina where the trunks reach a diameter of two metres. There are few existing trees comparable with these venerable types in the impression they produce of the lapse of ages and the vicissitudes of a dwindled race.

CUPRESSINEAE. *Cupressus* occurs in North America including the Californian coast and Mexico, in S.E. Europe, temperate Asia, China, and Japan. *Chamaecyparis* extends to the Sitka Sound and flourishes in China, Japan, and Formosa. *Libocedrus,* one of the few genera met with in both hemispheres, has a discontinuous distribution; it occurs in California, Chile, Japan, Australia, New Zealand, New Guinea, and New Caledonia. *Thuya* flourishes over a wide area in North America and occurs in the Far East. *Juniperus* is characteristic of temperate regions in both the old and new world and is represented in the Canaries, the Azores, Somaliland, and Mexico. *Fitzroya* is confined to Patagonia and Chile; *Diselma* to Tasmania. *Thujopsis* is exclusively Japanese. *Taxodium* is a native of Texas and Mexico, while *Glyptostrobus* is a closely allied genus in China. The monotypic *Cryptomeria* lives in China and Japan; *Cunninghamia* in China and Formosa. *Taiwania* and *Fokienia* have recently been described from Formosa and East China respectively. *Athrotaxis* is confined to Tasmania and *Tetraclinis* to North Africa.

CALLITRINEAE. *Callitris* occurs in Australia and New Caledonia; *Widdringtonia* grows in equatorial and South Africa and in Madagascar; *Actinostrobus* is restricted to West Australia.

SCIADOPITINEAE. *Sciadopitys* is confined to South Japan.

SEQUOIINEAE. *Sequoia* is confined to the Pacific coast of North California; *S. sempervirens* the species with 'the stronger hold upon existence' extends into Oregon, while *S. gigantea* forms groves in the valleys of the Sierra Nevada.

PODOCARPINEAE. *Podocarpus*, one of the more successful genera, is essentially a southern type: in Africa it extends from

[1] Wieland (16), p. 224.

Cape Colony through East Africa to Abyssinia; it occurs in S. America from Patagonia to Brazil and replaces *Pinus* on the mountains of Costa Rica[1]; in the West Indies, Malaya, in the Himalayas, China, Japan, Formosa[2], Tasmania, New Zealand, New Caledonia, the Fiji Islands and New Guinea. *Dacrydium* has also a fairly wide range in the southern hemisphere, but like the other members of the family, except *Podocarpus*, it does not cross the equator; it is abundant in the Malay Archipelago[3] and occurs in New Zealand, Tasmania, New Caledonia, New Guinea, and one species grows in the Chilean swamps. *Saxegothaea* is a monotypic genus in Chile and *Acmopyle* an imperfectly known New Caledonian genus. *Pherosphaera*[4] occurs in New South Wales, Victoria, and Tasmania; *Microcachrys*, like *Athrotaxis*, is Tasmanian. *Phyllocladus* has a wide range in Tasmania, New Zealand, Borneo, New Guinea, and the Philippines.

Some questions of exceptional interest from the point of view of the geographical distribution of Conifers in the Pacific region are ably discussed by Mr Guppy in the second volume of his admirable book *Observations of a Naturalist in the Pacific between 1896 and 1899*. He deals especially with *Agathis*, *Podocarpus*, and *Dacrydium*, and his remarks illustrate the importance of taking into account palaeobotanical data in any general discussion of the problems suggested by the present and often discontinuous range of existing genera. 'If,' he says, 'there is a real difficulty in applying our canons of plant-dispersal to the distribution of *Dammara* [*Agathis*], it is merely the same difficulty that has so often perplexed the botanist with other Coniferous genera in continental regions, such as, for instance, the occurrence of *Pinus excelsa* on the far-removed mountains of the Himalayas, and the existence of the Cedar in its isolated homes on the Atlas, the Lebanon mountains, and the Himalayas. Such difficulties largely disappear if we regard the present distribution of the Coniferae as the remnant of what it was in an ancient geological period[5].'

TAXINEAE. *Taxus* is chiefly a northern hemisphere genus; it occurs also in North Africa, Persia, India, the Philippines and the Far East, and extends from Newfoundland to Pacific North

---

[1] Harshberger (11) p. 304.   [2] Dümmer (12).   [3] Stapf (96).

[4] Groom (16).   [5] Guppy (06) p. 300.

America, Mexico, and Florida. *Torreya* has a more restricted and less continuous range in China, Japan, Florida, and California. *Torreya taxifolia* is almost extinct; it is separated by over 3000 miles from the other American species *T. californica* and the Pacific separates the latter from the two species in China and Japan[1]. *Cephalotaxus* lives in central China, Japan, and India.

Anatomical features.

The anatomy of Conifers, more especially from the point of view of the identification of families and genera, has long occupied the attention of botanists, and although much has been done in the direction of more intensive study, the limits within which anatomical features may be safely used are still but vaguely defined. Jeffrey[2] goes so far as to claim for the anatomical characters of plants a taxonomic value equal to that assigned by zoologists to the anatomical features of animals. Though often extremely useful, in many cases anatomical characters do not reveal more than an affinity between a fossil specimen and a group of recent genera. Statements are often based on insufficient data and many authors have not appreciated the range of variation in the vegetative shoots of a single tree. Attention has been drawn to the fact that anatomical features are especially variable in branches, and several authors have shown that characters to which importance has been attached are much less constant than has usually been supposed: many features, frequently accepted as trustworthy criteria from the point of view of identification, occur sporadically in other genera than those with which they are usually associated. In the following summary attention is directed to the comparative value of different characters, and prominence is given to possible sources of error in inferences based on anatomical features.

The wood of a Conifer consists only of tracheids, with or without resin-canals, and xylem-parenchyma and is characterised by narrow medullary rays usually one-cell broad. For convenience in description it is proposed to speak of the wood of the Conifer type as pycnoxylic[3] in distinction to the Cycadean type of wood which is styled manoxylic. The presence or absence of

---

[1] Berry (08[2]) p. 648.   [2] Jeffrey (05) p. 1.
[3] πυκνός, compact; μανός, porous, loose in texture.

well defined rings of growth should be noted and attention paid
to the breadth of the late summer ('autumn') wood: Goeppert[1]
considered the breadth of annual rings a character of importance,
but Kraus[2] and others have shown that this is of little significance.
In the Cupressineae (in the more restricted sense) it is probably
true that the rings are generally though not invariably narrower
than in Abietineae: in roots the later wood is smaller in amount
and there is a more sudden transition to the spring-wood than in
stems[3]. Though as a rule there is a considerable difference in
the thickness between the walls of the spring and summer tracheids,
in *Podocarpus Nagi*[4] the difference is slight. In some species
of *Araucaria* the rings are absent or feebly marked, a fact noticed
long ago by Nicol[5], and in other Conifers, *e.g.*, *Widdringtonia
juniperoides* and *Tetraclinis*[6] there may be no definite rings; in
Libocedrus macrolepis[7] there is but little difference in the thickness
of the spring and summer tracheids. It is, however, impossible
to say to what extent this is an inherent tendency and how far it
reflects the influence of external conditions: it may be that the
frequent absence of rings in Araucarian wood is explicable on the
hypothesis that this family is the oldest and most closely related
to Palaeozoic types, which are almost invariably characterised by
an absence of rings: the habit of forming well defined spring- and
late summer-wood may have been acquired at a later stage[8]. The
interest of annual rings is rather biological than taxonomic and it
is chiefly in connexion with fossil plants as tests of climate that
attention has been directed to this feature[9].

The genus *Taxus* is peculiar in having no resin-ducts in the
cortex or stele of stem and root or in the leaves. In some genera
resin-canals are a constant feature in the secondary wood, *e.g.*,
Pinus, Picea, Larix, Pseudotsuga; while in other Abietineae
canals do not usually occur in the xylem. This distinction is,
however, by no means constant and, as Jeffrey[10] has shown, the

[1] Goeppert (50).　　　　　　　　　[2] Kraus (64) p. 146.
[3] Gothan (10) p. 11; Penhallow (07) p. 31; von Mohl (62).
[4] Fujioka (13).　　　　　　　　　[5] Nicol (34) A. p. 139.
[6] Conwentz (90) A. p. 33.
[7] Fujioka (13) p. 213.　　　　　[8] Thomson (13) p. 33; Gothan (07) p. 25.
[9] Seward (92) B; Gothan (08²); Antevs (16); (17).
[10] Jeffrey (03); (05); etc.; Penhallow (07) pp. 123 *et seq.*; Jones (13²).

great majority of Conifers which are normally without resin-canals in the wood have the power of producing them in response to traumatic stimuli. In *Cedrus*, *Pseudolarix*, and *Tsuga* resin-canals are usually confined to the primary xylem of the root but wounding induces the development of canals in other parts of the wood. In *Cedrus*, however, both horizontal and vertical traumatic canals may occur whereas in other Abietineae the traumatic canals are only vertical[1]. Resin-canals may occur in the first-year wood of some species of *Abies* (fig. 690, B) and in *Sequoia gigantea* they are present in the first-year wood of vigorous branches and in the peduncles of cones, but do not normally occur in the later wood. In *S. sempervirens* canals are as a rule absent and are developed only after wounding (fig. 690, A). In the Araucarineae

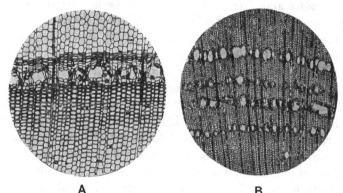

A   B

FIG. 690. A, *Sequoia sempervirens*. B, *Abies* sp. showing traumatic canals in the wood. (After Jeffrey.)

resin-canals are absent nor are they produced in injured stems: this failure to produce canals in response to disturbances set up by wounds is considered by Jeffrey to be an indication of the relatively late evolution of the family. *Pinus*, with abundant canals, is regarded as one of the more primitive types; *Abies*, with very few canals in healthy specimens but readily producing them on wounding (fig. 690, B), is regarded as a slightly later product of evolution, while *Sequoia sempervirens* (fig. 690, A) in which traumatic canals alone occur is still further removed from the original stock, and the Araucarineae, which are considered to

[1] Jeffrey (05) p. 25.

have lost the power of reversion retained by *Sequoia* and *Abies*, are placed higher in the evolutionary series. The vestigial significance of resin-canals is by no means generally admitted. Penhallow[1] holds, and I believe rightly, that they are not primitive features; their occurrence in the young shoots of certain species and in the peduncles of cones but not in the older wood may, as Gothan[2] suggests, be correlated with a greater need of protection. Kirsch[3] considers that the development of canals in young wood and in peduncles may be connected with the relatively greater abundance of food in those regions which, in his opinion, would induce a greater production of parenchyma and secretory passages. Moreover, if the occurrence of canals in the axis of a female cone of *Sequoia gigantea* is attributed to the retention of an ancestral character, why do not canals also occur in the axis of the microstrobili? The facts demonstrated by Jeffrey and his pupils are of great interest, but considered by themselves they may equally well be interpreted as favouring the greater specialisation and more recent development of those genera in which the production of resin-canals is a normal character.

The structure of the epithelial cells is employed as a taxonomic character though, as Conwentz[4] suggests, it is not a very satisfactory criterion and in petrified tissues it is often difficult to distinguish between true thick walls and walls thickened by secondary deposits. In *Pinus* the walls of the cells lining the canals are frequently thin[5], but in some species thick; *Larix* and *Picea* have thick-walled epithelial cells. The occurrence of tyloses,— the parenchymatous cells that invade the cavities of water-conducting elements,—has generally been regarded as the monopoly of Angiosperms: though unknown in recent Ferns they occur in some extinct types. Chrysler[6] has shown that tyloses are produced in the tracheids of *Pinus*, apparently as a consequence of wounding. Tyloses have also been found in some fossil coniferous woods.

The arrangement of the bordered pits on the radial walls of the tracheids is the character to which most attention has been given. In the Abietineae they form either single or double,

---

[1] Penhallow (07) p. 150.   [2] Gothan (07) p. 40.
[3] Kirsch (11); Thomson (13) p. 38.   See also Burlingame (15²).
[4] Conwentz (90) A, p. 45.   [5] Groom and Rushton (13).
[6] Chrysler (08) B. p. 204.

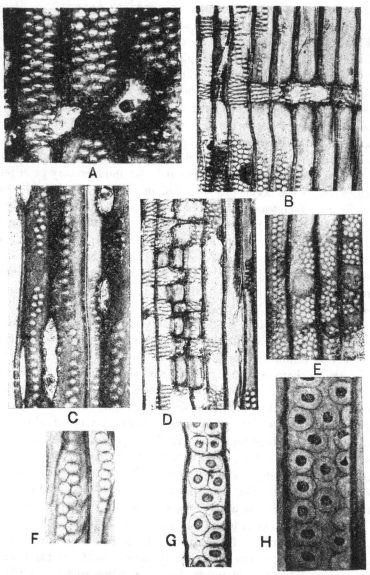

FIG. 691. A, *Araucaria Bidwillii*, tracheids from the cone-axis. B—D, F, *Agathis bornensis*; B, wood of branch, multiseriate pits and, at the ray, scalariform pits; C, tangential section; D, radial section of branch showing transition from alternate and opposite to scalariform pits; F, pits and rudimentary Sanio's rims. E, *Araucaria Cookii*, radial section of root. G, H, *Larix americana*, radial section of root. (After Thomson.)

and occasionally three or even four, rows (fig. 691, H); they are circular and not contiguous and the pits of a double row are as a rule on the same level; they are opposite and not alternate. This type of pitting occurs also in all the other families except the Araucarineae though in *Agathis* opposite pits are not unknown[1] (fig. 691, D) and Conwentz[2] states that he has seen separate and circular pits in the tracheids of recent species. The occasional occurrence of pits in clusters and not in opposite pairs has been described in *Pinus Merkensis*[3]: this is a feature characteristic of the tracheids of some fossil types, *e.g.*, *Cedroxylon transiens* Goth. In *Agathis* and *Araucaria* there may be 1—3 rows and as many as 5 rows on the tracheids of cone peduncles (fig. 691, A). The pits are contiguous and flattened, and those of adjacent rows are alternate and hexagonal[4]. Thomson has called attention to the occasional occurrence, especially in the region of the rays, of transversely elongated or scalariform pits in the tracheids of *Araucaria*. A single series of flattened pits and even the occasional occurrence of alternate hexagonal pits are not infallible criteria of an Araucarian affinity: in *Dacrydium* the pits of a double row may be alternate though rarely contiguous, and this is the case in some other genera, while in *Saxegothaea*[5] (fig. 692) the pits are as a rule uniseriate and often flattened. Worsdell[6] describes circular and separate pits in the cone-scales of *Araucaria* and Thomson records alternate biseriate pits in the cone-axis and early wood of the Abietineae. Flattened pits are described in *Podocarpus polystachya*[7] and I have seen similar pits in the wood of *Torreya californica* and several other conifers other than the Araucarineae. The size of the bordered pits though worthy of notice is not in itself a feature of much value. As Nicol first pointed out, in *Araucaria* they are larger than in *Taxus*; in *Pinus* they are larger than in *Araucaria*: Kraus[8] speaks of the

FIG. 692. Tracheids of *Saxegothaea conspicua.* (After Stiles.)

---

[1] Jeffrey (12) Pl. VI. fig. *b*.   [2] Conwentz (92) p. 35.
[3] Groom and Rushton (13).
[4] For good figures, see especially Thomson (13).
[5] Stiles (08).   [6] Worsdell (99).   [7] Gerry (10).   [8] Kraus (83).

Araucarian pits as small (9—$12 \cdot 8\,\mu$) in contrast to the broader pits (up to $21\,\mu$) of the Abietineae, those in Cupressineae being inter-mediate in size. There may, however, be considerable difference in the size of the pits in a single type[1]. The occurrence of spiral thickening bands in addition to bordered pits is characteristic of the Taxineae, but spiral bands occur sporadically in the secondary tracheids of other Conifers, e.g., *Phyllocladus, Larix leptolepis,* species of *Abies* and other Abietineae[2], also in some species of *Cupressus*[3]. In *Pseudotsuga* spiral bands may occur in all the tra-cheids of an annual ring. Some authors assert that the arrangement and grouping of the bands in a tracheid constitute a character of generic value, but there is not complete agreement on this point[4]

The walls of tracheids frequently exhibit well marked spiral patterns[5], due to an entirely different cause, which, especially in some petrified woods, closely simulate spiral bands. In the process of decay enzyme-action may etch into prominence the striation or spiral method of wall-construction; but the spirals are steeper than those of the true thickening bands. The presence of xylem-parenchyma, though of diagnostic value, is too uncertain and variable a character to be used with great confidence. In young shoots of *Sequoia* xylem-parenchyma may be absent though it is present in older branches[6]. Such parenchyma occasionally occurs in Abietineous wood[7], but it is generally considered a charac-teristic feature of the Cupressineae though in the wood of some members of that family it is not always obvious. The presence of drops of resin in the cells which form vertical series in different parts of the wood, or only in the late summer wood, may render the xylem-parenchyma conspicuous both in transverse and longi-tudinal sections. Rows of parenchyma occur in the wood of *Abies pectinata*[8], also in *Podocarpus* and *Dacrydium*. In *Taxodium*[9] the thick horizontal walls of the cells are a characteristic feature. Wood-parenchyma is rare in the Araucarineae and, as Penhallow[10]

---

[1] Schenk in Schimper and Schenk (90) A. p. 848.
[2] Bailey (09).   [3] Jones (12); (13).
[4] Gothan (05) p. 54; Penhallow (07) p. 41; Burgerstein (08) p. 104; Kraus (83) p. 103; Nakamura (83).
[5] Kraus (88); Gothan (05).   [6] Conwentz (92) p. 35.
[7] Burgerstein (06); Bailey (09).   [8] Kny (10).
[9] Schroeter (80) p. 30.   [10] Penhallow (04).

pointed out, tracheids with horizontal patches of resin may be mistaken for resiniferous parenchyma; but true parenchyma occasionally occurs[1].

Attention has been called to the diagnostic value of the horizontal thickening bands which on staining, and often in fossil wood, stand out as conspicuous features on the tracheids of the great majority of Conifers (fig. 693, C). Many authors speak of these bands as bars of Sanio[2], apparently overlooking the fact that this term (Sanio's 'Balken') was used by Müller[3] for the horizontal bars previously described by Winkler[4] on the tracheids of *Araucaria brasiliensis* (fig. 693, I). Groom and Rushton[5] have also called attention to the inaccurate use of the term Sanio's bars and they suggest the more appropriate expression Sanio's rims for the persistent margins of the primordial pit-areas which appear as horizontal lines between the bordered pits. An American author goes so far as to claim that 'by far the most reliable criterion for diagnosing coniferous wood is the occurrence of the bars [rims] of Sanio[6].' But if, as Jeffrey and his pupils assert, Sanio's rims are present on the tracheids of all Conifers except the Araucarineae the diagnostic value of this feature is exceedingly small. Jeffrey[7] has shown that in the first-year wood of *Araucaria* and in the cone of *A. Bidwillii* the pits are not always contiguous and rims of Sanio may then be present. Moreover, as Thomson[8] states, the darkly stained lines between contiguous pits on some Araucarian tracheids (fig. 691, F) may be regarded as feebly marked rims of Sanio. It is not surprising that in the case of tracheids with 2—3 series of contiguous hexagonal pits, which leave no free surface[9], Sanio's rims are not represented[10].

The most recent contribution to our knowledge of the rims of Sanio is by Mr Sifton[11] who describes them in petioles of *Cycas*

[1] Jeffrey (12) p. 536.   [2] Gerry (10).
[3] Müller (90).   [4] Winkler (72).
[5] Groom and Rushton (13). See also Rushton (16).
[6] Holden (13) p. 252; (13²).   [7] Jeffrey (12) Pl. VI. fig. *b*
[8] Thomson (13) p. 22.   [9] Gothan (10) p. 32.
[10] In a recent paper entitled 'Gliding growth and bars of Sanio' (Grossenbacher. *Amer. Journ. Bot.* vol. I. no. 10, 1914) the expression 'bars of Sanio' is employed in an unusual sense and not in accordance with the ordinary usage of the term bars, or rims, of Sanio.
[11] Sifton (15).

revoluta. Jeffrey regards the occurrence of Sanio's rims in the cone-axis of *Araucaria* as a vestigial phenomenon. He failed to find any rims of Sanio in the cone-axes of Cycads and this negative evidence was regarded as favourable to his view that the rims in the Araucarineae are derived from the more fully developed rims in the Abietineae. Sifton shows that the rims on the tracheids of *Cycas revoluta* agree closely with those in the xylem of the Araucarian cone-axes and with those in the cone-axis and root of certain Pines. On the assumption that roots and cone-axes are likely to retain ancestral characters, the resemblance of their rims to those found in the Araucarineae supports the view that the Abietineae are descended from ancestors which had rims of Sanio of the Araucarian or Cycadean type. The conclusion is that the shorter rims in the Araucarineae and on the tracheids of the cone-axis and root of the Abietineae represent the primitive form, the broader rims met with in the Abietineae and most other Conifers being later developments.

The pitting on the walls of medullary-ray cells has in recent years received special attention: in some Conifers the horizontal and tangential walls are strongly pitted (fig. 693, A, G), and this feature is clearly seen in both radial and tangential sections as also, in the case of the horizontal walls, in transverse sections (fig. 693, D, E, F). In most of the Abietineae the pits on the horizontal and tangential walls are a prominent feature while on the other hand in some Abietineae the pitting of these walls is feebly developed: to this type of pitting Gothan[1] has given the name Abietineous pitting. In the great majority of recent genera other than members of the Abietineae the horizontal and tangential walls are smooth (fig. 693, L, O); but there are exceptions. The ray cells in the cone-scales of *Agathis* are pitted and species of *Juniperus*[2], *Libocedrus decurrens* and *Fitzroya* also exhibit a form of Abietineous pitting. Gothan points out that in some Junipers and a few other Cupressineae the pits in the tangential walls differ in detail from the typical Abietineous form and that the pits in the horizontal walls are much less distinct than in the Abietineae: there is, however, no very clear distinction between

---

[1] Gothan (05) p. 43.
[2] *Ibid.* pp. 43, 45, fig. 7; Stopes (15) p. 63.

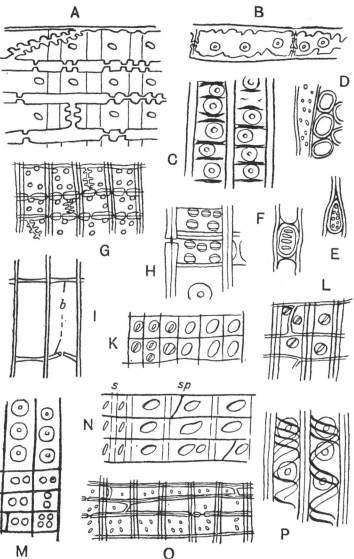

FIG. 693. A, *Abies Veitchii*, medullary ray. B, *Pinus silvestris*, medullary-ray
tracheid. C, *Abies balsamea*, pits and Sanio's rims. D, E, Pits in medullary-ray
cells in *Abies homolepis*. F, Pits in tangential wall of ray cell of *Juniperus
virginiana*. G, H, K, L, M, N, O, Pits in medullary-ray cells (radial view) in
Cedrus atlantica (G); *Taxodium distichum* (H); *Podocarpus andina* (K); *P. salici-
folia* (L); *Glyptostrobus* (M); *Sciadopitys* (N); *sp*, spring wood; *s*, summer wood;
Thuya gigantea (O). I, Bars, *b*, in tracheids of *Araucaria brasiliensis*. P. Spiral
bands in tracheids of *Torreya nucifera*. (A, N, after Nakamura; B. D—H,
K, L, P, after Gothan; C, after Gerry; I, after Winkler; M, after Kraus;
O, after Penhallow.)

his Juniperoid and Abietineous types. Each medullary-ray cell is longer in a radial direction than the breadth of a single tracheid and statements as to the number of pits on the radial wall of a ray cell have reference to the area bounded laterally by the vertical walls of a tracheid: this area may be designated the field ('Kreuzungsfeld'; 'aire mitoyenne'). It is, however, incorrect in many cases to speak of pits on the radial walls of medullary-ray cells, and if pits occur they are never bordered: in the Abietineae the walls are pitted, but in most other Conifers the pits seen in the field belong to the tracheids in contact with the rays. But in view of the general use of the expression medullary-ray pitting it would be inconvenient to discontinue the current terminology. There is a difference of opinion as to the value of medullary-ray pitting as a criterion of affinity, and it is probable that Gothan over-estimates the taxonomic significance of this character. Within certain limits the pitting on the walls of medullary-ray cells is undoubtedly important, but a comparison of sections of the wood of a collection of genera shakes one's confidence in the conclusions based by some authors on the form and number of the pits in the field. In the Abietineae a single large simple pit ('Eipore') occupies the field in some species of *Pinus* but the same type occurs also in *Sciadopitys*[1], species of *Podocarpus*[2] (fig. 693, K, N), *Microcachrys*, *Dacrydium*, and *Phyllocladus*[3]. The Araucarineae are usually described as having several (2—6) oblique pits in each field and as a rule no pits in the horizontal and tangential walls of the ray cells, but Thomson[4] states that the bordered pits in the field of *Araucaria* and *Agathis* are confined to the tracheids and an examination of macerated tissue confirms the absence of pits on the walls of the ray cells. A similar absence of pits characterises some other Conifers. Gothan has suggested the term Cupressoid pitting for Conifers in which the field in the spring-wood contains bordered pits with a fairly broad pore in a more or less horizontal position, a type of pitting found in some Cupressineae as also in *Sequoia* and certain other genera. He applies the name Podo-carpoid pitting to woods in which the field shows bordered pits

---

[1] Nakamura (83); Saporta (84) Pl. cxxxviii.; Fujioka (13).
[2] Kleeberg (85). .
[3] Schenk in Schimper and Schenk (90) A. p. 855.
[4] Thomson (13) p. 30.

with a narrower pore occupying an obliquely vertical position, a form of pit well shown in some species of *Podocarpus*. In the Podocarpineae and in most of the Cupressineae the tangential and horizontal walls of the ray cells are unpitted. An examination of sections of species of *Thuya* and some species of *Cupressus* reveals the presence of pits in the field with an almost vertical pore, and the variation in the breadth of the border and in the position of the pore is too great to admit of more than a restricted and cautious use of this anatomical feature as a means of distinguishing genera or even families. It is by no means easy even in sections of recent woods to observe with accuracy the structure of the ray pits: in many cases they are more or less bordered, but the greater distinctness of the pore often leads to the neglect of the fainter border. Moreover the small medullary-ray pits may be converted into large pits by the action of fungal hyphae. The large pits of some Pines, *Sciadopitys*, etc, represent one extreme; intermediate types are represented by *Cedrus*, *Taxodium*, and *Glyptostrobus*, while in *Juniperus* and several other genera the pits are smaller and more numerous[1].

The depth of the rays as seen in tangential section is a feature to which much attention has been paid, but this is a very variable and comparatively unimportant character[2]. In a single species of *Abies* the depth varies from 1 to 63 cells[3]. Many authors in describing fossil wood state the number of rays per square millimetre of a tangential section. Characters such as these may undoubtedly be useful in certain cases if used in conjunction with others, due allowance being made for the range of variation within the limits of a single stem. A more important feature is the occurrence of broad rays containing horizontal resin-canals such as those of *Pinus*, *Picea*, *Larix*, and *Pseudotsuga*. Another useful criterion is afforded by the association of horizontal tracheids (fig. 693, B) with the parenchyma of a ray usually at the upper and lower margin but sometimes, *e.g.*, *Pinus canariensis*[4], in the middle. The occurrence of such tracheids was formerly regarded as a trustworthy distinguishing feature of the Abietineae with the exception of *Abies* and *Pseudolarix*, but they are now known to

[1] Penhallow (07).   [2] Essner (86); Barber (98).
[3] Fujioka (13).   [4] Strasburger (91) p. 21.

occur in *Abies*[1] and several members of other families. Chrysler[2] states that *Cedrus* differs from *Abies* in having ray tracheids mixed with marginal parenchyma, and at the limit of an annual ring the marginal tracheids may be replaced by shadowy cells or ghosts of cells. The tracheids may have smooth walls as in *Cedrus, Tsuga, Larix, Pinus Strobus*, or, as in other species of *Pinus* ('Hard Pines'), their walls are characterised by irregular ingrowths or pegs. Wettstein[3] states that in *Picea omorica* horizontal tracheids though common in the main stem do not occur in the rays of branches. De Bary[4] recognised tracheids in the rays of *Sciadopitys*; they have been recorded also in *Juniperus, Cupressus, Thuja, Sequoia*[5] and, as the result of wounding, in *Cunninghamia*[6].

The occurrence of idioblasts in the form of irregular thick-walled elements is characteristic of the pith and cortex of *Araucaria*, but similar cells are found in the pith of *Torreya nucifera, Podocarpus neriifolia, Dacrydium cupressinum*[7] and *Cryptomeria*. In some cases, *e.g., Abies magnifica*[8], *Picea omorica*, horizontal rows of thick-walled cells form diaphragms in the pith. The structure of the secondary phloem has received relatively little attention and owing to its comparatively rare preservation in fossils it is less important to the palaeobotanist. In the Cupressineae the regular alternation of tangential rows of hard and soft bast is a characteristic feature, while in the Abietineae the phloem consists of sieve-tubes and parenchyma with a few scattered stone-cells. The absence of albuminous cells in the medullary rays of the phloem region in the Araucarineae is noteworthy and Thomson[9] states that he found none in *Podocarpus*.

The structure of the stomata in the leaves of Conifers is fairly uniform: their distribution should be noted though this in itself is not of much value as a distinguishing feature. They may be confined to regular bands (*Abies*, etc.) or grooves (*Torreya*, etc., fig. 694, B), or irregularly distributed. The position and number of resin-ducts is often a useful guide: to quote one example only, in the leaves of *Agathis* and *Araucaria* (fig. 694, C) the ducts

---

[1] Thompson (12).   [2] Chrysler (15).
[3] Wettstein (90) p. 511.
[4] De Bary (84) A. p. 490. See also Tassi (05) quoted by Vierhapper (10).
[5] Gordon, M. (12); Jones (13²).   [6] Jeffrey (08).
[7] Kubart (11²)   [8] Jeffrey (05).   [9] Thomson (13) p. 31.

occur between the veins, but in the leaves of some species of
Podocarpus (fig. 695, D), externally indistinguishable from those
of *Agathis*, the ducts are below the veins[1].   Caution must be exercised in using the number of resin-ducts as a diagnostic character.
Schroeter[2] draws attention to the occasional absence of ducts in
Picea excelsa leaves while in others 1 or 2 are present.   In most
leaves there is some mechanical tissue immediately below the
epidermis either as scattered fibres or a continuous layer, but in

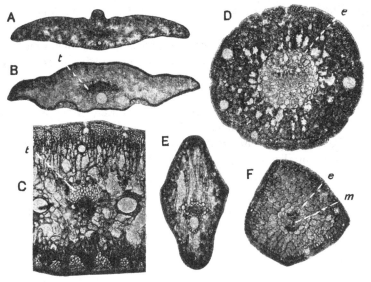

FIG. 694.   Leaves in transverse section.   A, *Cephalotaxus Fortunei*. B, *Torreya
myristica*; *t*, transfusion-tissue. C, *Araucaria imbricata*; *t*, transfusion-tissue
D, *Pinus monophylla*; *e*, endodermis. E, *Cryptomeria japonica*. F, *Cedrus
Libani*; *e*, endodermis; *m*, medullary ray.

Taxus and *Torreya* there is none.   The occurrence of branched
idioblasts is a striking feature in the mesophyll of *Agathis, Araucaria, Sciadopitys*, and *Podocarpus*.   The infoldings of the walls
of the chlorenchyma are especially characteristic of Pine leaves
(fig. 694, D) and they occur also in *Cedrus* (fig. 694, F), *Pseudolarix* and some other genera.   The structure, extent, and position
of short isodiametric tracheids in association with the conducting
tissue is an important feature.   This tissue, the elements of which

[1] Seward and Ford (06) B.   [2] Schroeter (97).

are usually termed transfusion-tracheids[1], is regarded by some authors as homologous with the centripetal wood of Cycadean leaves (Vol. III. p. 31) and Bernard definitely adopts the term centripetal xylem. Jeffrey[2] and, more recently, Takeda[3] do not attach a similar morphological significance to the short tracheids, and they are probably justified in their sceptical attitude. The transfusion tracheids are often reticulately pitted: in many leaves they form conspicuous flanges on the sides of the vascular bundles (fig. 694, B, C, *t*) or they may more or less encircle the vein. In some leaves, *e.g.*, *Araucaria*, it is noticeable that the amount of transfusion tissue (fig. 694, C, *t*) increases as the vein is traced towards the leaf-apex until the long and narrow elements may be entirely replaced by a group of short transfusion-tracheids. Another type of accessory tracheid is occasionally met with, namely elongated tracheids traversing the mesophyll between the veins and the edge of the lamina. This is seen in the long and narrow leaves of some Podocarps[4].

The anatomy of cone-scales is too wide a subject for adequate treatment in this sketch. The general rule is that in each scale there are two sets of vascular strands, a lower set of normally orientated bundles and an upper inversely orientated series. In some cone-scales, *e.g.*, *Araucaria*, concentric vascular strands are a prominent feature. The taxonomic significance of the anatomical characters of cone-scales has been discussed by Radais and more recently by Miss Aase[5], to whose accounts the student is referred.

It has been shown that characters which it has been customary to associate with a definite type of wood may occur sporadically in several other Conifers; but this does not invalidate conclusions based on the prevalent occurrence of such features in a given specimen. It is untrue to say that contiguous and alternate pits are the monopoly of the Araucarineae and it is incorrect to assert that in Araucarian wood the pits are never separate. Similarly

---

[1] So named by von Mohl; see Wordsell (97); Bernard (04) B.; Carter (11).
[2] Jeffrey (08²).
[3] Takeda (13); see also Thomson (13).
[4] For further details with regard to leaf-anatomy, see Thomas, F. (66); Bertrand, C. E. (74); Mahlert (85); Strübing (88); Daguillon (90).
[5] Radais (94); Sinnot (13); Aase (15).

the pits on the medullary-ray cells, whatever relative value we may assign to this character, are in many cases of considerable assistance even though we are not prepared to follow Gothan to the full extent of his trust in the taxonomic importance of medullary-ray pitting.

Dr Groom[1] has described the northern evergreen Conifers as architectural xerophytes having xeromorphic leaves with a xerophytic structure. He discusses in his remarks on the Ecology of Conifers the correlation of the characteristic wood-structure and the xeromorphic leaves, the cause of the survival of the Coniferae in competition with Dicotyledons, and other questions of interest to the student of the evolution and past history of the group.

Short summary of the characteristics of recent Conifers[2].

ARAUCARINEAE. There is a close agreement in the structure of the wood in the two members of this family. Attention has already been called to the normal type of pitting of the tracheids and to certain exceptional forms. The occasional tendency towards a scalariform type of pitting (fig. 691, D) is an interesting point. Pits are fairly abundant on the tangential walls of the xylem-elements. In the wood of *Agathis robusta*[3] vertical rows of parenchyma are said to be fairly abundant. Annual rings not infrequently absent or feebly developed. Medullary rays one-cell broad, rarely double, usually 7—15 cells deep; but in *Araucaria* the depth may reach 26 cells: the ray cells occasionally present a distended appearance in tangential sections of the wood (cf. *Ginkgo biloba*). The persistence of the leaf-traces in the old wood of *Araucaria* is a striking feature[4] considered by Lignier to possess diagnostic importance: Jeffrey[5] states that traces are much less persistent in some young stems. In *Araucaria* each trace arises as a single strand, but in *Agathis* it leaves the perimedullary region as a double bundle[6].

Agathis (fig. 695). Leaves sessile or slightly petiolate, opposite, subopposite or, on the main axis, spiral; ovate, broadly lanceolate (*A. loranthifolia*, 13 × 5 cm.; *A. macrophylla* 17 × 5 cm.). Resin-canals between the veins; transfusion-tracheids fairly abundant, but less prominent than in *Araucaria*. The almost spherical megastrobili (fig. 696) are very characteristic; they

---

[1] Groom (10).
[2] In addition to the text-books on Conifers by Beissner, Veitch, and the account in *Die Natürlichen Pflanzenfamilien* the student should consult the Report of the Conifer Conference, *Journal of the Royal Horticultural Society*, Vol. XIV. 1892.
[3] Noelle (10).
[4] Thiselton-Dyer (01²); Seward and Ford (06) B.
[5] Jeffrey (12) p. 565.   [6] Thomson (13) p. 15.

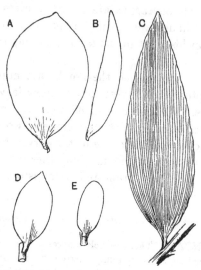

Fig. 695. A, *Agathis Moorei*. B, E, *Agathis australis*. C, *Agathis vitiensis*. D, *Podocarpus Motleyi*. (After Seward and Ford; ½ nat. size.)

A B

Fig. 696. *Agathis Moorei* (A) and *Agathis australis* (B) cones. (After Seward and Ford; A, ½ nat. size; B, ¾ nat. size.)

bear a close superficial resemblance to Cedar cones but the latter are relatively narrow and often more or less flattened at the apex[1]. Bommer[2] calls attention to the resemblance of *Agathis* cones to those of the Dicotyledon *Dammaropsis kingiana* (Moraceae).

Araucaria. (Frontispiece; figs. 678—681, 689, etc.) The falcate tetragonal leaves of *A. excelsa* illustrate one type of leaf that is seen in its smallest form in *A. Balansae* (4—5 × 2·5 mm.). In *A. Bidwillii* (fig. 697) the leaves are subsessile and the flat lamina may reach a length of 7 cm.: in *A. Housteinii*

Fig. 697.   *Araucaria Bidwillii.*   (After Seward and Ford; nat. size.)

the ovate-lanceolate leaves may be 10 cm by 1 cm. Dimorphism[3] in the foliage of a single shoot is not uncommon. The striking difference in some species between the juvenile and adult foliage is illustrated in fig. 698. In the broad-leaved species (*Colymbea* section) resin-canals occur between the veins (fig. 694, C) and in the *Eutacta* section, e.g., *A. excelsa*, the canals are scattered. Strobili and cone-scales of *Araucaria* are described in an earlier part of this chapter (see page 113).

---

[1] Fliche (96).
[2] Bommer (03) B. Pl. x. figs. 164, 165.
[3] Masters (91); Bommer (03) B. Pl. v. fig. 23; Siebold (70) Pl. cxl.

CUPRESSINEAE. The absence of resin-canals in the xylem is a feature shared by other families; but in the occurrence of xylem-parenchyma in different regions of the wood the Cupressineae differ as a rule from the Abietineae, though this is not a constant distinguishing character. The pits in the field vary from 1 to 6 or 8 in some genera, *e.g.*, *Taxodium* and *Glyptostrobus*[1]: Gothan[2] applies the term Cupressoid to medullary-ray pits character-

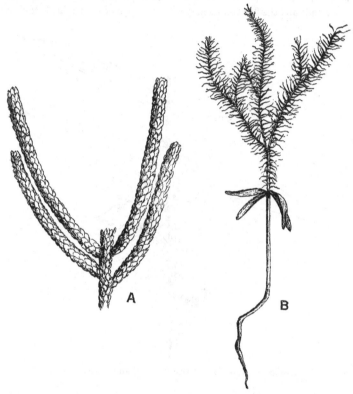

FIG. 698. *Araucaria excelsa.* A, Mature form. B, Seedling. (¾ nat. size.)

ised by an upper and lower border (fig. 693, H) but, as pointed out on a previous page, the position of the pore is by no means constant and in *Taxodium* the ray pits are fairly large and almost simple.

Cupressus and *Chamaecyparis*. By Beissner[3] and many other authors both names are used in a generic sense, though Masters regards *Chamaecyparis* as a subsection of *Cupressus*. One distinguishing feature is the presence of more

[1] Kleeberg (85).   [2] Gothan (05) p. 47.   [3] Beissner (91).

than 2 seeds on each cone-scale of *Cupressus*. The leaves are whorled, in 4 ranks, and appressed to the axis (fig. 699). The young foliage of *Cupressus*, *Thuya*, and other genera (fig. 700), which in the adult state has the form of scale-leaves, consists of spreading linear leaves: individuals in which this juvenile foliage persists are often spoken of as species of the 'genus' *Retinospora*. Cones oblong or globular, composed of a few pairs of scales with peltate distal ends either smooth or provided with a prominent umbo (fig. 699).

FIG. 699.   *Cupressus Macnabiana*.   (From Rendle and the *Gardeners' Chronicle*.)

Thuya. Very similar to *Cupressus* in the habit of its bi- or tri-pinnate shoots. The cones of 8—10 decussate scales are distinguished from those of *Cupressus* by their elongated, oblong, form and by the upwardly directed scales with thickened apices in contrast to the more horizontal and peltate scales of *Cupressus*: there are 2 seeds to each scale, winged (sect. *Euthuja*) or wingless (sect. *Biota*).

Libocedrus. Foliage shoots often broader than in *Thuya*; in *L. decurrens* the appressed, flat, leaves are characterised by a long decurrent portion and

10—2

in *L. Doniana* the whorled arrangement is less obvious. There is a large canal below the midrib of the leaf as in *Thuya* and *Cupressus*. The cones (4—6 valves) are longer and relatively narrower than in *Thuya*; seeds un-

FIG. 700. A, Young plant of *Libocedrus decurrens*; *c*, cotyledons; *pr*, primordial leaves; *tr*, transitional leaves. B, Branch with adult (*ad*) foliage. (After Rendle, from Veitch.)

equally winged. *Juniperus*. The polymorphism of the shoots is especially striking; the leaves, 2—3 in each whorl, being small and appressed, spinous and spreading, or flat and linear. The very narrow leaves of *J. Sabina* differ widely from the broader, flat and sharply pointed leaves of *J. drupacea*. There is a resin-canal near the lower surface[1] or, in some species, *e.g.*, *J. oxy-cedrus*, there may be no canal. The more or less globular or elongated fleshy strobilus consists of 1—4 whorls of megasporophylls.

Fitzroya and *Diselma*. In some examples of *F. patagonica* the leaves are short and crowded and slightly falcate, in others more spreading; while in *Diselma Archeri*[2] the shoots resemble those of some Lycopodiums. The leaves are in alternate ternary whorls and there is a single canal below the vein. The cones consist of 3 alternate trimerous whorls and the fertile scales bear a variable number of winged ovules. *Thujopsis*. Shoots similar to those of *Thuya* but the decussate leaves are rather larger[3]. Cones narrow, consisting of 8 clavate scales with 5 small winged seeds on each scale.

Taxodium. The slender deciduous shoots, 8—10 cm. long, bear 2-ranked linear leaves 2 cm. long, acute and lanceolate, also small leaves 10—17 mm. long and barely 1 mm. broad. Circular depressed branch-scars are a charac-teristic feature of the leafless shoots. Globular cones[4] composed of a few peltate scales with an irregular crenulate upper margin (fig. 686, D—F) each scale bearing 2 slightly winged seeds. The almost leafless spikes of micro-strobili are a noteworthy feature. The wood of *Taxodium* closely resembles that of *Sequoia sempervirens*: the thicker horizontal walls of the xylem-parenchyma are quoted by Gothan as a characteristic feature[5] of *Taxodium*. Lingelsheim[6] says that this distinction is not valid; he states that simple pits occur in the cross-walls of the xylem-parenchyma of *Taxodium* but not in the cells of *Sequoia*. Pits are, however, present in the parenchyma of both these genera. *Glyptostrobus*. Similar in habit to *Taxodium*, but the leaves are not 2-ranked[7]; the cones are more pyriform; the seeds are smooth and narrow with lateral wings. The comparatively large circular pits on the radial walls of the ray cells are characteristic[8]

Cryptomeria. Foliage-shoots (fig. 677) as in *Araucaria excelsa*: a large resin-canal occurs below the midrib of the laterally compressed leaves (fig. 694, E). The cones have 20—30 peltate scales characterised by the deeply cleft comb-like upper portion (fig. 684, M). The xylem-parenchyma is said by Fujioka to be confined to the region between the spring and summer wood: the pits in the field are variable in position and the breadth of the pore may be vertical or oblique.

Cunninghamia. Leaves densely spiral and spreading, narrow, lanceolate, acuminate and serrate, 2·5—5 cm. long, with a narrow decurrent base: a large canal below the midrib. Cones ovoid-globular, 4 cm. long, composed of broad thin scales with a serrate edge and a fimbriate membrane on the abaxial

---

[1] Kirchner, Loew, and Schröter (06) pp. 293, 314.
[2] Hooker, J. D. (60) Pl. XCVIII.   [3] Fujioka (13).
[4] *Gard. Chron.* Nov. 25, 1893, p. 657.
[5] Gothan (06); (09).   [6] Lingelsheim (08).
[7] Masters (00).   [8] Kleeberg (85).

side of the 3 winged seeds (fig. 684, K).   Branch-scars occur on the stems[1].
Taiwania.   This genus[2] has the habit of *Cryptomeria* and cones recalling
those of *Cunninghamia* and *Tsuga*: each scale bears 2 seeds.   *Fokienia.*   The
single species[3] is, in certain respects, intermediate between *Cupressus* and
Libocedrus; the cones are globose like those of *Chamaecyparis* and each
scale bears two unequally winged seeds; the foliage is nearly identical with
that of *Libocedrus*

Fig. 701.   *Athrotaxis cupressoides.*

Athrotaxis.   (Figs. 684, N, O;   701.)   Leaves short, loosely spreading
and slightly imbricate (*A. selaginoides*[4]), similar to those of *Araucaria excelsa,*

[1] Lotsy (11) p. 51.   [2] *Gard. Chron.* Feb. 4, 1911.
[3] A second species has recently been described by Hayata (17)
[4] Baker and Smith (10) p. 305 (and photograph).

more appressed and smaller, decussate or in close spirals (*A. cupressoides*) or, in *A. laxifolia*[1], more like the foliage of *Sequoia gigantea*. The apex of each cone-scale has a sharp point at the distal end. *Tetraclinis* (fig. 703, B). This genus, which occurs in Algeria and Morocco, and is usually placed next to *Callitris* and *Widdringtonia*, has recently been transferred by Saxton[2], as the result of his work on the life-history of *T. articulata*, to the Cupressineae: he believes that the Callitrineae were derived from the Northern Cupressineae through some type resembling *Tetraclinis*. The flattened foliage-shoots resemble *Salicornia* and bear 4-ranked small leaves; the cones consist of 4 nearly equal decussate scales, each scale having 2 seeds with unequal wings.

SCIADOPITINEAE. *Sciadopitys.* The single species is characterised by the long and rigid 'double needle' reaching a length of 12 cm. with a broad furrow on the upper surface and a deeper groove on the lower face. Mr Boodle[3] has recently described an example of concrescence in needles of *Pinus Laricio* Poir. var. *nigricans* Parl. which points to the probability that a morphological similarity exists between the double needle of *Sciadopitys* and abnormal, fused, leaves of *Pinus*. Reference is made by Boodle to other possible views that have been advanced with regard to the morphological nature of *Sciadopitys* needles. The oblong woody cones, 7 × 4 cm.[4] are fairly easy to identify by the rounded and reflexed upper margins of the scales (fig. 684, S); there are 5—15 seeds, with a narrow wing, on each cone-scale.

SEQUOIINEAE. *Sequoia.* Attention has already been called to some of the anatomical features. The oval pits on the field (2—6) have an upper and a lower border though the pore is not infrequently obliquely vertical. Anatomically the wood of *S. sempervirens* is considered by Gothan[5] to agree more closely with that of *Taxodium* than with *S. gigantea*. The leaves of *S. gigantea* (= *Wellingtonia*) are 3-angled and decurrent, ·5 mm. long; those on the fertile shoots are broader, shorter and imbricate. *S. sempervirens* bears 2-ranked linear, sessile, but not decurrent leaves with an abruptly spinous apex. There is a resin-canal below the midrib and transfusion-tracheids form conspicuous lateral groups. The cones of the two species are of the same type, but those of *S. gigantea* are larger and occasionally reach a length of 9—10 cm. (fig. 702).

CALLITRINEAE. The foliage-shoots, similar to those of some Junipers, are characterised by short decurrent scale-like leaves in alternate ternary or decussate whorls; the ovate or globular cones are composed of a few valvate scales. *Actinostrobus*[6] (fig. 703, A). The very small leaves in ternary whorls have a free apex and the slender shoots closely simulate those of some species of *Veronica* and *Thamnea depressa* (Bruniaceae). The 6 cone-scales are oblong, acute, and the base of the cone is invested by 6 rows of crowded scales which gradually pass into the foliage-leaves: each scale has 2—3 winged

---

[1] *Gard. Chron.* Jan. 31, 1891, p. 147.   [2] Saxton (13²); (13³).
[3] Boodle (15).   [4] Siebold (70) Pl. CII
[5] Gothan (06); (09). See also Jeffrey (03); Gordon (12).
[6] Saxton (13).

seeds. *Callitris.* Similar in habit to *Tetraclinis* (fig. 703, B): in some forms (*C. arborea*) the small leaves are closely appressed to the axis; *C. glauca*[1] shows a considerable range in the form of the leaves, and in *C. rhomboidalis* the

FIG. 702 A.   *Sequoia gigantea.*   Shoot with cones.   (Nat. size.)

shoots are especially slender. Saxton[2] points out that the tracheids have a single row of separate pits and draws attention to the occurrence of horizontal

[1] Baker and Smith (10) p. 118.
[2] Saxton (10²); (10³). The wood of the Callitrineae requires more thorough investigation.

bands above and below each pit.   Cones usually spherical, reaching 2·5 cm.
in length, with 6 smooth or tuberculate valves.   In *C. Macleayana* the cones
are pyramidal and have 6—8 valves[1].

Widdringtonia.   The adult foliage-shoots bear rather longer leaves in
decussate pairs[2]; the twigs are cylindrical and not flat as in *Thuya*.   In
quickly growing shoots the leaves may be spiral; they have no hypodermal

Fig. 702 B.   *Sequoia gigantea.*   An unusually large cone.
(British Museum; nat. size.)

layer like that in the leaves of *Callitris*.   Spirally thickened tracheids occasionally occur in the wood.   The cones consist of 4 decussate thick and warty
valves, each scale bearing 7—8 winged seeds.

ABIETINEAE.   *Pinus.*   The needle-like leaves on short shoots are a
striking feature.   In *P. silvestris* and many other species each short shoot

---

[1] Baker and Smith (10).   [2] Rendle (96).

FIG. 703. A, *Actinostrobus pyramidalis*. B, *Tetraclinis articulata*. (After Saxton.)

FIG. 704. *Pinus excelsa*, cone and foliage-spur.
(From a photograph by A. Howard.)

bears 2 needles plano-convex in section. In *P. monophylla* (fig. 694, D) each shoot has usually a single sharp-pointed cylindrical leaf 4 cm. long. Three-needled Pines occur in N. America and the Himalayas but not in Europe. In *P. Cembra, P. peuce* (the only European examples), *P. Strobus, P. excelsa* (fig. 704), *P. koraiensis*, etc., the short shoots bear 5 needles and each is triangular. Thomson[1] regards the dwarf-shoot of *Pinus* as a derivative of a longer shoot with spirally disposed needles: he has recently described examples of dwarf-shoots in recent Pines bearing an abnormally large number of leaves. In *P. silvestris* shoots with 3 leaves are not uncommon: in *P. excelsa* wounding caused the development of as many as 15 needles on a single dwarf-shoot. Jeffrey, on the other hand, regards the spur-shoot as a primitive attribute of the coniferous stock. The length of the leaf varies considerably and may reach 30 cm.; the margin is generally entire, but in *P. Cembra* the apical portion is finely serrate. The structure of Pine leaves[2] is well known, but reference may be made to the twin bundles in hard Pines, a distinction from the single strand in the soft Pines[3], the presence of a well-defined endodermis (fig. 694, D, *e*) and the infoldings of the chlorenchyma[4]. There are two types of cone, that of the *Pinaster* group in which the distal end of the woody seminiferous scale is more or less pyramidal and has a central umbo (cf. fig. 785), prolonged in some species, *P. ponderosa*[5], *P. Jeffreyi*[6], etc., into a short recurved spine or, in *P. Coulteri*, into a large curved spinous process. Another type, illustrated by the *Strobus* group, is distinguished by the flatter imbricate scales with the umbo at the tip of the free rounded margin (fig. 704). The cones of the latter type resemble those of *Picea*. The variation in the size and form of cones from the same tree in *Pinus excelsa* is worthy of notice from a palaeo-botanical standpoint[7].

Cedrus. The more slender needles of *Cedrus*, most of which occur in clusters on short shoots, are approximately triangular in section (fig. 694, F) and have a single bundle and two canals next the lower epidermis. The persistent leaf-bases resemble those of *Tsuga* and *Picea* (fig. 706, B, D). Canals, though not normally present in the wood, are induced by wounding; xylem-parenchyma may occur in the late summer-wood and ray-tracheids are present, but less prominent than in Pines and some other genera. The deciduous nature of the mature cone-scales would lead one to expect their preservation as fossils. A characteristic feature of the cones is the presence of radially disposed lines normal to the edge of the sporophylls[8] *Larix.* The narrow linear decurrent leaves, keeled on the lower surface, have a resin-canal close

---

[1] Thomson (14).

[2] Zang (04) attempts to classify Pine leaves on the basis of anatomy.

[3] For anatomical features of hard and soft Pines, see Holden, R. (13³); Jeffrey and Chrysler (06).

[4] Kirchner, Loew, and Schröter (06) figs. 188, 222, etc.

[5] *Gard. Chron.* Nov. 15, 1890, pp. 561, 569.

[6] *Ibid.* March 23, 1889.   [7] Bommer (03) B. Pls. II., IV.

[8] Fliche (96) p. 86.

to the epidermis at each angle. There is a single vascular bundle but no thick-walled hypoderm. Bailey[1] notes the occurrence of wood-parenchyma on the outer face of the summer-wood. The persistent subglobose or more elongated cones, reaching a length of 10 cm. (fig. 705, G), are in most species characterised by the large size of the bract-scales like those of *Abies* and *Pseudotsuga*. Bommer calls attention to the superficial resemblance of Larch cones to the fruit of *Petrophila diversifolia*[2] (Proteaceae). *Pseudolarix* resembles *Larix* in habit (fig. 705, A, B); the male strobili are umbellate; the leaves linear lanceolate and up to 7 cm. long. The cones are ovate and bear loosely imbricate, pointed deciduous scales; the bract-scale is shorter than in *Larix*. *Picea*. The short and narrow leaves are tetragonal or more or less flat in section: a canal occurs below the vein or there may be two lateral resin-ducts. The persistent bases of the leaves form prominent pegs (fig. 706, D). The cylindrical or oval cones, reaching 16 cm. in length, consist of leathery, concave imbricate scales; the bract-scales are concealed[3]. The stem of a variety of *P. excelsa* (var. *tuberculata*)[4], characterised by large conical tubercles of cork, bears a close resemblance to that of *Xanthoxylum* (Rutaceae)[5]. *Tsuga*. The leaves are flat, decurrent (fig. 706, B), similar to those of *Taxus*, *Abies*, and some species of *Picea*; but there is a single resin-canal below the vein and the lamina is petiolate and not sessile as in *Sequoia sempervirens*. In some cases the lamina is finely serrate. The cones closely resemble those of *Picea* except in their smaller size; the bract-scales are usually concealed, but in *T. Pattoniana*[6] they are longer and reach a length of 7·5 cm.

Pseudotsuga. The leaves resemble those of *Tsuga*, but there are two lateral canals and the lamina is attached by a narrow base to a slightly prominent leaf-cushion (fig. 706, C). The wood is like that of *Larix* but the tracheids often possess spiral bands, and xylem-parenchyma occasionally occurs. Resin-ducts are present in both normal and injured stems. The cones, 5—10 cm. long, resemble those of *Tsuga* but differ in the 3-pronged conspicuous bract-scale; they are pendulous and the scales are persistent. *Keteleeria*. Shoots similar in habit to those of *Abies*; leaves flat, a canal at each angle. Cones similar to those of *Pinus Cembra*: microstrobili umbellate. *Abies*. Leaves usually flatter than in *Picea* and often larger; the apex may be notched though this is not a constant feature[7]. There is no persistent leaf-base (fig. 706, A). Two lateral canals[8] are generally present in the leaf, but in some species the canals are median[9]. Transfusion-tracheids may form a ring round the phloem (*A. magnifica*). Cones large (in *A. nobilis* 25 × 10 cm.) and in most species the bract-scales are conspicuous (fig. 705, C, F). The cones

[1] Bailey (09).   [2] Bommer, *loc. cit.* Pl. IX. figs. 167, 168.
[3] For figures illustrating the range in form of cones, see Kirchner, Loew, and Schröter (06) pp. 152, 153.
[4] Schröter (97) figs. 24, 26.
[5] Barber (92).   [6] *Gard. Chron.* June 3, 1893, p. 659.
[7] Kirchner, Loew, and Schröter (06) fig. 26, p. 89.
[8] *Ibid.* p. 90.   [9] Elwes and Henry (06) p. 717.

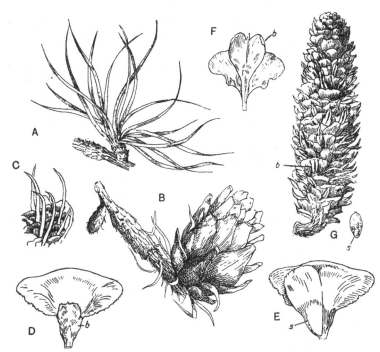

Fig. 705. A, B Short shoot and cone of *Pseudolarix Kaempferi.* C, *Abies bracteata,*
showing the long bract-scales. D, E, *Abies concolor,* cone-scale; *b,* bract-
scale; *s,* seed. F. *Abies Fraseri,* cone-scale; *b,* bract-scale. G, *Larix Griffithi;*
b, bract scale; *s,* seed. (C—F, from the *Gardeners' Chronicle.*) M. S.

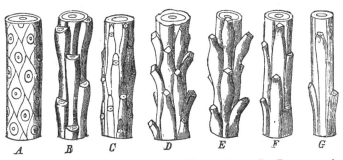

Fig. 706. Branches of Abietineae. A, *Abies pectinata.* B, *Tsuga canadensis.*
C, *Pseudotsuga Douglasii.* D, *Picea excelsa.* E, *Cedrus Libani.* F, *Larix
europaea.* G. *Pseudolarix Kaempferi.* (After Rendle from Veitch and after
Engler and Prantl.)

of some species with concealed bract-scales, e.g., A. violacea[1], A. Webbiana[2], A. concolor (fig. 705, D, E), are hardly distinguishable from those of Picea and the larger cones of Tsuga, while those of Abies amabilis resemble Cedar cones.

PODOCARPINEAE. The wood agrees with that of the Cupressineae in the presence of xylem-parenchyma, but Gothan[3] states that the pores of the bordered pits on the radial walls of the medullary rays in the region of the spring-wood are narrow, elliptical, and vertical (fig. 693, K), while in the Cupressineae the pore is horizontal (cf. fig. 693, H), a feature that is not sufficiently constant to afford a trustworthy criterion. *Podocarpus.* Leaf-lamina linear and short (*P. nivalis*, etc.) as in *Taxus*, longer and broader (in *P. elatus* reaching a length of 29 cm. and 2 cm. broad), as in *Cephalotaxus* (fig. 694, A) and the pinnae of *Cycas* small appressed and scale-like (*P. cupressina*) or ovate and provided with several veins (*P. Nagi*, 6 × 2 cm., *P. Wallichiana*, 13—15 × 3·5 cm.) The leaves are opposite in the section *Nageia* (fig. 707), scattered in other species. Tison[4] has shown that in the broad-leaved species only one vascular bundle is given off from the stele and this branches in the petiole, not in the cortex as in *Agathis*. The leaf of *P. macrophylla*[5] (5—6 cm. × 7 mm.) represents a fairly common type: short reticulate tracheids occur on each side of the vein and elongated tracheids extend from the midrib to the leaf-edge. There are three canals near the lower surface, but in some species only one is present. *P formosensis*[6] (fig. 707) has leaves like those of *P. Nagi*, but more rigid, thicker, and smaller; the epidermal cell-walls are thick and elongated and 4—5 cells surround each stoma (fig. 707, *a*, *b*). The presence of hypodermal mechanical tissue distinguishes *Podocarpus* leaves from those of *Torreya*, and another characteristic feature is afforded by the two kinds of accessory tracheids in place of the ordinary transfusion-tissue in the great majority of leaves. Reference has already been made to the 'cones.' *Dacrydium.* The dimorphism of the foliage-shoots is illustrated in fig. 708. The very simple type of megastrobilus is a striking feature (fig. 684, P). *Microcachrys.* Leaves small, appressed, in decussate pairs[7]. As in *Dacrydium* there is a single canal below the vein. Gothan notices the frequent occurrence of a single large pit in the field of the rays. The fleshy mulberry-like cone is a peculiar feature. *Saxegothaea* (fig. 687). Lindley[8], the author of the genus, describes this Conifer as having the male flowers of a Podocarp, the cones of an *Agathis*, the fruit of a Juniper, and the habit of *Taxus*. There is a canal below the vein, also lateral transfusion-tracheids but less numerous than in *Podocarpus*. The occurrence of pits in the horizontal walls of the ray cells is a character in which *Saxegothaea* differs from the genera comprised under the general term *Cupressinoxylon*[9]. The cones have already been described.

[1] *Gard. Chron.* Dec. 27, 1890.   [2] *Ibid.* Oct. 3, 1891.
[3] Gothan (05) p. 47.   [4] Tison (12) Pl. IV. fig. 2.
[5] Stiles (12).   [6] Dümmer (12).
[7] Hooker, W. J. *Icones Plantarum*, 1843.·
[8] Veitch (00) p. 158; for figures, see *Gard. Chron.* June 22, 1889, p. 782.
[9] See p. 186.

FIG. 707. *Podocarpus formosensis.* A, Foliage-shoot. *a*, *b*, Epidermis with stomata. (A, after Dümmer; ½ nat. size.)

Acmopyle. This generic name was given to an imperfectly investigated New Caledonian Conifer formerly known as *Dacrydium Pancheri.* The sessile, decurrent, falcate leaves on the lateral branches (1—1·6 cm. × 2·5 mm.) are a characteristic feature, those on the main axis being small and scale-like[1].

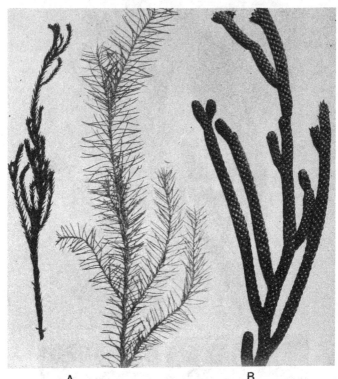

A   B

FIG. 708. A, *Dacrydium elatum.* B, *Dacrydium araucarioides.* (From specimens in the British Museum.)

Pherosphaera. The vegetative shoots resemble those of *Microcachrys* and some Lycopods. The genus is peculiar among the Podocarpineae in the absence of an epimatium. Sinnott[2] regards the megastrobili, composed of 2—5 one-seeded scales, as 'the last step in reduction,' but the difficulty is to discriminate between simplicity as a primitive and as a reduction-phenomenon. *Phyllocladus* (fig. 675). The cladodes have a central rib and several vascular

[1] Pilger (03) p. 117. Mr Sahni (Emmanuel College, Cambridge) has in preparation an account of *Acmopyle* based on material collected in New Caledonia by Mr Compton.

[2] Sinnott (13) p. 73; see also Masters (93) p. 6; Hooker. J. D., *Icon. Plant.*, 1882, Pl. 1383; Baker and Smith (10).

strands, and stomata occur on both surfaces[1]. The small megastrobili consist of spiral or decussate scales, each with one seed. Miss Young[2] mentions the occurrence of tracheids with spiral bands: a single large pit occupies the field of a ray-cell as in *Dacrydium Franklini*[3]. The seed in the early stage of development shows signs of a ruminated endosperm.

TAXINEAE. The occurrence of spiral bands in the secondary tracheids is a family-character but similar tracheids are not uncommon in other Conifers. *Taxus.* The difference in leaf-form and in the arrangement of the foliage is well illustrated by Kirchner and Schröter[4]. The leaves have recurved margins, a mucronate acute apex, and a small median vein. *Torreya.* In habit this genus resembles *Taxus* but the leaves are longer (6 cm. in *T. californica*); there is a resin-canal below the vein and two conspicuous stomatal grooves on the lower surface with papillose epidermal cells (fig. 694, B): the midrib is not prominent. *Cephalotaxus.* Leaves linear, more or less falcate, reaching a length of 12 cm. in *C. Henryi* (fig. 709), with a prominent midrib and one canal

FIG. 709. *Cephalotaxus Henryi.* (British Museum; ½ nat. size.) M. S.

(fig. 694, A); there are no stomatal grooves. Seeds like those of *Torreya* in size and in the thick fleshy sarcotesta and inner shell; the endosperm is not ruminated. Rothert[5] has described an interesting departure from the usual structure in the stem of *C. koraina* (according to Beissner = *C. pedunculata* var. *fastigiata*): a resin-canal occupies the centre of the pith and several short tracheids occur internal to the edge of the xylem-cylinder. There do not appear to be any anatomical characters apart from the spiral bands of the secondary tracheids by which the Taxineae can be distinguished from some Cupressineae and other Conifers: xylem-parenchyma is said to be present in both spring and summer wood of *Cephalotaxus drupacea*, while it is unrepresented in *Taxus cuspidata* and *Torreya nucifera*[6]; the pits in the field vary in number and may be simple or bordered and in the latter case the pore is obliquely vertical.

[1] Robertson (06); Bernard (04) B.; Stiles (12).
[2] Young, M. S. (10). [3] Gothan (05) p. 55.
[4] Kirchner, Loew, and Schröter (06) p. 69.
[5] Rothert (99). [6] Fujioka (13)

Sources of error in the determination of fossil Conifers.

The determination of fossil Conifers is one of the most difficult tasks of the palaeobotanist. It is comparatively seldom that well-preserved cones are found in organic connexion with the twigs that bore them and the cones rarely exhibit those features which are the best guides to affinity. Excessive trust in superficial similarity has frequently led to the employment of generic names suggesting relationships which are thoroughly misleading. In comparing fossil and recent forms authors are apt to confine their attention to the better-known types, forgetting that it is often with the less familiar and geographically restricted genera that extinct plants are most closely allied. Even the data supplied by petrified wood are often insufficient to enable the student to do more than refer a specimen to some comprehensive genus based on characters shared by several recent genera. Though it is as a rule easy to distinguish between the wood of a Conifer and that of Cycads and Dicotyledons, the agreement between the xylem elements of many Cycads and those of the Araucarineae is sufficiently close to afford opportunity for error. The homogeneous structure of the secondary wood of some Magnoliaceous genera, *e.g.*, *Trochodendron* and *Drimys*[1] (fig. 710), closely simulates that of a Conifer, but the medullary rays are approximately equal in breadth to the tracheids and the cells are more elongated

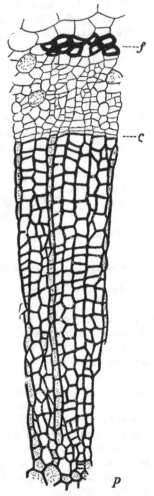

FIG. 710. *Drimys Winteri.* Transverse section of part of a stem. *p*, pith; *f*, pericycle fibres; *c*, cambium.

[1] Groppler (94); Solereder (99) p. 34; (08) p. 5.   See also Jeffrey and Cole (16).

vertically than in the rays of a Conifer. Attention has already been
called to the difficulty of distinguishing between the foliage-shoots
of some Conifers, Dicotyledons and Lycopodiaceous plants. The

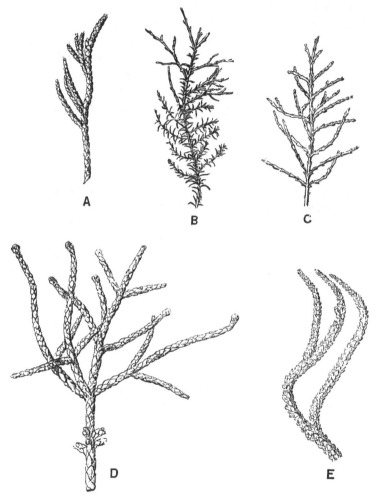

FIG. 711. A, *Veronica Hectori*; B, *Callitris calcarata*; C, *Veronica cupressoides*;
D, *Athrotaxis cupressoides*; E, *Crassula lycopodioides*. M. S.

twigs reproduced in fig. 711 are examples of misleading resemblances,
and similar instances are cited by Bommer. The Conifer *Podocarpus
dacrydioides* was described by Banks and Solander as *Lycopodium*

11—2

arboreum; *Dacrydium Bidwillii*[1], *D. Franklini* and *D. araucarioides* (fig. 708, B) recall some species of *Lycopodium*, and *D. cupressinum*[2], a dimorphic species, may simulate *Lycopodium tetragonum*. The long, linear, distichous leaves of some species of *Podocarpus* and *Cephalotaxus* might, as fossils, be confused with the pinnate leaves of *Cycas*; further, as Bommer points out, the leaves of *Podocarpus Blumei* resemble those of *Agathis* and the seeds are similar to those of *Dehaasia media* (Lauraceae). The *Cupressus* type of shoot occurs in *Baccharis scolopendra* as in other Dicotyledonous plants (fig. 711). The presence of a midrib in a linear *Taxus*-like leaf though usually easy to recognise is not always obvious, *e.g.*, in *Torreya nucifera* the midrib cannot always be distinguished on the upper face of the leaves. The recurrence of a similar habit in many Conifers renders difficult the identification of vegetative shoots, particularly as in fossil specimens the precise method of attachment of the leaves, their texture, and other features are frequently unrecognisable. The tendency to dimorphism in many genera is another difficulty: examples of dimorphic shoots are afforded by *Dacrydium laxifolium, D. Kirkii, D. elatium* (fig. 708, A), *Juniperus chinensis, Araucaria excelsa, Callitris glauca, Podocarpus cupressina*[3], *P. imbricata, Thuya occidentalis*, etc. Allusion has been made to the considerable variation in the length of leaves on a single branch of different Conifers: in such a form as *Cryptomeria japonica* var. *spiralis* Sieb. and in similar varieties of other genera the spirally twisted leaves, reminding one on a small scale of the 'wind-blown' Acanthus leaves on a Byzantine capital, constitute a feature which might be regarded as of taxonomic importance. The investigation of the cuticular membranes of Conifer leaves, as yet but little attempted, may supply useful criteria as in the case of Cycadean fronds.

[1] Pilger (03) fig. 4 A.   [2] Kirk (89) Pls. xviii., xix.
[3] Bennett and Brown (52) Pl. x

CHAPTER XLIV.

CONIFERALES (FOSSIL).

THE task of deciphering the fragmentary remains of Conifers is particularly difficult and no branch of palaeobotanical research makes greater demands upon the patience and self-control of the student. As Saporta says, 'Aucune étude n'ouvre des perspectives plus étendues, mais aucune aussi n'exige plus de réserve et de tâtonnements[1].' The determination of impressions of ill-preserved vegetative shoots is often impossible and it is regrettable that many authors have been too ready to employ generic names denoting identity or close relationship with recent types on wholly inadequate grounds. A recent writer thus sums up the situation created by an excessive faith in superficial resemblances and a lack of familiarity with existing representatives of the group:— 'Where a knowledge of reproductive parts is lacking, chaos reigns supreme.' It may be added that impressions or casts of cones in many cases do not afford any real assistance. A comparison of the various forms of foliage-shoots and strobili met with among recent Conifers demonstrates the danger of placing confidence in external resemblance as a guide to affinity. It is seldom that reproductive organs are well enough preserved to enable us to recognise features of primary systematic value. Though little has so far been done to test the value of epidermal characters as aids to identification, such results as have been obtained[2] favour the conclusion that this line of investigation promises to be less fruitful for Conifers than for Cycadean plants. Petrified wood of the Conifer type is abundant in plant-bearing strata from the later Palaeozoic rocks upwards, and considerable pains have been taken to utilise to the full this source of information. Within wide limits anatomical characters are undoubtedly valuable, but the recent tendency to subdivide comprehensive genera, which are

---

[1] Saporta (62) p. 309.   [2] Holden, R. (15[2]).

recognised as embracing several recent genera, into genera implying
a limitation of affinity within narrower bounds has, I venture to
think, been carried too far. The investigation of fossil coniferous
wood, in spite of the disappointing quality of the data from the
systematist's point of view is well worth attention. An examination of fossil wood from different geological horizons brings to
light many striking instances of a mixture in single plants of
features now characteristic of distinct genera. It is the generalised
forms that throw light on the nature of the changes produced in
anatomical structure in the course of evolution. The older and
more generalised types are of special importance to the student of
phylogeny. The very difficult question as to the stock from which
the Conifers are derived is too wide to be adequately discussed in
a general treatise. It is probable that the Coniferales are mono-
phyletic, the Araucarineae being the oldest representatives of the
group while the Podocarpineae are a closely related series. The
widely held view that the Araucarineae are descended from Cordaitalean ancestors is by no means definitely established; it rests
mainly on anatomical evidence and the arguments based on a
comparison of the reproductive shoots are far from convincing.
On the other hand those who favour a Lycopodiaceous ancestry
for the Coniferales are confronted with difficulties which, though
I venture to think they are not insurmountable, have not been
adequately met[1]. The suggested linking up of the Cordaitales,
through types in which the cylinder of secondary xylem is sup-
plemented by separate primary strands of vascular tissue, with
Lyginopteris and other Pteridosperms leads to the inclusion of the
Coniferales among the descendants of an ancient Filicinean stock,
but here too the chain of evidence is incomplete particularly as
regards the lack of data as to the nature of the reproductive organs
of several Palaeozoic genera founded on anatomical characters.

   The problem is still unsolved: the discovery of additional
types and a more thorough comparative study of such data as we
possess may enable us to see more clearly the paths along which
evolutionary tendencies have operated, but the absence of records
of the vegetation of pre-Devonian times deprives us of the means
of following to their common source the different phyla of vascular

---

[1] For a useful summary of arguments see Burlingame (15).

plants which in the Permo-Carboniferous era had already advanced far beyond the simple ancestral forms which the botanist seeks but rarely finds.

The various examples of fossil genera founded on the anatomical features of vegetative organs are dealt with in a separate section and not included with impressions in the descriptions of the several families, partly on the ground that it is rarely possible to demonstrate a connexion between the two sets of records and in part with a view to give a more connected account of the results so far derived from a study of petrified wood. Cross-references to anatomical structure are given in the descriptions of vegetative and reproductive organs when there appear to be sound reasons for assuming a generic or family connexion. The classification of woods is at best provisional and the generic characters are far from constant. The main point is that the student cannot afford to neglect this line of enquiry if he desires to obtain a comprehensive view of the changing combinations of structural features preceding their distribution among existing genera.

A comparison of recent Conifers and Cycads with their Mesozoic representatives brings out very clearly the fact that while on the one hand the modern Cycads differ widely from the Cycadean type which played a prominent part in Mesozoic floras, recent Conifers on the other hand agree closely in their main features with their Mesozoic ancestors. The Cycads as we know them now are a more recent product of evolution than the Conifers though it by no means follows that the Conifers in the wide sense are the more ancient group.

FOSSIL GYMNOSPERMOUS WOOD (Coniferales).

The earliest attempts to identify petrified wood are summarised by Goeppert[1], Knowlton[2], and other authors. Luidius (Lhwyd)[3] at the end of the seventeenth century employed the general designation *Lithoxylon*, and the termination *-xylon* is still used in generic names applied to fossil wood in conjunction with some prefix implying agreement in the more important anatomical features with some recent genus or family. For woods exhibiting a combination of characters unknown in existing genera a distinctive

[1] Goeppert (50).　　[2] Knowlton (89[2]); Gothan (05).　　[3] Luidius (1699) A.

prefix is employed, *e.g.*, *Xenoxylon,* and some authors make use of a name, *e.g.*, *Woodworthia*, which does not indicate that the diagnosis is based on anatomical features. Conwentz[1] adopted the method of adding the prefix *Rhizo-* to generic terms for wood believed to belong to roots, and Felix[2] and Lignier[3] have employed the prefixes *Cormo-* and *Clado-* for stem- and branch-wood respectively. It is, however, seldom that such differentiation is possible and it is questionable whether it is wise to attempt refinements of this kind. Barber[4], in his critical paper on a species of *Cupressinoxylon*, calls attention to Strasburger's description of an old moribund stem of *Larix* with root-like characters and Gothan[5] speaks of a branch of *Pinus silvestris* with root-attributes. The differences between branches and the main stem are not sufficiently known even in the more familiar types to justify the use of the prefixes *Cormo-* and *Clado-* in descriptions of fossil specimens.

The scientific study of fossil wood began with Nicol[6] and Witham whose work was rendered possible by methods of section-cutting first employed, according to Nicol, by a Mr Sanderson, a lapidary. Opinions expressed by Nicol on methods of investigating petrified wood are still pertinent after a lapse of 80 years:— 'To pronounce with certainty whether a fossil Conifer be essentially different from any known individual of the recent kind, it would be requisite to have a thorough knowledge of the structure at least of all the different tribes of recent Coniferae; and yet several distinct fossil genera have been indicated by a person who has examined, and that too very superficially, only three slices of three recent Pines, differing not essentially from one another.' In recent years the tendency has been towards a more detailed study of anatomical characters such as the distribution and form of the pits on medullary-ray cells. The facts recorded in the Chapter on Recent Conifers illustrate the difficulty of arriving at a thoroughly satisfactory classification of anatomical features that may serve as criteria in the identification of recent genera: even in the case of well-preserved fossil wood we have as a rule to rest content with a generic name denoting a combination of characters met with in more than one existing genus. Moreover, as already

[1] Conwentz (80) A.   [2] Felix (82).   [3] Lignier (07²).
[4] Barber (98).   [5] Gothan (05) p. 19.   [6] Nicol (34) A. p. 141.

shown, recent work has tended to reduce the taxonomic value of
certain characters such as the occurrence of ray-tracheids which
are more widely distributed than has generally been supposed.
In this connexion a word may be added with regard to some
common sources of error in anatomical investigation. There is
the obvious danger of confusion between features due to petrifying
agents or to decay before petrification and those present in the
living tree: the thickening of cell-walls, *e.g.*, those of medullary
rays, has been shown in some cases to be a pathological phenomenon[1]. The partial obliteration of bordered pits by decay may
cause them to appear separate though originally in contact (fig.
475, B, Vol. III. p. 257). The recognition of pits on the tangential
and horizontal walls of medullary-ray cells is often very difficult, and
negative evidence may be misleading. It is by no means always
a simple matter to distinguish between true canals and canal-like
spaces formed by the destruction of groups of tracheids (*e.g.*,
Pityoxylon eiggense; fig. 725). In one case it has been shown
that leaf-traces traversing broad medullary rays were mistaken
for horizontal resin-canals[2]. The spiral lines frequently seen on
the walls of petrified tracheids caused by the directive influence on
the structure of the membrane of the course of enzyme-action may
simulate the spiral bands characteristic of *Taxus, Torreya*, and
Cephalotaxus. These are a few of the pitfalls in the path of the
palaeobotanist, but despite the difficulties and the frequency with
which imperfect preservation prohibits complete diagnosis, the
investigation of fossil wood is well worth the attention of students
equipped with an intimate knowledge of recent Conifers. The
unpromising nature of the material may be a deterrent, though
lignitic and other specimens not thoroughly petrified are amenable
to special treatment[3].

In the account of recent Conifers attention is called to the significance of rings of growth: the subject has recently been exhaus-
tively treated by Antevs[4] and students should consult his memoir
in the *Progressus rei botanicae* for references to the literature. The

---

[1] Gothan (07), p. 25.
[2] Penhallow (00) p. 76; Thomson and Allin (12).
[3] For methods, see Jeffrey and Chrysler (06); Hollick and Jeffrey (09) B.
Gothan (09); Sinnott (09).
[4] Antevs (17).

subject is interesting and beset with difficulties but well worthy of more thorough treatment than it has so far received. Though petrified Coniferous stems are usually represented by the secondary wood only, the phloem and cortical tissues are sometimes preserved and afford useful information. Examples of petrified phloem and other extra-xylem tissues are described by Lignier and other authors. In his description of silicified plants from Franz Josef Land, of Lower Cretaceous or Upper Jurassic age, Solms-Laubach[1] includes some pieces of Coniferous bark showing patches of periderm alternating with secondary phloem consisting of sieve-tubes, phloem-parenchyma, and fibres, also some stone-cells. Some of the sieve-tubes are shown in fig. 718, B, with well-preserved sieve-plates, a feature very rarely preserved. There is not enough wood associated with the phloem and periderm to serve as a means of identification, but Solms-Laubach speaks of the bordered pits on the tracheids and the pits of the medullary rays as indicating *Pityoxylon* or *Cedroxylon*.

A new generic name *Vectia* has been instituted by Dr Marie Stopes[2] for a mass of petrified phloem which she compares more especially with the phloem of recent Conifers: while recognising that the specimen cannot be assigned with confidence to a particular group of Gymnosperms I venture to think it is almost certainly a portion of a Cycadean stem.

Reference has already been made in the section devoted to the anatomy of recent Conifers to the relative importance and constancy of different characters from a taxonomic point of view and this question need not be further considered. The method of classifying coniferous wood in general use is based on a scheme proposed by Kraus[3] A modified form of this scheme was published by Schenk[4] and more recently Penhallow[5], Jeffrey, Lignier, Gothan, Dr Stopes, and other authors have considerably extended our knowledge. Dr Gothan[6], whose memoir on the anatomy of Conifers contains much valuable information, employs several generic names denoting identity with recent types, and while admitting the great advance made by him and other workers in

[1] Solms-Laubach (04).   [2] Stopes (15) p. 247. See p. 419, Vol. III.
[3] Kraus in Schimper (72) A. p. 363.
[4] Schimper and Schenk (90) A. p. 860
[5] Penhallow (07).   [6] Gothan (05).

this field, it is difficult to avoid a suspicion of overstraining the significance of certain anatomical minutiae beyond the limits of safety.

The great abundance of petrified wood in strata ranging from the late Palaeozoic through the Mesozoic and Tertiary formations,

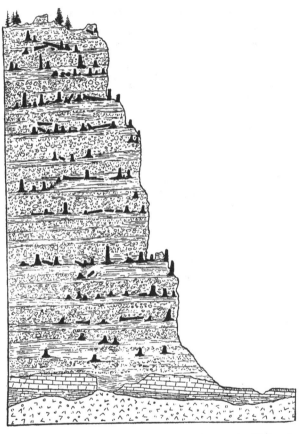

FIG. 712.　Section of the north face of Amethyst Mountain, Yellowstone Park.
(After Holmes.)

often in places which have yielded few other plant fossils, is in itself a strong incentive to research in a department of palaeobotany that has suffered from superficial work and hasty conclusions. Petrified tree-trunks are among the most impressive monuments

of former ages: the petrified forests in the desert east of Cairo[1];
the vast accumulation of Triassic stems, some reaching a length of
200 ft. and 7—10 ft. in diameter, over an area of 10 square miles
in Arizona County[2]; the succession of Tertiary forests in 2000 ft.
of volcanic sediment exposed on the sides of Amethyst Mountain
in the Yellowstone Park[3] (fig. 712); and on a smaller scale the
Jurassic trees in the Portland quarries are a few of many striking
examples of the wealth of material.

CLASSIFICATION OF FOSSIL CONIFEROUS WOOD.

A.   *Araucarian pitting on the tracheids.*

I. **DADOXYLON.**   (Including *Araucarioxylon* of Authors.)

Bordered pits on the radial walls of the tracheids, if uniseriate
flattened above and below, when in two or more rows alternate
and polygonal; separate and circular pits, though rare, may occur.
Rims of Sanio usually absent though their occasional occurrence
on the secondary tracheids in the cone-axis of recent species of
the Araucarineae shows that they are not entirely foreign to wood
of the Araucarian type.

Xylem-parenchyma absent or rare and may be represented by
resiniferous tracheids.   Medullary rays uniseriate, rarely double;
horizontal and tangential walls smooth; there may be 1—15
small pits in the field, though whether they actually belong to the
walls of the ray cells or to the adjacent tracheids has not been
definitely determined in fossil species.   The pits are simple or
bordered, circular or elliptical.

Resin-canals are absent both from normal and wounded wood.
Palaeozoic to Recent.

B.   *Pitting on the radial walls of the tracheids of the common
Coniferous type ; the pits are separate and circular and, if in two
or more rows, opposite.   Contiguous and more or less flattened pits
occur spasmodically on the tracheids of the wood of the genera included
in this section.   Well-developed rims of Sanio usually occur on the
tracheids.*

[1] Unger (59).
[2] A good example of an Arizona tree-trunk is exhibited in the Plant-Gallery of
the British Museum.   See Ward (00[3]).
[3] Knowlton (99); Holmes (78); Seward (11[3]) p. 60, fig. 6.

II. CUPRESSINOXYLON. (Including *Glyptostroboxylon* and *Taxodioxylon*.)

Xylem-parenchyma scattered through the wood and not confined to any particular region, often containing resin; the transverse walls may be thick and pitted. Medullary rays uniseriate, horizontal and tangential walls smooth (unpitted); there are generally several small pits in the field though in some species referred to this genus there may be a single pit. In the region of the spring-wood the pore of the apparently bordered ray-pits is more or less horizontal; but the form and position of the pore are variable.

Resin-canals absent except in wounded parts of the wood.

Jurassic to Recent.

III. TAXOXYLON.

The same anatomical features as in *Cupressinoxylon* except that the tracheids of the secondary xylem have spiral thickening bands.   Tertiary to Recent.

IV. MESEMBRIOXYLON. Gen. nov. (Including *Podocarpoxylon*, *Phyllocladoxylon* and *Paraphyllocladoxylon*.)

Xylem-parenchyma usually present and scattered, but it is not so characteristic a feature as in *Cupressinoxylon*.

Medullary rays usually uniseriate; the pitting is confined to the radial walls as in *Cupressinoxylon* but in the region of the spring-wood the pore is oblique or more or less vertical; in the summer-wood the pits in the field are indistinguishable from those of *Cupressinoxylon*. There are often several pits in the field but in some species there may be one or two large simple pits in the field.

Resin-canals present only in wounded parts of the wood.

Jurassic to Recent.

V. PARACEDROXYLON.

A genus of doubtful affinity. Bordered pits on the tracheids usually separate; no rims of Sanio.

Xylem-parenchyma confined to wounded regions. Medullary-ray cells pitted only on the radial walls except in the injured parts of the wood where the other walls may be pitted. There are 4—6 circular, apparently bordered, pits with an oblique pore in the field. In the characters of the normal wood this genus agrees most closely with *Cupressinoxylon* and *Mesembrioxylon*.

Cretaceous.

VI. CEDROXYLON.

In some species included in this genus contiguous and flattened bordered pits are fairly common on the radial walls of the tracheids.

Xylem-parenchyma, if present, confined to the late summer-wood.

Medullary rays uniseriate; all the walls are pitted; tracheids may be present in the rays though they are usually absent. There are 1—6 or rarely more pits in the field, either simple or apparently bordered. Resin-canals confined to wounded regions.

Jurassic (Triassic species doubtful) to Recent.

VII. PITYOXYLON. (Including *Piceoxylon* and *Pinuxylon*.)

Though in the great majority of cases the tracheids of the secondary xylem have no spiral bands, the presence of such bands in the recent genus *Pseudotsuga* shows that this feature may occur in wood of the *Pityoxylon* type.

Resin-canals present in the normal wood. Medullary rays of two kinds, uniseriate rays consisting of parenchyma and, in many cases, ray-tracheids, also fusiform rays with horizontal resin-canals. The walls of the ray-tracheids are either smooth or irregularly dentate. All the walls of the medullary-ray cells are pitted; there may be one large simple pit in the field or several small, apparently bordered, pits.

Jurassic to Recent.

VIII. PROTOPICEOXYLON.

Similar to *Pityoxylon* except in the absence of horizontal resin-canals in the normal wood; such horizontal canals as occur are regarded as traumatic.

Cretaceous to Tertiary.

C. *Genera in which Araucarian features, especially as regards the tracheal pitting, occur in association with characters met with in recent Abietineae. The genera included in this section afford examples of generalised types and do not resemble recent forms so closely as do the other genera.*

IX. WOODWORTHIA.

Araucarian tracheal pitting; annual rings feebly marked. Short shoots and a subtending leaf are present in the secondary wood. Resin-canals absent.

Medullary rays uniseriate; pits confined to the radial walls.

Triassic.

X. ARAUCARIOPITYS.

Similar to *Woodworthia* in the possession of short shoots and in the pitting of the tracheids though separate and circular pits also occur. Vertical resin-canals abundant in wounded regions.

Medullary rays uniseriate; all the walls pitted.

Cretaceous.

XI. PROTOCEDROXYLON. (Including *Metacedroxylon*.)

Bordered pits on the radial walls of the tracheids in 1—3 rows, usually of the Araucarian type but separate pits also occur. No rims of Sanio and no resin-canals.

Xylem-parenchyma usually absent. Medullary rays generally uniseriate; all the walls pitted; 1—3 circular, simple, pits in the field.

Jurassic.

XII. XENOXYLON.

Tracheal pits large, generally flattened above and below though not always, often transversely elongated. Resin-canals absent.

Medullary rays uniseriate; pitting confined to the radial walls; usually one large simple pit in the field.

Jurassic (Triassic?).

XIII. ANOMALOXYLON.

When uniseriate the pits on the tracheids are usually contiguous and flattened; if biseriate the pits are opposite; separate pits also occur.

True resin-canals absent, but canal-like spaces lined with small cells occur in some large medullary rays and constitute a characteristic featuie.

Medullary rays uniseriate; pits confined to the radial walls; 2—3 circular simple pits in the field.

Jurassic.

XIV. THYLLOXYLON.

Tracheal pitting partially Araucarian.

Xylem-parenchyma at the end of a year's growth. No. true resin-canals, but the central parenchyma of some of the broader rays is replaced by a canal-like space often filled with tyloses.

Medullary rays uniseriate; all the walls pitted.

Jurassic.

XV. PLANOXYLON.

Wood with well-marked annual rings, resin-canals usually absent; tracheids with 1—3 rows of alternate hexagonal bordered pits on the radial walls and in the late wood there may be a single row of separate pits. Xylem-parenchyma occurs only between the spring elements and the last-formed tracheids of the previous year. Medullary rays almost entirely uniseriate; all the walls pitted.

Lias to Cretaceous.

I. DADOXYLON [and *Araucarioxylon*].

The anatomical characters implied by the expression 'wood of the Araucarian type' are enumerated in the Chapter on Recent Conifers, and in Chapter XXXIII. reference is made to the impossibility of drawing a clear line of division between the wood of Araucarian plants and that of certain members of the Cordaitales[1]. The name *Araucarites* was used by Goeppert[2] for fossil wood of the Araucarian type, but the previous use of this name by Presl for impressions of foliage-shoots and cones renders unsuitable its application to wood apart from the fact that *Araucarites* at once commits an author to a determination implying an affinity which in many cases cannot be demonstrated. Endlicher's non-committal genus *Dadoxylon*[3] has been widely used, especially for Palaeozoic wood having the characters of *Araucaria* or *Cordaites*. This designation leaves open the question of precise systematic position. In 1882 Kraus[4] instituted the genus *Araucarioxylon*, a name which has been widely adopted for fossil wood both from Palaeozoic and later formations. The practice of limiting *Dadoxylon* to Palaeozoic species and reserving *Araucarioxylon* for Mesozoic and Tertiary

---

[1] See p. 248, Vol. III.   [2] Goeppert (45).
[3] Endlicher (47).   [4] Kraus in Schimper (72) A. p. 370.

wood has already been criticised[1]: it is pointed out that the application of an age-test is scientifically unsound and cannot fail to be misleading. Although it is probably true that species of *Dadoxylon* from strata later than the Rhaetic series are in the majority of cases Araucarian, we have no adequate grounds for definitely naming such types *Araucarioxylon* in the sense of membership of the Araucarineae. Similarly some Palaeozoic species of *Dadoxylon* may well be more closely allied to the Araucarineae than to *Cordaites*: *Dadoxylon valdajolense* (Moug.)[2] and *D. Rhodeanum* (Goepp.)[3] have both been referred to *Walchia*: the latter species is included by Tuzson in the genus *Ullmannites*. We do not know the lower geological limit of the Araucarineae, nor do we know when the Cordaitales became extinct. Tuzson[4] subdivides wood of the *Dadoxylon* type into several genera including *Pycnophyllum,* to which is referred *Dadoxylon Brandlingii* (Lind. and Hutt.), *Ullmannites* applied to the type recently named *Eristophyton Beinertianum* by Zalessky. and previously included by Scott in *Calamopitys,* also to *Dadoxylon Rhodeanum* and *D. saxonicum, Pagiophyllites,* including *P. keuperianus* (Goepp.), and for Tertiary species the term *Araucarites* is used. This nomenclature, based partly on age and in part on a supposed connexion between the wood and foliage-shoots, is opposed to sound principles and in some cases is at variance with the true character of the species. When evidence is available in support of a reference either to the Araucarineae or to the Cordaitales the qualifying terms *Cordaioxylon* or *Araucarioxylon* may be added after *Dadoxylon.* Such a species as *Dadoxylon permicum* Merck.[5] is one of many examples of a *Dadoxylon* that cannot be more precisely identified. *Dadoxylon australe* Crié[6] from Triassic rocks in New Caledonia must be included in the same category. This species is founded on wood showing well-marked rings of growth; there are two or more rows of alternate polygonal pits on the radial walls of the tracheids and the medullary rays are uniseriate and 3—15 cells in depth. Crié's species should not be confused with

---

[1] See p. 249, Vol. III. See also Gothan (05) p. 14; Potonié (02) p. 229.
[2] Fliche (03); Mougeot (52) A.
[3] Potonié (99) B. p. 294; Gothan (07) p. 17; Tuzson (09) p. 34.
[4] Tuzson (09) p. 17.   [5] Mercklin (55) p. 53.
[6] Crié (89) p. 5, Pls. I., III., V.

Dadoxylon australe Arber[1] based on Palaeozoic wood from
Australia: the substitution of *Dadoxylon* for *Araucarioxylon*, the
name used by Crié for his New Caledonian species, necessitates
a new specific designation for Arber's type, which it is proposed
to rename *Dadoxylon Arberi*. From Liassic beds in Yorkshire
Miss Holden[2] has described a species as *Araucarioxylon* sp. which
she suggests may be one of the oldest representatives of the Araucarineae or perhaps a Jurassic example of *Cordaites*: the latter
identification is supported by a reference to the recorded occurrence
by Lignier of an *Artisia*[3] in French Jurassic beds, a test of affinity
that cannot be accepted as satisfactory.

There remains for consideration the debated question as to the
value to be attached to the occurrence of contiguous and flattened
pits as an index of Araucarian affinity when this feature is associated
with a type of medullary-ray pitting foreign to *Dadoxylon*. In
Gothan's genus *Xenoxylon*[4] the tracheids have usually large
flattened pits, but the pits on the radial walls of the medullary-
ray cells are very different from those characteristic of Araucarineous wood. It is, moreover, not uncommon to find instances of
contiguous and alternate pits on the tracheids of a stem in which
the more usual type is the Abietineous arrangement. Gothan[5]
lays greater stress on the nature of the pitting on the walls of
medullary cells, but Jeffrey[6] has discovered typical Abietineous
ray cells in the cone-axis of *Araucaria* and *Agathis*. Miss Holden[7]
goes so far as to maintain that the only feature which holds
absolutely is provided by the rims of Sanio: these are invariably
absent in Conifers with Araucarian affinities except on the first
few tracheids of the cone-axis of *Araucaria* and *Agathis*. This
author records as *Araucarioxylon* sp.[8] a wood (described on a later
page) from New Jersey possessing opposite pits on some of the
tracheids, also rims of Sanio. We cannot lay down any definite
rules with regard to the sporadic variation in tracheal or medullary
pitting or as to the relative value to be assigned to one or other
character. The statement by Thomson[9] that the ray cells of such

[1] Arber (05) B. p. 191, text-figs. 40—43.
[2] Holden, R. (13²) p. 540, Pl. XL. fig. 28.   [3] See p. 248, Vol. III.
[4] See page 248, Vol. III.   [5] Gothan (05).   [6] Jeffrey (12).
[7] Holden (14) (13²) p. 544. See also Sifton (15).
[8] Holden (14) p. 171.   [9] Thomson (13).

Araucarias as he examined have no pits even on the radial walls leads one to suspect the accuracy of some of the many recorded instances of fossil Araucarian wood with medullary-ray pits; but this does not affect the value to be attached to the presence or absence of well-defined pits on the vertical or horizontal walls of medullary-ray cells. It is generally true that in the ray cells of Abietineae these walls are pitted while in the Araucarineae they are unpitted, and it is equally true that the predominance of alternate and contiguous pits on the tracheids is evidence of Araucarian affinity. Though generally absent from Araucarian wood xylem-parenchyma occasionally occurs in stems otherwise identical with the usual Araucarian type. The genus *Araucariopsis*[1] was instituted by Caspary for specimens distinguished from most examples of *Dadoxylon* by the presence of scattered xylem-parenchyma but, as Gothan[2] points out, a distinctive name is superfluous; the type-species of *Araucariopsis, A. macractis*, should be transferred to *Dadoxylon*; in the possession of xylem-parenchyma it agrees with *Dadoxylon septentrionale* Goth. from Spitzbergen. The importance attached by Jeffrey and other American authors to the presence of Sanio's rims on the tracheal walls is, I venture to think, greatly overestimated. The determination of fossil wood is to a large extent a question of relative values. There is clear evidence, and it would be surprising were it not so, that in several extinct types there are combinations of character pointing to less sharply defined boundaries than those which delimit existing families and genera. It is in the conclusions drawn from generalised types that authors differ. An outstanding fact is the predominance in Palaeozoic stems of the Araucarian form of tracheal pitting which is unquestionably a much older type than that characteristic of the Abietineae. The following definition is based on specimens agreeing in the sum of their characters with recent Araucarineae, but there are various genera described by Jeffrey, Miss Holden, and other authors and believed by them to be more or less closely allied to the Araucarineae which are not provided for in the definition. These genera are treated separately as generalised types and the decision as to the nature of the evidence

[1] Caspary and Triebel (89) p. 81. Pls. XIV., XV.
[2] Gothan (10) p. 9.

they afford with regard to the phylogeny of the Araucarineae and
the Abietineae must be left to the individual student.

Annual rings occasionally well marked but not infrequently
absent or indistinct. Tracheids with uniseriate and more or less
flattened bordered pits or with two or more rows of alternate
polygonal pits on the radial walls. The alternate disposition,
even if unaccompanied by flattening and the polygonal contour
of the pits, is a *Dadoxylon* feature if it is the dominant arrangement
and not a sporadic occurrence. Bordered pits occasionally occur
on the tangential walls but they are smaller and comparatively
rare. Rims of Sanio absent. Xylem-parenchyma usually absent
or feebly developed; resiniferous tracheids occasionally present.
Medullary rays uniseriate, very rarely double, homogeneous, 1—30
or as many as 50 cells deep; walls comparatively thin and without
pits on the horizontal and vertical walls; the radial walls may show
1—15 small pits, the oblique pore being occasionally enclosed in a
feebly developed border. In view of the entire absence of pits on
the ray cells of at least some recent Araucarias the structure of the
ray cells in fossil stems requires careful revision[1].

In the following brief descriptions of species of *Dadoxylon* a few
examples are chosen to illustrate the wide geological and geographical range of fossil wood of this type, but it must be remembered that in many cases no positive statement can be made with
regard to the nature of the parent-plant beyond the facts afforded
by the anatomical characters of the stem. Evidence bearing on
the geological age of the Araucarineae is discussed in the course
of the description of genera founded on vegetative shoots and
reproductive organs. Species of *Dadoxylon* from Carboniferous
and Permian strata have already been described in Chapter XXXIII.
as more probably referable to *Cordaites* or at least to the Cordaitales, and it is not by any means impossible that some of the
Dadoxylons recorded from Triassic or even higher strata may
belong to Cordaitalean species rather than to the Araucarineae.
The evidence afforded by petrified wood in conjunction with that
derived from vegetative remains lends probability to the view that
Araucarian plants existed at least as early as the later Palaeozoic

---

[1] See, in addition to Kraus and other authbrities, Lignier (07[2]).

age. The lack of satisfactory knowledge with regard to the morphology of the reproductive organs of such genera as *Walchia*, *Voltzia*, and other plants closely resembling living Araucarias in the habit of their foliage-shoots precludes any definite statement as to the precise degree of relationship between these and other types and existing Araucarineae, though it is certain that the Araucarineae were at least foreshadowed before the close of the Palaeozoic era.

Dadoxylon keuperianum (Goepp.).

This species[1], from Franconia and Würtemberg, is considered by Schimper[2] on the ground of association to be the wood of a *Voltzia*, and Tuzson[3] adopts the generic name *Pagiophyllites* implying relationship with *Pagiophyllum*; he includes in *D. keuperianum* the species *Araucarioxylon würtembergicum* Kr. and *A. thuringicum* Born[4]. The tracheids of *D. keuperianum* have one or more rows of contiguous and more or less flattened pits; the medullary rays are uniseriate and 2—50 cells in depth; Tuzson figures 2—4 circular simple pits in the field. Though possibly belonging to *Voltzia* or *Pagiophyllum* this wood is best retained in *Dadoxylon*. Other Triassic Dadoxylons are described by Wherry[5] from Pennsylvania: he records *Araucarioxylon virginianum*, a species described by Knowlton[6] from Potomac beds, and *A. vanartsdaleni*: in both forms the tracheal pits are compressed and alternate, the rings of growth indistinct, and the medullary-ray cells are said to have no pits. Reference has already been made to a Triassic species from New Caledonia, *D. australe* (Crié).

Dadoxylon septentrionale Gothan.

This species[7], founded on material believed by Gothan to be Triassic in age, has the following characters:—Annual rings often distinct macroscopically but microscopically showing little contrast between spring- and summer-wood; bordered pits in a single row and separate or polygonal and in two alternate rows; they are often arranged in stellate clusters as in some species of *Cedroxylon*.

[1] Goeppert (81) p. 42.
[2] Schimper (72) A. p. 384.
[3] Tuzson (09), p. 30, fig. 5.
[4] Schimper (72) A. p. 384.
[5] Wherry (12).
[6] Knowlton (89²), Pl. VII. figs. 2—5.
[7] Gothan (10) p. 8, Pl. I. figs. 4—8, Pl. II. fig. 1.

Medullary rays uniseriate, reaching a depth of 30 cells, 2—4 elliptical oblique pits in the field; wood-parenchyma, often with dark contents, is not uncommon. Gothan emphasises the abundance of xylem-parenchyma as a character in which this species differs from typical Dadoxylons.

Dadoxylon mahajambjense (Fliche).

Fliche[1] described this species, from Liassic strata in Madagascar, under the name *Araucarioxylon*: the radial walls of the tracheids have two rows of contiguous and alternate pits; the medullary rays are uniseriate, usually 8—16 cells deep, and small circular pits occur on the radial walls.

Dadoxylon divescence (Lignier).

An Oxfordian species from Normandy characterised by leaf-traces larger than those of *Araucaria imbricata* and, as Lignier[2] states, suggesting leaves comparable in size with those of *Cordaites*. The tracheids have 1—4 rows of pits and the medullary rays are 8—11 cells in depth.

Dadoxylon argillicola (Eichwald), recorded from Moscow[3], is another example of a similar type of wood, and many other instances might be quoted in illustration of the wide distribution of vegetative organs in Jurassic beds agreeing anatomically with the Araucarineae.

Dadoxylon (Araucarioxylon) novae zeelandiae[4] (Stopes).

A Cretaceous species from Amuri Bluff, New Zealand[5], founded on a piece of decorticated stem 8 cm. in diameter preserved partly in silica and in part in a calcareous medium. The small pith, not more than 1 mm. in diameter, is imperfectly petrified: the tracheids of the secondary xylem, which shows well-marked annual rings, have biseriate, alternate, hexagonal, pits; there is no xylem-parenchyma. An interesting feature is the occurrence of tracheids on each side of the medullary rays with thicker walls and containing discs of resin: these resiniferous tracheids, similar to those described by Thomson[6] and other authors, afford particularly good

---

[1] Fliche (05) p. 350, Pl. x. fig. 1.

[2] Lignier (07²) p. 257, Pl. xvii. figs. 10—13.   [3] Eichwald (68) Pl. v. fig. 12.

[4] The pecific name is given by Dr Stopes as *novae zeelandii*.

[5] Stopes (14²) Pl. xx.   [6] Thomson, R. B. (13).

examples of this type of element. The uniseriate medullary rays are usually 3—4 cells deep and are described as having 5—6 bordered pits with oblique pores in the field. In view of Thomson's conclusion with regard to the absence of pits on the ray cells in recent Araucarineae and considering the form of the pits as shown in Dr Stopes' drawing it may be doubted whether the pits actually belong to the walls of the ray cells.

Dr Stopes comments on the scarcity of Araucarian remains recorded from New Zealand and adds a brief account of some Tertiary wood described by Ettingshausen[1] as *Araucaria Haasti*, a species founded on foliage-shoots which, without any adequate reason, it is surmised belong to the petrified wood. The latter is poorly preserved: Dr Stopes examined the type-specimen and found that the bordered pits on the tracheids are circular and not compressed; she expresses some doubt as to its Araucarian affinity, but renames the species *Araucarioxylon Ettingshauseni*. A specimen of wood from Amuri described by Ettingshausen as *Dammara Oweni*[2] appears to be undoubtedly Araucarian.

Dadoxylon sp. (Holden).

An interesting type of stem-wood has been described by Miss Holden[3] as *Araucarioxylon* sp. from the Cretaceous lignites of Cliffwood, New Jersey, which shows Araucarian characters in combination with certain anatomical features not usually associated with *Araucarioxylon*. The tracheids for the most part have alternate compressed pits; the medullary rays consist of thin-walled cells and there is no xylem-parenchyma. Near the inner edge of the wood the tracheids are characterised by bordered pits in opposite pairs and rims of Sanio occur between adjacent pairs. The specimen is said to supply the only missing link in the chain of evidence pointing to the derivation of the Araucarineae from the Abietineae. Opposite pits are figured by Miss Holden in tracheids from the cone-axis of *Araucaria Bidwillii* and, as stated elsewhere, Sanio's rims are not unknown in Araucarian wood. I have adopted the name *Dadoxylon* because the characters as a whole are consistent with that designation though it might be contended that a new name is desirable to indicate the occurrence of unusual

[1] Ettingshausen (87) p. 154, Pls. II. figs. 1, 2; VI. figs. 10—12.
[2] *Ibid.* p. 16, Pl. VI. figs. 13—15.   [3] Holden, R. (13²).

features.   The occurrence of opposite pits and rims of Sanio in the younger portion of the xylem is regarded as evidence in support of the view that the stem, while in the main Araucarian, exhibits features indicative of the origin of the Araucarineae from the Abietineae.   The presence of opposite pits in wood in which the normal arrangement is alternate is not surprising if it is admitted that the Coniferous pitting is derived from an earlier scalariform type.   Even in stems in which the alternate or opposite pitting is well established it is not very uncommon to find occasional departures from the normal pattern.   This Cretaceous stem is one of many generalised types, and the arguments based on the admixture of characters in favour of the greater antiquity of the Abietineae do not present any insuperable difficulty to the opposite view namely that the Araucarineae preceded the other families of the Coniferales.

Dadoxylon (Araucarioxylon) breveradiatum (Lignier).

This species from the Cenomanian of Normandy[1] affords one of the few examples of the preservation of phloem, cortex, and pith.

Lignier adopted the generic name *Arauca-riocaulon*: he describes the tracheids as having 1—3 rows of pits, crowded and alternate but not flattened.   The medullary rays are very short, 1—3 and rarely 4 cells in depth; there are said to be 8—15 pits in the field (fig. 713).   Resiniferous parenchyma is abundant as in the wood referred by Caspary to a special genus *Araucariopsis*[2] and in *Dadoxylon septentrionale* Goth.   The phloem includes well-preserved sieve-tubes and the

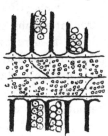

Fig. 713. *Dadoxylon (Ar-aucarioxylon)breveradiatum.* (After Lignier.)

cortex is characterised by numerous sclerites, an Araucarian feature.   This species appears to be an aberrant type the position of which is by no means clear.

Among other Cretaceous species are *Dadoxylon albianum*[3] (Fliche) from L'Aube; *Dadoxylon Dantzii* Pot.[4] from beds probably of Upper Cretaceous age in East Africa, without definite rings

---

[1] Lignier (07²) p. 290. Pl. XIX.   [2] See page 179.
[3] Fliche (97) p 8, figs. 2—4.   [4] Potonié (02) Pls. I., II.

of growth and with a single row of contiguous pits on the tracheid walls; *Dadoxylon virginianum* (Knowlton)[1] from the Potomac series; *Dadoxylon barremianum* (Fliche)[2] from the Lower Cretaceous of France; *Dadoxylon noveboracense* (Holl. and Jeff.)[3] from the Middle Cretaceous beds of Staten Island; *Dadoxylon Zuffardii* Negri[4] from middle Cretaceous rocks in the Gulf of Tripoli; *Dadoxylon tankoense* (Stopes and Fujii)[5] from Upper Cretaceous beds in Japan; *Dadoxylon madagascariense* (Fliche)[6] from Madagascar.

Dadoxylon (Araucarioxylon) kerguelense sp. nov.

In 1881 Goeppert[7] mentioned some wood received from Kerguelen Island from Baron von Schleinitz, probably of Tertiary age, under the name *Araucarites Schleinitzi et Hookeri*, but, as Gothan[8] points out, it is not clear whether he refers to one or two species, and as there are no figures or full description Goeppert's designation cannot stand.

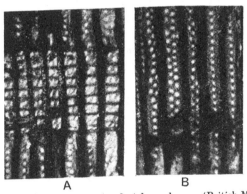

A   B

Fig. 714.   *Dadoxylon (Araucarioxylon) kerguelense.*   (British Museum.)

The sections on which this Kerguelen species is founded are in the British Museum[9]; they show the following characters:—

Annual rings narrow, often 15—20 tracheids broad, the summer wood being frequently represented by only two rows of elements. There are 1 or 2 rows of bordered pits on the radial walls of the

---

[1] Knowlton (89²) Pl. VII.
[3] Hollick and Jeffrey (09) B. p. 58, Pl. XXI.
[5] Stopes and Fujii (10) Pl. III. figs. 17, 18.
[6] Fliche (00²) p. 472, fig. 1.
[7] Goeppert (81).   [8] Gothan (08) p. 13.

[2] Fliche (00) p. 18.
[4] Negri (14).

[9] Sections V. 8388, V. 8390.

tracheids, contiguous, alternate, and often slightly flattened (fig. 714, B). The medullary rays have usually 5—8 elliptical pits in the field (fig. 714, A).

Dadoxylon (*Araucarioxylon*) *pseudoparenchymatosum* Gothan.

A species from Tertiary or possibly Upper Cretaceous rocks in Seymour Island (S. lat. 64° 16′)[1] agreeing closely with the wood of recent Araucarineae. The annual rings are distinct; there are 1—2 rows of pits on the tracheids 10—12 μ in diameter; the medullary rays, 2—10 cells deep, are usually uniseriate and there are several small oblique pits in the field. Cross-bars[2] (Müller's 'querbalken') like those described in *Araucaria brasiliensis* occur in some of the tracheids.

Dadoxylon Doeringii Conwentz[3] is a Patagonian species of Sub-Oligocene age characterised by distinct annual rings; rays up to 40 cells in depth, with 1—2 pits in the field. Among other Tertiary species are *Dadoxylon aegyptiacum* Unger[4], recorded from several localities in the Libyan desert; *Dadoxylon Robertianum* (Schenk)[5] of Tertiary or possibly Cretaceous age from the province of Nagpur, India; *Araucarioxylon koreanum* (Felix)[6] from Korea, characterised by the occurrence of a single row of continuous pits on the tracheids, is referred by Gothan to the genus *Xenoxylon* and regarded as identical with *X. latiporosum*[7].

II.　CUPRESSINOXYLON.　Goeppert.

The name *Cupressinoxylon*[8] or, as written by Kraus, *Cupressoxylon*[9], is usually applied to fossil wood exhibiting the following features:—Annual rings well defined, often narrow; vertical rows of parenchyma, often containing resin and recognisable by their dark contents even in transverse section (fig. 715, A), scattered through the spring- and summer-wood. Bordered pits on the tracheids usually separate and circular and if in more than one row opposite; medullary-ray cells generally characterised by the presence of several small pits in the field. Used in this sense

[1] Gothan (08) p. 10, Pl. I. figs. 12—16.
[2] See page 135.　　　　　　　　　　　[3] Conwentz (85) p. 16.
[4] Unger (59); Schenk (80) p. 3, Pls. I., II.
[5] Schenk (82²).　　　　　　　　　　　[6] Felix (87) p. 518.
[7] Gothan (10) p. 23.　　　　　　　　　[8] Goeppert (50) p. 196.
[9] Kraus in Schimper (72) A. p. 374.

Cupressinoxylon denotes wood similar to *Cedroxylon* except in the greater abundance of xylem-parenchyma and its occurrence in the spring- as well as in the summer-wood. The medullary rays afford another distinction which according to some authors is more trustworthy than the presence or distribution of the xylem-parenchyma. Gothan[1], who uses *Cupressinoxylon* in a more restricted sense, lays stress on the pitting of the medullary cells as a distinctive feature: the pitting is confined to the radial walls, or in other words there is no Abietineous pitting in *Cupressinoxylon* in the stricter sense. The medullary-ray pits have a broadly elliptical pore which is more or less horizontal at least in the spring-tracheids—the Cupressoid type in contrast to the Podo-carpoid type in which in the spring-wood the pore is narrower and more vertical, though in some Podocarps the bordered pits are replaced by large simple pits: in the summer-wood the difference between the Cupressoid and Podocarpoid type disappears. In *Cupressinoxylon* the medullary rays are uniseriate and not very deep though the depth is a variable character. Lignier[2] states that 60—150 rays occur in 1 square millimetre, another feature of doubtful value. In some species included in *Cupressinoxylon* the pitting of the tracheids is partly Araucarian as it also is in certain types of *Cedroxylon*. The presence of Sanio's rims, though not mentioned by many authors, is regarded by Jeffrey and some other American botanists as an important character to be expected in all Coniferous wood other than that of the Araucarineae. It is clear that unusually good preservation is essential for the recognition of such features of the medullary-ray cells as Gothan includes in his definition of the genus; unless the tissues are well preserved the generic separation of Coniferous types except within very wide limits is impossible. The name *Cupressinoxylon* may conveniently be restricted to wood having the usual type of tracheal pitting though pits of the Araucarian type may occur locally, with medullary rays in which the pitting is confined to the radial walls and generally with several fairly small and apparently more or less definitely bordered pits in the field, the pores in the spring-wood being elliptical and more or less horizontal. Resin-canals absent except in wounded regions

---

[1] Gothan (05) p. 39.   [2] Lignier (07²) p. 245.

of the wood; xylem-parenchyma normally abundant and not restricted to the end of the year's growth. The photographs

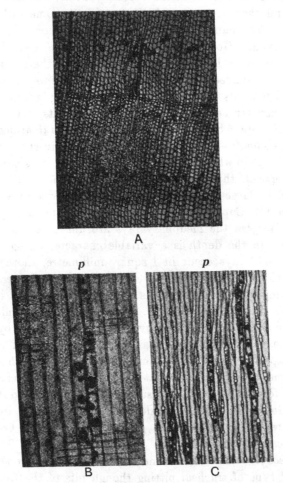

FIG. 715. *Cupressinoxylon* sp. Lough Neagh, Ireland; *p*, xylem-parenchyma.
(British Museum; A. V. 8248, B. V. 8253, C. V. 8257.)

reproduced in fig. 715 of sections of a piece of petrified wood from Lough Neagh[1] in Ireland, probably of Upper Tertiary age, illustrate

[1] See vol. I. p. 80.

very clearly the appearance of xylem- or resin-parenchyma as seen in transverse, radial and tangential section. The presence of dark contents renders these cells conspicuous in transverse section (A) and both contents and cross-walls are seen in the longitudinal sections (B, C). In fig. 715, B, the opposite and scattered bordered pits are shown on the radial walls of the tracheids with an occasional tendency to an alternate arrangement. Goeppert founded his species *Pinites Pritchardti*[1] on wood from Lough Neagh, the generic name being altered by Kraus to *Cupressinoxylon*[2].

Kraus[3] enumerates 46 recent species of Conifers with wood of the *Cupressinoxylon* type and Beust[4] mentions 37 species. As used by most authors *Cupressinoxylon* includes members of the Cupressineae (in the wide sense as used on page 124), Podocarpineae, *Abies Webbiana*, and some other recent genera. As already stated Gothan has essayed the difficult task of defining the genus in such a way as to restrict the wood so named to a smaller number of recent Conifers, recognising as distinct genera certain species previously included in *Cupressinoxylon*, e.g., *Taxodioxylon* (= *Cupressinoxylon Sequoianum* Merck.[5]), *Podocarpoxylon* (= *C. Hookeri* Arb.[6]), *Xenoxylon* (= *Cupressinoxylon Barberi* Sew.[7]): these and other species formerly included in *Cupressinoxylon* are distributed among genera not always well defined but instituted with a view to increase the value of fossil species considered in relationship to recent types.

[*Paracupressinoxylon*. Holden.]

Miss Holden[8] has proposed a new generic name, *Paracupressinoxylon* for wood which agrees with *Cupressinoxylon* in having xylem-parenchyma scattered through the year's growth but differing in the Abietineous pitting of the medullary-ray cells and in the absence of Sanio's rims. In *Paracupressinoxylon* are included two species, *P. cedroides* and *P. cupressoides*, both Jurassic and the latter also Cretaceous in age, which appear to differ too much from one another to be referred to one genus. Both species

[1] Goeppert (50) p. 220.   [2] Kraus in Schimper and Schenk (72) A. p. 376
[3] Kraus (64).   [4] Beust (85).
[5] Mercklin (55) Pl. xvii.   [6] Arber (04); Gothan (08) p. 7.
[7] Seward (04) B. p. 60, Pl. vii.
[8] Holden, R. (13²) p. 537; (14) p. 173.

are regarded by the author of the genus as undoubted representatives of the Araucarineae, a determination that is hardly consistent with the affinity implied by the generic name. The species
P. cupressoides, from Yorkshire and the Cretaceous lignites of
New Jersey, is characterised by the restriction of pits to the radial
wall of the medullary-ray cells, the absence of Sanio's rims, the
presence of scattered xylem-parenchyma, and by the occurrence
both of scattered and crowded pits on the tracheal walls. Moreover in this species the phloem shows an alternation of hard and
soft elements. The affinity suggested by these features would
seem to be to *Cupressinoxylon*. On the other hand, *Paracupressinoxylon cedroides*, founded on material from the Yorkshire coast,
is characterised by the Abietineous pitting of the medullary-ray
cells, an admixture of scattered and compressed bordered pits on
the radial walls of the tracheids, scattered xylem-parenchyma, no
alternation of hard and soft bast, and by the absence of any
sclerous cells in the pith. This species also illustrates the occurrence of resin-canals in wounded regions of the wood. The reference of both these species to the Araucarineae, chiefly because of
the absence of Sanio's rims, though consistent with the principle
that this character is all important, implies the neglect of other
characters, more especially the nature of the medullary-ray pitting,
which in the case of recent Conifers are unquestionably of taxonomic importance. The species *P. cedroides* should not, in my
opinion, be included with *P. cupressoides* in one genus; it is probably more closely allied to the Abietineae than to any other
family. It should, however, be remembered that pitting of the
horizontal and tangential walls of medullary-ray cells is a feature
that is not confined to the Abietineae; it occurs also in some recent
Junipers and the extinct genera *Protocedroxylon* and *Thylloxylon*.

Cupressinoxylon liasinum Lignier.

This Liassic species[1] from Orne, France, is founded on the wood
of a pentarch and hexarch root: the bordered pits on the radial
walls of the tracheids are usually in one row, occasionally in two
opposite series; smaller pits occur on the tangential walls; medullary rays 1—5 cells deep with a few ovoid-oblong pits, often simple

[1] Lignier (07²) p. 306, Pl. xxi. figs. 58—61; Pl. xxiii. fig. 83.

and rarely bordered, in the field. Resin-cells are abundant. Lignier suggests that if the genus *Cupressinoxylon* is subdivided this species might be referred to *Glyptostroboxylon*, though the medullary-ray pitting is not consistent with the characters of that genus.

Cupressinoxylon vectense Barber.

Founded on both stem and root wood from the Lower Greensand of the Isle of Wight and described by Barber[1] with a thorough-

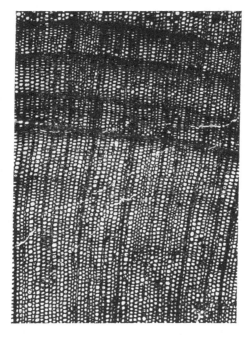

FIG. 716. *Cupressinoxylon vectense.* Transverse section of branch-wood showing a compound ring. (After Barber.)

ness of detail rarely met with in accounts of petrified wood. Dr Stopes[2] has recently re-described this species, adding a figure of the medullary-ray pitting. Annual rings well marked, often illustrating the occurrence of compound rings of growth (fig. 716) which are discussed in detail in the original account; bordered pits in a

[1] Barber (98) Pls. XXIII.—XXIV.
[2] Stopes (15) p. 169, text-fig. 50; also text-figs. 48, 49, and Pl. xv.

single row, rarely double in the roots, free and circular in branches, often contiguous and compressed in roots; tangential pits common Medullary rays usually uniseriate, 1—16 cells deep, pits confined to the radial walls, usually 1 but sometimes 2—4 oval and oblique pits in the field. Resin-canals absent; resin-parenchyma in vertical rows, abundant and scattered. The pith consists of pitted parenchymatous cells separated by intercellular spaces; in the roots the rows of tracheids pass directly into the cells of the pith; in the branches they terminate in small groups of cells irregularly arranged.

Cupressinoxylon McGeei Knowlton.

This is one of several species from the Potomac lignites included by Knowlton[1] in *Cupressinoxylon*. The annual rings are well marked; the tracheids have 2—3 rows of opposite and circular pits on the radial walls and small bordered pits are abundant on the tangential walls. The uniseriate medullary rays, 2—49 cells deep, have 1—2 oval apparently simple pits in the field and resin-parenchyma is abundant. Gothan[2] has described some wood of Lower Cretaceous age[3] from King Charles Land as *Cupressinoxylon, cf. C. McGeei,* agreeing with Knowlton's type in the medullary-ray pitting; there are 2—4 simple pits in the field, elliptical and horizontal; an indication of a border was seen in some of the pits in the region of the summer-wood, but the general absence of a border, if an original feature, is a difference between this wood and that of recent genera included in *Cupressinoxylon*. There is no Abietineous pitting on the ray cells.

The species *Cupressinoxylon luccombense* described by Dr Stopes[4] from the Lower Greensand of the Isle of Wight closely resembles *C. vectense*, but it has stone-cells in the pith, the tracheids are larger and there are usually 3—4 pits in the field in place of 1 or sometimes 2—3 in *C. vectense*; moreover in the latter species the pits of the ray cells are more uniform in size.

[1] Knowlton (89²) p. 46, Pl. II. fig. 5; Pl. III. figs. 1—5.
[2] Gothan (07²) p. 19, fig. 10.
[3] Gothan speaks of the King Charles Land fossils as Jurassic, but the beds have since been shown to belong to the Cretaceous system See Burckhardt (11).
[4] Stopes (15) p. 180, text-figs. 51—53.

Cupressinoxylon cryptomerioides Stopes.

A species[1] founded on a small branch from the Lower Green-sand of Kent showing the following features:—the primary xylem, composed of spiral and scalariform elements, also tracheids

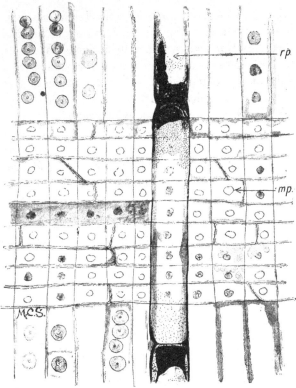

FIG. 717. *Cupressinoxylon Hortii.* Radial section showing the pits of the medullary rays, *mp*; the pits on the tracheids, and the wood-parenchyma, *rp.* (After Stopes.)

with small circular pits, forms projecting wedges in the perimedullary region; there are no resin-canals and resiniferous parenchyma is abundant; the medullary rays, generally 2—3 cells in depth, have two circular pits in the field, a feature regarded by Kraus as indicative of affinity to *Glyptostrobus*[2]. The cortex contains large

----

[1] Stopes (15) p. 186, Pls. XVI., XVII., text-figs. 54, 55.
[2] See page 198.

resin-canals. Dr Stopes draws attention to certain features, particularly the pitting of the medullary-ray cells and the structure of the cortex, in which this species resembles the genus *Cryptomeria*.

Cupressinoxylon Hortii Stopes.

This Lower Greensand species[1] from Bedfordshire is distinguished from other types by the very numerous medullary rays, often separated from one another by a single row of tracheids, which are both uniseriate and multiseriate, the same ray exhibiting both forms at different heights. The rays attain a depth of 80 cells and there is generally one large oval or circular pit in the field (fig. 717) though two are occasionally present. This species is hardly a typical *Cupressinoxylon* and Dr Stopes points out that the large single pit in the field is suggestive of *Podocarpoxylon*: it affords another illustration of the impossibility of identifying the majority of fossil woods within narrow limits.

Cupressinoxylon Holdenae sp. nov.

A species of Eocene age from the London Clay of Faversham in the Cambridge Botany School collection characterised by well-defined annual rings and the presence in some but not all the bands of summer-wood of resin-canals (fig. 718, C) which vary in size, some of the larger being formed by the confluence of smaller adjacent canals. The presence of rather thick-walled, pitted, cells lining the canals or in close association with them is a characteristic feature. Tyloses occur in some of the canals. The large number of canals in each row suggests their development in response to traumatic stimuli. The bordered pits occur in single, separate, rows or in double and opposite rows with rims of Sanio occasionally preserved. In a few places the pits of a single row are in contact and slightly flattened. Resin-parenchyma occurs in vertical rows in both spring- and summer-wood. Medullary rays uniseriate, reaching occasionally 30 cells in depth; the tangential and horizontal walls are unpitted and on the radial walls the pits are preserved only in a few places; there appear to be 2—4 fairly large simple pits in the field. The crowded series of canals (fig. 718, C) are identical with the traumatic ducts described in *Sequoia sempervirens*[2]

Stopes (15) p. 194, Pl. XVIII. text-figs. 56—58.   [2] Jeffrey (03).

and *Abies* (cf. fig. 690, B, p. 130). The absence of Abietineous pitting in the ray cells, the distribution of the canals, and the presence of scattered rows of xylem-parenchyma are features indicating affinity to *Sequoia sempervirens*. The preservation is not sufficiently good to warrant any definite statement with regard to the pits on the radial walls of the ray cells: the absence of a border is in contrast to the pits in *Sequoia*, but the apparent lack of a border may be due to imperfect preservation or to decay. Miss Holden, who carefully examined the sections, called my attention to the occasional occurrence of obscure and narrow cells of unequal breadth on the edge of some medullary rays bearing a resemblance to the ghost-like ray-tracheids described by Thompson[1].

Cupressinoxylon Koettlitzi sp. nov.

Silicified wood is by no means uncommon in the Franz Josef Archipelago and several specimens have been found in talus-heaps and in basaltic lavas. The age may be Upper Jurassic, Cretaceous, or possibly Tertiary. A radial longitudinal section of a piece of wood is figured, though not named, by Newton and Teall[2] and without a full description. The following account is based on sections cut from the same block in the possession of the Geological Survey, which was collected at Northbrook Island by members of the Jackson-Harmsworth Expedition. The species is named after Dr Koettlitz, the geologist of the Expedition[3]. Annual rings narrow and distinct: there are no resin-canals and no clearly preserved xylem-parenchyma, though in a few places there are indications of what appear to be elongated cells containing a dark substance. It is noteworthy that in some recent Cupressineae resin-parenchyma is not invariably present. The bordered pits on the radial walls of the tracheids are variable in their arrangement; they occur in single rows (fig. 718, E), contiguous and sometimes slightly flattened or more or less widely scattered, also in double rows with an opposite or occasionally an alternate disposition. There are a few pits on the tangential walls of the tracheids and rims of Sanio are seen in places on the radial faces. The medullary rays are 1—25 cells deep, uniseriate and very rarely two-cells

---

[1] Thompson (12).
[2] Newton and Teall (97) p. 508, Pl. xli. fig. 11.   [3] Koettlitz (98)

broad; 2—4 small circular or oval (simple?) pits in the field, but none on the tangential or horizontal walls.

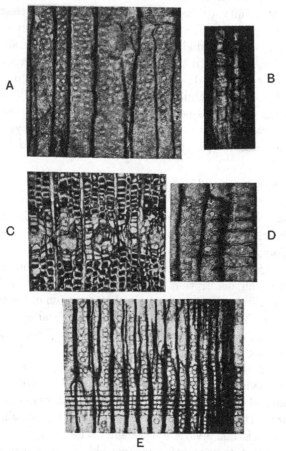

FIG. 718. A, D, *Cupressinoxylon polyommatum.* B, Phloem showing sieve-plates (Franz Josef Land). C, *Cupressinoxylon Holdenae.* E, *Cupressinoxylon Koettlitzi.* (A, D, Dublin Museum; B, E, Geological Survey Museum, Jermyn Street; C, Cambridge Botany School.)

Cupressinoxylon polyommatum Cramer.

The petrified wood on which this species[1] was founded was obtained from Banksland, from Miocene rocks, N. lat. 74° 40',

[1] Cramer (68) p. 172, Pls. xxxiv., xxxv., xxxvii.

long. 122° W., during the voyage of H.M.S. Investigator in 1851. Through the kindness of Prof. Johnson I was able to have sections cut from Dr Cramer's specimen in the Dublin Museum.

Annual rings clearly marked; summer-wood narrow. The most striking feature is the irregular distribution and unusually large number of small bordered pits on the radial walls of the tracheids. The pits vary in size and have an average diameter of 13·77 μ; they are usually separate but occasionally those of a single row are in contact. There are frequently as many as 3—4 rows (fig. 718, A) and occasionally 5 opposite and separate pits, a feature suggesting comparison with *Pinus Merkusii*[1], but in that species the pits of the three opposite rows are in contact. Rims of Sanio are occasionally present. Rows of narrow parenchyma occur in different regions of the wood. Medullary rays uniseriate, 1—16 cells deep, with 2—4 simple large oval pits in the field (fig. 718, D); there are no pits on the other walls of the ray cells. The pitting in the field is very similar to that in *Taxodium* and *Cupressinoxylon* (*Taxodioxylon*) *Taxodii* (fig. 720, A, B).

Cupressinoxylon taxodioides Conwentz.

Under this name Conwentz[2] describes some wood, probably of Pliocene age, from California which he compares with *Sequoia sempervirens*. He speaks of one stem 22 metres long and with a maximum diameter of 3—4 m.; the bordered pits are in 1—2 rows on the radial walls and small pits occur on the tangential walls; the medullary rays are usually two-cells broad and have 3—4 generally elliptical pits in the field apparently simple and arranged in a horizontal row. The rays are usually 15—20 cells deep but may reach a depth of 56 cells. Resin-parenchyma occurs in vertical rows but it is not stated whether it is confined to any definite region of the wood.

Schmalhausen's species *Cupressinoxylon* (*Glyptostrobus* ?) *neosibiricum*[3], characterised by medullary rays 13—20 or even 40—48 cells deep, and 1—2 circular or oval pits in the field, though compared by him with *Glyptostrobus*, cannot safely be regarded as more nearly allied to that genus than to some other members of the Cupressineae.

[1] Groom and Rushton (13) p. 484.   [2] Conwentz (78) Pls. XIII., XIV.
[3] Schmalhausen (90) Pl. II. pp. 44—49; Gothan (05) p. 50.

Many of the species included by authors are not described in sufficient detail to satisfy modern requirements with regard to the structure of the medullary rays and other characters. From the point of view of geographical distribution reference may be made to *Cupressinoxylon antarcticum* described by Beust[1] from Kerguelen Land.

[*Glyptostroboxylon* Conwentz.]

This generic name was first employed by Conwentz[2] for wood from sub-Oligocene beds in Argentina which he described as *Glyptostroboxylon Goepperti*: no figures accompany the description and it is hardly possible to determine with accuracy the precise affinity of the specimen. The annual rings are said to be distinct, the pits on the tracheids uniseriate and contiguous; resin-parenchyma occurs, and the medullary-ray cells have large circular pits on the radial walls. It is suggested by Gothan[3] that this species should be transferred to *Podocarpoxylon*, though in the medullary-ray pitting it differs from typical representatives of the Podocarpineae. Kraus[4] described the pits in the radial walls of the ra cells of *Glyptostrobus* as large and circularyand in a Tertiary specimen from Niederwöllstadt, named *Glyptostrobus tener* Kr., he figures the pits in the field as simple (fig. 719); these are said to be 1—8 in number and they are arranged in horizontal series, a feature characteristic of *Taxodium*.

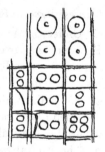

FIG. 719. *Cupressinoxylon tener* (Kraus). (After Kraus.)

Sections of wood of *Glyptostrobus heterophyllus* which I have examined show 2—8 pits in the field but there is a distinct border and the pore is narrow and oblique or in the spring-wood the pore may be broader and almost horizontal. Xylem-parenchyma is scattered through the wood and the thick transverse walls are pitted. Gothan believes that the medullary-ray pitting of *Glyptostrobus* is distinguishable from the Cupressoid type by the increase in the breadth of the pore as the ray cells are followed from the summer- to the spring-wood and by the fact

[1] Beust (85) Pl. IV.   [2] Conwentz (85) p. 13.
[3] Gothan (08) p. 9.   [4] Kraus (64) p. 195, Pl. V. fig. 12.

that the pits in the region of the spring-wood may be simple. The same author[1] points out that this type of medullary-ray pitting occurs in *Cunninghamia*. The impression produced by an examination of the recent *Glyptostrobus* is that this so-called Glyptostroboid pitting is not a sufficiently well-defined type to serve as a trustworthy diagnostic character. The medullary-ray pitting in the region of the summer-wood is similar to that in some species of *Podocarpoxylon* (= *Mesembrioxylon*), while in the spring-wood the pits in the field are rather of the type associated with *Cupressinoxylon* and the scattered xylem-parenchyma is another characteristic of the latter genus. There would seem to be no adequate grounds for regarding the two fossil species referred to *Glyptostroboxylon* as more nearly related to *Glyptostrobus* than to certain other recent genera. The retention of the name *Glyptostroboxylon* is inadvisable in that it implies an affinity which is not supported by satisfactory evidence.

CUPRESSINOXYLON, sub-genus **TAXODIOXYLON** Felix.

The generic name *Taxodioxylon* was applied to a Tertiary species from Hungary, originally referred by Felix[2] to *Rhizotaxodioxylon*, on the ground of a resemblance in structure to the wood of the recent genus *Taxodium*. Schenk[3], who examined the type-specimen, confirmed this comparison. *Taxodioxylon* has been adopted for fossils agreeing with the wood of *Taxodium* and *Sequoia sempervirens*: *Sequoia gigantea*, on the other hand, agrees more closely with typical species of *Cupressinoxylon*. *Taxodioxylon*, while similar in most respects to *Cupressinoxylon*, is said to differ in the medullary-ray pitting, the pits in the field being almost simple and elliptical with their long axis horizontal in contrast to the more definitely bordered pits of the Cupressoid type. This distinction is, however, not entirely satisfactory: in the wood of *Taxodium* the pits in the field are rather large and, though often simple, they occasionally present the appearance of pits with a well-developed border and the pore may be almost vertical or horizontal. In the recent species, as in some fossil examples, the tendency of the pits to arrange themselves in one or two

[1] Gothan (05) p. 49.   [2] Felix (84) p. 38.
[3] Schimper and Schenk (90) A. p. 872.

horizontal rows is a characteristic feature. The characters of
Taxodioxylon may be summarised as follows:—Annual rings
distinct; bordered pits on the radial walls of the tracheids in 1—4
rows, circular and separate and, if in two or three rows, opposite;
rims of Sanio present. Medullary-ray cells pitted only on the
radial walls; pits in the field, 2—8 in number, often arranged in
horizontal rows (fig. 720, A), sometimes fairly large, simple, or
bordered and horizontally elliptical. The thick walls of the xylem-
parenchyma (fig. 720, B) are characteristic of *Taxodium* and
Taxodioxylon and this character has been quoted as a trustworthy

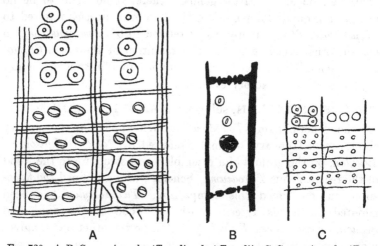

A   B   C

FIG. 720. A, B, *Cupressinoxylon* (*Taxodioxylon*) *Taxodii*. C, *Cupressinoxylon* (*Taxodioxylon*) *Sequoianum*. (A, B, after Gothan; C, after Mercklin.)

distinction between *Taxodium* and *Sequoia sempervirens* though,
as already pointed out, this is not a safe test[1]. In *Glyptostrobus
heterophyllus* the transverse walls of the xylem-parenchyma are
also thick and pitted and this tissue in some species of *Cupressus*
exhibits precisely similar features. It is in certain inconstant
features that *Taxodioxylon* differs from *Cupressinoxylon*. In
several instances the occurrence of wood referred to *Taxodioxylon*
in beds containing impressions of foliage-shoots like those of
Taxodium and *Sequoia sempervirens* strengthens the conclusions
based on anatomical characters.

[1] See page 149.

Cupressinoxylon (*Taxodioxylon*) *Taxodii* Gothan.

In his description of the Tertiary beds at Senftenberg Potonié[1] identified some of the wood as *Taxodium* and compared the deposits with those of a *Taxodium*-swamp. Gothan[2] has given a fuller description of the wood, which is characterised by xylem-parenchyma with thick transverse walls (fig. 720, B) and fairly large elliptical pits in the field; in the region of the spring-wood the medullary-ray pits have a horizontal pore and in the late wood the pore is said to be vertical. The bordered pits on the vertical walls of the xylem-parenchyma have a vertical pore in the summer-wood but it is horizontal in the spring-wood.

Cupressinoxylon (*Taxodioxylon*) *Sequoianum* Mercklin.

Gothan[3] recognised this species, originally described by Mercklin[4] from an unknown locality, in Tertiary beds in Germany associated with foliage-shoots agreeing with *Sequoia sempervirens*. The walls of the xylem-parenchyma are thinner than in *C. Taxodii*; there are 2—7 apparently simple pits in the field (fig. 720, C) in one or two horizontal rows. There may be three rows of opposite pits on the radial walls of the tracheids. It may be that *C. Sequoianum* is the wood of a *Sequoia*. A closely allied species is represented by *Cupressinoxylon uniradiatum* Goepp.[5] from Brühl afterwards recorded by Conwentz[6] as *Rhizocupressinoxylon* from Tertiary beds in Silesia. Schmalhausen[7] has described some interesting specimens of wood from Tertiary beds in Russia as *Cupressinoxylon Sequoianum* characterised by the possession of elliptical simple pits in the ray cells (from 2 to 6 in the field) of the transversely elongated form characteristic of *Taxodioxylon*. Xylem-parenchyma is fairly abundant and the tracheids are peculiar in having three or rarely four bordered pits on the walls as in some recent Pines.

Another example of wood of the *Taxodioxylon* type is afforded by *T. palustre* described by Felix[8] from Tertiary rocks in Hungary and recorded also from sub-Oligocene beds in Silesia. Gothan[9]

---

[1] Potonié (96); see also Eberdt (94).
[2] Gothan (06) p. 164.
[3] *Ibid.* p. 165; (09) p. 518.
[4] Mercklin (55) p. 65, Pl. XVII.
[5] Goeppert (50) p. 203, Pl. XXVII.
[6] Conwentz (80) A. p. 25, Pls. IV., V.
[7] Schmalhausen (83) Pl. XII.
[8] Felix (82) p. 278; (84) p. 38.
[9] Gothan (10) pp. 40, 43, Pl. VII.

assigns to *Taxodioxylon* two specimens from Tertiary strata in Spitzbergen and it is probable that the wood may belong to plants which bore the twigs described by authors as *Taxodium distichum* and *Sequoia Langsdorfii*.

The name *Taxodioxylon* is retained as a section of *Cupressinoxylon* and not as a separate genus on the ground that the characters on which it is based do not appear to be sufficiently distinctive or constant to warrant its recognition as a well-defined generic type.

III. **TAXOXYLON.** Unger.

Unger[1] gave this name to fossil wood characterised by the presence of spiral bands in the secondary tracheids, a feature especially associated with the recent genera *Taxus*, *Torreya*, and *Cephalotaxus*, but by no means unknown in other Conifers[2]. The name, in the form *Taxoxylum*, was substituted for *Taxites* employed by Goeppert[3] for some species of Tertiary wood. Apart from the presence of spiral bands *Taxoxylon* agrees with *Cupressinoxylon*, though according to Lignier the medullary rays are deeper in the latter genus.

Taxoxylon scalariforme (Goeppert).

This Tertiary species, originally described by Goeppert from Hungary as *Taxites scalariformis*, was renamed by Unger *Taxoxylum Goepperti*. According to Schenk[4] it is the only species among those recorded by Goeppert which should be retained in *Taxoxylon*, the spiral pattern of the tracheids being due to the presence of true bands and not, as in the other species, the result of enzyme action on the wood which produces a spiral striation closely simulating spiral bands. The bordered pits on the tracheids are circular and separate; the medullary rays are uniseriate and from 1 to 10 cells in depth; there are no resin-canals.

Goeppert's species *Taxites Ayckii* (after Herr Aycke) of Tertiary age was retained by Kraus[5] as an example of that genus but afterwards transferred to *Cupressinoxylon* on the ground of the absence of true spiral bands in the tracheids. Lingelsheim[6] also

---

[1] Unger (47) p. 33.   [2] See page 134.
[3] Goeppert (40).   [4] Schimper and Schenk (90) A. p. 859.
[5] Goeppert (40) p. 188; (50) p. 244; Kraus (64) p. 197.
[6] Lingelsheim (08) p. 27.

states that there are no true spiral bands in Geoppert's supposed *Taxoxylon*. The Permian species *Taxoxylon ginkgoides* Ren.[1] and Grand'Eury's Upper Carboniferous species *T. stephanense*[2] are probably founded on wood of the Araucarian type in which the tracheids show spiral striation.

An Aptian species *Taxoxylon anglicum* Stopes[3] is referred to that genus because of the occurrence of a spiral marking on the tracheids which the author of the species believes to indicate the presence of true spiral bands in the wood of recent Taxaceae, and because of the groups of 3—4 bordered pits in the fields of the medullary rays.

There are no resin-canals in the wood; the tracheids have a single row of circular pits on the radial walls and occasional rims of Sanio are preserved. There are 1—6 pits in the field and the presence of a border is regarded by Dr Stopes as an argument in favour of the tracheal nature of some of the medullary-ray elements, though the appearance of the cells does not afford any substantial ground for interpreting them as other than parenchymatous elements.

An examination of the type-specimens convinced me that the spiral markings on the tracheids are not true bands like those of recent Taxineous wood and the pitting of the medullary-ray cells is in itself by no means a trustworthy criterion. There are, I venture to think, no good reasons for referring this wood to the genus *Taxoxylon*.

IV. MESEMBRIOXYLON. Gen. nov.

This generic name[4] is proposed for fossil wood exhibiting certain features associated with several recent genera which have a southern distribution. It is intended to replace Gothan's two genera *Podocarpoxylon* and *Phyllocladoxylon*, types differing from one another in features which, as Dr Stopes[5] points out, are too inconstant to justify the retention of both designations. Moreover the use of Gothan's names implies affinities to recent genera which there are no adequate reasons for assuming. In this instance, as in many others, the anatomical characters do not enable us to

---

[1] Renault (85) p. 163

[2] Grand'Eury (90) A. p. 317; Gothan (05) p. 68.

[3] Stopes (15) p. 204, Pl. xix. text-fig. 59.

[4] μεσημβρινός, southern.   [5] Stopes (15) p. 210.

assign fossil species to a position within the Coniferales sufficiently
definite to be denoted by the use of a name implying close relationship to a particular genus as distinct from a group of allied
types.

Podocarpoxylon. This name[1] has been applied to wood agreeing
in structure with recent species of *Podocarpus* and *Dacrydium*
more closely than with other Conifers. As generally understood
the genus stands for wood without resin-canals, possessing xylem-
parenchyma not necessarily confined to a particular region of
the year's growth. In *Podocarpoxylon aparenchymatosum* Goth.
xylem-parenchyma is absent. The bordered pits on the tracheids
are in 1—2 rows and, if in two series, the pits are opposite or sub-
opposite (*Podocarpoxylon Schwendae* Kub.); rims of Sanio are
present. There is no Abietineous pitting in the ray cells; the
pits in the field are typically Podocarpoid, that is there are few in
the field and these appear to be bordered and characterised by
an elliptical or linear pore which is oblique or more or less vertical.
The medullary-ray pitting next the summer-wood does not afford
a satisfactory means of separating *Podocarpoxylon* and *Cupressinoxylon,* but in the region of the spring-elements the Podocarpoid
type is a distinguishing feature of *Podocarpoxylon,* though as
stated in the account of recent Conifers the position of the pore
is by no means a constant character. On the other hand, the pits
in the field may be large and simple as in *Sciadopitys,* some species
of *Podocarpus,* in *Phyllocladoxylon, Xenoxylon* and some other
genera.

Phyllocladoxylon. This name was given by Gothan[2] to wood
similar to *Podocarpoxylon* but differing chiefly in the occurrence of
large, simple pits in the field ['Eiporen'], a feature shared with
Sciadopitys and some species of *Podocarpus,* by *Microcachrys,*
Dacrydium, and *Pherosphaera.* The tracheids have 1—2 rows of
bordered pits on the radial walls, scattered and circular, but not
infrequently contiguous and flattened, and if in two rows they may
be alternate. There are no clear indications of Sanio's rims in
the specimens figured by Gothan and Schenk. There are no resin-
canals and no xylem-parenchyma. *Phyllocladoxylon* agrees closely
with *Xenoxylon*[3], but in *Phyllocladoxylon* the tracheal pits are often

---

[1] Gothan (05) p. 48.   [2] *Ibid.* p. 55; (10) p. 37.   [3] See page 238.

separate and are smaller than in *Xenoxylon*. The pits on the medullary-ray cells are also smaller, though it is doubtful if this is a constant character. Miss Holden[1] has instituted a new generic name, *Paraphyllocladoxylon*, for two specimens of wood from Jurassic rocks on the Yorkshire coast which do not appear to differ from *Mesembrioxylon* in any respect calling for generic recognition. In *Paraphyllocladoxylon eboracense*, from the Oolite of Scarborough, the tracheids have usually scattered and circular pits on the radial walls and pits are also abundant on the tangential walls: in *Paraphyllocladoxylon araucarioides* the pits on the radial walls are always closely compressed and flattened. There is no Abietineous pitting and there may be one or occasionally two large simple pores in the field like those in *Xenoxylon* and *Mesembrioxylon* (= *Phyllocladoxylon* of Gothan) but smaller than those of *Xenoxylon*. Xylem-parenchyma is absent, but some tracheids have apparent cross-walls that are believed to be resin-plates. Miss Holden recognises the close resemblance of her species to Gothan's *Phyllocladoxylon*, but a new name is employed on the ground that the absence of Sanio's rims shows that the wood of the Yorkshire plants is Araucarian. The absence of Sanio's rims cannot be confidently regarded as an original feature and, assuming this negative character to be a real one, it does not differentiate the specimens from those described by Gothan; Gothan's figures afford no evidence of the presence of Sanio's rims in his species of *Phyllocladoxylon*. If the Yorkshire stems are Araucarian so too are those from King Charles Land and Seymour Island[2]. In one of Miss Holden's species the tracheal pitting is not of the Araucarian type, while in the other it is Araucarian; the pitting of the medullary rays is opposed to an affinity to any recent Araucarian Conifer. Both of the Yorkshire species are therefore transferred to *Mesembrioxylon*: their anatomical characters indicate that they are generalised types which cannot legitimately be included in any family based solely on existing Conifers. While recognising that it is not always easy to draw a definite distinction between *Xenoxylon* and *Mesembrioxylon* the two names may be conveniently retained, the former being used in a much more restricted sense than the latter. *Mesembrioxylon* is applied to woods in which the general

---

[1] Holden, R. (13²) p. 536, Pl. xxxix. figs. 7—10.   [2] Gothan (07²); (08).

features are similar to those associated with *Cupressinoxylon*, but the xylem-parenchyma may not be always present and the medullary-ray cells have one or two large simple pits, or two or more smaller bordered pits, in the field, the pore being rather vertical than horizontal. *Mesembrioxylon* undoubtedly includes species which if additional data were available would be assigned to distinct genera. Apart from the probability that anatomical characters were even less restricted in their range through different types in former periods than they are in existing genera, the impossibility of discriminating between certain closely allied recent Conifers points to the advisability of employing designations for fossil woods in a wide sense and thus avoiding the danger of misleading students in search of material on which to base conclusions with regard to the relative antiquity of existing genera.

Mesembrioxylon sp. (= ? *Podocarpoxylon* sp. Gothan).

This wood from Bathonian rocks in Russian Poland[1] affords an example of the difficulty of distinguishing clearly between *Podocarpoxylon* and *Glyptostroboxylon*: the bordered pits on the radial walls of the tracheids are separate or contiguous and slightly flattened; xylem-parenchyma occasionally occurs; there are usually two fairly large circular simple pits in the field (fig. 722, C). The systematic position of this wood cannot be regarded as well established.

Miss Holden[2] has recently described two specimens from the Jurassic beds on the Yorkshire coast as *Podocarpoxylon* sp. but the evidence in support of affinity to the Podocarpineae is not by any means conclusive. In one specimen there is no xylem-parenchyma and in the other parenchyma occurs at the end of the year's growth. The pits on the medullary rays are described as piciform, 1—2 in the field. The anatomical features described hardly afford adequate reasons for assigning the wood to *Podocarpoxylon* rather than to *Cupressinoxylon*.

? *Mesembrioxylon* sp. (Thomas).

A specimen of imperfectly preserved wood from Jurassic rocks in the Izium district, in South Russia, described by Thomas[3] as

---

[1] Gothan (06²) p. 456, fig. 5.   [2] Holden, R. (13²) p. 542, Pl. XL. figs. 31, 32.
[3] Thomas, H. H. (11) p. 80, Pl. v. figs. 5—7.

Phyllocladoxylon sp. may perhaps be included in *Mesembrioxylon* though the preservation is hardly sufficiently good to admit of accurate determination. The bordered pits on the tracheids are circular and separate, about 15μ in diameter; the medullary rays have one or rarely two large simple pits in the field.

Mesembrioxylon woburnense (Stopes).

An Aptian (Lower Greensand) species from Bedfordshire founded on two blocks of secondary wood and referred to *Podocarpoxylon*[1]. The tracheids have 1—2 rows of bordered pits, the pits in two series being opposite; Sanio's rims are present. Resin-parenchyma is abundant all through the wood; the medullary rays are for the most part 3 cells deep but vary from 1 to 25; there is one large circular or oval pit, or sometimes two, in the field, and a narrow border is occasionally preserved. This species is near to *M. Schwendae* but there are fewer pits in the field in the English type.

Mesembrioxylon bedfordense (Stopes).

This Aptian species[2] is especially characterised by the arrangement of the bordered pits on the radial walls of the tracheids; the pits are uniseriate and occur in chains of 3—10, the border being flattened above and below by contact (fig. 721, *t*): the narrower parts of the xylem-elements are often without pits. Xylem-parenchyma is scattered through the wood and the medullary-ray cells have an oval or nearly circular large pit, sometimes with a border (fig. 721, *m*, *p*), in the field. The contiguous pits constitute an Araucarian feature though similar pits occur in *Cedroxylon* and in some other genera.

Mesembrioxylon Gothani (Stopes).

Dr Stopes regards this species[3], from the Aptian of the Isle of Wight, as highly suggestive of the genus *Phyllocladus*. The medullary rays are generally 2—4 cells deep and there are 1—2 large oval simple pits in the field. Xylem-parenchyma is sparsely scattered through the wood, and stone-cells occur in the pith.

---

[1] Stopes (15) p. 211, Pl. xx. text-figs. 60—63.
[2] *Ibid.* p. 223, Pl. xxi. text-fig. 64.
[3] *Ibid.* p. 228, text-figs. 65, 66.

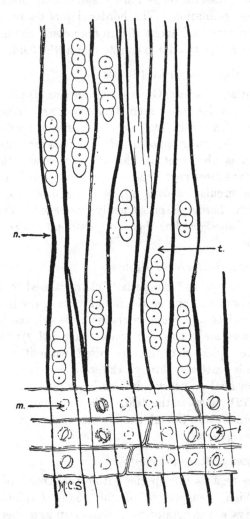

Fig. 721. *Mesembrioxylon bedfordense.* Radial section showing groups of adjacent pits on the tracheids, *t*; the medullary-ray pits, *m*, and a few with a border, *p*; *n*, narrow part of a tracheid. (After Stopes.)

Mesembrioxylon Schwendae (Kubart).

This species is recorded from Attersee in Upper Austria and though probably of Tertiary age it may be derived from Cretaceous strata[1]. Xylem-parenchyma is present; the bordered pits on the radial walls of the tracheids are in 1—2 rows, usually separate but if contiguous not flattened; if in 2 rows opposite or sub-opposite; the medullary rays reach a depth of 13 cells; there are generally 1—3 pits (fig. 722, A, B) but occasionally as many as 5 in the field;

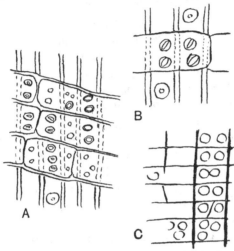

Fig. 722. A, B, *Mesembrioxylon Schwendae*. C, *Mesembrioxylon* sp. (A, B, after Kubart; C, after Gothan.)

they are bordered and the pore is obliquely vertical, though this feature is inconstant and in some places the pore is circular or the bordered pit may be replaced by a large simple pit. Sclerous cells occur in the pith.

Mesembrioxylon aparenchymatosum (Gothan).

In this Tertiary species, included by Gothan in *Podocarpoxylon*, from Seymour Island[2] there is no xylem-parenchyma; the medullary-ray cells have 1—2 elliptical-circular pits in the field and the major axis is oblique. The evidence as to affinity is far from conclusive.

[1] Kubart (11²) Pl. III. text-figs. 1—10.
[2] Gothan (08) p. 8, Pl. I. figs. 9—11.

Mesembrioxylon sp. (Gothan).

Gothan[1] described some wood as *Phyllocladoxylon* sp. which he originally stated to have been derived from King Charles Land, but Nathorst pointed out that it came from Scoresby Sound in East Greenland, N. lat. 70° 50′. The pits on the tracheids are in 1—2 rows, about 16 μ high, and smaller than in *Xenoxylon*; they are scattered or contiguous and flattened, but more often separate; the medullary rays are composed of a small number of cells and there are 1—2 simple pores in the field.

Mesembrioxylon antarcticum (Gothan).

In this Tertiary species from Seymour Island, included by Gothan in *Phyllocladoxylon*[2], the bordered tracheal pits are generally uniseriate and separate; the pitting of the medullary-ray cells is like that in *Mesembrioxylon* sp. (fig. 722, C). In the account of this wood Gothan points out that the similar pits on the walls of the medullary rays of *Sciadopitys* are smaller than in the fossil type; in contrast to the tendency towards a vertical elongation of the ray pits in some recent Podocarpineae those of *Mesembrioxylon* are usually more horizontally stretched as in *Phyllocladus* and some species of *Pinus*. Gothan compares *Cupressinoxylon Hookeri* Arb.[3] with *Mesembrioxylon antarcticum*, but the former is distinguished by the smaller pits on the ray cells and by the occurrence of xylem-parenchyma. It is noteworthy that, as Gothan states[4], the only recent South American Conifer possessing large simple pits in the medullary-ray cells is *Podocarpus andina*, while on the other hand there are several Australian genera agreeing closely with the Seymour Island species in the character of the medullary-ray pitting: from this it is concluded that in Tertiary times there was a closer connexion between the South American and Australian regions than at the present day, an inference which, though not improbably correct, rests on slender evidence in this particular case.

Mesembrioxylon Mülleri (Schenk).

This species from Pliocene strata in New South Wales, was described by Schenk[5] as *Phyllocladus*: the pits on the tracheids

---

[1] Gothan (07²) p. 9, fig. 2; (08) p. 6 (footnote).
[2] *Ibid.* (08) p. 4, Pl. I. figs. 4—8.　　[3] Arber (04).　　[4] Gothan (08) p. 25.
[5] Schenk in Schimper and Schenk (90) A. p. 873, figs. 424, 425.

are represented by Schenk as widely separated and the single pits in the field are narrower and more oblique than in other examples of the genus. The impression made by these species, formerly referred to *Phyllocladoxylon* and distinguished from one another by no very well-defined characters, is that they agree with certain recent Podocarpineae and with *Sciadopitys* more closely than with any other recent Conifers; but in the absence of any definite evidence with regard to foliage or reproductive organs it is impossible to select any one existing genus as the modern representative of the Arctic and South American fossil species. If the absence of Sanio's rims is accepted as a criterion of affinity, some species of *Mesembrioxylon* would be included in the list of types allied to the Araucarineae, but even assuming that the preservation of the wood is such as to admit of their recognition, were they present, their absence does not nullify the evidence afforded by the tracheal and medullary-ray pitting.

? *Mesembrioxylon Hookeri* (Arber) (= *Cupressinoxylon Hookeri* Arber).

A species[1] founded on a splendid specimen of silicified wood nearly 9 ft. long and with a diameter of 3 ft. from Tasmania exhibited in the Geological Department of the British Museum. The stem was discovered early in the nineteenth century in Tertiary basaltic lava on the Macquarie plains. Dr Arber quotes Sir Joseph Hooker who gives an interesting account of the method of preservation of the decorticated wood. The annual rings are well marked and narrow; the tracheids have usually a single row of circular and scattered bordered pits on their radial walls and smaller pits are abundant on the tangential walls. Sanio's rims are clearly shown on some of the tracheids. The medullary rays are generally uniseriate and in exceptional cases reach a depth of over 20 cells. Arber speaks of the occurrence of a small simple pit on the radial wall of the ray cells; an examination of the sections in the British Museum showed that for the most part the pitting on the ray cells is not preserved but in some places a single, fairly large, simple pit occurs in the field. Resin-parenchyma is present in both spring- and summer-wood.

---

[1] Arber (04).

It has been suggested by Gothan[1] that this species should be referred to *Podocarpoxylon*: the pitting of the medullary-ray cells, though seldom preserved, seems to differ from the typical *Cupressinoxylon* form. It is therefore referred though with some hesitation to *Mesembrioxylon*.

V. PARACEDROXYLON. Sinnott.

Paracedroxylon scituatense Sinnott. The generic name *Paracedroxylon*[2] was instituted for this Cretaceous species from Scituate, Massachusetts, in order to indicate its resemblance to *Cedroxylon* as defined by Kraus, more especially as regards the pitting of the tracheids and the absence of resin-canals and xylem-parenchyma in the normal wood, and to show that in some features it differs from that genus, namely in the absence of Sanio's rims and in the smooth and thin unpitted horizontal walls of the medullary-ray cells. In typical species of *Cedroxylon* xylem-parenchyma is present and the tracheal pitting alone is not a distinguishing feature. The annual rings are broad and not well defined; resiniferous parenchyma is restricted to wounded regions. Groups of thin-walled cells, which it is suggested may be abortive resin-canals, though there is no evidence that this is the case, and thick-walled parenchyma occur in the wounded tissue. The tracheids have a single row of bordered pits, generally circular. The medullary rays, 2—12 or more cells in depth, are pitted only on the radial walls and there are 4—6 circular pits with an oblique slit-like pore in the field. The occurrence of bands of much thickened and pitted parenchyma is regarded as evidence of wounding: canal-like spaces occur in the traumatic tissue. Moreover in the affected regions the medullary-ray cells often show pitting on their horizontal and tangential walls.

Sinnott regards the absence of Sanio's rims as indicative of Araucarian affinity while the traumatic phenomena are interpreted as Abietineous characters. The genus rests on a slender basis: except for the absence of xylem-parenchyma the normal wood differs very slightly from *Cupressinoxylon* and it is not distinguished by any well-marked features from *Mesembrioxylon*.

Jeffrey[3] has described the axis of a *Geinitzia* cone from the

[1] Gothan (08) p. 7.   [2] Sinnott (09).   [3] Jeffrey (11).

Mataram formation as exhibiting the features of *Parace-droxylon*.

VI. CEDROXYLON. Kraus.

This generic name was instituted[1] for fossil wood agreeing with *Cupressinoxylon* in the arrangement of the pits on the tracheids and in the absence of resin-canals, but differing in the scarcity or absence of xylem-parenchyma. As defined by Schenk[2], *Cedroxylon* stands for fossil wood agreeing generally with that of recent species of *Cedrus, Abies,* and *Tsuga* with or without tracheids in the medullary rays. Brongniart's genus *Eleoxylon*[3] is included by Schenk as a synonym of *Cedroxylon*. The chief distinguishing character of *Cedroxylon* as compared with *Cupressinoxylon*, as used by some authors, is the more restricted occurrence of xylem-parenchyma; in *Cedroxylon* it is confined to the end of each year's wood whereas in *Cupressinoxylon* the parenchyma is not so limited in its distribution. A closer examination of different types of wood included in *Cedroxylon* shows that the xylem-parenchyma is an unsafe guide: Barber[4] states that he found more xylem-parenchyma per square millimetre in *Cedrus* wood than in *Crypto-meria* (a genus included in the general term *Cupressinoxylon*) and Lignier[5] speaks of the absence of parenchyma in some species of *Cedroxylon*. Gothan[6], who has discussed the distinctive features of these and other genera in considerable detail, points out that in *Abies Webbiana* xylem-parenchyma is abundant as in *Cupress-inoxylon*, while in some Cupressineae the parenchyma is so scarce that it is often difficult to discover. It is clear that a test based on the presence or distribution of xylem-parenchyma is unsatis-factory; the application of such a test would lead to the inclusion of both Abietineous and Cupressineous genera in one generic type. In typical cases the distribution of xylem-parenchyma is none the less a useful character, but Conifers with parenchyma scattered through the year's growth are not confined to types usually in-cluded in the comprehensive genus *Cupressinoxylon*: many Podo-carps and some other genera not members of the Cupressineae possess abundant parenchyma in the wood. The structure of the

[1] Kraus in Schimper (72) A. p. 370.   [2] Schimper and Schenk (90) A. p. 862.
[3] Brongniart (49) A. p. 76.   [4] Barber (98) p. 332.
[5] Lignier (07²) p. 245.   [6] Gothan (05) p. 45.

medullary-ray cells affords an important distinguishing feature. In *Cedroxylon* they are characterised by the occurrence of pits on all the walls as in *Abies, Tsuga, Cedrus, Larix, Picea, Pseudolarix,* while in wood of the *Cupressinoxylon* type there is no Abietineous pitting but only pits on the radial walls. Another distinguishing feature, mentioned by Lignier,—whether important or not is open to question—is based on the greater number of medullary rays per square millimetre in *Cupressinoxylon*.

The characters of *Cedroxylon* may be briefly summarised as follows:—Annual rings well marked; bordered pits on the radial walls of the tracheids usually circular and separate and if in more than one row, opposite, but in some species the Araucarian type of pitting also occurs (fig. 723), the pits being contiguous and alternate or sometimes arranged in stellate clusters. Xylem-parenchyma typically confined to the end of an annual ring, but sometimes absent; medullary rays generally uniseriate and composed exclusively of parenchyma though horizontal tracheids may occur; pits on all the walls of medullary-ray cells as in the Abietineae; on the radial walls there may be 1—6 apparently simple circular pits in the field. There are no resin-canals except as the result of injury.

Cedroxylon transiens Gothan.

This species from Upper Jurassic rocks in Spitzbergen[1] and from Lower Cretaceous beds in King Charles Land[2] illustrates the admixture of opposite, separate, and contiguous, alternate, pits in the same wood (fig. 723, A, B); stellate groups also occur (fig. 723, A), an arrangement occasionally seen in some recent Pines, e.g., *Pinus Merkusii*[3]. The Araucarioid type of pitting is characteristic of the spring-wood where it is associated with the more usual opposite and separate arrangement. Smaller separate pits occur on the tangential walls of some of the tracheids. Resin-canals are absent, though a solitary example is recorded by Gothan. The medullary rays have several small pits in the field, also pits on the tangential and horizontal walls; the xylem-parenchyma is confined to the end of the year's wood; the cross-walls are pitted (fig. 723, B).

[1] Gothan (10) p. 38, Pl. VI. figs. 11—13.
[2] *Ibid.* (07²) p. 26, figs. 14, 15; Pl. I. fig. 1.
[3] Groom and Rushton (13) Pl. XXV. figs. 47, 48.

Gothan makes no mention of Sanio's rims. The Abietineous features predominate over the Araucarian, the latter being limited to the local occurrence of polygonal and alternate bordered pits.

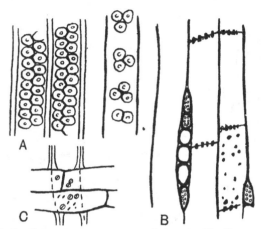

FIG. 723. A, B, *Cedroxylon transiens.* C, *Cedroxylon blevillense.* (A, B, after Gothan; C, after Lignier.)

Wood of similar type was described by Schroeter[1] from King Charles Land as *Pinus (Larix) Johnseni*: resin-canals, possibly due to wounding, occur in the summer-wood. The medullary rays are 1—18 cells deep and there are 1—3 simple circular pits in the field; all the walls of the ray cells are pitted. As in *Cedroxylon transiens* the Araucarian type of pitting is represented on some of the tracheids.

Cedroxylon Hornei Seward and Bancroft.

An Upper Jurassic species from Helmsdale[2], Sutherland, a locality from which Hugh Miller recorded numerous specimens of fossil wood which is still abundant on the beach immediately north of Helmsdale. The annual rings are well defined: the bordered pits are usually in a single row on the radial walls of the tracheids, occasionally in contact and flattened; double rows of opposite pits are not uncommon. Xylem-parenchyma is confined to the late wood. Medullary rays, 1—26 cells deep, generally 8—12,

---

[1] Schroeter (80), Pl. I. figs 1—8.
[2] Seward and Bancroft (13) p. 883, text-fig. 5; Pl. II. figs. 22—25.

and uniseriate; there are 2—4 simple or faintly bordered circular pits in the field and pits occur also on the tangential and horizontal walls.

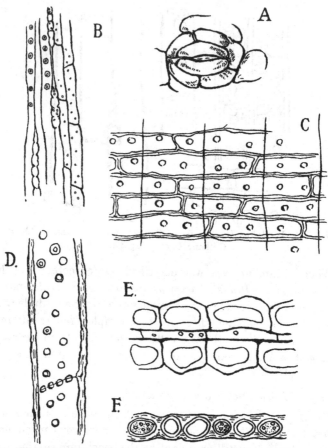

Fig. 724. A, *Brachyphyllum eathiense*; stoma. B—F, *Cedroxylon Hornei*.
(After Seward and Bancroft.)

Cedroxylon cedroides Gothan.

In this species from King Charles Land[1], probably Lower Cretaceous, the pitting of the tracheids is in the main of the usual Coniferous type and not Araucarioid, though in places the pits are

[1] Gothan (07[2]) p. 23, figs. 11—13.

alternate and form stellate groups. The medullary rays, reaching a depth of 30 cells, show very clearly the Abietineous pitting and there are 4—5 simple circular pits in the field. The occasional occurrence of single large pores in the field would seem to be due, at least in part, to the destructive action of fungi. Xylem-parenchyma occurs in the summer-wood. The structure agrees with that of the wood of *Cedrus*, *Pseudolarix*, and *Tsuga*; the Abietineous features are relatively more conspicuous than in *Cedroxylon transiens* in which the tracheal pitting is more Araucarian.

Cedroxylon maidstonense Stopes.

In this wood[1], from the Lower Greensand of Kent, the rings of growth are well marked; the bordered pits on the radial walls of the tracheids are usually uniseriate and Sanio's rims are distinctly preserved. Xylem-parenchyma is absent or very rare and there are no resin-canals. The medullary rays are seldom deeper than 10 cells; there are 4—6, or occasionally more, oval or circular pits in the field and some have a slit-like pore and are bordered; pits are clearly shown on the tangential walls of the ray cells.

Cedroxylon pottoniense Stopes.

This species[2] of the same geological age, from Potton in Bedfordshire, differs from *C. maidstonense* in the comparative abundance of xylem-parenchyma: the medullary-ray cells show very clearly the Abietineous type of pitting.

Cedroxylon blevillense Lignier.

In this species from the gault of Bléville (Seine-Inférieure)[3] the tracheal pits are usually uniseriate and separate but if in two rows they may be either opposite or alternate. The pits in the field are small, numerous, and have an oblique pore (fig. 723, C). The tangential walls of the ray cells are sometimes pitted. There is no resiniferous parenchyma. The characters afford another illustration of the impossibility of drawing any clearly defined line between *Cedroxylon* and allied generic types.

[1] Stopes (15) p. 149, Pl. XII. text-figs. 41—43.   [2] *Ibid.* p. 154, text-fig. 44.
[3] Lignier (07²) p. 267, Pl. XVIII. figs. 15—17; Pl. XXI. fig. 66; Pl. XXII. fig. 72; Pl. XXIII. fig. 87.

A Triassic species described by Wherry[1] as *Brachyoxylon penn-sylvanicum* may, as that author suggested, belong to *Cedroxylon*: the tracheids have 1—2 rows of pits, usually separate but sometimes alternate and hexagonal as in *Cedroxylon transiens* and *C. Hornei*; no description is given of the medullary-ray pitting or of any xylem-parenchyma. There are no adequate grounds for referring this Triassic wood to *Cedroxylon*. Several species of wood from Triassic and higher horizons have been assigned to *Cedroxylon*, but in many cases the descriptions fall short of modern standards and accurate determination is impossible. Crié[2] describes a species, *C. australe*, from the Trias of New Caledonia though his figures and descriptions do not afford satisfactory evidence in support of this reference. Schenk[3] mentions *Cedroxylon pertinax* (Goepp.) as the oldest representative of the genus and speaks of it as Rhaetic, while Gothan refers the species to a Jurassic horizon. A species founded by Goeppert and described by Mercklin[4] from Jurassic rocks of Russia, *Pinites jurassicus*, may be a *Cedroxylon*: the bordered pits are usually separate and opposite but sometimes in contiguous groups. Mercklin states that small thick-walled cells, often with dark contents, occur at the outer limit of each ring. A specimen described by Felix[5] as *Cormocedroxylon jurense* from the Braun Jura of Galicia is compared by him with *Pinites jurensis*. Fliche[6] records, though without complete diagnoses, some French Lower Cretaceous species: the tracheids of *Cedroxylon reticulatum* Sap., from the Albian of L'Aube, are characterised by pits which are usually separate but may be contiguous and flattened. Cones closely resembling those of *Cedrus* occur in the same beds. This author gives partial descriptions of *C. barremianum* Fliche[7] from the Lower Cretaceous of Haute Marne and a Cenomanian species *C. manekildense* Fliche[8], but in neither case are the data adequate.

C. matsumurae Stopes and Fujii[9] is an Upper Cretaceous Japanese species with 1—2 rows of tracheid-pits, generally opposite

---

[1] Wherry (12) Pl. IV.   [2] Crié (89) Pls. II.—v.
[3] Schimper and Schenk (90) A. p. 871.
[4] Mercklin (55) p. 48, Pl. VIII. figs. 6—10.   [5] Felix (82) p. 264.
[6] Fliche (97) p. 7.   [7] *Ibid.* (00) Pl. II. fig. 1.
[8] *Ibid.* (96) Pl. XV. fig. 3.
[9] Stopes and Fujii (10) p. 42, Pl. I. fig. 10; Pl. IV. figs. 20—23.

but sometimes alternate though not contiguous. The medullary rays, 5—12 or rarely 20 cells deep, are imperfectly preserved. Another species, *C. Yendoi* St.[1] and Fuj. from the same locality is also founded on material that is insufficient for accurate determination. Sporadically occurring resin-ducts are regarded as traumatic.

Among Tertiary species reference may be made to *Cedroxylon affine* Kraus[2] from Sicily, without resin-parenchyma and characterised by usually two large simple pits in the field; *C. Hoheneggeri* Felix[3] from the Eocene of Moravia figured by Schenk as from Cretaceous strata; *C. Hermanni* Sch.[4], an incompletely described species from Assam, probably of Tertiary age.

VII. PITYOXYLON. Kraus.

Kraus[5] included in this genus some of the species previously referred by Goeppert to *Pinites*; others he assigned to *Cedroxylon*. *Pityoxylon* is distinguished from *Cupressinoxylon* and *Cedroxylon* by the normal occurrence of resin-canals in the wood and by the presence of horizontal tracheids in some of the medullary rays. Within the limits of the genus the following differences occur in the characters of the medullary rays and the resin-canals:—the walls of the ray-tracheids are smooth or provided with dentate ingrowths; the pits on the medullary-ray cells are large and simple or smaller and apparently bordered, and there may be one or several pits in the field; the parenchyma of the resin-canals has thin or thick walls. As generally employed *Pityoxylon* includes species exhibiting anatomical features met with in *Pinus, Picea, Larix, Pseudotsuga*, and some other Abietineae. Gothan[6] makes use of two generic names, *Piceoxylon* and *Pinuxylon*, to denote the possibility of more precise comparison with recent types than is implied by Kraus's more comprehensive term. *Piceoxylon* is characterised by thick-walled epithelial cells lining the resin-canals, by small pits in the ray cells, spiral tracheids in the summer-wood, the absence of teeth in the ray-tracheids, clearly marked Abietineous

[1] Stopes and Fujii (10) Pl. IV. figs. 24—26.
[2] Kraus (83).　　　　　　　　　　　　[3] Felix (82) p. 268.
[4] Schenk (82²) p. 355.　　　　　　　　[5] Kraus in Schimper (72) A. p. 377.
[6] Gothan (05) p. 102.

pitting in the ray cells, and by the occurrence of numerous pits in the tangential walls of the summer tracheids.

Pinuxylon is used by Gothan in preference to *Pinoxylon*, the name adopted by Knowlton[1] for wood in which there are no resin-canals in the medullary rays. In *Pinuxylon* the walls of the epithelial cells are thin, rarely thick; the medullary rays have large simple pits in the spring-wood; there are no spiral bands in the tracheids. The horizontal tracheids have smooth or dentate walls and the Abietineous pitting is much reduced. The distinctions on which these two genera are based are thus not very clearly defined and it is only in particularly well-preserved material that the two generic types can be recognised with certainty. Dr Stopes[2] follows Jeffrey and Chrysler[3] in regarding Gothan's twofold division as unnecessary.

In the majority of species referred to *Pityoxylon* the published information is insufficient for a sub-division in Gothan's sense and as a rule the generic name stands for wood of an Abietineous type which cannot be assigned with confidence to any one recent genus. The question of the antiquity of the Abietineae has been confused by the too liberal use of the term *Pinites* by Goeppert and some other authors for stems which have no claim to be placed in the genus *Pityoxylon*. Jeffrey and Chrysler[4], who follow previous authors in quoting *Pinites Conwentzianus* Goepp.[5], described as a Carboniferous species from Waldenburgh, as evidence of a Palaeozoic Pinus-like wood, state that the species receives 'full confirmation from the description of a similar type, *Pityoxylon chasense* Pen.[6] from the Permian of Kansas.' Goeppert and Stenzel state that *Pinites Conwentzianus* was found on a rubbish-heap ('Halde'), but Goeppert apparently entertained no doubt as to its Carboniferous age. Through the courtesy of Prof. Frech of Breslau I was able to examine the original sections and convinced myself that the wood is Abietineous: the rings of growth are well defined; horizontal tracheids occur in some of the rays and the tracheal pits, 1—2 rows, are widely separated, though occasionally the pits

[1] Knowlton in Ward (00) B. p. 420.   [2] Stopes (15).
[3] Jeffrey and Chrysler (06).   [4] *Ibid.* p. 13.
[5] Goeppert and Stenzel (88) p. 54, Pls. XI., XII.
[6] Penhallow (00) p. 76; Thomson and Allin (12).

of a single row are in contact and slightly flattened. Feeling
sceptical as to the Carboniferous age of the wood I wrote to Dr
Conwentz who confirmed my doubts with regard to the value of the
evidence as to the geological horizon. Thomson and Allin have
shown that Penhallow's *Pityoxylon* cannot be accepted as trustworthy evidence of the occurrence of a Palaeozoic Abietineous
type. *Pityoxylon chasense* is not an Abietineous species; it is
founded on *Dadoxylon* wood devoid of annual rings and without
resin-canals traversing the medullary rays.

The fragments of wood from the Muschelkalk of Recoaro
figured by Schleiden and Schenk[1] as *Pinites Goeppertianus* afford
no evidence of Abietineous affinity beyond the occurrence of
separate bordered pits on the walls of the tracheids.

Pityoxylon eiggense (Witham).

The petrified wood first named by Witham[2] *Pinites eiggensis*
and afterwards[3] referred by him to the genus *Peuce* was originally
recorded by Macculoch in 1814 from below the massive and pre-
cipitous ridge of pitchstone which forms a striking feature above
the basaltic lavas of the Sgurr of Eigg in the Inner Hebrides.
Lindley and Hutton[4] and Nicol[5] also gave short descriptions of the
structure and Miller[6] in the *Cruise of the Betsey* alludes to a fossil
trunk as 'an ancient tree of the Oolites.' The wood occurs with
fragmental sedimentary rocks below the pitchstone and not
actually *in situ*; Mr Harker's thorough examination of the island
led him to the conclusion that the wood and associated rock-
fragments are derived from Jurassic (Oxfordian) strata and were
carried up by volcanic agency[7]. Mr Harker tells me that he has
never seen the *Pityoxylon* with any undoubted matrix adherent;
it occurs with wood of a different type (*Dadoxylon*) which is embedded in a white sandstone agreeing exactly with the Great
Estuarine Sandstone of Eigg in which similar wood has been found
in place. It is, however, possible that *Pityoxylon* did not come

[1] Schenk (68) Pl. v. figs. 4—7.   [2] Witham (31).
[3] *Ibid.* (33) A. Pls. xiv., xv.   [4] Lindley and Hutton (33) A. Pl. xxx.
[5] Nicol (34) A. p. 154.   [6] Miller (58) p. 37.
[7] Harker (06) p. 55; (08) p. 52. In these memoirs Mr Harker discusses the
earlier conclusions of Sir Archibald Geikie as to the geological history of Eigg and
gives references to previous notices of the fossil wood; Seward (11²) p. 652.

from the same source as the *Dadoxylon* wood. Though probably Jurassic, a Tertiary source is by no means ruled out.

Kraus[1] transferred Witham's species to the genus *Pityoxylon* and that name is used by Schroeter[2] and Schenk[3]. The specimens on which the following account is based are most of them in the British Museum. Annual rings clearly defined, usually 1—1·5 mm.

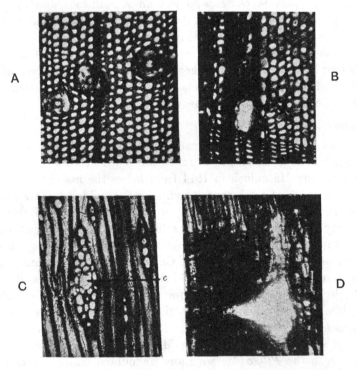

Fig. 725. *Pityoxylon eiggense*; *c*, resin-canal in a fusiform medullary ray. (British Museum, 51427, 51641, 51727.)

broad; the reduction in diameter of the summer-tracheids extends over several rows, the transition being much more gradual than in some types of Coniferous wood. A characteristic feature is the occurrence of more or less circular patches where the tracheids have been destroyed with the exception of a single tracheid or a

[1] Kraus in Schimper (72) A. p. 378.   [2] Schroeter (80) p. 13.
[3] Schimper and Schenk (90) A. pp. 855, 874.

small group in the centre of a clear crystalline matrix. Some of these patches simulate resin-canals, a fact which led Schenk to deny the existence of true canals. All stages of decay are shown, from the partial obliteration of a circular group of tracheids to the destruction of the group, one central element being left, or to the formation of a canal-like cavity (fig. 725, A). It is often difficult to decide whether a clear space in the wood is a canal or the result of *post-mortem* changes, but there is no doubt as to the occurrence of some true secretory canals in different regions of the wood. There is very little parenchyma accompanying the canals. The medullary rays are of two kinds, uniseriate, 1—13 cells in depth, though usually about 6 cells in depth, and lenticular rays with a central canal (fig. 725, C, c) identical with those in a modern Pine. Fig. 725, D shows part of a vertical canal with some parenchymatous lining in continuity with a horizontal canal in a broad medullary ray. Several small pits occur on the tangential and horizontal walls of the ray cells, and the radial walls, which are less clearly preserved, occasionally show 1—3 elliptical pits. In radial section the upper and lower cells of a medullary ray are often distinguished by their less uniform breadth and resemble in this respect ray tracheids. A careful examination of sections revealed the existence of bordered pits in the tangential walls of these elements and confirmed their tracheal nature[1].

The bordered pits are generally single and sparsely scattered on the radial walls of the tracheids; occasionally the pits are in contact and a few double and opposite rows occur. An interesting feature is the occurrence in some tracheids of a biconcave patch of some brown substance agreeing closely with resinous deposits described by Penhallow[2]. If this species is from Jurassic strata its close resemblance to recent types of *Pinus* is a fact of considerable interest.

Pityoxylon Ruffordi Seward.

This species[3] is founded on a specimen of wood obtained by Mr Rufford from Wealden beds at Ecclesbourne on the Sussex

---

[1] I am indebted to Miss Ruth Holden for calling my attention to this feature and for other assistance in the examination of the sections.

[2] Penhallow (04) p. 526.

[3] Seward (95) A. p. 199; (96) p. 417, Pls. II., III.

coast: the sections on which the description is based are in the British Museum. Annual rings well marked, varying in breadth from 1 to 3 mm.; resin-canals are abundant both in the spring- and

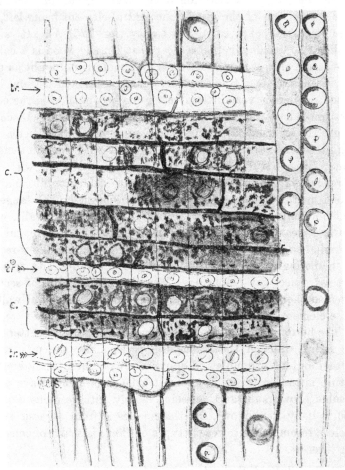

FIG. 726. *Pityoxylon Sewardi.* Radial section showing tracheids, *tr*, in a medullary ray; *c*, parenchyma of the ray with simple pits. (After Stopes.)

autumn-wood and horizontal canals occur in the lenticular medullary rays: some of the canals are occupied by large rounded cells like tyloses. The bordered pits on the radial walls of the tracheids form single or double rows; in the latter case the pits are generally

opposite but stellate groups also occur as in *Cedroxylon transiens*
Goth. (fig. 723, A) and several other species. The uniseriate
medullary rays reach a depth of 30 cells: there are usually 2—4
oval or circular pits in the field.

A similar type of stem is represented by *Pityoxylon Nathorsti*[1]
(Conw.) from the Lower Cretaceous of Sweden.

Pityoxylon Sewardi Stopes.

This species[2] is founded on a petrified branch, not less than 18
cm. in diameter, from the Lower Greensand of Kent. It exhibits
the usual features characteristic of the genus; the wood contains
horizontal and vertical canals with thin-walled epithelial cells. The
medullary rays are larger and more abundant than in most Coniferous woods and horizontal tracheids (fig. 726, *tr*) occur interspersed with the parenchymatous cells, *c*, as well as on the upper
and lower margins, an arrangement in which the fossil bears a
striking resemblance to the recent species *Pinus monticola*[3].

Pityoxylon Benstedi Stopes.

In this Lower Greensand species[4] from Kent the resin-canals
often contain tyloses as in *P. Nathorsti* Conw. and the epithelial
cells have very thick walls, a feature suggesting comparison with
the genus *Larix*. The medullary rays show well-marked Abietineous pitting (fig. 727, *a*) and ray-tracheids (*rt*) occur. Rims of
Sanio are shown in fig. 727 between the circular bordered pits,
tr. The difference between the tracheal and ordinary parenchymatous elements of the rays, as represented in fig. 727 from a
drawing by Dr Marie Stopes, is not very clearly defined and in the
upper ray shown in the figure part of a tracheid is seen abutting
laterally on parenchymatous ray cells, the only difference between
them being in the form of the pits, a criterion which is largely
dependent for its value on the state of preservation. Dr Stopes is
inclined to regard this species as most nearly allied to *Larix*.

Pityoxylon statenense Jeffrey and Chrysler.

A species from the Middle Cretaceous of Staten Island[5] found
in association with the short shoots described by Jeffrey and

[1] Conwentz (92) p. 13, Pls. I.—III., VI., VII.
[2] Stopes (15) p. 95, Pls. IV., V. text-figs. 23, 24.   [3] *Ibid.* text-fig. 25, p. 103.
[4] *Ibid.* (15) p. 105, Pls. V.—VII. text-figs. 26, 27.
[5] Jeffrey and Chrysler (06).

Hollick as *Pinus triphylla*, etc. The anatomical features are as
follows:—annual rings narrow, not clearly marked owing to the
walls of the summer-tracheids being thinner than in recent species
of *Pinus*; xylem-parenchyma confined to the periphery of the

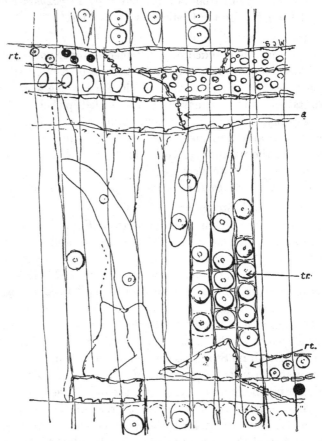

Fig. 727. *Pityoxylon Benstedi.* Radial section showing the tracheal pitting, *tr*;
    rt, ray-tracheids; *p*, ray cells with large pits; *a*, typical Abietineous pitting
    of end-walls of medullary-ray cells. (After Stopes.)

resin-canals which occur in any part of the wood and are often
filled with tyloses. The tracheids have a single row of pits not
contiguous or flattened; the pits on the tangential walls are
confined to the summer-tracheids; both linear and fusiform

medullary rays occur, the latter with horizontal resin-ducts. There are no ray-tracheids. There is usually one circular or elliptical pit in the field. A second species from the same locality, *Pityoxylon scituatense*, differs only in some unimportant features from *P. statenense*. These fossils differ from recent Pines as also from *Picea*, *Pseudotsuga*, and *Larix* in the absence of ray-tracheids. In the restriction of bordered pits to the tangential walls of the tracheids of the summer-wood they agree with the soft Pines, but though this character is generally lacking in hard Pines, Jeffrey and Chrysler point out that in some hard Pines without pits on the tangential walls of the tracheids of vegetative shoots the summer elements of the cones have tangential pits. The occurrence of bordered pits on the tangential walls of the late wood and the absence of ray-tracheids are regarded by the authors of the species as ancestral features.

Pityoxylon protoscleropitys (Holden).

A Middle Cretaceous species[1] from New Jersey, referred by Miss Holden to *Pinus*, showing the following features:—annual rings well developed; linear and fusiform medullary rays, horizontal and vertical resin-canals, bordered pits uniseriate and scattered on the radial walls of the tracheids; none on the tangential walls. Rims of Sanio are present. There are 1—2 pits in the field with a lenticular pore and circular border; the other walls of the ray cells are abundantly pitted. Ray-tracheids occur on the margins of the medullary rays and rarely interspersed with the parenchyma; their walls are denticulate as in recent hard Pines.

The presence of horizontal tracheids in the medullary rays is an important character: in *Pityoxylon scituatensiformis* (Bailey)[2], another Middle Cretaceous species, ray-tracheids are present but they have smooth walls and are not met with in the first 10—15 rings of wood, whereas in *P. protoscleropitys* they occur even in the wood of the first year. In this connexion the presence of ray-tracheids in *Pityoxylon eiggense* is noteworthy at least if that species is from a Jurassic source. *Pityoxylon protoscleropitys* is considered by Miss Holden to be 'probably the earliest form with all the characters of a modern hard Pine, yet retaining certain ancestral

[1] Holden (13³).   [2] Bailey (11).

features, as the association of primary and fascicular leaves, the latter borne on brachyblasts subtended by a foliar trace.'

Miss Holden[1] has also described from New Jersey two species which she compares with *Prepinus*, namely *Pityoxylon foliosum* and *P. anomalum*.

Pityoxylon Nathorsti (Conwentz).

Under the name *Pinus Nathorsti* Conwentz[2] described in considerable detail specimens of petrified wood from the Senonian Holma sandstone of Sweden, also a cone and two detached needles. The rings of growth agree with those of stems and older branches in the gradual increase in the thickness of the tracheid-walls in passing from the spring to the late summer elements. The bordered pits on the radial walls of the tracheids are uniseriate and separate and none were found on the tangential walls. Both horizontal and vertical resin-canals occur, several of them with well-preserved tyloses; the epithelial cells are thin-walled and unpitted as in *Pinus silvestris*. It may be that the abundance of tyloses is connected with the presence of fungal mycelia as in wood of *Hevea* stems recently described by Mr Brooks[3]. The medullary are uniseriate, generally 5—7 cells deep, also fusiform and with resin-canals: the preservation is not sufficiently good to admit of any definite statement as to the occurrence of horizontal tracheids.

Pityoxylon zezoense (Suzuki).

This Upper Cretaceous Japanese species was described as *Abiocaulis zezoensis*[4], but in view of the occurrence of features suggesting comparison with *Pinus* as well as with *Abies* the more comprehensive generic name is preferable. In the presence of pits on the tangential and horizontal walls of the medullary-ray cells and in the pitting of the tracheids this wood conforms to the Abietineous type. There are no ray-tracheids: the pitting on the radial walls of the ray cells agrees in part with that in *Abies* and in some of the cells there are large circular pits like those of *Pinus* Normal resin-canals are present in the second ring only, others being interpreted as traumatic. Xylem-parenchyma is

[1] Holden (13³).   [2] Conwentz (92) p. 13, Pls. I.—III., VI., VII.
[3] Brooks and Sharples (14).   [4] Suzuki (10).

sparsely distributed. There would seem to be little difference of importance between this species and wood referred to *Cedroxylon*. The number of Cretaceous examples of *Pityoxylon* might be considerably extended: for an account of French species reference should be made to Lignier and to Fliche.

Pityoxylon Pseudotsugae (Gothan).

Gothan described this species, from South Nevada and probably of Tertiary age, as *Piceoxylon Pseudotsugae*[1]: it is interesting as a type of *Pityoxylon* agreeing closely with the recent genus *Pseudotsuga* in the presence of spiral bands in the tracheids. There are both vertical and horizontal resin-canals and the ray cells have Abietineous pitting. Xylem-parenchyma occurs next the summer-wood and the epithelial cells have thick walls, features in agreement with Gothan's genus *Piceoxylon*. Bailey[2] points out that in the absence of spiral bands in the ray-tracheids the fossil species resembles *Pseudotsuga Douglasii*, while in *Pseudotsuga macrocarpa* the tracheids of the rays have spiral bands.

Fritel and Viguier[3] have described a species from Eocene beds in the Paris Basin as *Piceoxylon Gothani* in which some of the xylem-tracheids have spiral bands.

Pityoxylon pulchrum (Cramer).

A Tertiary species[4] originally described from material collected by Sir Leopold MacClintock in Banksland as *Cupressinoxylon pulchrum*. A piece of wood in the Dublin Museum labelled 'from Ballast Bay, Baring Island, given by Sir L. MacClintock' agrees very closely with Cramer's type-specimen, and as the resemblance extends to most of the anatomical characters, I believe it to be the material on which *C. pulchrum* was founded. The chief difference is that the Dublin wood has resin-canals as in the specimen described by Cramer as *Pinus MacClurii* (?) Heer[5]; in *C. pulchrum* no resin-ducts are recorded. On the other hand in the sum of its characters the Dublin specimen agrees much more closely with *C. pulchrum*. Annual rings well marked;

---

[1] Gothan (06³).   [2] Bailey (09) p. 54.

[3] Fritel and Viguier (11) p. 63.

[4] Cramer (68) p. 171, Pl. xxxiv. fig. 1; Pl. xxxvi figs. 6—8.

[5] *Ibid.* Pls. xxxv, xxxvi.

bordered pits large, approximately 25μ in diameter, in 1—2 opposite rows, sometimes in contact and slightly flattened. Medullary rays uniseriate, 1—14 cells deep, also fusiform rays containing a horizontal canal; 2 or 3 large oval pits occur on the radial walls of the ray cells and in a few cases pits on the tangential walls. Ray-tracheids with bordered pits occur on the edges of the medullary rays.

Among other Tertiary species reference may be made to *Pityoxylon parryoides* Goth.[1] from the Braunkohle of Rheinland, so named from its resemblance to the North American *Pinus Parrya,* characterised by horizontal tracheids with smooth walls and thin-walled epithelial cells; also *Pityoxylon pineoides* Kraus[2] a Sicilian Tertiary species without ray-tracheids.

Pityoxylon succinifer (Goeppert).

This species from the Oligocene amber beds of the Baltic coast was first named *Pinites succinifer*[3] and several years later fully described and admirably illustrated as *Pinus succinifer*[4]. It affords a striking illustration of the possibilities of amber as a petrifying agent and shows several features of anatomical interest. The roots are represented by pieces of wood in a pathological state: the tracheids have 1—3 rows of pits on their radial walls and some of them contain tyloses; the walls of the ray-tracheids have dentate ingrowths. The stem and branch wood is more complete. Sieve-tubes and sieve-plates are exceptionally well preserved and both cortex and pith tissues are represented. The tracheids have 1—2 rows of separate pits; a spiral sculpturing on the walls of the tracheids was mistaken by Menge for the spiral bands characteristic of the Taxineae and he named the species *Taxoxylum electrochyton.* Conwentz describes tyloses in the tracheids, also a crescentic patch of parenchyma in the wood passing into a mass of resin[5], a feature occasionally seen in recent wood. The medullary rays have 1—4 pits in the field; both ray-tracheids and horizontal resin-canals occur and in some cases

[1] Gothan (09) p. 523, figs. 3—5.
[2] Kraus (83) p. 83, Pl. I. figs. 1—3.
[3] Goeppert (41) p. 39; Goeppert and Berendt (45) A. p. 61.
[4] Conwentz (90) A. p. 26, with numerous plates.
[5] *Ibid.* (89); (90) A. p. 48; cf. Hollick and Jeffrey (09) B. Pl XXI. fig. 4.

rays are said to consist exclusively of tracheal tissue[1]. Pine needles and cones have been obtained from the amber beds.

The Fossil forests of the Yellowstone Park include examples of Pityoxylon trees some of which have been described by Knowlton[2] and Felix[3], but unfortunately the anatomical details are not as a rule well preserved. The most striking exposure of the Tertiary (probably Miocene) trees is on the slopes of Amethyst mountain (fig. 712), where a succession of forests is represented throughout the 2000 ft. of strata. Felix describes a species, *Pityoxylon fallax*, chiefly interesting from the point of view of a comparison between the stem and root wood of the same tree: the elements of the root are in general larger than those in the stem. Knowlton gives an account of *P. Aldersoni* and *P. amethystinum*, species which may be identical: the pits on the tracheids and medullary rays are seldom preserved, but the occurrence of both vertical and horizontal resin-canals is clearly shown.

VIII. PROTOPICEOXYLON. Gothan.

Protopiceoxylon exstinctum Gothan. The generic name *Protopiceoxylon* was proposed for some Lower Cretaceous wood from King Charles Land[4] possessing Abietineous characters, intermediate between *Cedroxylon* and *Pityoxylon* in having only vertical resin-canals, at least in uninjured wood. The anatomical features of the type-specimen are complicated by the occurrence of additional resin-canals in wounded portions of the stem. It is difficult to determine the precise extent of the traumatic influences, but the presence of callus-wood healing a wound leaves no doubt as to the correctness of Gothan's conclusion that certain features are abnormal and due to the effects of wounding. In the species *P. exstinctum* are also included specimens from Spitzbergen[5] and some of the material on which Cramer[6] founded his species *Pinites cavernosus*: the later specific name is not retained on the ground that the original diagnosis is incorrect and it was only after examining sections of the type-specimen that Gothan recognised the true nature of Cramer's species.

[1] Conwentz (90) A. Pl. IX. fig. 2.
[2] Knowlton (99) p. 763, Pls. CVI.—CVIII., CXII.—CXV., CXVIII., CXIX.
[3] Felix (96) p. 254.   [4] Gothan (07²) p. 32, figs. 16, 17; Pl. I. figs. 2—6.
[5] Gothan (10) p. 15, Pl. II. figs. 5—8; Pl. III. figs. 1—4, 6—8.   [6] Cramer (68).

Protopiceoxylon exstinctum shows the following characters:—
annual rings well marked; vertical resin-canals occur in the wood
but there are no canals in the medullary rays except a few of
unusually large diameter in wounded areas; there is no xylem-
parenchyma apart from the resin-canals. Tracheids with 1—2
rows of bordered pits on the radial walls, separate and circular,
also contiguous and flattened, opposite, or sometimes alternate:
in the occurrence of the Araucarian type of pitting on some
tracheids this species agrees with several types of Mesozoic wood.
Medullary rays uniseriate, characterised by well-developed Abietineous pitting; on the radial walls there are 2—4 circular and
bordered pits in the field. No undoubted ray-tracheids were
noticed; numerous small pits occur on the horizontal walls of parenchymatous cells associated with the resin-canals. The pith consists
of parenchyma with thin sclerenchymatous diaphragms.

The horizontal canals, presumably traumatic, in some of the
medullary rays resemble in their large size those in *Anomaloxylon*
but in that genus there is no Abietineous pitting on the medullary-
ray cells; similar canals are described by Jeffrey[1] in wounded
wood of *Cedrus* and other Conifers. In the occurrence of vertical
canals only in the normal wood *Protopiceoxylon* is intermediate
between *Cedroxylon*, which has no canals, and *Piceoxylon* and
Pinuxylon of Gothan (= *Pityoxylon* of Kraus), the fossil representatives of such recent genera as *Larix*, *Picea*, and *Pinus*, in which
both vertical and horizontal ducts occur. Gothan holds, and
probably with good reason, that vertical canals preceded those
in the medullary rays and regards the fossil species as a primitive
type.

A species from the Black Hills described by Knowlton[2] as
Pinoxylon dacotense agrees with *Protopiceoxylon* in having only
vertical canals, but it is not clear whether they are normal or
traumatic: *Piceoxylon* would seem to be the more appropriate
designation for Knowlton's species.

Protopiceoxylon articum sp. nov.

This species is founded on a specimen from Cape Flora, Franz
Josef Land, probably Oxfordian in age. Annual rings are distinct

---

[1] Jeffrey (03); (05).   [2] Knowlton in Ward (00) B. p. 420, Pl. CLXXIX.

and narrow; several oval or circular spaces are conspicuous in transverse section, some being true canals and others the result of decay. There are 2—3 opposite rows of bordered pits on the radial walls of the tracheids. Partially destroyed rows of resin-parenchyma occur which probably belong to secretory canals. The uniseriate and comparatively deep medullary rays, 20—30 cells, are characterised by rather thick and pitted horizontal and

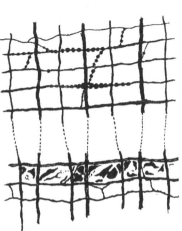

FIG. 728. *Protopiceoxylon arcticum.*
(Cambridge Botany School.)

FIG. 729. *Protopiceoxylon Edwardsi.* Longitudinal view of the thick-walled, pitted, epithelial cells of the resin-canals. (After Stopes.)

vertical walls (fig. 728); 4—5 small simple pits occur on a few of the cells and on the upper and lower edges of some of the rays are empty elements of unequal breadth which in all probability are ray-tracheids. The wood agrees in the presence of vertical canals only and in the structure of the medullary rays with *Protopiceoxylon exstinctum* Goth. In Gothan's species there are 2—4 bordered pits in the field, but the absence of a border in the Franz Josef Land wood may be a consequence of imperfect preservation.

Protopiceoxylon Edwardsi Stopes.

Founded on a branch from the Lower Greensand of Sussex, showing 17 annual rings, having the following characters[1]: a large pith nearly 3 mm. in diameter composed of parenchyma without stone-cells; tracheids with usually one row of circular bordered pits; vertical canals in the summer-wood and associated with a small amount of resiniferous parenchyma; medullary rays uniseriate, with Abietineous pitting and 2—4 more or less circular pits in the field. The small size of the resin-canals is a characteristic feature, also their thick-walled pitted epithelial cells (fig. 729). The species differs from Gothan's *P. exstinctum* in the smaller diameter of the canals, the absence of traumatic horizontal canals, and in the greater number of the vertical secretory passages.

IX. **WOODWORTHIA.** Jeffrey.

Woodworthia arizonica Jeffrey. This genus[2] is founded on specimens from the Triassic petrified forest of Arizona characterised by the occurrence of short shoots in the secondary wood comparable with those in the stem of *Araucariopitys*. In the type-specimen the annual rings are not very clearly defined: the pitting on the tracheids is definitely Araucarian. The medullary rays are uniseriate, 2—9 cells deep: they appear to have pits only on the lateral walls.

On the surface of the wood are several small scars and a few larger ones, the former representing short shoots subtended by a leaf-trace; the shoots are not infrequently branched as they pass through the secondary xylem, a feature recorded also in *Ginkgo*[3]. Jeffrey describes the short shoots as having a limited existence and disappearing in the wood at a comparatively short distance from the pith; they have no rings of growth, a character associated with short-lived leaf-spurs in recent species but a feature in which they differ from those of *Ginkgo*. The leaf-traces subtending the short shoots, in contrast to those of *Araucaria*, are not persistent throughout the secondary wood. Jeffrey regards this fact as an argument against the view that the persistence of the traces in *Araucaria* is a primitive character; but it is worthy of note that

---

[1] Stopes (15) p. 81, Pl. III. text-figs. 17—22.
[2] Jeffrey (10²), Pls. XXXI., XXXII.   [3] Tupper (11).

the leaves accompanying the shoots of *Woodworthia* are not strictly comparable with those of the foliar organs of recent Araucarias which have no short shoots in their axils. Jeffrey regards the short shoot as a primitive attribute of the coniferous stock and its occurrence in the stems of *Woodworthia* and *Araucariopitys* is held to be evidence in support of the interpretation of the seminiferous scales of Abietineous genera as metamorphosed short shoots, an interpretation which is open to question. The presence of short shoots is not a monopoly of the Abietineae and their presence in a stem may be regarded as a point of contact with *Ginkgo* as well as with Abietineous plants. Attention is called elsewhere to the probability that foliar spurs like those of *Pinus* are specialised forms of ordinary shoots. However we may interpret the characters exhibited by *Woodworthia*, the genus is an interesting example of an extinct type illustrating the combination with Araucarian characters of a morphological feature that is no longer represented in the Araucarineae.

X. ARAUCARIOPITYS. Jeffrey.

A genus founded by Jeffrey[1] on a stem from the Middle Cretaceous beds of Staten Island, New York, showing on its decorticated surface scars of short shoots and in the structure of the wood both Abietineous and Araucarian features.

Araucariopitys americana Jeffrey. The bordered pits on the radial walls of the tracheids are often contiguous and flattened though in places separate and circular, usually arranged as a single row. All the walls of the ray cells are pitted as in the Abietineae. The large number of vertical resin-canals (fig. 758, C, D, page 323) in a single tangential row is regarded as evidence of traumatic origin. This conclusion is based on the fact that when canals are present in wood that is known to have been wounded they occur in crowded tangentially arranged rows in contrast to their sparser distribution in the normal wood. There are no canals in the uninjured tissues of *Araucariopitys*. Diaphragms of sclerous cells occur in the pith as in some recent Abietineae. The short shoots are shown in tangential section of the stem and stated to be accompanied by a subtending leaf.

---

[1] Jeffrey (07) Pls. XXVIII—XXX.

It is suggested by Jeffrey that *Araucariopitys* may be the stem of Heer's genus *Czekanowskia*[1], but there is no proof of a connexion and evidence afforded by *Czekanowskia* favours an alliance with the Ginkgoales. *Araucariopitys* is regarded by Jeffrey as 'unquestionably Araucarian' though 'nearer in structure to the Abietineae than any other known Araucarian genus living or extinct.' The chief Araucarian feature would appear to be the occurrence of flattened tracheal pits, but Jeffrey points out that not only are the pits in a single row and sometimes separate, but even when in double rows the pits may be opposite. The other characters, *e.g.*, the pitting of the medullary-ray cells and the presence of resin-canals, are more Abietineous than Araucarian[2]. In view of the occasional occurrence in Abietineous genera of contiguous and alternate pits on the tracheids it is open to doubt whether there are adequate grounds for assuming a definite Araucarian affinity. *Araucariopitys* is one of several genera described by Jeffrey and other American authors exhibiting features shared by recent Araucarineae and Abietineae which are claimed as evidence of the greater antiquity of the Abietineous type. In this genus the balance of evidence would seem to be in favour of an Abietineous alliance, the tendency towards an Araucarian pitting of the tracheids being reminiscent of ancestral types in which that character was more pronounced.

XI. **PROTOCEDROXYLON.** Gothan.

Protocedroxylon araucarioides. Gothan. The type-species of the genus founded on Upper Jurassic wood[3] from the Esmarks Glacier, Spitzbergen, is one of the most striking examples of a group of generalised types from Upper Jurassic strata especially from the Arctic regions. The generic name emphasises the Abietineous characters while the specific term gives expression to the presence of Araucarian features. The following account is based chiefly on the description by Gothan, and a few additional facts are taken from an account of some specimens from Liassic and Oolitic rocks on the Yorkshire coast by Prof. Jeffrey[4] and Miss Holden[5].

---

[1] See page 63.　　　　　　　　　　　　　[2] See also Gothan (10) p. 30.
[3] Gothan (10) p. 27, Pls. v., vi.　　　　　[4] Jeffrey (12) p. 533, Pl. i. figs. *a, b*.
[5] Holden (13²) p. 538, Pl. xl. figs. 17—21.

Annual rings well marked; bordered pits on the radial walls of the tracheids in 1—3 rows; in the Spitzbergen wood these are from 20 to 24 μ in height, dimensions larger than in recent Araucarineae, and in the type-specimen the pits are always contiguous, more or less flattened and alternate—that is Araucarian; in the English specimens the pits when in a single row are often separate and circular but equally often contiguous. Jeffrey points out that the alternate pits when in more than one series are less crowded than in Araucarian wood. There are no Sanio's rims. The medullary rays are uniseriate and characterised by Abietineous pitting on the horizontal and tangential walls; on the radial walls there are 1—3 circular, apparently unbordered, pits in the field. Xylem-parenchyma is practically absent. An interesting feature is the abundance both in the Spitzbergen and Yorkshire material of tyloses in many of the tracheids, a feature occasionally met with in recent Conifers[1] as in some other fossil species. The pitting of the tracheids in the type-specimen may be described as exclusively Araucarian, but in the English specimens separate pits also occur though on the whole the Araucarian type is dominant. The pitting of the medullary rays is on the other hand definitely Abietineous. The American authors, particularly Miss Holden[1], consider that the absence of Sanio's rims suffices to tip the balance on the Araucarian side. On most of the tracheids the crowding of the pits precludes the occurrence of Sanio's rims and in other cases their absence is not necessarily an original feature. Abietineous pitting is recorded by Jeffrey in the cone-axis of an *Agathis* and it has also been found in *Araucaria*; but in the Araucarineae it is very exceptional: its occurrence as a constant feature in *Protocedroxylon* may be regarded as an indication of Abietineous relationship. No substantial assistance is afforded by impressions in Spitzbergen rocks: the abundance of *Elatides* is consistent with the occurrence of Araucarian wood, but impressions of Abietineous Conifers afford at least as strong an argument in favour of the occurrence of Abietineous wood.

Protocedroxylon scoticum (Holden). This species, described by Miss Holden[2] under the generic name *Metacedroxylon* from Corallian beds on the Sutherland coast of Scotland, is founded on a piece

---

[1] See page 178.   [2] Holden, R. (15).

of stem showing 75 rings of growth. There are no resin-canals and no xylem-parenchyma; the tracheids have uniseriate bordered pits on the radial walls only and they are almost invariably compressed by mutual contact; the presence of a torus is a feature characteristic of the Abietineae and not of the Araucarineae. Bars of Sanio are present but there are no rims of Sanio. Tyloses are abundant in the tracheids. The medullary rays are 2—20 cells deep, generally uniseriate though occasionally biseriate; the pitting is of the Abietineous type. In the abundance of tyloses and in other characters the wood resembles *Protocedroxylon araucarioides* Goth. a species transferred by Miss Holden to *Metacedroxylon*, but it differs in the absence of tangential tracheal pits and in the occurrence of biseriate medullary rays.

Some fossil wood, which is not very well preserved, from Middle Cretaceous rocks near Iefren in the Gulf of Tripoli is described by Negri as *Protocedroxylon Paronai*[1]. The tracheal pits where biseriate are often alternate and compressed; rims of Sanio are absent: the latter feature, deduced from negative evidence, and considering the state of preservation, is surely of little value.

The presence of Araucarian pitting on the tracheids in several Jurassic species is far from surprising in view of the prevalence of that type of pitting in Palaeozoic stems; moreover an admixture of characters is a natural result of progressive development. It is a matter of opinion with regard to the relative value of tracheal or medullary-ray pitting whether *Protocedroxylon* should be placed nearer to the Araucarineae or to the Abietineae. Miss Holden[2] discards the name *Protocedroxylon* for *Metacedroxylon* on the ground that the former implies Abietineous affinity, a fine shade of difference that hardly gives adequate expression to her conclusion that '*Metacedroxylon araucarioides* cannot be other than an Araucarian Conifer.'

XII. XENOXYLON. Gothan.

Gothan[3] instituted this generic name for some Upper Jurassic wood, originally described by Cramer[4] from Green Harbour,

[1] Negri (14) p. 340, Pl. v. figs. 1—6; Pl. vi. figs. 1—3.
[2] Holden, R. (14) p 538   [3] Gothan (05) p. 38.
[4] Cramer (68) Pl. xl.; Schroeter (80) p. 7.

Spitzbergen, as *Pinites latiporosus* in order to give expression to the combination of distinctive features in both the tracheal and medullary-ray pitting. The most striking characteristics are the very large size of the bordered pits of the tracheids, the occasional (*X. phyllocladoides*) or constant (*X. latiporosum*) occurrence of contiguous and vertically flattened and transversely elongated pits on the radial walls, the absence of pits on the transverse and tangential walls of the medullary-ray cells, and the presence of large simple pores on the lateral walls. There is no definite evidence as to the nature of the foliage, though Nathorst[1] has suggested a possible connexion between *Elatides* and *Xenoxylon*. *Xenoxylon* would seem to have been widely distributed in later Jurassic floras.

Xenoxylon conchylianum Fliche.

Fliche[2] refers to *Xenoxylon* a piece of wood from the Muschelkalk of the Vosges characterised by (i) the occurrence of uniseriate bordered pits compressed above and below and occupying the whole breadth of the tracheids and (ii) a single elliptical pit in the field. The regularity of the pores in the medullary-ray cells and the fact that they are most clearly shown where the preservation is best favour the conclusion that they are an original feature. The medullary rays are usually from 5 to 10 cells in depth. There are no resin-canals and no xylem-parenchyma. Fliche states that the pits on the tracheids are rather less flattened than in *X. latiporosum* and they occupy a greater breadth of tracheal wall. The photographs accompanying the description are unfortunately too small to show the important characters. The annual rings are faintly marked and the summer elements are confined to 4—5 rows

Despite the resemblance between this Triassic species and those previously described it is by no means certain that Fliche's species is generically identical with the younger types. Large simple pores occur in the medullary-ray cells of recent Conifers belonging to different families, and it is not uncommon to find the bordered pits on the radial walls of tracheids in contact and slightly flattened in wood normally characterised by circular and separate pits. In

[1] Nathorst (97) p. 42.
[2] Fliche (10) p. 232, Pl. XXIII. figs. 4—5.

this connexion it is noteworthy that Fliche states that the pits in
his wood are occasionally circular.

Xenoxylon latiporosum (Cramer).

Gothan's examination of the specimens on which Cramer
founded this species[1] enabled him to confirm the main points of
the original description: he regards Cramer's species *Pinites
pauciporosus* as identical with the type-species with which he also
identifies *Araucarioxylon koreanum* Felix[2]. *Xenoxylon latiporosum*
is characterised by the large size of the pits on the radial walls of

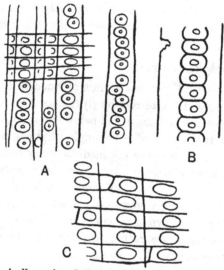

FIG. 730. A, *Xenoxylon phyllocladoides.* B, C, *Xenoxylon latiporosum.*
(A, after Gothan; B, C, after Cramer.)

the tracheids, 20—40μ broad and 15—20μ high, their vertically
flattened form (fig. 730, B) and their occurrence in one or two contiguous rows, the pits of double rows being generally opposite. The
medullary rays are uniseriate, reaching 17 cells in depth, characterised by the narrow form of the cells, the absence of pits on the
horizontal and vertical walls and by the presence of large simple
pores on the lateral walls, usually one pore in the field (fig. 730, C)
which it almost fills, or occasionally two; there is no xylem-

[1] Gothan (10) p. 23, Pls. IV., v.   [2] Felix (87) Pl. xxv. fig. 1.

parenchyma but tylose-like cross-walls occur in some of the xylem elements.

The nature of the pitting led Kraus[1] to include Cramer's species in *Araucarioxylon* and Miss Holden[2], who records this species from the Yorkshire coast, regards the absence of Sanio's rims as evidence of Araucarian affinity. The medullary-ray pitting is, however, very different from that in recent Araucarineae and the absence of Sanio's rims may well be a natural consequence of the crowded arrangement of the tracheal pits.

Xenoxylon phyllocladoides Gothan.

This species, founded on material from the Bathonian of Russian Poland[3], differs from *X. latiporosum* in the not infrequent occurrence of separate and circular pits on the tracheids: in it are included specimens from Liassic rocks at Gallberges near Salzgitter in Germany described by Conwentz[4] as *Araucarioxylon latiporosum* (Cram.) and, with some hesitation, *Cupressinoxylon Barberi* Sew.[5] from the Yorkshire coast. The tracheal pits are uniseriate, flattened or separate and circular (fig. 730, A), or in two rows, generally though not invariably opposite; they vary in size from $22 \times 30\mu$ to $24 \times 36\mu$; the medullary rays are generally less than 10 cells deep and in pitting agree with those of the type-species. This species is recorded from Poland, Spitzbergen, King Charles Land[6], Yorkshire, and Germany.

Though similar to *Araucaria* and *Agathis* in the flattened contiguous pits, *Xenoxylon* differs in the elliptical form of the border and pore, also in the occurrence of separate and circular pits and in the occurrence of opposite pairs. In the form of the pits on the tracheids *Xenoxylon* resembles the Palaeozoic species *Dadoxylon protopityoides* Fel.[7] and pits of similar form occur in the wood of the recent Magnoliaceous plant *Drimys Winteri*[8]. From the Abietineae the genus is distinguished by the restriction of the medullary-ray pitting to the radial walls, though the large pores

[1] Kraus in Schimper (72) A. p. 384.
[2] Holden, R. (14) p. 536, Pl. xxxix. figs. 5, 6.
[3] Gothan (06²) p. 454, fig. 4; (10) p. 36, Pl. vi. figs. 9, 10, etc.
[4] Conwentz (82) p. 170
[5] Seward (04) B. Pl. vii.; Holden, R. (14) p. 535.
[6] Gothan (08²) p. 10, figs. 3—9.
[7] Felix (86) A. Pl. v. fig. 4.   [8] Groppler (94) Pls. i., ii.

(fig. 730, C) in the field resemble those of some Pines and other Abietineae as also those of *Sciadopitys* (fig. 693, N): in *Xenoxylon* there are no resin-canals and no xylem-parenchyma. Gothan considers that while differing in the sum of its characters from any other type of Conifer, *Xenoxylon* shows most resemblance to Gothan's genus *Phyllocladoxylon* (= *Mesembrioxylon* Sew.)[1], a genus including fossil species which suggest affinity not only with *Phyllocladus* but with other members of the Taxaceae. In his account of *Xenoxylon phyllocladoides* from King Charles Land Gothan describes instructive examples of the effect of the action of fungal hyphae on the structure of tissues. The genus may be described as a generalised type exhibiting features shared by the Araucarineae and Taxaceae.

XIII. ANOMALOXYLON. Gothan.

Anomaloxylon magnoradiatum Gothan. Gothan proposed this name for some Upper Jurassic wood from Spitz-bergen[2] which cannot be definitely assigned to a family-position: its most striking feature is the occurrence of large spindle-shaped medullary rays containing a large 'canal,' or spaces lined by a single layer of cells (fig. 731). The rings of growth are well marked, the summer-wood being composed of a very few rows of tracheids in abrupt juxtaposition to the larger spring elements, a character associated with roots. There are no vertical resin-canals and no regular or typical horizontal canals. Xylem-parenchyma is rare or absent. The bordered pits on the radial walls of the tracheids are in 1—2 rows, separate and circular or, more frequently, con-tiguous and more or less flattened but, if in two rows, not alternate. The medullary rays are uniseriate, generally 5—7 cells deep or in places forming broad and deep spindle-shaped areas either empty or containing a large circular canal-like passage. These peculiar rays, as seen in a

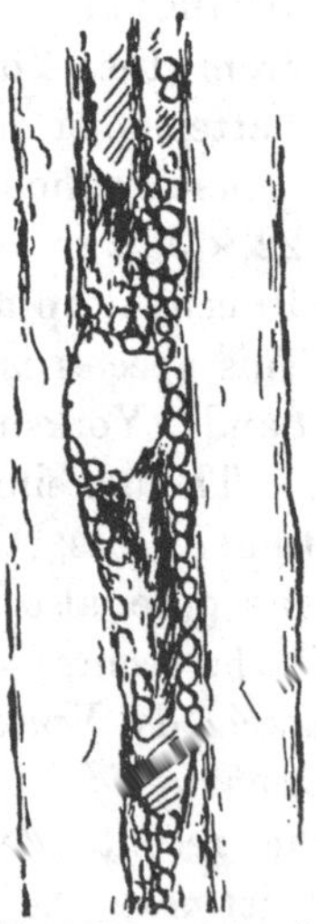

FIG. 731. *Anomalo-xylon magnoradi-atum*. Medullary ray showing the small cells and a resin-canal. (After Gothan.)

tangential section of the wood, are a conspicuous feature and are

---

[1] See page 303.   [2] Gothan (10) p. 10, Pl. I. figs. 9—11; Pl. II. figs. 2, 3.

often lined by a layer of small cells (fig. 731). In the presence of these large medullary rays *Anomaloxylon* resembles *Thylloxylon*, but in the latter genus the rays are smaller and more uniform in size. Gothan discusses the nature of these medullary rays and inclines to the view that they agree more closely with abnormal or traumatic formations in certain Conifers than with any normal structures. There are no pits on the horizontal or tangential walls of the ray cells and there are 2—3 simple circular pits in the field.

The general impression gained from an examination of Gothan's photographs is that no true canals occur, and that the peculiar medullary rays owe their form to partial decay of abnormal patches of parenchyma possibly produced as the result of wounding.

Though on the whole nearer in structure to the Taxodineae[1] than to any other family *Anomaloxylon* is a type which cannot be assigned to a definite position.

XIV. THYLLOXYLON. Gothan.

Thylloxylon irregulare Gothan. The generic name *Thylloxylon* was given to a single species of Upper Jurassic age from Spitzbergen[2] on account of the occurrence of tüllen-like parenchyma in horizontal canal-like spaces in some of the larger medullary rays. The wood is characterised by separate bordered pits in the summer tracheids and 1—2 rows of alternate contiguous, Araucarioid, pits on the spring elements; xylem-parenchyma occurs only at the end of the year's growth. The medullary cells have Abietineous pitting and there are 2—3 small circular, apparently simple, pits in the field, or occasionally only one in the region of the late wood. The rays are uniseriate or 2—3 cells broad and some medullary rays closely resemble those of certain Abietineae possessing horizontal resin-canals; but in *Thylloxylon* there are no true canals. The central parenchyma of some of the broad rays is replaced by a canal-like space and these spaces are often filled with spherical tüllen-like tissue, a feature shared with *Anomaloxylon*, but in that genus there is no Abietineous pitting on the medullary-ray cells. There are no vertical resin-canals.

[1] That is *Athrotaxis, Cryptomeria, Sequoia, Taxodium*, and *Sciadopitys*, genera which are now (see page 126) assigned to different families.
[2] Gothan (10) p. 34, Pl. VI. figs. 2—8.

XV. **PLANOXYLON.** Stopes.

Dr Marie Stopes[1] has recently instituted this generic name for a piece of Coniferous wood from Middle (or Upper?) Cretaceous rocks in New Zealand and in it she also includes the Liassic species *Araucarioxylon Lindleii* (Witham). The genus is a striking example of a combination of Araucarian and Abietineous characters[2], and, as Dr Stopes points out, it resembles in this respect *Cedroxylon transiens* Goth. and other generalised types.

Planoxylon Hectori Stopes. The type-specimen, from Amuri Bluff, New Zealand, is part of a stem 150 years old or more. The rings of growth are well marked; the tracheids have 1—3 rows of alternate and hexagonal bordered pits on the radial walls and there may be a single row of separate pits on the elements at the end of an annual ring. The medullary rays are nearly always uniseriate, 1—24 cells deep but usually from 3 to 9 cells in depth; all the walls of the ray cells are pitted and there are 1—2 vertical rows of three pits in the field in the neighbourhood of the spring tracheids and generally a single vertical pair in the region of the late wood. Xylem-parenchyma appears to occur only between the spring tracheids and the latest formed wood of the previous year. Like many other fossil stems this species indicates the existence of Conifers with typical Araucarian pitting on the tracheids and equally well defined Abietineous pitting on the medullary-ray cells. It is especially interesting as showing the presence in the southern hemisphere of a type very similar to *Cedroxylon transiens* and other species recorded from high northern latitudes.

Planoxylon Lindleii (Witham).

This Liassic species from Whitby was originally referred by Witham[3] to the genus *Peuce*; subsequently included in *Araucarioxylon*[4] it has recently been transferred by Dr Stopes to her new genus *Planoxylon*[5]. The pitting of the tracheids is essentially Araucarian; there are 1—3 rows of alternate hexagonal pits on the radial walls, but the pitting of the medullary-ray cells, as Dr Stopes has shown, is typically Abietineous.

[1] Stopes (16)

[2] πλανάομαι, to wander; suggesting that 'the forms comprising the genus were moving from one position to another in a systematic sense.'

[3] Witham (33) A. p. 58, Pls. IX., XV.   [4] Seward (04) B. p. 56, Pls. VI., VII.

[5] Stopes (16) pp. 118, 120, text-figs. 6, 7.

CHAPTER XLV.

CONIFERALES.

SHOOTS, CONE-SCALES, ETC.

THE majority of the vegetative and fertile shoots, cone-scales, seeds, etc., selected in illustration of the past history and geographical distribution of the Coniferales are described under the different families enumerated on page 124. Under each family are included not only specimens which, with a fair amount of confidence, can be assigned to a family-position but also genera of doubtful affinity which it has been contended afford evidence of greater or less value in favour of an alliance with the family under which they are described. The inclusion of certain genera in a chapter or section devoted to a particular family does not necessarily mean that they show clear evidence of relationship to that family: many of the genera might with equal propriety be relegated to Chapter L, which is devoted to *Coniferales incertae sedis.* On the other hand some genera included in that category would by other authors be given a place in the Araucarineae or some other family. I have endeavoured to state the different views expressed by authors with regard to the affinity of imperfectly known genera, but in many instances the available data do not afford any trustworthy evidence of relationship to existing types.

ARAUCARINEAE.

DAMMARITES. Presl.
PROTODAMMARA. Hollick and Jeffrey.

The distinctive characters of the recent genus *Agathis* (*Dammara*) are briefly described in Chapter XLIII. Fossil records bearing on the past history of *Agathis* are more meagre and more difficult of interpretation than those relating to *Araucaria*. The evidence at present available points to the greater antiquity of *Araucaria* at least as regards the type of cone characteristic of that genus.

On the other hand the type of foliage-shoot represented by existing
species of *Agathis*—without taking into account the Palaeozoic
leaves assigned to *Cordaites*, some species of which bear a close
superficial resemblance to those of certain examples of the recent
genus—is widely represented in Rhaetic and Jurassic floras by
Podozamites[1]. There is, however, no proof that *Podozamites* was
nearly related to *Agathis*, and, indeed, such information as we have
with regard to the reproductive organs of that genus does not
point to any very close Araucarian affinity. Fossil wood gives
no help towards a distinction between the two members of the
Araucarineae nor do impressions of vegetative shoots materially
aid us.

Palaeobotanical literature contains a few records of leaves
referred to *Dammara* or *Dammarites* but in no case is there any
conclusive evidence of generic identity of the fossils with the
recent genus. Leaves from Lower Cretaceous rocks in Bohemia
described by Velenovský and by Frič and Bayer[2] as *Dammaro-
phyllum striatum* and *D. bohemicum* exhibit a close agreement in
shape and venation with those of some species of *Agathis*, though
they differ but slightly from some forms of *Podozamites*, e.g.,
P. Reinii Geyl. (fig. 814, p. 456). Other leaves that may belong to
plants similar to *Agathis* are represented by *Dammarites caudatus*
and *D. emarginatus* Lesq. from the Dakota series[3]: these, probably specifically identical, forms present, as Lesquereux says, a close
resemblance to *Agathis robusta*. It is impossible without additional
data to determine the true position of these and similar leaves
though it is permissible to regard them as possible examples of
the foliage of Conifers closely allied to *Agathis*. Similarly, some
detached leaves from Cretaceous and Tertiary strata referred to
Podozamites may well be more akin to *Agathis* especially in view
of the fact that *Podozamites* is essentially a Rhaetic and Jurassic
genus. The leaves figured by Saporta[4] from Lower Cretaceous
beds in Portugal as *Podozamites ellipsoideus* agree closely with
those of *Agathis*. In the case of separate linear leaves like those
described by Hollick[5] from the Cretaceous of Long Island as

[1] See page 447.   [2] Frič and Bayer (01) B. p. 96.
[3] Lesquereux (91) p. 32, Pl. i figs. 9—11.
[4] Saporta (94) B. Pl. xxxiii. fig. 5; Pl. xxxv. fig 12.
[5] Hollick (18) Pl. 163, figs. 2, 3.

Podozamites lanceolatus the term *Desmiophyllum* would be a more appropriate generic designation, the name *Dammarites* being adopted for broader foims. This distinction is purely arbitrary and it must be admitted that there is no substantial justification for the use of a generic name implying affinity with *Agathis*. Unless there are adequate grounds for assuming generic identity of detached Tertiary and Cretaceous leaves with *Podozamites* it is inadvisable to make use of that designation. As Schenk[1] points out Velenovský's Tertiary species *Podozamites miocenicus* may be a leaf of *Agathis* or possibly a *Podocarpus*.

Dammarites Bayeri Zeiller.

This name was given to some oval-lanceolate leaves from Upper Cretaceous beds in Bulgaria varying in length from 10 to 12 cm. and from 15 to 30 mm. broad agreeing closely with Heer's *Podozamites marginatus* from the Cenomanian of Greenland but wisely excluded by Zeiller[2] from that genus, though on grounds which are no longer cogent if the interpretation of *Podozamites* impressions as shoots and not pinnate leaves is accepted.

Ettingshausen[3] records two species of *Dammarites* from Tertiary rocks in New Zealand: *Dammarites Oweni* includes in addition to leaves a cone-scale, the impression of a cone, and some petrified wood of the Araucarian type. There is no proof that these *disjuncta membra* belong to the same plant though it is not improbable that they are parts of a Conifer closely allied to *Agathis*. Ettingshausen's second species *D. univervis* is founded on a leaf and a supposed cone-scale of doubtful value.

The data furnished by leaves alone are of little value. In addition to the cone described from New Zealand by Ettingshausen other examples are recorded as species of *Dammarites* but without any satisfactory evidence of affinity to the recent genus, *e.g.*, *Dammarites albens* Presl.[4] from the Quadersandstein of Bohemia and *D. crassipes* Goepp.[5] These two species are united by

[1] Schimper and Schenk (90) A. p. 279.
[2] Zeiller (05²) p. 17, Pl. vii. figs. 8—11.
[3] Ettingshausen (87) p. 15, Pl. i. figs. 20—24.
[4] Sternberg (38) A. Pl. lii.; Corda in Reuss (46) B. Pl. xlviii.; Goeppert (50) p. 237; Schimper and Schenk (90) A. p. 279, fig. 292 *b*
[5] Goeppert (50) Pl. xlv. fig. 6; Corda in Reuss (46) B.

Velenovský under a single type which he calls *Krannera mirabilis*[1]
from a name suggested by Corda: additional examples superior
in preservation to those previously figured are illustrated in
Velenovský's memoir on the Bohemian Cretaceous Gymnosperms.
Velenovský regards the supposed cones as stems bearing crowded
woody scales which originally had long *Cordaites*-like leaves attached to a transverse ridge just internal to the thickened distal
ends: there appears to be no absolute proof in support of this
connexion between scales and foliage-leaves, but one specimen
figured shows portions of leaf-like organs attached to two of the
scales, though these may be petiolar and not pieces of laminae.
Reference is made elsewhere to the *Krannera* leaves.   It is probable
that, as Velenovský believes, the supposed cones are stems similar
to tuberous Cycadean species but it is doubtful if they were provided with leaves of the type included in *Krannera mirabilis*.
Schmalhausen[2] figures a Tertiary cone from Russia as *Dammara
Armaschewskii* which in the form of the distal ends of the scales
resembles *Araucaria brasiliensis* and species of *Pinus*, but some
detached scales agree closely in shape and in the possession of a
single seed with those of *Agathis*. Small detached cone-scales
of Tertiary age, described by Schmalhausen[3] as *Dammara Tolli*
from the New Siberian Islands,
(fig. 732), may be allied to *Agathis*;
they agree generally with those of
D. borealis and other western types.
It is, however, from detached cone-
scales obtained from Cretaceous
strata in Greenland and some
European localities but especially

FIG. 732.   *Dammarites Tolli*. (After
Schmalhausen; nat. size.)

from the Eastern United States that the most promising information has been gained.   Hitchcock first recorded these scales
from Martha's Vineyard and spoke of them as 'seed-vessels' of
some Coniferous plants[4], but it was Heer[5] who compared them
with the cone-scales of *Agathis*. The latter author described
several examples from Cenomanian strata in West Greenland as
Dammara borealis, D. microlepis, etc.

[1] Velenovský (85) B. p. 1, Pls. I., IV.   [2] Schmalhausen (83) p. 313, Pl. xxxvi.
[3] *Ibid.* (90) p. 14, Pl. I. fig. 19.   [4] Hollick (06) p. 38.   [5] Heer (82).

Dammarites borealis (Heer). Though it is clearly impossible to define with any precision the limits of species based on detached scales varying considerably in size and shape, several types have been recorded, particularly from different localities on the Atlantic Coastal plain of North America[1]. The larger forms may conveniently be included in *Dammarites borealis* Heer and smaller forms are illustrated by *Protodammara speciosa* Holl. and Jeff.[2] It is probable that these two types are generically identical, but the name *Protodammara* implies the presence of certain structural features while Heer's species is founded on

casts or impressions. A specimen of the latter species from Greenland is shown in fig. 733, the scale is 22 mm. broad and is characterised by several parallel lines, either vascular bundles or resin-canals, and the white patches represent some exuded resinous material. Other Greenland examples are more elongated basally and are identical in shape with the smaller scales from Staten Island seen in fig. 758, E, F,

FIG. 733. *Dammarites borealis*. Cone-scale from Igdlokungnak, West Greenland. (Stockholm Museum; nat. size.)

page 323. Fossils of similar form were described by Heer from the same locality as *Eucalyptus Geinitzii*[3]. Krasser[4] and some other writers have retained the generic name *Eucalyptus* on the ground of association with *Eucalyptus*-like leaves. Hollick speaks of scales like *D. borealis* as 'among the most abundant and characteristic remains found in the Cretaceous deposits of America and Europe[5]': he adds that the name *Dammarites* is chosen for the sake of convenience rather than from a conviction that it represents their true generic relationship. Newberry in describing this type from the Amboy clays states that some of the scales have grooves, corresponding to the dark lines in fig. 733, filled with amber[6], and anatomical evidence derived from *Protodammara* supports the view that the cone-scales were rich in resinous substance. Both

[1] Hollick (97) Pl. xi. figs. 5—8; (06) p. 37; Newberry and Hollick (95) p. 46.
[2] Hollick and Jeffrey (06) p. 199, Pl. i. figs. 5—13; Pl. ii. figs. 1—5.
[3] Heer (82) p. 93.
[4] Krasser (96) B. Pl. xvi. fig. 6.
[5] Velenovský (89) Pl. i. figs. 28, 29.
[6] Newberry and Hollick (95) p. 47 and see also Berry (07).

large and small cone-scales are recorded by White[1], Berry[2], Knowlton, and especially by Hollick from several places on the Atlantic Coastal plain. Knowlton's species *Dammarites acicularis*[3] is probably identical with *D. borealis*. As examples of smaller forms reference may be made to *D. northportensis*[4], *D. minor*[5], and the scales now included in *Protodammara*. These numerous scales occur as detached specimens and without seeds, but their resemblance to the cone-scales of *Agathis* and the anatomical features exhibited by the lignitic examples described by Hollick and Jeffrey afford strong arguments in favour of an affinity to *Agathis*. We have no proof as to the nature of the vegetative shoots of the parent-plants. Newberry states that in the Amboy clays the scales occur in association with shoots like Heer's *Juniperus macilenta*, in some cases, apparently, attached. On the other hand Krasser considered the association of the specimens from Moravia, which he called *Eucalyptus Geinitzii*, with dicotyledonous leaves as evidence of original connexion.

Protodammara speciosa Hollick and Jeffrey.

It is probable that these scales[6] are generically identical with the larger forms referred to *D. borealis* and other species, *e.g.*, *D. cliffwoodensis*[7], but until anatomical evidence is obtained a distinction should be recognised. The type-specimens were found at Kreischerville, Staten Island, New Jersey, in Middle Cretaceous beds, and the species is thus defined:—kite-shaped cone-scales from 4 to 6 mm. long by 4—6 mm. broad above, abruptly narrowed from about the middle to the base (fig. 758, E, F, page 323), rounded, incurved, and apiculate above; resin-ducts five or more, extending down the lower surface of the limb; seed-scars three in number, crescentically arranged above the middle and approximately in the broadest part of the scale, with the central one higher up than the laterals. Near the base of a scale there is a single vascular bundle with the xylem uppermost: at a higher level a single strand with reversed orientation is given off and the original

---

[1] White (90) p. 97. [2] Berry (03) Pl. XLVIII. figs. 8—11.
[3] Knowlton (05) Pl. XV. figs. 2—5.
[4] Hollick (04) p. 405, Pl. LXX. figs. 1, 2. [5] *Ibid.* (12) Pl. II. figs. 35—37.
[6] Hollick and Jeffrey (06); (09) B. p. 46, Pls. IV., X., XIV.—XVI.
[7] Hollick (97) Pl. XI. figs 5—8.

bundle divides into three. In the lower portion of the scale there are seven resin-canals and above these is a band of transfusion-tracheids surrounding and connecting the vascular bundles. In median longitudinal section a scale shows a terminal spinous process similar to that in the scales of *Conites Juddi*[1] (fig. 734); on the adaxial side of this the scale is swollen and internal to the swollen part is a small pit marking the position of the middle of the three seeds. The upper surface of the scales is covered with periderm and stone-cells occur in the ground-tissue. In anatomical characters *Protodammara* resembles the scales of recent Arauca-rineae more closely than those of any other Conifers, and in the absence of a definite ligule and in the relation of the seeds to the scale the fossil scales are similar to those of *Agathis*. In the description of a Scotch Upper Jurassic cone, *Conites Juddi*, attention is called to a close resemblance in anatomical features to *Protodammara*. The American scales occur in association with shoots of the type represented by *Brachyphyllum macrocarpum* Newb.[2] (=*B. crassum*) (fig. 758, G), an association noted also in other localities than Kreischerville. The structure of these shoots is described under the genus *Brachyphyllum*[3], but as regards the scales the important point is that if this association means original connexion, the habit of the parent-plant was wholly different from that of any recent *Agathis* or *Araucaria*. *Brachyphyllum macrocarpum* is regarded by Hollick and Jeffrey as certainly Araucarian. Wood of the Araucarian type is also found in associa-tion with the *Protodammara* scales and the *Brachyphyllum* shoots. Considering the cone-scales by themselves, their position would seem to be next to *Agathis* though they differ in bearing three seeds in place of the single seed in the recent genus: the number of seeds borne on the larger scales such as *D. borealis* is not known. *Protodammara* affords an interesting illustration of the co-existence of characters now characteristic of the Araucarineae with others no longer exhibited by members of that family: assuming a con-nexion between *Brachyphyllum macrocarpum* and the cone-scales, the habit of the vegetative shoots furnishes a further illustration of a wider range in the morphological features of fossil Conifers allied to existing Araucarineae.

[1] See page 252.   [2] Hollick and Jeffrey (09) B. p. 33.   [3] See page 322.

Conites Juddi Seward and Bancroft.

This name was given to partially petrified cones of Upper Jurassic age collected by Hugh Miller on the North-east coast of Scotland[1]: the fossils though differing in size and to some extent in form are included under one specific term but distinguished as

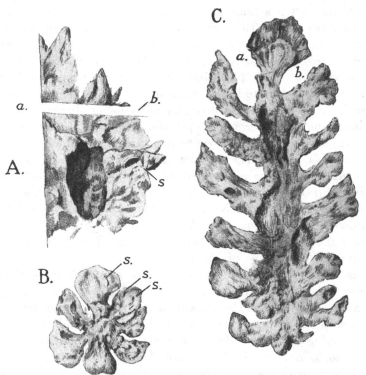

FIG. 734. *Conites Juddi.* A, *forma* γ; *ab*, space where a section was cut; *s*, seed. B, *forma* α; *s*, spaces, probably resin-canals, filled with crystalline materials and simulating seeds. C, *forma* β; *ab*, ridge.

forma α, β, γ, and δ (fig. 734). The type-specimens are in the Royal Scottish Museum, Edinburgh. Spirally disposed thick scales are attached by a comparatively narrow base to a thick axis and the individual scales agree closely in shape with those described as *Dammarites borealis* and with smaller forms referred

[1] Seward and Bancroft (13) p. 873, Pl. I. figs. 9—12; Pl. II. figs. 14—21.

by Hollick and Jeffrey to *Protodammara*. The parenchymatous tissue of the cone-scales contains several thick-walled idioblasts and the resin-canals and spaces form a conspicuous feature. A series of vascular bundles runs radially through the scale, but no evidence has been obtained of the occurrence of a double set of vascular strands like those in *Protodammara*. The seeds—the number of which, whether one or more, cannot be determined— lie in a depression near the proximal end of the scales and there

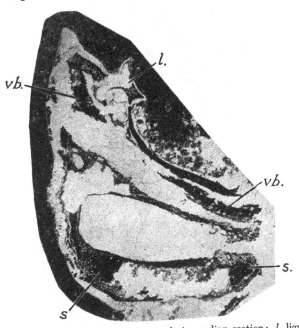

FIG. 735. *Conites Juddi, forma* a; cone-scale in median section; *l*, ligule; *s, s,* (?) resinous material; *vb,* vascular bundles. (× 6.)

is a ligule on the abaxial side of the seed or seeds (fig. 736, B, *l*). The uppermost scales on the two cones shown in fig. 734, B and C, illustrate the striking similarity to such detached scales as those of *Dammarites borealis*: the raised patches, *s, s,* simulating seeds, are formed by a crystalline substance filling cavities in the scales and probably corresponding to the resin-ducts which form a charac-teristic feature in the scales of *Dammarites* and *Protodammara*. A cone-scale from the specimen represented in fig. 734, B shows

in longitudinal section (fig. 735) a large cavity in the lower part of
the scale containing at each end a dark patch of some secreted
substance *s, s*; above this is a vascular strand *vb*, extending into
the distal end of the scale near which is a ligular outgrowth *l*, and
below this is a depression on the upper face of the scale in which

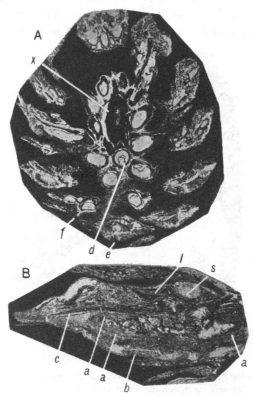

Fig. 736. *Conites Juddi, forma* δ.   A, Longitudinal section of cone; *x*, xylem,
d, seed, *e, f*, cone-scales.   B, Scale in longitudinal section; *c*, vascular bundle;
a, periderm; *b*, palisade-tissue; *l*, ligule; *s*, seed.   (After Seward and Bancroft.)

a seed was originally situated.   The cones shown in fig. 734, B, C,
have lost their seeds and indicate a persistent habit in contrast
to the cones of *Agathis* and, presumably, the cones which possessed
scales like *Dammarites borealis*.   The cone seen in fig. 734, A, is
probably younger; the scales are more crowded and in one of

them is the cast of a seed, s. One of the scales, 1·7 cm. long, of this cone is represented in section in fig. 736, B: a vascular bundle, c, runs through the length of the tissues towards the blunt spinous distal end above which is a prominent hump and next to this a ligule, l, close to the depression in which the seed, s, was situated. Below the vascular bundle, c, a band of periderm surrounds a central area of decayed tissue, a, a. Next the lower surface at b are a few layers of palisade cells, a characteristic feature. The specimen shown in section in fig. 736, A, is described as *forma* δ: the secondary xylem, x, of the axis is not sufficiently well preserved to throw any light on the nature of the tracheal pitting. The large (white) cavities at first sight suggesting seeds near the axis are bounded by periderm and correspond to the partially destroyed tissue in fig. 736, B: a delicate structure, the nature of which could not be determined, occurs in the cavity d, fig. 736, A. In the scale e several smaller cavities are seen near the upper face above the vascular strands and below the latter is a larger cavity; similar cavities are shown in the scale f (fig. 736, A).

Cones similar to *Conites Juddi* are described by Velenovský[1] as *Fričia nobilis* and *Sequoia fastigiata*, both from Lower Cretaceous strata, but it is impossible to say whether the resemblance has any significance. In several anatomical characters the scales of *Conites Juddi* resemble those of *Protodammara* described by Hollick and Jeffrey from Kreischerville. The only indication of pits on the xylem tracheids in the Scottish cones was seen in the scale shown in fig. 736, B: the pits are for the most part uniseriate but occasionally contiguous though generally not actually in contact. Large idioblasts and resin-cavities occur in both the Scottish and American cones, but in the former the occurrence of a ligule is a distinguishing feature in which they agree with cone-scales of recent Araucarias. In *Araucaria* the seeds are embedded in the substance of the scales while in the fossil species they are situated in a depression on the upper face, a feature in which *Conites Juddi* agrees more closely with the cones of *Agathis*. In the sporophylls of *Conites Juddi*, which anatomically are close to those of recent Araucarineae, characters occur which are now shared between *Araucaria* and *Agathis*. The apparently small size of the seeds

1 Velenovský (85) B. Pl. III. fig. 6; Pl. VIII. fig. 13.

and their relation to the ligular outgrowth, as well as the occurrence of separate bordered pits on the tracheids suggest comparison with the recent genus *Cunninghamia*, though the structure of the scales is more akin to that of Araucarian sporophylls. The combination of features which are now distributed among different genera is to be expected in extinct types belonging to evolutionary stages anterior to the divergence of characters along independent lines. The main conclusion is that the affinities are Araucarian though the morphological characters are such as to indicate a combination of features no longer found in a single genus.

Cones exhibiting a close resemblance to those of Araucaria.

ARAUCARITES. Presl.

Araucarites sphaerocarpus Carruthers.

This species (fig. 737), from Inferior Oolite rocks at Bruton, Somersetshire[1], affords a good example of a large Araucarian cone 13 cm. in diameter very similar in form to some recent species (*cf.* fig. 680 and fig. 681). The rhomboidal scales, 2 cm. broad at the distal end, are laterally winged as in *Araucaria Cookii* (fig. 638, A) and bear a single seed embedded in the middle of the upper surface: on the exposed distal ends is a transverse groove and on some of the more complete examples a short rounded umbo is seen below the groove; in some scales a transverse row of pits marks the position of vascular bundles just below the transverse depression.

Araucarites ooliticus (Carruthers).

This species was originally described by Carruthers as *Kinda-carpum ooliticum*[2] from the Great Oolite of Northamptonshire and regarded as an inflorescence of some Pandanaceous plant. Zigno[3] transferred it to *Pandanocarpum*. An examination of the type-specimen in the Northampton Museum led me to refer the cone to *Araucarites*[4]. The type-specimen (fig 738) is a portion of a cone 9 cm. long consisting of a stout central axis covered with spirally disposed deep pits bounded by a crystalline reticulum; the pits

[1] Carruthers (66) Pl. xv; Seward (04) B. p. 131; (11³) p. 116, fig. 18.
[2] Carruthers (68) p. 156, Pl. ix.   [3] Zigno (85) p. 3.
[3] Seward (96²) p. 216; (04) B. p. 135.

FIG. 737. *Araucarites sphaerocarpus.* (British Museum, 41,036; $\frac{2}{3}$ nat. size.)

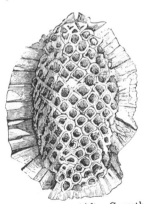

FIG. 738. *Araucarites ooliticus.* (After Carruthers; $\frac{2}{5}$ nat. size.)

being cavities in the proximal portion of the scales in which the
seeds were embedded.   Numerous imbricate scales are attached
laterally to the central region and partially hidden in the matrix.
The scales are approximately 1·7 cm. broad and slightly winged.
The single seed on each scale, the general form of the cone, the shape
of the individual scales, and the occurrence of sterile scales at the
base of the axis are features in which the fossil is practically
identical with recent forms.   Fig. 739 shows a piece of a smaller

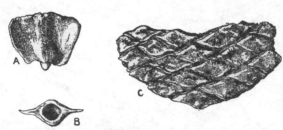

Fig. 739.   *Araucarites ooliticus.*   A, Scale in surface-view showing the projecting
end of a seed.   B, Scale seen from the proximal end showing the seed-cavity.
C, part of cone.   (Northampton Museum;   nat. size.)

cone (in the Northampton Museum), of the same type; this specimen
shows the appearance of the scales in end-view (C), in surface-view
(A), and as seen from the proximal end with the seed-cavity (B).
In fig. 739, A, the base of a seed is seen projecting from the middle
of the laterally expanded scale.   An oblong-ovate cone described
by Carruthers[1] from the Coralline Oolite at Malton, Yorkshire, as
Araucarites Hudlestoni is probably another example of this species:
in one of the specimens of *A. Hudlestoni* in the York Museum a
broad central region is occupied by a mass of pisolite to which
numerous cone-scales are attached.   The scales are shown in
section and in several of them there is a single seed lying in a cavity
occupying the proximal end of the scale precisely as in *A. ooliticus.*
It is possible that *A. ooliticus* is specifically identical with *A. sphaerocarpus*;   it is at least a closely allied type.

The specimen figured by Lindley and Hutton[2] as *Strobilites
Bucklandi* appears to be indistinguishable from *A. ooliticus.*
Similar cones are illustrated by *Araucarites Cleminshawi* Mansell-

[1] Carruthers (77);   Seward (04) B. p. 133.
[2] Lindley and Hutton (34) A. Pl. cxxix.

Pleydell[1] from the Inferior Oolite of Dorsetshire, *A. sphaericus* (Carr.)[2], originally referred to *Cycadeostrobus*, and other species.

Araucarites pippingfordensis (Ung.)[3] is a Wealden species first described by Fitton and named by Unger *Zamiostrobus pipping-fordensis* but recognised by Carruthers as a cone closely allied to *A. sphaerocarpus*. Detached scales which may belong to this species have recently been figured from the Wealden beds on the Sussex coast. The cone described as *Araucarites (Conites)* sp. from the same locality is possibly identical with *A. pippingfordensis*.

Araucarites Brodiei Carruthers.

The specimens from the Stonesfield slate[4] on which this species was founded are rather larger than the scales of *A. Phillipsi*: the scale shown in fig. 740, 4, is 3 cm. long and 1·7 cm. broad; a raised edge just beyond the single seed no doubt corresponds to the so-called ligule on an Araucarian scale and the distal spinous process is another feature shared with recent types.

Araucarites (Sarcostrobus) Paulini (Fliche).

A specimen described by Fliche[5] from Lower Cretaceous rocks in the Haute-Marne, France, is made the type of a new genus *Sarcostrobus* on the ground that the seeds are not so completely covered by the tissues of the scales as in recent Araucarias. The elliptical cone is 5·5 cm. long and 3·8 cm. in diameter; in form, in the stout axis, and in the shape of the single-seeded cone-scales it closely resembles the megastrobili of *Araucaria excelsa* and Jurassic species such as *Araucarites ooliticus* (Carr.) (fig. 738): the small seeds are sunk in a cavity at the proximal end of the scale, but Fliche states that they are not covered on their upper side by the substance of the scale. He is no doubt correct in assigning the cone to the Araucarineae, but the slight differences between the relation of seeds to scales referred to by Fliche do not appear to be sufficiently important to justify the creation of a distinctive generic name; moreover the preseivation of the specimens renders accurate description of details very difficult.

[1] Mansell-Pleydell (85).

[2] Carruthers (67³) p. 105; Seward (04) B. p. 138.

[3] Seward (13) p. 104.

[4] Carruthers (69²) p. 3, Pl. v. figs. 1—6; Seward (04) B. p. 137, Pl. III. fig. 5; Pl. XII. fig. 2.	[5] Fliche (00) p. 11, Pl. I.

Araucarites hespera Wieland. This type is described by Wieland[1] from a specimen obtained from Upper Cretaceous rocks in South Dakota consisting of half an eroded cone bearing scales with small seeds. Wieland's description, though brief and lacking details, and the photograph support his conclusion as to the Araucarian affinity of the specimen.

The generic name *Doliostrobus* was instituted by Marion[2] for specimens of foliage-shoots from Oligocene beds in the South of France agreeing with *Araucarites Sternbergii*, on the ground that the reproductive organs exhibit features more like those characteristic of *Agathis* than *Araucaria*. Laurent[3] refers a small piece of a foliage-shoot from the Aquitanian beds in the Puy-de-Dôme to *Doliostrobus Sternbergii*, though there is no information with regard to the cones. Gardner[4] describes branches from the Bembridge marls in the Isle of Wight as *Doliostrobus Sternbergii* (Goepp.) and accepts Marion's conclusion as to the intermediate character of the genus: the foliage-shoots, though rather more slender than those from Bournemouth referred by him to *Araucaria Goepperti*, are exactly similar in habit and cannot be distinguished by any feature of importance. Gardner reproduces a drawing communicated by Marion of a foliage-shoot bearing a terminal cone-axis from which the scales have fallen, also several detached cone-scales, agreeing closely in size and shape with scales of *Araucaria excelsa* and other recent species, and an impression described as a seed with a lateral wing. The supposed seed has, however, a terminal wing and moreover it is as large as the detached scales: it is permissible to suggest that it may be an imperfectly preserved cone-scale, but without examining the actual specimen any definite assertion would be hazardous. Gardner states that a reason for comparing the scales with those of an *Araucaria* is that in *Agathis* the scales are persistent, but as pointed out elsewhere[5] cones of *Agathis* very readily fall to pieces and the scales easily become detached from the axis. Having regard to the nature of the sterile shoots, the form of the cone, as shown in a drawing published by Gardner of a specimen sent to him by Ettingshausen from Häring, and an

---

[1] Wieland (08²) p. 4, fig. 1.
[2] Marion (84).
[3] Laurent (12) Pl. v. fig. 7.
[4] Gardner (86) p. 93, Pls. xxii., xxiii.
[5] Seward and Ford (06) B. p. 359.

unconvincing specimen of a winged seed figured from Marion's drawing, there would seem to be no valid reason for drawing a distinction between *Doliostrobus* and *Araucarites* or for regarding Marion's and Gardner's fossils as intermediate between *Araucaria* and *Agathis*. Attention has been called on a previous page[1] to the danger of placing too much confidence in the resemblance of foliage-shoots of fossil specimens to those of recent types, but in this case the presence of cones and scales like those of *Araucaria* supplies confirmatory evidence.

Pseudo-Araucaria. Fliche.

The generic name *Pseudo-Araucaria* was given by Fliche[2] to several cones from the Lower Cretaceous beds of the Argonne which he described under three specific names, *Pseudo-Araucaria Loppinetti*, *P. major*, *P. Lamberti*. Externally they are similar to those of some recent Araucarias and in shape agree with cones of *Cedrus*: a stout axis bears deciduous scales with two seeds, the seeds of each pair being separated from one another by a median ridge of the cone-scale which covers them laterally. The seeds appear to bear a relation to the scale similar to that between the single seed and the cone-scale of an *Araucaria*. The cone-scales are slightly expanded láterally as in the *Eutacta* section of the recent genus. Fliche's descriptions are unfortunately inadequately illustrated and it is difficult to obtain a very clear impression of the structural features. The most interesting peculiarity of these cones is the occurrence of two seeds in each cone-scale agreeing in their position on the sporophyll with the single seed of *Araucaria*: the author of the genus regards it as a type intermediate between the Abietineae and the Araucarineae.

Araucarian cone-scales.

The question of the lower geological limit of cones or cone-scales of the Araucarian type is one which cannot be settled with any certainty: there are many examples of vegetative organs very similar in habit to *Araucaria excelsa* and allied species recorded from Triassic, Permian, and to a less extent from Upper Carboniferous strata, also others which agree in the broader form of the

---

[1] See page 162.

[2] Fliche (96) p. 70, Pl. vi. figs. 3—5; Pl. vii. figs. 1, 2.

leaves with *Araucaria Bidwilli* and *A. imbricata*; but the majority
of these shoots are referred to such genera as *Voltzia*, *Walchia*,
Albertia, and *Ullmannia*.   It is pointed out in the description of
these genera that there are reasons for believing them to have
Araucarian affinities, though there is no definite evidence that any
of them bore cones exhibiting the same order of resemblance to
those of recent Araucarineae as is the case with Jurassic and
Cretaceous types.

Araucarites Delafondi Zeiller.

One of the very few Palaeozoic species of seed-bearing scales
that can reasonably be referred to the genus *Araucarites* is
A. Delafondi founded by Zeiller[1] on some detached scales from
Permian beds at Charmoy; the scales are broadly triangular
10—12 mm. long and 8—10 mm. broad, the base is cuneate and
truncate, the apical margin is rounded and has a small median
depression instead of the usual spine.   In the middle of the scale
is a shallow depression which contained a single seed 8—10 mm.
long and 2 mm. broad.   As Zeiller says, there is no absolute
certainty as to the affinity of this species but the scales are unquestionably very similar to those of Mesozoic and recent species
of *Araucarites* and *Araucaria*.   It is suggested that the vegetative
shoots of *Ullmannia frumentaria* (fig. 750) from the same beds may
belong to the plant which bore cones with scales of *A. Delafondi*.

The occurrence of widely distributed Jurassic cone-scales,
bearing a single seed and agreeing very closely in their shape and
size, as also in the laterally expanded borders and in many cases
in the presence of a distal spinous process, with those of recent
species of *Araucaria* especially those belonging to the section
Eutacta, bears striking testimony to the former extended geographical distribution of Araucarian plants.   It has been pointed
out in a previous chapter that a single seed occasionally occurs on
the seminiferous scales of recent Pine cones (fig. 686, B), but in the
scales now under consideration the occurrence of a single seed is
a constant feature and moreover the form of the scales is identical
with that of such species as *Araucaria excelsa* and *A. Cookii*.   The
number of names given to the fossil scales is but a rough index of

---

[1] Zeiller (06) B. p. 215, Pl. L. fig. 1.

the number of actual species: it is obviously impossible to decide with any assurance how much value should be attached to differences in size or to slight variations in form, but the main point is that cones and cone-scales of the Araucarian type are among the most familiar Jurassic fossils. The following selected examples are chosen in illustration of this statement and reference to others will be found in some of the sources quoted in the footnotes.

Araucarites Phillipsi Carruthers.

Carruthers[1] described this species from the Middle Jurassic rocks on the Yorkshire coast: the type-specimen is in the Leckenby collection in the Sedgwick Museum, Cambridge. The scales are

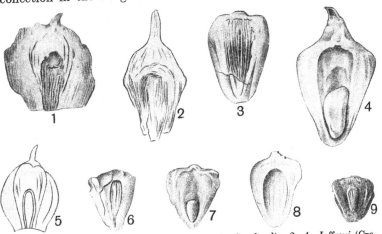

FIG. 740. 1, *Araucarites Milleri* (Upper Jurassic, Scotland); 2, *A. Jeffreyi* (Cretaceous, N. America); 3, *A. Rogersi* (Wealden, S. Africa); 4, 7, *A. Brodiei* (Middle Jurassic, England); 5, *A. Häberleinii* (Middle Jurassic, Germany); 6, *Araucarites* sp. (Middle Jurassic, Australia); 8, *A. cutchensis* (Middle Jurassic, India); 9, *A. Phillipsi* (Middle Jurassic, England). (Slightly reduced; 1, 3, 4, 6, 7, 9, after Seward; 2, after Berry; 5, after Thiselton-Dyer; 8 after Feistmantel.)

cuneate, nearly as long as broad (fig. 740, 9), and in shape similar to those of the cone already described as *Araucarites ooliticus* (Carr.).

Examples of French Jurassic cone-scales are afforded by *Araucarites Moreauana* Sap.[2], from Corallian beds near St Mihiel

[1] Carruthers (69²) p. 6, Pl. II. figs. 7—9; Seward (00) B. p. 285, Pl. x. fig. 4.
[2] Saporta (84) p. 425, Pls. CLXXXIV., CLXXXV.

and other localities, similar to those of *A. Brodiei* (fig. 740, 4; 7)
but reaching a length of 4·5 cm., kite-shaped and provided with a
terminal spine; also *Araucarites microphylla* Sap.[1] represented by
foliage-shoots and cone-scales; the shoots bear linear-lanceolate
leaves similar to those of *Araucaria Bidwilli* but smaller, and the
scales are of the *Eutacta* type.   *Araucarites Falsani* Sap.[2] is founded
on twigs similar to those of *Araucaria excelsa* and scales characterised by stout terminal spines.   Both *Araucarites Falsani* and *A.
microphylla* are from Kimeridgian strata in Ain.

German cone-scales hardly distinguishable from some of the
British and French examples are described by Sir William Thiselton-Dyer[3] from Solenhofen as *Araucarites Häberleinii* (fig. 740, 5).
Salfeld figures some detached scales from the Malm of South-
West Germany as *Araucaria?* which are undoubtedly Araucarian
cone-scales agreeing closely with *A. Milleri* from Scotland (fig.
740, 1).

Araucarites Rogersi Seward.   The scales of this species (fig.
740, 3)[4] from the Uitenhage (Wealden) series of Cape Colony reach
a length of 3 cm. and the straight distal margin bears a median
spine; there is no indication of a ligule.   A specimen in the British
Museum collected by Atherstone and referred to by Tate[5] shows
several scales still in their natural position.   This type bears a
striking resemblance to some of the Indian specimens described
by Feistmantel and is similar to the North American Neocomian
species *A. wyomingensis* Font.

Araucarites macropterus Feistmantel and *A. cutchensis* Feist.

Several examples of typical Araucarian cone-scales are figured
by Feistmantel from Upper Gondwana rocks in India.   The
scales described as *A. macropterus*[6] are distinguished by their
large size; specimens from the Rajmahal series reach a breadth of
5 cm.   The scales of *A. cutchensis*[7] recorded from the Cutch flora
(fig. 740, 8) and elsewhere are smaller, but in some cases it is

[1] Saporta (84) p. 431, Pls. CLXXXVI., CLXXXVII.
[2] *Ibid.* p. 439, Pls. CLXXXVI., CLXXXVII.
[3] Thiselton-Dyer (72).
[4] Seward (03) B. p. 37, Pl. VI. figs. 4—7.
[5] Tate (67) p. 147.   [6] Feistmantel (77²) p. 186, Pl. VIII. figs. 9—12
[7] *Ibid.* (76²) p. 62, Pls. VII.—IX., XII.; (82) Pl. III.

impossible to draw any sharp line between the two species; they agree very closely with both British and French Jurassic types.

Araucarites cutchensis Feist. is recorded by Halle[1] from the Upper Jurassic flora of Graham Land. The scales exhibit a considerable range in size and shape and more than one type may be represented. They are always more or less cuneate and have a narrow truncate base; some of them show broad lateral wing-like extensions; the distal end is nearly truncate and bears a narrow linear appendage. As Halle says, the scales closely resemble those of *A. Brodiei* Carr.

Cone-scales from Jurassic rocks in Victoria[2], Australia, described as *Araucarites* sp., A and B, demonstrate the occurrence of cones with scales almost identical with *A. Phillipsi* and other European forms. There is a comparative scarcity of Araucarian cone-scales in Jurassic and Cretaceous strata in North America but some examples are recorded. *Araucarites wyomingensis* Font.[3] from the Lower Cretaceous of the Black Hills is represented by broadly cuneate scales 1·2 cm. long and with a maximum breadth of 9 mm. and a broad beak at the apex bearing seeds 4—5 mm. long. Larger cone-scales are described by Berry[4] from Middle Cretaceous rocks in North Carolina as *Araucarites Jeffreyi* (fig. 740, 2). These scales are associated with the foliage-shoots referred to *Araucarites bladensis* and the two may belong to one plant. The same author also figures a specimen from the Upper Potomac series as *Araucarites patapscoensis*[5] from Virginia.

Foliage-shoots.

There are numerous examples of foliage-shoots among Mesozoic, and to a less extent Palaeozoic, strata which bear a striking resemblance to branches of recent species of *Araucaria*, especially species of the *Eutacta* section, but in many cases confirmatory evidence such as would be afforded by reproductive shoots is lacking. The practice adopted by some authors of referring

[1] Halle (13²) p. 72, Pl. VIII. figs. 3—10.
[2] Seward (04²) B. p. 181, figs. 42, 43.
[3] Fontaine in Ward (99) B. p. 669, Pl. CLXIII. figs. 1—9.
[4] Berry (08) p. 258, Pl. XVI.; (14) p. 20.
[5] *Ibid.* (11) p. 399, Pl. LXXVII. fig. 5.

impressions of vegetative branches to the genus *Araucarites* solely
on the ground of similarity in habit and leaf-form to the recent
genus is not in accordance with sound principles, though in some
instances the implied relationship may be a reality. Pending
more satisfactory evidence many of the sterile *Araucaria*-like
shoots are referred to *Pagiophyllum*, while branches of similar
habit bearing oval cones are included in the genus *Elatides*.

An example of a Palaeozoic fossil which has been assigned to
Araucarites on slender grounds is afforded by *Araucarites Oldhami*
Zeiller.

Araucarites Oldhami Zeiller.

The specimen from the Lower Gondwana rocks in India to
which this name is applied[1] consists of an axis bearing spirally
disposed lanceolate-acuminate leaves reaching 4·5 cm. in length,
slightly contracted at the base and longitudinally striated:
portions of the axis show rhomboidal and feebly convex areas
separated by narrow scars where the laminae have been broken off.
As Zeiller says, the resemblance of the shoot to a branch of *Araucaria imbricata* is very close, but considering the age of the beds
and the absence of any Araucarian cone-scales from rocks at this
horizon in India it is questionable whether it is wise to adopt
the name *Araucarites*. It is not unlikely that a small specimen
figured by Feistmantel[2] from the Karharbari coal-field as possibly
a Fern rhizome is a portion of a leafless axis of Zeiller's species.

Araucarites ovatus Hollick.

This species from Cretaceous strata in New Jersey[3] is founded
on fragments of sterile branches bearing elliptical-ovate leaves
resembling the foliage of *Araucaria imbricata* and the fossil species
Araucarites Nathorsti Dus. A very similar type is represented by
Araucarites bladenensis, described by Berry[4] as *Araucaria bladenensis*, from the Upper Cretaceous rocks of Carolina and Alabama:
the leaves are decurrent, ovate-lanceolate, about 1·6 by 0·8 cm.
with a cuspidate apex and rounded base; there are 14—16 parallel

[1] Zeiller (02) B. p. 36, Pl. VII. fig. 6.   [2] Feistmantel (79²) Pl. XIII. fig. 6.
[3] Hollick (97) p. 128, Pl. XII. figs. 3 *a*, 4.
[4] Berry (08) p. 255, Pls. XII.—XIV.; (14) pp. 19, 105, Pl. III. figs. 6, 7; Pl. XIX.
figs. 1, 2.

veins and imperfectly preserved stomata occur in rows on the lower surface. Berry compares the species with *Araucaria Toucasi* figured by Saporta[1] from Turonian rocks in the South of France. Another species founded on a sterile shoot is *Araucarites Hatcheri* described by Wieland[2] from Upper Cretaceous rocks in Wyoming.

These and other examples that might be quoted, though referred to the Araucarineae on evidence that cannot be considered conclusive, are probably correctly determined; the comparison with *Araucaria Bidwilli* and *A. imbricata* suggested by the striking resemblance of the leaves is supported by the occurrence of Araucarian cone-scales in some of the localities.

Araucarites Sternbergii Goeppert.

This species was founded on sterile branches, from the rich Eocene flora of Häring in the Tyrol[3], practically identical in habit with foliage-shoots of *Araucaria excelsa* and other recent species. From the same locality Goeppert[4] figured an imperfectly preserved cone approximately 6 cm. long and 3 cm. in diameter characterised by imbricate, spirally disposed scales with reflexed apices which he compares to a male cone of *Araucaria imbricata* incorrectly spoken of as *A. excelsa*: Goeppert suggests the possible specific identity of *A. Sternbergii* and *A. Goepperti* Sternb.: the latter species was founded by Sternberg[5] on a Tertiary cone from Häring in the Tyrol. Ettingshausen[6] subsequently figured several good examples of vegetative shoots of this type from Häring and described a subglobose cone, figured by Gardner[7], which he refers to the same species: this author also records *A. Sternbergii* from Bilin in Bohemia[8], but under the generic name *Sequoia*: in his account of the occurrence of the species in Carinthia[9] he adopts the designation *Araucarites*. Ettingshausen figures a single cone-scale from Eocene beds in Styria as *Araucarites schoeneggensis*[10] and compares it to the scales of

[1] Saporta (79) A. p. 198, Pl. xxvii., 2.
[2] Wieland (08²) p. 6, fig. 2.
[3] Goeppert (50) p. 236, Pl. xliv. fig. 1.
[4] *Ibid.* Pl. xliv. fig. 2.
[5] Sternberg (38) A. Pl. xxxix. fig. 4.
[6] Ettingshausen (55) p. 36, Pls. vii., viii.
[7] Gardner (86) p. 96.
[8] Ettingshausen (67²) p. 116, Pl. xiii. figs. 3—8.
[9] *Ibid.* (85).
[10] *Ibid.* (90) Pl. i. fig. 93.

A. Sternbergii. The latter species is recorded also by Massalongo[1] from Eocene rocks in Italy and on imperfect evidence by Heer[2] from Switzerland. Gardner[3] describes several good specimens of vegetative shoots from the Eocene flora of Bournemouth which he names *Araucarites Goepperti* Sternb. though the specific name

Sternbergii would be more appropriate as that designation was first applied to similar branches from Häring and *A. Goepperti* was founded on a detached cone. Two small pieces of larger specimens in the British Museum from Bournemouth are represented in fig. 741 in illustration of the very close resemblance of the leaves to those of recent species. Gardner draws attention to the similarity of some of the fossil examples to deciduous shoots of *Araucaria Cunninghamii*: with reference to the absence of cones or cone-scales he quotes the fact, communicated to him by an observer in Madeira, that the foliage

Fig. 741. *Araucarites Sternbergii.* (British Museum, V, 523; nat. size.)

of *A. Cunninghamii* requires two or three days to sink while mature seeds do not begin to sink before the fifth or sixth day, so that in moving water shoots and seeds would necessarily be deposited separately.

Some of the fragments of branches described by Gardner as *Athrotaxis* (?) *subulata*[4] may well belong to *Araucarites*. It must be admitted that in the case of the English specimens, as in many others, the use of the generic name *Araucarites* is based on the evidence of vegetative branches only, but Gardner correctly states that in the shoots of similar habit referred to *Cryptomeria* the leaves are straighter, and moreover the presence on some of the shoots of the latter of persistent cones like those of *Cryptomeria japonica* constitutes a clear distinction. Having regard to the very striking resemblance of the widely spread Tertiary specimens

[1] Massalongo (59) Pls. v.—vii. [2] Heer (55) A. Pl. xxi. fig. 5.
[3] Gardner (86) p. 55, Pl. xi. fig. 1; Pl. xii. [4] *Ibid.* Pl. xi.

included in *Araucarites Sternbergii* or *A. Goepperti* to those of
A. Cunninghamii and other species the probability of generic
identity is such as to justify the retention of the designation
Araucarites.

Araucarites Haastii (Ettingshausen).

Ettingshausen[1] described this species as *Araucaria Haastii*
from beds at Shag Point, New Zealand, believed to be of Eocene
age; it is represented by sterile branches bearing crowded ovate-
lanceolate, acuminate, leaves apparently of leathery texture
reaching a length of 5 cm. and 2 cm. or less in breadth. As
Ettingshausen says, they agree very closely with the leaves of
Araucaria imbricata but like those of *A. Nathorsti* Dus. they have
a less spinous apex than in the recent species. Some petrified
wood from Malvern Hills in New Zealand is referred by Ettingshausen to the same species but without any evidence of connexion between the wood and the foliage-shoots. The same
author describes a branch similar in habit to *Araucaria excelsa*,
from Shag Point, as *Araucaria Danai*[2], but the specimen is too
imperfect to warrant the use of the designation *Araucarites*.

Araucarites Nathorsti (Dusén).

This species, described as *Araucaria Nathorsti*, is recorded by
Dusén[3] from Punta Arenas on the Magellan Straits: the age of the
beds is believed to be Oligocene though the precise horizon has
not been determined. The material consists of fragments of
foliage-shoots bearing short and relatively broad leaves of leathery
texture, varying from linear to ovate; they agree closely with the
leaves of *Araucaria imbricata*, differing chiefly, as Dusén states,
in their blunter apices.

Araucarites imponens (Dusén).

Nathorst[4] first suggested a reference to *Araucaria* of the single
leaf on which this species was founded[5]: it was collected in a
marine volcanic tuff in Seymour Island and is probably of Lower
Tertiary age. The leaf is linear, 6 cm. long, and tapers gradually
towards an incomplete apex; it agrees in form and size with

---

[1] Ettingshausen (87) p. 154, Pl. II.   [2] *Ibid.* p. 155, Pl. I. fig. 18.
[3] Dusén (99) p. 105, Pl. XII.   [4] Nathorst (04²) B.
[5] Dusén (08) p. 11, Pl. I. figs. 16, 17.

leaves of *A. Bidwilli* and *A. brasiliensis*, but the single impression
is hardly sufficient to demonstrate the existence of *Araucarites*
in this southern flora (lat. 64° 16′ S.). On the other hand the
occurrence of wood of the Araucarian type[1] in Seymour Island
in beds that are either Lower Tertiary or Upper Cretaceous
supports the conclusion of Nathorst and Dusén.

ELATIDES. Heer.

Heer[2] proposed this name[3] for some Jurassic Coniferous
remains from Siberia characterised by spirally disposed falcate
leaves (figs. 742, 743) and cones similar externally to those of
Picea, Abies and other Abietineae. The genus is based primarily
on the form of the cones and cone-scales. In the new genus
were included three species, *E. ovalis*, represented by oval cones
2·7 cm. long and 6--7 mm. broad, *E. Brandtiana* characterised by
cylindrical cones, and *E. falcata* founded on vegetative branches
very similar to those of *E. Williamsonis*. Nathorst[4] has included
these three species in *Elatides curvifolia* (Dunk.) a Wealden
species abundantly represented in the plant-beds of Spitzbergen
(fig. 743). In the absence of cones it is impossible to draw any
satisfactory distinction between foliage-shoots belonging to *Elatides*
and those referred by authors to *Sequoia, Pagiophyllum,* and other
genera. It is therefore only in cases where cones are present
that the designation *Elatides* is admissible. The vegetative
characters of *Elatides* are those of *Araucaria Cunninghamii, A.
excelsa*, and allied species while the cones consist of flat imbricate
scales with narrower and more or less pointed or spinous distal
ends. There is some reason to believe that the cone-scales were
monospermic but the evidence is not conclusive and rests on a
single species. The data are insufficient to fix definitely the position of the genus, though it is in all probability a member of the
Araucarineae. *Elatides* is characteristic of Rhaetic, Jurassic, and
Wealden floras.

Elatides Sternbergii (Nilsson).

A Rhaetic species originally described by Nilsson from Rhaetic
rocks in the South of Sweden as *Abies Sternbergii*, subsequently

---

[1] Gothan (08).   [2] Heer (77) ii. pp. 77—79, Pl. XIV.   [3] ἐλάτη, Fir.
[4] Nathorst (97) pp. 35, 58, Pls. I., II., IV., VI.

included by Nathorst[1] in *Palissya Braunii* but afterwards recognised by him as a distinct species[2] and recently transferred to *Elatides*[3]. An examination of specimens in the Stockholm Museum leads me to agree with the substitution of the designation *Elatides*. Nathorst has also pointed out that some of the cones from the Rhaetic of Franconia referred by Schenk[4] to *Palissya Braunii* are of the *Elatides* type and distinct from cones of *Palissya* which are characterised by their more open habit and by other more important morphological features. *Elatides Sternbergii*, though similar in the habit of the vegetative shoots to *E. Williamsonis* from Jurassic strata, differs in the narrower and straighter leaves which may reach a length of 2—3 cm. and are either straight or slightly curved in contrast to the stouter and strongly falcate leaves of *E. Williamsonis* and *E. curvifolia*. A cone figured by Nathorst[5] is practically identical in external form with one of *E. Williamsonis* illustrated in volume I. of the *Jurassic Flora of the Yorkshire Coast*[6]. We have no knowledge of the structure of the reproductive shoots and no evidence other than the habit of the foliage-shoots with regard to systematic position: it is, however, probable that this Rhaetic species is closely allied to the later Jurassic and Wealden types.

Elatides Williamsonis (Brongniart).

This Jurassic species described by Brongniart[7] as *Lycopodites Williamsonis*, was figured by Phillips[8] as *L. uncifolius* and by Lindley and Hutton[9] under Brongniart's name. The specimens figured by the English authors are in the York and Manchester Museums respectively. Schimper transferred the species to *Pachyphyllum* and it has usually been assigned to that genus or to *Pagiophyllum*[10], the name substituted by Heer for Pomel's *Pachyphyllum*, a designation now reserved for sterile shoots and therefore inapplicable to the present species which possesses cones of the *Elatides* type. The vegetative shoots are monopodially

[1] Nathorst (78[2]) B. p. 28, Pl. IV. figs. 1—3.
[2] *Ibid.* (86) p. 107, Pl. XXIII. figs. 8—12; Pls. XXIV., XXV.
[3] *Ibid.* (97) p 34; (08).
[4] Schenk (67) A. Pl. XLI. fig. 7.   See also Solms-Laubach (91) A. p. 73.
[5] Nathorst (86), Pl. XXV. fig. 8.   [6] Seward (00) B. Pl. X. fig. 3.
[7] Brongniart (28) A. p. 83.   [8] Phillips (29) A. Pl. VIII. fig. 3
[9] Lindley and Hutton (33) A. Pl. XCIII.   [10] Seward (00) B. p. 291.

branched, the smaller branches being given off at an acute angle; the leaves are crowded, fleshy, tetragonal and falcate (fig. 742), agreeing closely with the foliage of the *Eutacta* species of *Araucaria* and with *Cryptomeria*. The megastrobili are cylindrical, approximately 6 cm. long and 2 cm. in diameter, bearing imbricate, flat, scales with narrow pointed distal ends resembling the free portion

FIG. 742. *Elatides Williamsonis.* (British Museum; nat. size.)

of the foliage-leaves. No specimens have been described showing seeds attached to the scales. The more slender microstrobili, 2 cm. long, bear sporophylls at right angles to the axis with triangular upturned distal ends characterised by a median keel. In the vegetative shoots this species closely resembles the Liassic *Pagiophyllum peregrinum*[1] (fig. 744) (Lind. and Hutt.), but in the

---

[1] See page 276.

absence of cones the latter species is retained in *Pagiophyllum*. From the Wealden species *E. curvifolia, E. Williamsonis* differs in its stouter and more crowded leaves though the differences are slight both in the cones and vegetative shoots.   In all probability this species is represented in several Jurassic floras, but unless cones are present specimens should be referred to *Pagiophyllum*.

Elatides curvifolia (Dunker).

Dunker[1] first described this Wealden type (fig. 743) from North Germany as *Lycopodites* and it was referred by Ettingshausen[2] to *Araucarites*, the generic name, though probably correctly express-

FIG. 743.   *Elatides curvifolia*.   (After Nathorst; nat. size.)

ing the position of the fossil Conifer, being used without adequate reasons.   The identity of the cones discovered by Nathorst[3] in the Wealden or Upper Jurassic beds of Spitzbergen with those on which Heer founded the genus *Elatides* led to the adoption of that generic term.   Nathorst's discovery of several fertile branches justifies his reference of Heer's specimens from Spitzbergen described as *Sequoia Reichenbachii*[4] to *Elatides curvifolia*, as also the employment of Dunker's specific term for *Elatides ovalis* and *E. Brandtiana* Heer.   The cones of *E. curvifolia* are cylindrical or oval and it is suggested by Nathorst that these forms might be regarded as varieties, the oval form being spoken of as var. *ovalis*

[1] Dunker (46) A. p. 20, Pl. VII. fig. 9.
[2] Ettingshausen (52) Pl. II.
[3] Nathorst (97) pp. 35, 58. Pls. I., II., IV., VI.
[4] Heer (75) ii. Pl. XXXVI. figs. 1—8; Pl. XXXVII. figs. 1, 2.

after Heer's specific name and the cylindrical cones being distinguished as var. *Brandtiana*, but it is doubtful whether the retention of these varietal names is advisable. The cone-scales have pointed apices and agree closely with those of *E. Williamsonis*. On a specimen of a cone of this species from Kimeridge strata in Scotland[1] one scale afforded evidence of the occurrence of a single seed as in *Araucaria*. Nathorst regards some smaller cones on a branch from Spitzbergen as microstrobili and suggests that longitudinal striae on the sporophylls may represent long microsporangia like those of *Araucaria*; but the preservation is too imperfect to demonstrate the nature of the specimen. The vegetative branches bear falcate leaves rather more slender and as a rule less crowded than in *E. Williamsonis*: on older branches from which the free part of the lamina has fallen there are leaf-bases or in some cases an oval leaf-scar. This type is characteristic of Wealden strata in Spitzbergen, North Germany, and other European localities; it is no doubt represented by some of the impressions of branches assigned to *Sphenolepidium Sternbergianum*[2]; it is also recorded from Kimeridgian strata in the North of Scotland.

PAGIOPHYLLUM. Heer.

Heer[3] instituted this genus in place of *Pachyphyllum*, previously adopted by Pomel[4] for a section of his genus *Moreauia*, on the ground that the latter name had been applied to a member of the Orchidaceae. Some of the species referred to *Pagiophyllum* have also been included in *Araucarites* and *Brachyphyllum*. Tuzson[5] instituted a new genus. *Pagiophyllites* for petrified Mesozoic wood having Araucarian features, the type-species being *P. keuperianus* (Goepp.), but no evidence is furnished in support of a connexion of this wood with foliage-shoots of *Pagiophyllum*. Schimper and Saporta include in their diagnosis of the genus both vegetative and reproductive shoots and consider *Pagiophyllum* to be allied to *Agathis*, *Cunninghamia*, and *Araucaria*. Certain species have in recent years been transferred to *Elatides* because of the occurrence of cones conforming to Heer's genus.

It has been suggested[6] that *Pagiophyllum* may most conveni-

---

[1] Seward (11[2]) p. 684, fig. 10.   [2] *Ibid.* (95) A. p. 205; (11[2]) p. 685.
[3] Heer (81) p. 11.   [4] Pomel (49) p. 352.
[5] Tuzson (09) p. 30.   [6] Seward (12) p. 41.

ently be reserved for vegetative branches of Conifers (fig. 744) possessing foliage like that of *Araucaria excelsa* and allied species, which in the absence of cones cannot safely be referred to *Elatides* or other genera based, in part at least, on strobilar characters. *Pagiophyllum* is essentially an artificial genus: as Solms-Laubach says, 'it is only in accordance with old custom to distinguish the Ullmanniae of the Zechstein from *Pagiophyllum*[1],' and it is equally difficult to draw any clearly defined line between this genus and some forms included by authors in *Brachyphyllum*. A Triassic species from Raibl originally referred to *Voltzia heterophylla*[2] afterwards named *V. Foettleri* by Stur[3], *Pagiophyllum Sandbergi* by Schenk[4] and figured by Schütze[5] as *P. Foettleri*, has the habit of a *Brachyphyllum*. This is one of many examples of sterile shoots illustrating the arbitrary use of generic names for coniferous remains which afford no definite evidence of their systematic position. The Araucarian habit is in itself of little value as evidence of affinity, but the abundance of petrified wood with Araucarian features (*Dadoxylon*) in strata yielding *Pagiophyllum* shoots suggests an Araucarian alliance, and the fact that some *Pagiophyllum* shoots bear *Elatides* cones affording indications of Araucarian characters points in the same direction. It cannot be assumed that all *Pagiophyllum* shoots bore similar cones, and it is mainly on this account that the employment of *Pagiophyllum* as a provisional designation is recommended.

Pagiophyllum is widely distributed in Jurassic strata and extends into Cretaceous and Tertiary rocks: it occurs also in pre-Jurassic floras and has recently been described by Zeiller[6] from the Permian of France. It should be recognised that this extended use of the name is not in accordance with general practice, but it is adopted on the ground that, as in recent Conifers so in the case of extinct types, similarity in the habit of vegetative branches does not necessarily imply close relationship as regards the more important characters of the reproductive shoots.

[1] Solms-Laubach (91) A. p. 77.
[2] Bronn (58) p. 135, Pl. viii.
[3] Stur (68) p. 104.
[4] Schimper and Schenk (90) A. pp. 276, 290.
[5] Schütze (01) Pl. vi. fig. 1.
[6] Zeiller (06) B. p. 219, Pl. li.

Pagiophyllum peregrinum (Lindley and Hutton).

This species, first named by Lindley *Araucaria peregrina*[1], was founded on material from the Lias of Lyme Regis in Dorsetshire[2]. It is possible that the generic name chosen by Lindley correctly expresses the position of the species, but decisive evidence is lacking. Vegetative shoots bear crowded imbricate, spirally disposed, leaves tetragonal in section, broadly triangular, sometimes falcate and more or less appressed to the stem in the lower portion of the lamina (fig. 744). There is a distinct dorsal keel and occasionally rows of papillae are visible on the lamina; the apex is obtuse or acute. The leaves vary considerably in size and shape. Zeiller[3] describes the cuticle of the dorsal and ventral surfaces of some leaves on Permian specimens from Blanzy: the stomata occur in longitudinal rows on the lower face only, the guard-cells being usually at right-angles to the long axis of the leaf.

In habit this species agrees closely with *Elatides Williamsonis*, a Middle Jurassic type; it occurs in Jurassic rocks of England, France, Germany, Italy, and elsewhere, the oldest recorded examples being those described by

Fig. 744. *Pagiophyllum peregrinum.* (British Museum; nat. size.)

[1] Lindley and Hutton (33) A. Pl. LXXXVIII.
[2] Seward (04) B. p. 48, Pl. v.; Saporta (84) p. 383, Pls. 173—176.
[3] Zeiller (06) B. p. 219.

Zeiller from the Permian of France which he refers to *Pagio-phyllum* in preference to *Ullmannia*, pointing out that the leaves are relatively longer and less appressed to the axis than in the shoots known as *U. Bronni* Goepp. (fig. 750, D, E).   Triassic specimens from North Italy in the Bologna Museum named *Pagiophyllum Rotzoanum* appear to be indistinguishable from the English species.

It is unnecessary to describe other examples of the genus as the *Pagiophyllum* type is illustrated by many Mesozoic and Tertiary species referred by authors to *Sequoia, Geinitzia, Elatides, Sphenolepidium,* and other genera.   The important point is that in place of generic names connoting definite forms of cone, the designation *Pagiophyllum* should be adopted for all foliage-shoots of a certain habit which afford no satisfactory evidence as to the nature of the reproductive shoots.

PALAEOZOIC CONIFERS EXHIBITING CERTAIN FEATURES SUG-
    GESTIVE OF ARAUCARIAN AFFINITY BUT WHICH CANNOT BE
    DEFINITELY ASSIGNED TO THAT OR TO ANY OTHER FAMILY
    OF CONIFERALES ON THE EVIDENCE AT PRESENT AVAILABLE.

WALCHIA.   *Sternberg.*

The name *Walchia*[1] is applied to foliage-shoots, occasionally bearing terminal cones, from Permian and to a less extent Upper Carboniferous rocks, which present a striking agreement in habit with branches of *Araucaria excelsa* and other recent species of the section *Eutacta* of *Araucaria*.   Information with regard to repro-ductive shoots is very incomplete and we have little more than circumstantial evidence as to the anatomical features of the stem. In many cases the ultimate branches bear terminal cones similar to the megastrobili of *Elatides*[2], but it is only in a few specimens that seeds are preserved on the cone-scales: in some species, *e.g.,* *W. frondosa* Ren. and *W. fertilis* Ren. the fertile shoots appear to be of a distinct type though the evidence is not wholly satisfactory. It is probable, as several authors have suggested, that the species included in *Walchia*, were our information fuller, would be referred to more than one generic type.   The resemblance of branches of *Walchia* to the foliage-shoots of *Lepidodendron*, especially in the

---

[1] Sternberg (26) A. p. xxii.   [2] See page 272.

FIG. 745. *Walchia piniformis.* (British Museum; ½ nat. size.)

case of specimens too small to show the characteristic branching-habit, has led to confusion between the two genera. It is often very difficult to draw a definite line between *Walchia* and *Ullmannia,* and in the absence of sporophylls the genus *Gomphostrobus* may be easily confused with species of *Walchia.*

Foliage-shoots characterised by a pinnate arrangement of the ultimate branches (fig. 745) attached at right-angles or obliquely to an axis of higher order. Leaves spirally disposed, crowded and imbricate, short and ovate or linear and spreading, usually tetragonal and more or less falcate and decurrent. The dimorphism of branches and differences due to age or position on the tree render a satisfactory delimitation of species almost impossible though a few fairly well defined types can be recognised with reasonable certainty. As Bergeron[1] says, in the absence of strobili the separation of species represented only by sterile shoots is hardly possible. Further reference is made to the features exhibited by reproductive shoots in the appended account of a few selected types. Information with regard to the anatomical characters of *Walchia* is very scanty and is based on evidence afforded by the association of foliage-shoots and petrified wood or on inferences drawn from unconvincing considerations. Among specimens which may belong to this genus one of the more interesting is that on which Mougeot[2] founded the species *Araucarites valdajolensis.* The type-specimen, from the Permian of Val d'Ajol in the Vosges, has been refigured and critically discussed by Fliche[3] though no complete investigation of its structure has been made. The cylindrical piece of stem, 9 cm. in diameter, has a large pith and a broad zone of secondary wood composed of tracheids, with two rows of alternate bordered polygonal pits, and narrow medullary rays. It seems clear from Mougeot's brief account and from the description of other specimens by Fliche that the anatomical features are Araucarian though we have no information as to the structure of the inner edge of the xylem, a region of special importance as regards comparison with other types possessing a similar Araucarian pitting on the tracheids. The surface of Mougeot's specimen is characterised by numerous spirally disposed, elliptical projections 5—7 mm. long and 3—4 mm. wide

---

[1] Bergeron (84).   [2] Mougeot (52) A. p. 27, Pl. IV.   [3] Fliche (03).

which Fliche regards as leaf-bases and compares with those on Araucarian stems. While admitting the possibility that the wood belongs to *Gomphostrobus* or some Cordaitean species Fliche considers *Walchia* the most likely genus. The comparative closeness of the leaf-bases would seem to be a difficulty: in stems as large as the type-specimen of *W. valdajolensis* one would expect to find the leaf-bases more widely separated and tangentially stretched. It is by no means unlikely that the supposed surface-features may belong to a deeper zone of the cortex of a partially decorticated stem; but in any case they do not suggest a stem of *Cordaites* or *Mesoxylon*. It is impossible to assign the species with confidence to *Walchia* though Fliche may be correct in his opinion as to the likelihood of that being its true position. The generic name *Araucarites* implies a degree of affinity which has not been established and the designation *Dadoxylon* would be more in keeping with the facts.

Walchia is especially characteristic of Permian floras though it has been shown to occur in the Stephanian of several countries. In Britain *Walchia* is recorded from a very few Permian[1] and Upper Coal Measures[2] localities. We cannot speak with confidence as to the position of the genus: the striking resemblance in the system of branching and in the foliage-shoots to certain species of *Araucaria* at once suggests a possible affinity to the Araucarineae, and this slender basis of comparison receives support from the occurrence in a few instances of single seeds on the upper face of sporophylls and from the Araucarian type of pitting in wood associated with *Walchia* branches. It may fairly be said that although proof is lacking there is a strong presumption in favour of regarding this Permo-Carboniferous genus as more nearly allied to the Araucarineae than to any other family of Gymnosperms.

Walchia piniformis (Schlotheim).

This, the commonest species, was originally described by Schlotheim as *Lycopodiolithes piniformis*[3]: it occurs in both Permian and Stephanian strata[4]. The pinnately branched shoots

---

[1] Vernon (12) p. 607.   [2] Kidston (02) B.
[3] Schlotheim (20) A. p. 415, Pl. xxiii. fig. 1 a; Pl. xxv. fig. 1.
[4] *E.g.* Stefani (01) p. 111; Weiss, C. E. (72) p. 179; Heer (76) A. p. 57; Goeppert (65) p. 236; Grand'Eury (77) A. p. 514.

(fig 745) are characterised by the more or less oblique insertion of
the slender branchlets and by the comparatively long, narrow,
falcate, decurrent leaves. The ovoid or cylindrical megastrobili
terminal on the ultimate shoots bear imbricate ovate-lanceolate
sporophylls, but their preservation is not such as to throw any
light on the structure of the seed-bearing organs. Zeiller[1] mentions
a cone from Lodève (Permian) 10 cm. long and 1—1·2 cm. in
diameter, but the average length is less than this. A branch
figured by Potonié[2] from the Permian of Thuringia shows elliptical
leaf-cushions very like those on the larger stem described by
Mougeot as *Araucarites valdajolensis*. Potonié has drawn attention to the difficulty of distinguishing small specimens of this
species from *W. filiciformis* and *W. linearifolia*, and some forms
described as *W. imbricata* and *W. hypnoides* are by no means
clearly distinguished from *W. piniformis*. In *W. filiciformis*
(Schloth.) the leaves are characterised by the downward curve of
the lamina near the base though this in itself is hardly a decisive
criterion. In *W. linearifolia* Goepp. the leaves are rather more
delicate and less falcate, while in *W. imbricata* they are usually
shorter, relatively broader, and more strongly imbricate and
incurved. *W. hypnoides* (Brongn.) is a smaller form though, as
Kidston[3] suggests, this may not be a specific character. Renault's
species *W. fertilis*[4] represents a similar form but with smaller
leaves, and each branchlet ends in a long and narrow strobilus
which affords no indication of the nature of the sporophylls.

Walchia filiciformis (Schlotheim).

This species, recorded from Permian and Stephanian beds, and
differing but slightly from *W. piniformis* is important as supplying
more satisfactory evidence as to the nature of the megastrobili.
Zeiller[5] has described a fertile specimen from the Permian of
Brive in which the ovate-lanceolate cone-scales (sporophylls) bear
on their upper concave face single ovoid seeds, 7—8 mm. long.
The strobili appear to be lax in the arrangement of the monospermic
sporophylls.

1 Zeiller (06) B. p. 204, Pl. L. figs. 3, 5; Bergeron (84).
2 Potonié (93) A. p. 218.   3 Kidston (86) A. p. 15.
4 Renault (96) A. p. 359; (93) A. Pl. LXXX. fig. 2.
5 Zeiller (92) A. p. 99, Pl. XV. fig. 3.

Walchia imbricata Schimper.

In habit this type[1] closely resembles the foliage-shoots of *Araucaria Rulei* on a smaller scale, the crowded leaves being strongly incurved and imbricate: it is readily distinguishable from *Ullmannia Bronni* Goepp. (fig. 750, D, E). A good example is figured by Zeiller[2] in which the branches are unusually large, 8 mm.—1·2 cm. in diameter: the species occurs in Permian and Stephanian strata and is recorded by Kidston from the Upper Coal Measures of central England.

Walchia Schneideri Zeiller.

This species[3] from Charmoy is characterised by the long filiform leaves, ·6 mm. broad and 1·2 cm. long, usually straight and decurrent. The ultimate branches, some of which bear comparatively long and slender cones, are oblique and alternate as in *W. piniformis*. *W. foliosa* Eich.[4] from the Permian of Russia is a similar form but with less delicate leaves.

Walchia frondosa Renault.

A species from the Permian of Autun having slender leaves rather shorter and more falcate than those of *W. Schneideri*: some of the branches bear a terminal globular bud superficially resembling the ovuliferous shoot of *Taxus*. It is, however, not improbable that the buds are purely vegetative like those figured by Bergeron on a specimen of *W. piniformis*.

SCHIZODENDRON. Eichwald. (*TYLODENDRON* Weiss.)

Prof. C. E. Weiss[5] instituted the generic name *Tylodendron* for casts from Upper Carboniferous and Permian strata in Germany which he described as branches of a Conifer with spirally arranged rhomboidal raised areas or pulvini each of which has a median slit in its apical portion (fig. 746). The elongate, narrow, raised areas (*cf.* the medullary cast of a *Voltzia*, which shows precisely similar areas, represented in fig. 748) were regarded as casts of leaf-cushions and the slit was interpreted as a resin-canal.

[1] Schimper (72) A. p. 239.   [2] Zeiller (06) B. p. 211, Pl. xlix. figs. 1, 2.
[3] *Ibid.* p. 206, Pl. xlviii. figs. 4, 5.
[4] Eichwald (60) B. Pl. xix. fig. 1.
[5] Weiss, C. E. (72) p. 182, Pls. xix., xx.

Schizodendron speciosum (Weiss). One of the casts figured by Weiss and assigned by him to this species has a length of 70 cm. and at intervals of about 30 cm. shows periodic swellings where it assumes a barrel-shaped form. Pieces of wood attached to some of the casts were investigated by Dippel who found that they agreed anatomically with Araucarian stems. Weiss considered his specimens to be generically identical with casts figured by Eichwald[1] from Russia as species of *Schizodendron* and *Angiodendron*. Potonié, while uncertain as to the close agreement with some of Eichwald's fossils, regarded *Tylodendron* as identical with Eichwald's *Schizodendron*, and Zeiller[2], in view of this agreement, adopted the older name *Schizodendron*. The latter author[3] formerly believed *Tylodendron*, as described by Weiss, to be distinguished from *Schizodendron* by the apical occurrence of the slit on the so-called leaf-cushion in contrast to the basal slit in *Schizodendron*, but Potonié[4] proved that in both cases the median groove extends up the lower portion of each projecting area from its base and represents an out-going leaf-trace; he also

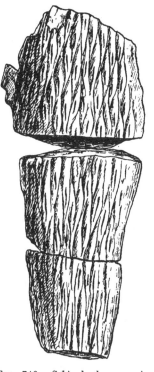

FIG. 746. *Schizodendron speciosum.* (After Potonié; ⅔ nat. size.)

demonstrated that *Schizodendron* is a pith-cast, the tapered areas being the inner ends of medullary rays. In some specimens the casts afford some indication of a discoid pith. The relation between the wood and the pith-casts is also very clearly shown in a section of a petrified stem of Permian age from Prince Edward Island described by Miss Holden[5]. The true nature of

[1] Eichwald (60) B. Pl. xviii.　　　　[2] Zeiller (92²) A. p. 102.
[3] *Ibid.* (80) p. 263, Pl. v.　　　　　[4] Potonié (88).
[5] Holden, R. (13) p. 245.

the *Tylodendron* casts is also shown in specimens from the Lower
Permian of Saxony in the Chemnitz Museum[1]. Casts similar to
those described as *Tylodendron* and *Schizodendron* were recorded
by Schleiden in 1846 and referred to a new genus *Endolepis*: he
believed them to be casts of the pith-cavity of some Dicotyledonous
stem. Examples of *Endolepis* have been described by Schenk and
more recently by Fliche[2] who discusses the history of the genus
and on the ground of priority adopts Schleiden's name in preference to *Schizodendron*. It is, however, preferable to retain
Schizodendron for the larger casts with periodic swellings. The
smaller type represented by *Endolepis* has in several instances
been found in connexion with the foliage of *Voltzia*[3] (fig. 748) and
it is questionable if a special designation is needed. In the form
of the raised areas on the surface of the cast *Schizodendron* and
Endolepis appear to be identical: while suggesting the advisability
of retaining the former name I recognise that the Permian and
Triassic casts may belong to stems which are closely allied or even
generically identical.

The structure of the wood of *Schizodendron speciosum* is of the
Araucarian type; the tracheids have 1—3 rows of contiguous and
alternate pits on the radial walls and the medullary rays are
usually uniseriate. Potonié compares the pith-casts of the Palaeozoic stems with those of recent species of *Araucaria* and *Agathis*:
the pith of the recent species is much smaller but in both fossil
and recent medullary casts there are periodic swellings where the
presence of scars, sometimes in a whorl or pseudowhorl[4], marks
the position of branches. The pith-cast of a recent Cycad (fig. 398,
p. 29, Vol. III.) bears a general resemblance to *Schizodendron*: in
Araucaria the medullary rays are narrower and so produce
narrower raised areas on a pith-cast.

Schizodendron Cowardi (F. E. Weiss).

Prof. F. E. Weiss[5] has recently described an interesting example
of *Schizodendron* which throws some fresh light on structural
features. The specimen was found in Cheshire but not *in situ*
and nothing is known as to its geological age; it consists of a

[1] Sterzel (00).   [2] Fliche (10) p. 212, Pls. xix., xx.
[3] Seward (90). See also page 290.
[4] Zeiller (92²) A. Pl. xv. fig. 5.   [5] Weiss, F. E. (13²).

petrified barrel-shaped piece of pith with portions of the inner edge of the xylem-cylinder. The surface-features agree with those of *S. speciosum*, each rhombic area being divided for a third of its length by a median groove. The pith is composed of thin-walled parenchyma with several secretory canals in the outer region; patches of xylem are preserved in the depressions between the lozenge-shaped areas showing the same anatomical characters as those described by Potonié : internal to the secondary xylem are very small groups of tracheids separated by 1—2 rows of parenchyma from the secondary elements, which pursue a sinuous longitudinal course. These tracheal strands are, as Weiss points out, at least superficially comparable with the primary xylem of such a type as *Pitys antiqua*. The innermost elements of the secondary xylem are usually scalariform and these pass gradually into tracheids with two alternate rows of bordered pits often slightly polygonal. The leaf-traces are formed of two endarch strands which coalesce as they pass downwards and eventually merge laterally with the secondary xylem.

Medullary casts with the external features of *Schizodendron* might well belong to stems which are not identical in anatomical characters, and from casts alone all that can be inferred is the presence in the vascular cylinder of medullary rays with fairly broad inner faces separated by prominent wedges of tracheids, also the spiral disposition of leaves each supplied with a single vascular bundle given off from the lower angle of the xylem-meshes. In the case of *Schizodendron Cowardi* the presence of small strands of primary xylem suggests comparison with such a genus as *Pitys* or *Mesopitys*, while in the other examples there is no indication of any xylem internal to the main cylinder. The characters of the secondary xylem point to an Araucarian or Cordaitean affinity and the pith agrees with that of *Araucaria*, though in *S. Cowardi* the presence of secretory canals is a Cycadean feature. Bain and Dawson[1], though they did not correctly interpret the surface-characters of *Schizodendron*, referred to it as representing decorticated branches of the Conifer *Walchia*. Sterzel recorded the association of *Schizodendron* with *Walchia* foliage-shoots in Saxony, and Zeiller, who noticed a similar association

---

[1] Bain and Dawson (85).

in French Permian rocks, expressed the opinion that the casts belonged to *Walchia* stems. In this connexion it is noteworthy that shoots of *Voltzia*[1] also possess medullary casts (fig. 748, A, B) with the superficial features of *Schizodendron*. Though we have no proof of a connexion between casts and leaf-bearing branches, it is probable that some forms of *Schizodendron* represent the pith-casts of *Walchia*: if this view is correct it affords another argument in favour of connecting *Walchia* with the Araucarineae, but how close the connexion is cannot be definitely settled without further evidence as to the reproductive shoots.

Schizodendron, though not confined to Permian rocks, is most abundant in beds of that age; it is recorded from several localities in Germany[2], from France, Russia[3], and Canada while the British specimen, though presumably from English rocks, was not found *in situ*.

HAPALOXYLON. Renault.

Renault[4] instituted this genus for a cylindrical stem 2 cm. in diameter from the Permian of Autun characterised especially by the parenchymatous structure of the secondary xylem. The type-species, *Hapaloxylon Rochei*, resembles *Araucarites valdajolensis*, a Permian species founded by Mougeot, in its spirally disposed leaf-scars each with an elongated groove marking the position of the leaf-trace. The solid parenchymatous pith is surrounded by a narrow zone of 2—3 layers of tracheids with a single row of bordered pits which Renault speaks of as primary xylem: this forms the inner edge of a broad cylinder of homogeneous parenchyma traversed by uniseriate medullary rays 1—3 cells deep. The secondary-xylem elements are rectangular 7—8 times as long as broad and without pits. Beyond the cambium is a broad zone of secondary phloem consisting of a regular alternation of well preserved sieve-tubes with lateral sieve-plates and parenchyma. The cortex contains some secretory sacs and is bounded by periderm.

The inference drawn by Renault is that the leaves were small like those of *Walchia* and each had a single vein. There is,

[1] Seward (90).   [2] Weiss, C. E. (72); (74); Potonié (88); (93) A. etc.
[3] Schmalhausen (87) Pl. VII. fig. 34.
[4] Renault (96) A. p. 360; (93) A. Pl. LXXVI.

however, no definite evidence as to the nature of the foliage: the stem structure represents a type previously unrecorded among Gymnosperms, but comparable with the structure of the stem of *Aeschynomene* a recent genus of the Leguminosae.

GOMPHOSTROBUS[1]　Marion.

This generic name was instituted by Marion[2] for Permian foliage-shoots from Lodève bearing vegetative leaves similar to those of *Walchia* and *Araucaria excelsa* but distinguished by the bifurcate form (fig. 747) of the relatively long sporophylls borne in a crowded cluster on the apical region of the axis. Marion referred *Gomphostrobus* to the Coniferales. Geinitz[3] had previously described detached sporophylls from the Lower Permian of Saxony, of the same type as those on which the genus *Gomphostrobus* was founded, as *Sigillariostrobus bifidus*. The Saxon specimens are represented in the drawings published by Geinitz as distally-forked scales bearing a single seed at the base. Potonié[4], who examined the original fossils in the Dresden Museum—and 1 am able to confirm his view—states that there are no undoubted seeds but only a faintly outlined area near the proximal end of each scale which no doubt marks the position of a seed or sporangium.

1873. *Sigillariostrobus bifidus*, Geinitz, Neues Jahrbuch Min. p. 700, Pl. v. figs. 5—7.
1890. *Dicranophyllum gallicum*, Schenk in Schimper and Schenk, p. 266.
1890. *Gomphostrobus heterophyllus*, Marion, Compt. Rend. cx. p. 892.
1891. *Psilotiphyllum bifidum*, Potonié, Ber. deutsch. Bot. Ges. Bd. ix. p. 256.
1892. *Gomphostrobus bifidus*, Zeiller, Bassin Houill. Perm. Brive, p. 101, Pl. xv. fig. 12.

Gomphostrobus bifidus (Geinitz).

Potonié in his account of the genus and type-species reproduces Marion's original drawings showing a *Walchia*-like axis with short falcate leaves bearing crowded linear and distally forked sporophylls reaching a length of 8 cm. in the apical region. The sporophylls, which are bent to one side, giving the impression of wind-blown foliage, consist of a simple lamina 8·5 to 25 mm. long,

---

[1] See vol. ii. p. 26.　　　　[2] Marion (90) A.　　　　[3] Geinitz (73).
[4] Potonié (93) A. p. 197 Pls. xxvii., xxviii., xxxiii.

with two divergent distal prongs varying considerably in the angle of divergence, a variation noticed also by Zeiller. On one example (fig. 747, B) Potonié records the occurrence of two scars; a lower scar, *a*, representing the attachment of the lamina and a second scar, *b*, which he attributes to a sporangium.

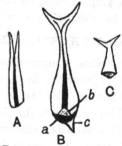

The species, represented usually by de-tached sporophylls only, is recorded from Lodève, Brive, and other French localities[1], also from Permian localities in Germany. It was referred by Schenk to *Dicrano-phyllum*, but in that genus it is the foliage-leaves that are forked and there is no reason to assume any close relationship between the two imperfectly known types. If the scar

Fig. 747. *Gomphostrobus bifidus*; *a*, leaf-scar; *b*, scar of sporangium?; *c*, torn piece of tissue. (A, B; after Potonié; C, Zeiller.)

at the base of the sporophylls marks the position of a seed a com-parison with the Araucarineae is suggested, and in this connexion it is noteworthy that Sterzel[2] records the association of *Gomphostrobus* with *Dadoxylon* wood. Potonié, who at first overlooked Marion's paper, proposed the name *Psilotiphyllum* to give expression to his opinion that the Permian plant is a Palaeozoic member of the Psilotales, a conclusion based on insufficient evidence. We have no definite information with regard to the nature of the organ borne on the sporophylls. The same author compares the sporo-phylls of *Gomphostrobus* with the leaves of *Sphenophyllum* though the verticillate disposition of the leaves of the latter genus is a well-defined difference. It would seem, as Zeiller says, that *Gompho-strobus* is probably allied to *Walchia* though its position cannot be precisely determined without further data.

A recent examination of some specimens from Lower Gondwana rocks in India described by Feistmantel[3] as *Voltzia* revealed the occurrence of some small distally forked leaves very similar to the sporophylls of the European *Gomphostrobus*.

[1] Zeiller (92²) A. p. 101, Pl. xv. fig. 12; (06) B. p. 213, Pl. L. figs. 6—8.
[2] Sterzel (00) p. 6.
[3] Feistmantel (79²) Pls. xxii. *et seq.*

VOLTZIA. Brongniart.

Brongniart[1] instituted this genus for foliage-shoots from the Bunter sandstones of the Vosges, the name being chosen in commemoration of Voltz; he compared the branches with those of *Araucaria excelsa* but added that the cone-scales bore three ovules. The leaves show considerable variation even on the same axis, a feature shared with *Walchia* and *Ullmannia*: the megastrobili are characterised by a lax disposition and the fan-like, lobed or crenulate form of the megasporophylls, which in the best preserved type, *V. Liebeana* (fig. 748, C—F), bear three ovate seeds on the upper surface. Many authors compare the Triassic genus with members of the Taxodineae, *e.g.*, *Cryptomeria*, and the Araucarineae: wood of the Araucarian type has been referred to *Voltzia* though without proof of connexion with the vegetative shoots. Gothan[2], who favours a Taxodineous alliance, points out that wood associated with *Voltzia* has Araucarian pitting on the tracheids, though he adds that the occurrence of typical Araucarian pitting in stems possessing other characters foreign to the recent Araucarineae justifies the conclusion that the presence of alternate polygonal pits on the tracheids is not necessarily proof of Araucarian affinity. An examination of some carbonised fragments attached to cone-scales of *V. Liebeana* in the British Museum from Gera revealed the occurrence of uniseriate pits both separate and in contact with one another. It is probable that *Voltzia* is related to the Araucarineae though in what degree is uncertain. A recent view[3] that *Voltzia* affords an illustration of a generalised type combining Araucarian and Abietineous features is in part based on an assumption that the cone-scales are double like those of the Abietineae. That the genus is a generalised type is probable, but the data are insufficient to warrant any definite statement as to which Coniferae are the nearest allies. The range of the genus is difficult to define: if we include the species *V. keuperiana*, also Heer's genus *Leptostrobus*, the geological range extends from the Permian to Middle Jurassic floras. The typical species are characteristic of Permian and Lower Triassic rocks. The similarity in habit of *Walchia*, some species of *Ullmannia*, and *Voltzia* renders

---

[1] Brongniart (28) p. 448, Pls. xv.—xvii. [2] Gothan (10) p. 31.
[3] Holden, R. (13)

exceedingly difficult the determination of sterile branches.
Further, the fact that the specimens of this presumably arborescent
genus are usually small branch-fragments sets a limit to our knowledge of the external features of the individual plants.

Voltzia heterophylla Brongniart.

The examination of numerous specimens from the Bunter
sandstone of the Vosges led Schimper and Mougeot to include

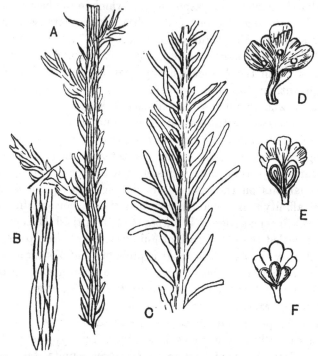

FIG. 748.   *Voltzia.*   A, B, *Voltzia heterophylla* (B, medullary cast).   C—F, *Voltzia
Liebeana.*   (A, B, after Seward;   C—F after Geinitz.)

under this name *Voltzia brevifolia* and *V. rigida* of Brongniart[1].
The Triassic species *Voltzia heterophylla* (fig. 748, A) is the best
known representative of the genus: the vegetative shoots agree
very closely with those of *Araucaria excelsa* and differ but little
from shoots of *Walchia*, some forms of *Ullmannia*, and *Pagiophyl-*

[1] Brongniart (28) p. 446; Schimper and Mougeot (44) A. p. 21, Pls. I., VI.—IX.

lum. Schütze[1] has given a long list of references to records o
V. heterophylla, but in making use of such lists it should be remem
bered that in the absence of reproductive organs the specific o
even generic determination of specimens resembling in habit
Araucaria excelsa is a hopeless task. The heterophylly of *Voltzia
heterophylla* is a striking feature: long linear obtuse leaves, 2—5
cm. long, occur in close association with falcate decurrent foliage.
Small oval strobili characterised by crowded imbricate appendages
are figured by Schimper and Mougeot as male cones, but in no case
have any sporangia been detected: similar strobili are also figured
by Leuthardt[2] from the Keuper of the Basel district. The megastrobili are longer and bear cuneate cone-scales, with 3—5 rounded
lobes on the upper surface, arranged in a lax spiral. The marginal
lobes of the scales are less deeply separated from one another
than in *V. Liebeana* (fig. 748, D—F). We have no satisfactory
information with regard to the nature or method of attachment
of the seeds. Saporta[3] figures a cone from Soultz-les-Bains
showing, as he asserts, the impressions of seeds, but the drawing
affords no definite evidence as to the relation of cone-scales and
seeds. Saporta regards the cone-scales as double, each consisting
of an ovuliferous scale and a bract-scale more or less completely
fused as in the recent genus *Taxodium.* The assumption that the
scales are double rests on a very slender basis, and even in the
much better preserved specimens of *V. Liebeana* 'there is nothing
to indicate that the scale was double[4].' In a recent paper Miss
Holden[5] speaks of *Voltzia* cone-scales as double in terms suggesting
a well-established fact, though this is by no means the case.

Blanckenhorn[6] and other authors have described specimens of
Voltzia heterophylla showing elongated leaf-cushions which they
compare with similar raised areas on the Permian casts on which
Weiss founded the genus *Tylodendron (Schizodendron[7]).* In the
latter genus the supposed leaf-cushions are casts of medullary
rays at the inner edge of the secondary xylem, and an examination
of *Voltzia* specimens in the Strassburg Museum[8] convinced me that

---

[1] Schütze (01).　　　　　　　　　　　[2] Leuthardt (03) p. 10, Pl. IV.
[3] Saporta (84) Pl. 154, fig. 4.　　　　[4] Solms Laubach (91) A. p. 68.
[5] Holden, R. (13) p. 251.
[6] Blanckenhorn (85) B. Pl. XX. figs. 17—20.
[7] See page 282.　　　　　　　　　　　[8] Seward (90).

the same explanation applies to *Voltzia heterophylla*. Triassic
Voltzia casts were referred by Schleiden to a distinct genus *Endolepis*: examples figured by Schenk[1] afford a good illustration of
their close resemblance to *Schizodendron*. The medullary casts
of the Triassic genus differ from those of *Schizodendron* in their
smaller diameter and in the absence of periodic swellings: narrow
slits in the elongated areas mark the position of out-going leaf-traces
(fig. 748, B). A similar though larger form of cast is figured by
Miss Holden[2] from Coburg and New Brunswick and referred to
Voltzia coburgensis. Fliche[3] has described a form from the
Muschelkalk of France with more slender shoots than in most
examples of *V. heterophylla*. Specimens from Swiss Triassic beds
figured by Heer[4] and Leuthardt[5] as examples of this species are
too incomplete to be identified with certainty. Feistmantel's
Indian specimens[6] referred to *V. heterophylla*, which I have recently examined, from Lower Gondwana strata, show a variation in
leaf-form suggestive of the European species, but the determination
is open to question. A supposed cone-scale figured by Feistmantel resembles in outline the lobed scales of *V. Liebeana*. Some
very incomplete branches regarded by Feistmantel as pieces of
Albertia shoots are probably identical with the impressions assigned
to *Voltzia*. On a few of the smaller Indian specimens I found
leaves 5 mm. long divided into two slightly divergent prongs,
a feature unknown in *Voltzia* but suggesting *Gomphostrobus* or
small leaves of *Dicranophyllum*. Some small seeds figured by
Zeiller[7] from the Karharbari beds of India as probably belonging
to *Voltzia* cannot be determined with confidence.

Voltzia walchiaeformis Fliche.

Fliche[8] gives this name to vegetative shoots from the Bunter
of the Vosges characterised by a *Walchia*-like habit of branching,
the pinnately disposed lateral branches being given off at about
40°; the leaves are elliptical, short and broad, more or less appressed
and less spreading than in *Walchia*. It is, however, impossible
without the confirmatory evidence of strobili to distinguish

[1] Schenk (68) p. 80, Pl. VI.
[2] Holden, R. (13).
[3] Fliche (10) Pls. XVIII., XIX.
[4] Heer (76) A. Pl. XXX.
[5] Leuthardt (03) Pl. IV. figs. 2—5.
[6] Feistmantel (79²) Pls. XXII.—XXV.
[7] Zeiller (02) B. Pl. VII. fig. 9.
[8] Fliche (10) p. 198, Pl. XXI.

clearly between certain forms of *Walchia* and *Voltzia*. A very similar type is figured by Schütze[1] as *Widdringtonites keuperianus* Heer[2] from Stuttgart but with no justification for the use of a generic name implying relationship with *Widdringtonia*. The fragments of foliage-shoots on which Heer founded this species are too small and of too common a type to be referred to a genus implying any definite position in the Coniferales.

Voltzia Liebeana Geinitz.

A Permian species[3], characteristic of the Zechstein copper-bearing beds of Gera and other localities, represented by foliage-shoots (fig. 748, C), well preserved cone-scales, and strobili. The vegetative branches closely resemble those of *V. heterophylla* and *Ullmannia selaginoïdes*: there is the same inconstancy in leaf-form as in the Bunter species. The strobili are also similar to those of *V. heterophylla*: the largest example figured by Geinitz is 2·5 cm. in diameter and 7 cm. long. The cone-scales (fig. 748, D—F) have five lobes, deeper than in *V. heterophylla*, and the central lobe is longer than the others. The occurrence of three seeds is a characteristic feature; these are ovate, 5 × 3 mm., and have a narrow marginal wing. From their close association with strobili Geinitz identified some spherical seeds, formerly described by him as *Cyclocarpon eiselianum*, as those of .*V. Liebeana*. This author figures an imperfectly preserved strobilus as a male catkin; it resembles those of *V. heterophylla*, but no microspores have been discovered. Heer describes some foliage-shoots and detached lobed cone-scales from the Permian of Hungary as *V. hungarica*[4]: the scales are similar to those of *V. Liebeana* but have slightly narrower lobes.

Voltzia keuperiana (Schimper).

This specific name[5] was given to fertile shoots characterised by long and lax strobili called by many authors *V. coburgensis* Schauroth[6], a designation first applied to a cast resembling *Lyginodendron* (*cf.* fig. 401, Vol. III. p. 37) from the Keuper of Coburg and having no proved connexion with *Voltzia*. The strobili and

---

[1] Schütze (01) Pl. x. [2] Heer (65) A. fig. 31; (76) A. Pl. xxx. figs. 4, 5.
[3] Geinitz (80) p. 26, Pl. v. [4] Heer (76²).
[5] Schimper (72) A. p. 243, Pl. LXXVI. [6] Schauroth (52) p. 540.

megasporophylls constitute the distinctive features of *V. keuperiana*: Schimper figures two strobili approximately 18 cm. long
characterised by fan-shaped scales; the lamina has a fairly long
stalk gradually passing into a broad rounded distal portion with
a crenulate edge, the sinuses between the numerous crenulations
being continued as grooves over the face of the expanded portion
of the scale. No information is available as to the seeds. The
similarity in the general plan of the strobili, apart from the clearly
marked distinguishing feature of the megasporophylls, points to
a generic affinity between this species and *V. Liebeana* and *V. heterophylla*. Schimper states that the strobili of *V. keuperiana* occur
in groups in contrast to the solitary cones of other types, and in
view of this distinction and the form of the cone-scales he employed
the generic name *Glyptolepis* for which Heer substituted *Glypto-
lepidium*[1] on the ground of the previous use of *Glyptolepis* for a
fossil fish. Schimper refers to this species the wood named by
Goeppert *Araucarites keuperianus* (= *Dadoxylon keuperianum*) but
there is no proof of actual connexion. Schenk[2] adopted the
generic name *Voltzia* and Potonié proposed a new term *Voltziopsis*[3]
to be used in a provisional and wide sense for *Voltzia keuperiana*,
Cheirolepis Escheri Heer, Heer's *Leptostrobus*[4] and Nathorst's
Swedenborgia[5] (fig. 749), including species ranging from the Keuper
to Middle Jurassic strata in contrast to the Lower Triassic and
Permian range of typical representatives of *Voltzia*. The species
Cheirolepis Escheri, included by Potonié in his genus *Voltziopsis*,
was founded by Heer on an imperfectly preserved scale from the
Lower Lias of Switzerland resembling the lobed cone-scales of
Cheirolepis Münsteri Schenk[6]. The genus *Cheirolepis* was instituted by Schimper[7] as a substitute for *Brachyphyllum* for the
Rhaetic species *B. Münsteri* Schenk, the new name being chosen
because of the digitate margin of the cone-scales which are said to
bear single seeds. My former employment of Schimper's generic
term for sterile branches originally named by Phillips *Brachyphyllum setosum*[8] was hardly justifiable in the absence of sporo-

[1] Heer (77) ii. p. 72.   [2] Schimper and Schenk (90) A. p. 290, fig. 199.
[3] Potonié (99) B. p. 304.   [4] Heer (77) ii. p. 72.
[5] Nathorst (78) B. p. 30, Pl. xvi. figs. 6—12.
[6] Schenk (67) A. p. 187 Pl. xliii. figs. 1—12.
[7] Schimper (72) A. p. 247.   [8] Seward (00) B. p. 294.

phylls. The species *Voltzia recubariensis* (Mass.) represented by
several vegetative shoots, imperfect cones, cone-scales, and seeds
in the Muschelkalk beds of Recoaro[1] illustrates the impossibility
in the case of sterile specimens of drawing any satisfactory line
between *Voltzia* and *Pagiophyllum*. Heer instituted the genus
Leptostrobus for strobili from Jurassic strata in Siberia agreeing
closely in habit and in the form of the megasporophylls with those
of *Voltzia*. The strobili, referred by Heer to three species, do not
exhibit any well-marked specific differences; the longest example,
L. crassipes, is 7 cm. in length and 1·5 cm. broad: the scales,
7—8 mm. broad, are entire at the distal margin or more or less
lobed and in some specimens the scales are hardly distinguishable
from those of *V. heterophylla*. Heer states that two-winged seeds
are borne on some of the scales though the evidence is not clear.
In a later account Heer[2] includes in *Leptostrobus* clusters of long
linear leaves apparently borne on short shoots and resembling the
needles of *Pityites Solmsi* Sew., *Schizolepis Braunii*, and Jeffrey's
Prepinus. These leaves, described as *L. rigida* and *L. angustifolia*, though in close association with strobili are not actually
connected with them; they differ considerably from the short,
triangular, imbricate leaves shown in one of Heer's figures immediately below the fertile portion
of an axis of *L. crassipes*[3]. It
would be difficult to draw any
generic distinction between *Leptostrobus* and *Voltzia* especially *V.
keuperiana*.

SWEDENBORGIA. Nathorst.

This genus[4] is founded on lax
oval strobili bearing small cone-
scales with long stalks and a
single seed (fig. 749, B). In the
type-species, *S. cryptomerides*, the
cones reach 7 cm. in length and

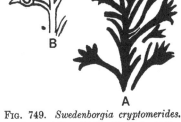

Fig. 749. *Swedenborgia cryptomerides.*
A, Strobilus. B, Fertile leaf. (After
Nathorst; nat. size.)

the scales are divided distally
into 4 or 5 acute digitate lobes. Nathorst compares the strobili

[1] Schenk (68) Pls. VII. *et seq.*   [2] Heer (82) p. 23, Pls. VII., VIII.
[3] *Ibid.* (77) ii. Pl. XIII. fig. 14.   [4] Nathorst (78) B. p. 30, Pl. XVI.

with those of *Cryptomeria* and *Voltzia* but mentions the presence
of a single seed as an important distinguishing feature.   To
unite *Swedenborgia* with *Voltzia* would be misleading, and there
is no valid reason for replacing Nathorst's term by Potonié's
genus *Voltziopsis*.

Strobilites. *Strobilites laxus* Seward.

This name was applied to a lax strobilus, 30 cm. long and 1·3 cm.
broad at the base, from Rhaetic beds on the Orange River, South
Africa[1], which may be allied to *Voltzia*, though in the absence of
seeds its position cannot be determined.   The sporophylls consist
of a lamina with a rounded distal edge and a radially folded surface
attached by a short horizontal stalk resembling the seed-bearing
scales of *Voltzia heterophylla*, *V. coburgensis*, and to some extent
Heer's Jurassic *Leptostrobus*.

ULLMANNIA. Goeppert.

Goeppert[2] in his description of *Ullmannia* refers to the extensive
literature on the fossils from the Permian copper mines of Frankenberg on which the genus was founded: the most complete of
these earlier accounts is that of Ullmann.   In habit similar to
Walchia, *Ullmannia* is represented by various forms of foliage-
shoots and impressions of buds and cones, but the data are insufficient to settle its position in the Coniferales.   *Ullmannia
Bronni* (fig. 750, D), the type-species, is practically identical in
leaf-form and habit with the Mesozoic genus *Pagiophyllum*, while
the species *U. frumentaria* (fig. 750, A) agrees closely with such
recent Conifers as *Araucaria excelsa* and *A. Bidwilli*.   The branches
bear spirally disposed crowded leaves with a median vein and
numerous longitudinal striations on the lamina.   The association
of impressions of foliage-shoots with wood having the Araucarian
type of pitting[3] affords contributory evidence, though by no
means proof, of Araucarian affinity.   In the absence of any definite
information as to the structure of the reproductive shoots *Ullmannia* must be left for the present as a Conifer which cannot be
assigned with certainty to a systematic position.   Tuzson[4] uses

---

[1] Seward (08) B. p. 101, text-fig. 7; Pl. v. fig. 3.   [2] Goeppert (50) p. 185.
[3] Solms-Laubach (84) Pl. iii. fig. 16; Schimper and Schenk (90) A. p. 275, figs.
190, 191.   [4] Tuzson (09) p. 23.

the generic name *Ullmannites* for wood having the Araucarian type of tracheal pitting which he believes to belong to *Ullmannia*, but under the former genus are included types of wood that are not generically identical and afford no evidence of connexion with *Ullmannia*. *Ullmannia* is essentially a Permian genus especially characteristic of the copper-bearing rocks of Frankenberg in Hessen and Ilmenau in Thuringia; it is recorded also from France and by Eichwald and Schmalhausen from Russia, the species

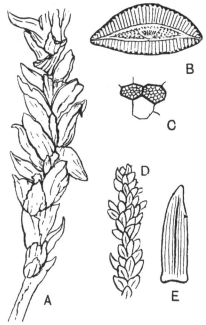

Fig. 750.   *Ullmannia*. A—C, *Ullmannia frumentaria*. D, E, *U. Bronni*. (A, after Geinitz; B, C, after Solms-Laubach; D, E, after Potonié.)

U. biarmica Eich.[1] being represented by both sterile and fertile branches. Imperfectly preserved impressions from the Permian of Durham[2] have been referred to *Ullmannia* but no satisfactory specimens have been discovered in English beds. The fragment figured by Lindley and Hutton[3] as *Voltzia Phillipsi* may belong to an *Ullmannia*.

---

[1] Schmalhausen (87) Pl. VI.
[2] Murchison and Harkness (64) p. 154; Kirby (64).
[3] Lindley and Hutton (37) A. Pl. 195.

Ullmannia Bronni Goeppert.

The fragments of foliage-shoots on which Goeppert[1] founded this species were described by some earlier authors as the Frankenberg ears of corn and by Schlotheim as *Poacites phalaroides*. Bronn, who first identified the fossils as Coniferous, named them *Cupressites Ullmanni*. We know nothing of the structure of the reproductive shoots, and the cones referred by Goeppert to this species have since been assigned to the genus *Strobilites*. In habit and leaf-form *Ullmannia Bronni* is indistinguishable from certain Triassic and Jurassic foliage-shoots referred by most authors to *Pagiophyllum*. The leaves are imbricate (fig. 750, D, E); the lamina is oval or broadly linear, elliptical, and characterised by longitudinal striations. Some imperfectly preserved leaves examined by Solms-Laubach showed clear indications of the presence of a midrib with lateral groups of transfusion-tracheids, as in *U. selaginoides* and *U. frumentaria*.

Ullmannia selaginoides (Brongniart).

This and the following species, *U. frumentaria*, are founded on vegetative shoots from the Permian of Ilmenau (Thuringia) and both were described by Brongniart as examples of *Fucoides*[2]. The leaves of *U. selaginoides* are longer than those of *U. Bronni*, linear and almost uniform in breadth, elliptical in section. There is a single vascular bundle accompanied by wings of reticulate transfusion-tracheids (*cf.* fig. 750, C) associated with parenchyma[3]. As in the leaves of recent Conifers the transfusion-tissue persists in the apical region of the lamina. There are 1—2 rows of hypodermal fibres below the epidermis with sunken stomata and the mesophyll consists largely of palisade-cells (*cf.* fig. 750, B). The stele of the shoot has a large pith with nests of dark cells enclosed by a cylinder of secondary xylem consisting of tracheids having a single row of separate circular pits on the radial wall and uniseriate medullary rays 1—6 cells deep.

Ullmannia frumentaria (Schlotheim).

This species, originally named *Carpolithes frumentarius*[4], is the commonest fossil in the Ilmenau mines. The leaves are lanceolate,

[1] Goeppert (50) p. 185, Pl. xx. For synonymy, see Solms-Laubach (84); Geinitz (80).   [2] Brongniart (28) A. Pl. ix.
[3] Solms Laubach (84) Pl. iii. figs. 1, 4, 6, 15.
[4] Schlotheim (20) A. Pl. xxvii. fig. 1 For figures, see Geinitz (80); etc.

acute, decurrent, and more or less falcate and like those of other species characterised by longitudinal striae on the dorsal face (fig. 750, A). Well-preserved impressions of this species in the Dresden Museum bear a close resemblance to shoots of *Araucaria Bidwilli* and *A. brasiliensis*. Some specimens show laterally attached oval cones, but it is not clear if these are reproductive shoots or vegetative buds and nothing is known as to the nature of the sporophylls. The seeds, *Cardiocarpus triangularis*[1], referred by Geinitz to this species on the ground of association, cannot be safely assigned to *Ullmannia*. In leaf-structure (fig. 750, B, C) *U. frumentaria* closely resembles *U. selaginoides* but the hypodermal fibres form oval strands instead of 1—2 layers. The species has also been recorded from the Permian of France[2].

Strobilites.   *Strobilites Bronni* (Goeppert).

Under this provisional name Solms-Laubach[3] described the problematical fossils from the Frankenberg copper mines which Goeppert believed to be the megastrobili and cone-scales of *Ullmannia Bronni*. In the absence of any satisfactory evidence of connexion with the vegetative shoots on which *U. Bronni* was founded it is better to follow Solms-Laubach in the adoption of the non-committal name *Strobilites*. These star-stones ('Sterngraupen') of the miners consist of more or less circular bodies bearing some resemblance to the peltate cone-scales of *Cupressus*; they occur either singly or in cone-like groups. No seeds have been found attached to the scales nor is there any proof that they were borne by a Conifer. The larger scales, 15—25 mm. in diameter, are characterised by 8—12 radial ridges and a central depression (? umbo), and to the under surface is attached a cylindrical stalk usually in the centre but occasionally excentric. The scales are sometimes found in almost spherical clusters and another form described by Solms-Laubach consists of a cylindrical cone-like aggregate 4 cm. × 2 cm. of rather smaller scales without radial ribs and characterised by a stout, rounded, peripheral border.

---

[1] Geinitz (80) Pl. III. figs. 11—15.
[2] Zeiller (06) B. p. 219, Pl. L.
[3] Solms-Laubach (84).

ALBERTIA. Schimper.

This generic name was given by Schimper to vegetative branches from the Bunter of the Vosges agreeing in habit with shoots of some species of *Agathis* but differing in the broader insertion of the lamina. For *Albertia* Endlicher[1] substituted *Haidingera.* Schimper and Mougeot[2] figure reconstructions of both male and female cones and a single cone-scale bearing a median seed. Schenk[3], who examined the original specimens, states that the supposed male cone is a young megastrobilus of *Voltzia,* and Solms-Laubach[4], who also examined the material in the Strassburg Museum, considers that Schimper's statement that the seed-scale and cones belong to *Albertia* is 'altogether arbitrary and unsupported.' There would seem to be no reason for connecting the cones figured by Schimper and Mougeot with the shoots referred by those authors to four species of *Albertia*[5]. *Albertia latifolia* is founded on branches bearing fairly large (2·5 × 1 cm.), obovate, slightly decurrent leaves with numerous longitudinal striations. The branches described as *A. elliptica* are not distinguishable by any clearly marked feature from *A. latifolia.* *Albertia Braunii* has larger obovate leaves and *A. speciosa* has broadly linear leaves reaching a length of 4·5 cm. There is a very close resemblance between the shoots from the Bunter beds and those of some forms of *Ullmannia* especially *U. frumentaria* (Schlot.)[6], and it is doubtful whether any useful purpose is served by the retention of the designation *Albertia*: the descriptions of the reproductive shoots are misleading and rest on no substantial basis and the sterile branches exhibit no characters by which they can be generically separated from *Ullmannia.* The important point is that there are no grounds for regarding the specimens usually referred to *Albertia* as Araucarian other than the uncertain and untrustworthy evidence afforded by a similarity to *Agathis.* The fragmentary impression from the Karharbari beds of India assigned by Feistmantel[7] to

[1] Endlicher (47) p. 303.
[2] Schimper and Mougeot (44) A. Pl. I.
[3] Schenk in Schimper and Schenk (90) A. p. 284.
[4] Solms-Laubach (91) A. p. 75.
[5] Schimper and Mougeot (44) A. Pls. I.—v.
[6] Geinitz (80) Pl. III.
[7] Feistmantel (79) p. 29, Pl. XXVI. fig. 2.

Albertia is too imperfect to be determined; it may be identical with those referred by the same author to *Voltzia*.

The abundance and wide distribution of wood with Araucarian features in Palaeozoic rocks though, for reasons already stated, not admissible as proof of the occurrence of members of the Araucarineae, at least shows the great antiquity and predominance of the Araucarian type. There can be no reasonable doubt that much of the wood described in Chapter XXXIII. as *Dadoxylon* belonged to Araucarian plants, more especially the examples furnished by Mesozoic and Tertiary strata. In considering the past history of the family the evidence of the wood must be taken into account.

The conclusions drawn from a survey of the fossil records are: (i) the type represented by *Araucaria* is older than that now illustrated by *Agathis*. In other words *Araucaria* possesses features, especially those associated with the megastrobili, which extend farther back without departing far from the existing type than is the case with *Agathis*. (ii) The Araucarineae foreshadowed in the later Carboniferous and earlier Permian periods were in all probability established as a family in Rhaetic times, and in the Jurassic and earlier Cretaceous periods the Araucarineae were almost cosmopolitan and represented by numerous forms. (iii) Such evidence as is afforded by Tertiary records, though meagre and often incomplete, points to the continued existence of the family in the Northern Hemisphere at least in the older Tertiary floras.

CHAPTER XLVI.

CUPRESSINEAE.

THE published records of fossil Conifers would seem to justify the conclusion that the Cupressineae were widely distributed and represented by a wealth of genera during the latter part of the Mesozoic era particularly in the later Jurassic floras, but on closer inspection of the material a student, having any familiarity with the external features of recent genera, cannot fail to recognise the wholly inadequate nature of the data on which the systematic determinations are based. It is undoubtedly true that in the later Jurassic and Lower Cretaceous floras Conifers agreeing generally in habit and in the possession of appressed imbricate leaves with such genera as *Cupressus*, *Chamaecyparis*, and *Thuya* were among the most characteristic types: some have the leaves in decussate pairs with an occasional tendency to a spiral phyllotaxis while others possess leaves of the same form but spirally disposed. Almost all are sterile and when cones are present the form and arrangement of the scales often suggest comparison with recent types other than the Cupressineae. In the Chapter on recent Conifers attention is called to the inconstancy of leaf-arrangement in certain species and to the close resemblance between vegetative shoots of plants belonging to different families. Fossil coniferous branches referred by authors to the Cupressineae afford a striking illustration of the insufficiency of the evidence on which sterile impressions have been named. This statement, though primarily concerned with Mesozoic records, applies also to many Tertiary species. The records of the rocks clearly show that European Tertiary floras contained a considerable number of Cupressineous types that are now confined to other regions, but a critical examination of the older fossils leads to the conclusion that in very many cases accurate determination of the affinities of sterile branches, superficially resembling existing members of the Cupressineae, is impossible without additional information.

In the absence of well-preserved cones or anatomical data it is possible that a comparative examination of cuticular membranes might furnish useful results. It is, however, only rarely that such information can be obtained and the only safe course to follow is to use, with greater freedom than has generally been the practice, provisional generic names which do not imply affinities to recent genera. Provisional names that have reference only to vegetative features should be superseded by designations denoting characters of greater taxonomic significance when the necessary information is available. As a preliminary to the description of a few selected types it may be useful to consider the sense in which some generic names have been employed and at the same time to state whether the retention of certain names or their use in a modified sense is advisable.

Thuytes Brongniart.

Brongniart[1] instituted this term for Branches like those of *Thuya*; fruit unknown.' The name, in the form *Thuyites*, had been used a few years previously by Sternberg and it was adopted by Unger[2] as *Thuites* for both sterile branches and cones similar to those of the recent genus. *Thuytes* has been widely used for vegetative branches agreeing generally in habit with those of *Thuya* or *Cupressus* and some other Cupressineae, but with the exception of some Tertiary species the designation has reference in nearly all cases to the form and arrangement of the leaves. The employment of *Thuytes* in this wide sense is open to criticism on the ground that in accordance with the usual practice specimens so named would be considered to be more nearly related to *Thuya* than to any other genus. It is therefore proposed to adopt the generic designation *Thuites* only for such specimens as afford evidence of close affinity to the recent genus and to discontinue its use for sterile shoots which suggest comparison not only with *Thuya* but also with *Cupressus*, *Libocedrus* and other genera.

CUPRESSINOCLADUS. Gen. nov.

Goeppert's term *Cupressites*[3] is retained for fossils which there is reason for associating with *Cupressus* and should not be employed

---

[1] Brongniart (28) A. p. 109; Seward (04) B. p. 140.
[2] Unger (50) A. p. 346.   [3] Goeppert (50) p. 183.

in a wider sense. Bowerbank[1] adopted the form *Cupressinites*
for some fossil cones from the London Clay in order to avoid the
implication of affinity only to *Cupressus* which is suggested by
Cupressites. This generic name would be convenient for Cupressineous branches had it not been restricted in the first instance to
cones: to avoid the revival of a term and its employment in a new
sense it is proposed to adopt the name *Cupressinocladus* for vegetative shoots agreeing in the habit of branching and in the predominance of a decussate arrangement of appressed leaves with recent
Cupressineae such as *Cupressus*, *Thuya*, *Libocedrus* and similar
types. When cones are present which throw any light on generic
affinity some other term should be adopted. It will, however, be
found in practice that the choice of the most appropriate name is
exceedingly difficult; and no sharp line can be drawn between
certain specimens which conform in part to *Cupressinocladus* and
in part to the characters of *Brachyphyllum*.

Palaeocyparis Saporta.

The published illustrations of Conifers included by Saporta[2] in
this genus afford examples of the inconstancy of leaf-arrangement
in a single type and demonstrate the impossibility of drawing any
definite distinction between this genus and *Thuytes* as used by
Saporta. With one exception all the specimens from Jurassic
rocks referred by Saporta to his genus are sterile and in habit
agree with several recent genera of the Cupressineae particularly
Cupressus, *Chamaecyparis*, and *Thuya*. In *Palaeocyparis* are included species previously referred to *Echinostrobus*, *Thuytes*, and
Athrotaxites. The branching is in one plane; the leaves are stated
to be usually though not invariably decussate, more or less triangular, appressed, and imbricate, rarely free at the apex of the
lamina. The supposed cone described in the case of *Palaeocyparis
elegans*[3], a species from Upper Jurassic beds in France, is only 7
by 9 mm. and it is not clear whether it is a true cone or a vegetative
bud or perhaps a male flower. The genus is practically founded
on vegetative characters only. An objection to the retention of
Saporta's term is that several of the specimens may legitimately
be included in a previously established genus *Brachyphyllum*.

[1] Bowerbank (40) p. 51.   [2] Saporta (84) p. 574, Pls. 202 *et seq.*
[3] *Ibid.* Pl. 214.

THUITES. Brongniart emend.

As stated on a previous page it is proposed to limit this name to fossils affording good evidence of close affinity to the recent genus *Thuya*. Among the few examples that appear to fulfil this condition are those described as *Thuya occidentalis* L. *succinea* Goepp.[1] and *Biota orientalis* Endl. *succinea* Goepp.[2] from the Oligocene amber deposits on the Baltic coast. Schlechtendal[3] records specimens from Oligocene beds at Weimar, consisting of shoots with opposite pairs of leaves, on which a resin-canal is seen below the apex, and the remains of male flowers, which he names *Thuya occidentalis* var. *thuringica*.

CUPRESSITES. Goeppert.

Cupressites MacHenryi Baily.

This name was given[4] to a piece of shoot of *Cupressus*-like habit from the Eocene beds of Antrim. Baily's figure, which Gardner says is inaccurate, shows the small scale-leaves as spiral, but the specimens subsequently described by Gardner[5] from the same locality leave no doubt as to the decussate arrangement of the foliage. Gardner adopts the name *Cupressus Pritchardi* on the assumption that the fossil wood from Lough Neagh described by Goeppert[6] as *Pinites Pritchardi* belongs to the species which furnished the Antrim specimens: the wood, subsequently referred to *Cupressinoxylon*[7], affords a typical example of that genus as is clearly shown by the photographs reproduced in fig. 715 (p. 188), but there is no proof of any connexion between it and the branches from the Antrim leaf-beds. The shoots are characterised by their slender pinnately arranged branchlets with small decussate leaves, and the cones, about 14 mm. long, bear 10 hexagonal scales of the *Cupressus* type very similar to those of *Cupressites taxiformis*. No seeds were found attached to the cone-scales. The abundance of specimens in the Irish beds indicates that this Cupressineous species was a common tree in the forests which

---

[1] Goeppert and Menge (83) A. p. 43, Pl. xv. figs. 199–206.
[2] *Ibid.* p. 42, Pl. xv. figs. 180—198.
[3] Schlechtendal (02) Pls. I., II.
[4] Baily, W. H. (69) p. 361, Pl. xv. fig. 5.
[5] Gardner (86) p. 82, Pl. xvi. figs. 8, 9; Pl. xviii. fig. 1; Pl. xix.
[6] Goeppert (50) p. 220.   [7] Kraus in Schimper (72) A. p. 376.

flourished on the western edge of Europe during the period of
volcanic activity responsible for the widespread sheets of lava in
the North-East of Ireland and the Western Isles of Scotland.

Cupressites taxiformis Unger.

This species was founded by Unger[1] on sterile and fertile shoots
from Eocene beds in the Tyrol.   Many of the sterile branches are
similar in the form of the linear leaves to *Taxus,* but other leaves
are appressed to the axis and free only at the apex, resembling on
a small scale those of *Sequoia gigantea.*   The cones consist of
polygonal, peltate, scales probably verticillate and superficially
similar to the strobili of *Cupressus*: they are borne on shoots with
scale-like leaves.   The species is recorded by De la Harpe[2] from
the Isle of Wight and several specimens are figured by Gardner[3]
from the Middle Bagshot beds of Bournemouth.   The material
from the latter locality consists of sterile shoots with linear *Taxus-*
like leaves from 5 to 15 mm. long and occasionally, on the same
axis, smaller decurrent leaves, though generally the two forms are
found on different twigs.   The cones occur only
in connexion with the shoots bearing small appressed leaves: the specimen shown in fig. 751
has been re-drawn from one of Gardner's figured
specimens[4]: it is characterised by peltate scales
with a central umbo and a wrinkled surface.
The evidence in favour of assigning all the
sterile shoots to the same species is not convincing, though a similar combination is met
with in the recent species *Glyptostrobus hetero-*
phyllus.   If we confine our attention to the cones

Fig. 751. *Cupressites
taxiformis.* (From
a specimen in the
British Museum;
nat. size.)

they may reasonably be retained in the genus *Cupressites* and
regarded as evidence of the existence in Western Europe in the
Eocene period of a type closely allied to the genus *Cupressus.*
The position of the sterile shoots cannot be determined without
further investigation.

    [1] Unger (47) p. 18, Pl. VIII. figs. 1—3; Pl. IX. figs. 1—4.
    [2] De la Harpe in Bristow (62) Pl. V. fig. 2.
    [3] Gardner (86) p. 26, Pl. I. figs. 1—13; Pl. V. figs. 13, 14; Pl. VII. fig. 8; Pl. IX.
figs. 22—26, 28—30.
    [4] *Ibid.* Pl. IX. fig. 27.

Another Tertiary representative of the recent genus *Cupressus* is that described from the Oligocene amber beds of East Prussia as *Cupressus sempervirens* L. *succinea* Goepp. and Menge[1]. This species is founded on fragments of sterile shoots with a well preserved male flower showing very clearly the form of the sporophylls.

CUPRESSINOCLADUS. Gen. nov.

The following examples of Cupressineous shoots that do not afford satisfactory evidence of relationship to any particular recent genus are given in illustration of the desirability of employing such a non-committal generic term as *Cupressinocladus*.

(i) *Species previously referred to* Libocedrus.

The Cretaceous specimens from the Atane beds of West Greenland described by Heer[2] as *Libocedrus cretacea* are unaccompanied by any cones and may equally well be compared with species of *Thuya*: specimens from the Amboy clays, believed to be identical with Heer's, are assigned to the latter genus by Newberry[3].

Cupressinocladus salicornoides (Unger).

A sterile piece of branch figured by Lindley[4] from Provence as *Thuya* is probably identical with Unger's species recorded by Saporta[5] from the same locality. The type-specimens on which Unger founded the species *Thuites salicornoides* (fig. 752) are from Eocene beds in Croatia[6]; they do not bear any mature cones and cannot be assigned with confidence to any recent genus. The flattened shoots bear appressed leaves in decussate pairs and the decurrent lamina may reach a length of 1·5 cm. By later authors this species, recorded from Styria[7], Bohemia[8], Switzerland[9], the Oligocene beds of East Prussia[10], Italy[11], the Miocene of France and from other localities[12], is spoken of as *Libocedrus*, but the few examples of cones that have been figured do not exhibit with

---

[1] Goeppert and Menge (83) A. p. 45, Pl. XVI. figs. 218—224.

[2] Heer (82) i. p. 49, Pl. XXIX. figs. 1, 2; Pl. XLIII. fig. 1 *d*.

[3] Newberry and Hollick (95) p. 53, Pl. X. figs. 1, 1 *a*.

[4] In Murchison and Lyell (29) p. 298, fig. B.

[5] Saporta (65²) p. 42, Pl. I. fig. 4.   [6] Unger (47) p. 11, Pl. II.

[7] Ettingshausen (70) p. 39; (88) p. 273.

[8] *Ibid.* (67²) p. 109, Pl. V. figs. 1—7, 14; Engelhardt (85) Pl. VIII. figs. 27 —30.

[9] Heer (55) A. p. 47, Pl. XXI. fig. 2.

[10] Goeppert and Menge (83) A. Pl. XV. figs. 175—177.

[11] Massalongo (59) p. 153, Pl. V. figs. 20—23.   [12] Marty (08).

308

CUPRESSINEAE

[CH.

sufficient clearness morphological features that justify the conclusion that the specimens agree more closely with *Libocedrus* than with *Thuya* or *Thujopsis*. The present discontinuous distribution of *Libocedrus* (page 126) is favourable to the view that it was formerly much more widely spread, but despite the very

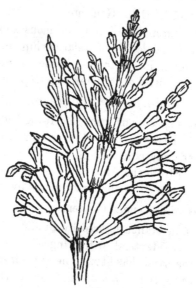

FIG. 752. *Cupressinocladus salicornoides.* (After Unger; nat. size.)

close resemblance between the sterile shoots of the Tertiary Conifer to those of some existing species of the genus it would be unwise to adopt the designation *Libocedrus* or *Libocedrites*.

Heer[1] described fragments from Miocene strata in Greenland as *Libocedrus Sabiniana* including a supposed cone-scale, which is too indistinct to be determined. To this species Beust refers some petrified wood from the same locality. The sterile specimens recorded by Gardner[2] from the Woolwich beds of Kent as *Libocedrus adpressa*, though similar to *L. salicornoides*, are no doubt specifically distinct: they are valueless as evidence of the existence of *Libocedrus*. Laurent[3] also records the species from Aquitanian beds in the Puy-de-Dôme.

[1] Heer (83) p. 58, Pls. LXX., LXXXVI., LXXXVII.; Beust (85) p. 40, Pls. III., v.
[2] Gardner (86) p. 25, Pl. II. figs. 17—20.　　　[3] Laurent (12) p. 69.

(ii) *Species previously referred to* Thujopsis.

Cupressinocladus massiliensis (Saporta). The small twigs described by Saporta from Provence as *Thujopsis massiliensis*[1] and compared by him with *Thuya occidentalis* are very similar to those named by the same author *T. europaea*[2]: in neither case is there any justification for the use of the generic name *Thujopsis*. Heer[3] records *T. europaea* from Miocene beds in Greenland and in a later account adopts the name *Biota orientalis* on the ground of the occurrence of imperfect cones and elongated cone-scales comparable with those of *Biota* (*Thuya*), but the figured specimens are too indistinct to warrant the employment of the generic term *Thuites* in the more restricted sense advocated above. Goeppert and Menge[4] refer some fragments from the Baltic amber beds to *Thujopsis europaea*, but no reproductive organs are figured.

(iii) *Species previously referred to* Thuya *or* Thuites.

Several Jurassic and Lower Cretaceous specimens formerly included in *Thuites* are now transferred to *Brachyphyllum* on the ground that the choice of one or other of these names has frequently been determined by characters that are both inconstant and of little morphological importance. There are, however, several examples of Coniferous shoots from Mesozoic and Tertiary strata that are clearly distinguished from such types as *Thuites expansa* (= *Brachyphyllum expansum*), in which the verticillate arrangement of the leaves is not a well marked or constant feature, by the very regular disposition of appressed leaves in decussate pairs as in recent species of *Libocedrus* and some other Cupressineae: for this form of shoot the generic name *Cupressinocladus* is now adopted.

Cupressinocladus valdensis Seward.

A species described as *Thuites valdensis*[5] from a single specimen from Wealden beds on the Sussex coast, characterised by decussate appressed leaves with a comparatively long basal portion concrescent with the axis of the branch and a free short triangular

---

[1] Saporta (65) p. 72, Pl. I. fig. 6; Pl. IV. fig. 2.
[2] *Ibid.* (65[2]) Pl. I. fig. 5.
[3] Heer (68) i. p. 90, Pl. L. fig. 11; (75) iii. p. 7. Pl. I. figs. 13¹—29.
[4] Goeppert and Menge (83) A. Pl. XVI. figs. 215—217.
[5] Seward (95) A. p. 209, Pl. XX. fig. 6.

apex, agreeing closely with some forms of *Libocedrus* and with the younger branches of *Frenelopsis*. Though accurate determination of the position of such specimens is impossible, they afford evidence of the fairly widespread occurrence of Conifers in Mesozoic and Tertiary strata exhibiting a striking resemblance in habit to recent Cupressineous genera.

Vegetative branches from Miocene beds in Spitzbergen and Greenland referred by Heer[1] to *Thuites Ehrenswaerdi* and *T. Meriana* respectively afford examples of specimens which would be more appropriately included in the genus *Cupressinocladus*. Similarly the fragments described by Goeppert and Menge as *Thuya Mengeana* Goepp.[2] afford no convincing evidence of generic identity with the recent genus.

(iv) *Species previously referred to* Juniperus *or* Juniperites.

Cupressinocladus hypnoides (Heer).

The slender sterile branches from the Lower Cretaceous beds on the West of Greenland described as *Juniperus hypnoides*[3] afford no substantial evidence of relationship to *Juniperus* rather than to some other member of the Cupressineae or Callitrineae. Specimens from the same locality which may be specifically identical with *Juniperus hypnoides* are described by Heer as *J. macilenta*[4]. The leaves are very small and occur on the slender axes in opposite pairs. Newberry records *J. macilenta*[5] from the Amboy clays and states that cone-scales of *Dammara* are associated with the foliage-shoots, though Hollick in a note to the description says that he was unable to find any such scales with the vegetative branches. Hollick and Jeffrey[6] figure specimens from the lignite beds at Kreischerville as *J. hypnoides* and believed them to be identical with those described by Newberry as *J. macilenta*; they also speak of the association of *Dammara* [*Agathis*] scales.

The Tertiary Greenland species *J. tertiarius* and *J. gracilis*[7], founded by Heer on sterile branches, are equally unsatisfactory as

[1] Heer (71) iii. p. 38, Pl. II. figs. 25, 26; (82) i. Pl. I.
[2] Goeppert and Menge (83) A. p. 44, Pl. XVI. figs. 211—214.
[3] Heer (82) i. p. 47, Pl. XLIV. fig. 3; Pl. XLVI. fig. 18.
[4] *Ibid.* Pl. XXXV. figs. 10, 11.   [5] Newberry and Hollick (95) Pl. X. fig. 7.
[6] Hollick and Jeffrey (09) B. p. 61, Pl. V. figs. 5, 6.
[7] Heer (83) p. 57, Pls. LXX., CII., CVI.

records of Conifers closely allied to *Juniperus*, and the same remark applies to *Juniperites eocenica* described by Ettingshausen[1] from Häring in the Tyrol. A single male flower figured by Goeppert and Menge[2] from the Baltic amber as *Juniperus Hartmannianus* may be correctly referred to that genus though other recent genera are not excluded.

Echinostrobus Schimper.

Proposed in the first instance by Schimper[3] for Unger's *Athrotaxites lycopodioides*, this term was adopted for several sterile shoots such as those named by Brongniart *Thuytes expansus*, characterised by the possession of decussate leaves like those of *Thuya* and *Cupressus* with others agreeing more closely with *Brachyphyllum*. As the name has reference to the spinous nature of the cone-scales, and as it is now agreed that Unger's earlier name *Athrotaxites* may be appropriately employed, *Echinostrobus* is discarded.

Phyllostrobus Saporta.

This generic name was given by Saporta[4] to an Upper Jurassic fertile shoot with whorled leaves of the *Thuites* form bearing a single cone compared with those of *Libocedrus*. The impression conveyed by Saporta's figures is that the preservation of the cone is too imperfect to warrant the institution of a new generic term.

Condylites Thiselton-Dyer.

This name applied to specimens from the Solenhofen slates[5] has reference to the elbow-like insertion of lateral branches: the foliage is like that of *Brachyphyllum*, and the cones, which are imperfectly preserved, are compared with those of *Thuya*. As in the case of Saporta's *Phyllostrobus* the cones are too obscure to admit of any satisfactory description.

Athrotaxites Unger.

This name was proposed by Unger[6] for a branched cone-bearing shoot from Solenhofen agreeing in vegetative characters with

---

[1] Ettingshausen (55). Pl. v. fig. 6.
[2] Goeppert and Menge (83) A. p. 39, Pl. xiv. figs. 156, 157.
[3] Schimper (72) A. p. 330.   [4] Saporta (84) p. 635.
[5] Thiselton-Dyer (72).   [6] Unger (49) Pl. v. figs. 1, 2.

specimens previously figured by Sternberg as *Caulerpites*. The specimen in the Munich Museum on which the type-species *Athrotaxites lycopodioides* was founded was examined and re-figured by Schimper[1] and by Saporta (fig. 753, C)[2]: the former author substituted for *Athrotaxites* a new genus *Echinostrobus* in order to avoid the implication of relationship with *Athrotaxis* which he was not prepared to accept, and without adequate reason altered Unger's specific name *lycopodioides* to *Sternbergii*. Saporta, who believed Unger's type to be intermediate between *Athrotaxis* and *Cryptomeria* as regards the features of the cones, retained Schimper's designation *Echinostrobus Sternbergii*. Unger[3] in subsequent accounts of Solenhofen plants extended the application of *Athrotaxites* to sterile shoots, and this course was also followed by Thiselton-Dyer[4] who expressed agreement with Unger as regards the resemblance of the fossil cones to those of the recent genus. In accordance with the principle advocated on a previous page it is suggested that Unger's generic name should be retained only for specimens which afford evidence, other than mere resemblance of foliage-shoots, of affinity to the recent genus *Athrotaxis*: this use of Unger's term has recently been adopted by Halle[5]. Most of the sterile specimens referred to *Athrotaxites* by Unger and other authors should be transferred to *Brachyphyllum*.

Athrotaxopsis Fontaine.

Fontaine[6] instituted this name for some fertile Coniferous shoots from the Potomac formation similar in habit to species included by Saporta in his genus *Palaeocyparis* and to the genus *Thu tes* but bearing cones different from those of *Athrotaxites lycopodioides*. Berry[7] subsequently reduced the number of Fontaine's species and pointed out that some of his specimens are indistinguishable from shoots included in *Sphenolepidium*, a conclusion to which I had been led[8] by a comparison of Wealden specimens of *Sphenolepidium Kurrianum* with Fontaine's figures. Some of the Potomac examples agree in vegetative characters with the genus *Brachyphyllum*. In the absence of any substantial

[1] Schimper (74) A. Pl. LXXV. fig. 21.
[2] Saporta (84) Pl. 199.   [3] Unger (52); (54²).
[4] Thiselton-Dyer (72).   [5] Halle (13) p. 40.
[6] Fontaine (89) B. p. 239.   [7] Berry (11⁴).   [8] Seward (95) A. p. 202.

grounds for assuming any direct relationship between the Potomac
Conifers and *Athrotaxis*, and in view of the fact that the American
specimens can be accommodated in previously instituted genera,
there are no good reasons for retaining Fontaine's name.

ATHROTAXITES. Unger emend.

The Tasmanian genus *Athrotaxis* (page 150) is one of the existing
Conifers of which our knowledge of morphological features is very

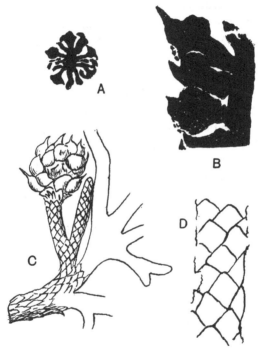

FIG. 753. A, B, D, *Athrotaxites Ungeri*. C, *Athrotaxites lycopodioides*. (A, B, D,
after Halle; A, nat. size. C, after Saporta; nat. size.)

incomplete: its restricted range and the striking resemblance of
the vegetative characters to those of many Jurassic species included in *Brachyphyllum* suggest antiquity, but palaeobotanical
records have furnished very little evidence of value in support of
this conclusion. It is however probable that like several other
plants now confined to the southern hemisphere *Athrotaxis* is a

survival of a type of Conifer which was widely spread in Jurassic
floras.

Athrotaxites lycopodioides Unger (= *Echinostrobus Sternbergii*
Schimp.).

Reference has already been made to the vegetative characters
of this Upper Jurassic species from Solenhofen[1]. In the crowded
imbricate leaves and in the blunt stiff branches it agrees very
closely with *Athrotaxis cupressoides*: the globular cones (fig. 753,
C), though incompletely preserved, also exhibit in the comparatively
small number of cone-scales and their thick spinous distal ends a
distinct similarity to those of the recent genus. Nothing is known
of the seeds.

Athrotaxites Ungeri Halle.

This species[2], founded on fertile specimens from the San Martin
flora of Patagonia, probably of Upper Jurassic or Wealden age,
is practically identical with Unger's type: the branches are more
slender and the globular cones, 10—15 mm. in diameter, appear to
be of the same type. The cone-scales have a cuneate base and a
thick spathulate distal end prolonged into a short pointed apex
(fig. 753, A, B); the thickening of the scales close behind the apex
recalls the form characteristic of recent cones (cf. fig. 684, N, p. 116).
As Halle says, the vegetative features (fig. 753, D) of this and the
preceding type are those of *Brachyphyllum*.

In his memoir on British Eocene Gymnosperms Gardner[3]
referred to *Athrotaxis* some of the foliage-shoots and cones from
Bovey Tracey in Devonshire which had previously been included
in *Sequoia Couttsiae* Heer: the reasons for the change of genus are
by no means adequate. Mr and Mrs Clement Reid[4] in their recent
investigation of the Bovey Tracey material, which they refer to
an Upper Oligocene age, made a careful examination of numerous
Sequoia fragments including a comparison of fossil cuticular
membranes with the epidermis of both *Sequoia* and *Athrotaxis*
leaves: they were unable to discover any evidence of the presence
of representatives of the latter genus. Gardner also assigns some

[1] See page 312; Unger (49).   [2] Halle (13) p. 40, Pls. ii.—v.
[3] Gardner (86) p. 90, Pl. vi. figs. 1—9; Pl. x. figs. 6—9.
[4] Reid, C. and E. M. Reid (10) p. 171.

pieces of vegetative organs and in one case a cone to *Athrotaxis*, but the evidence on which the species *Athrotaxis* (?) *subulata*[1] is founded has little value.

BRACHYPHYLLUM. Brongniart.

Brongniart[2] proposed this name for a Jurassic species, *Brachyphyllum mamillare*, founded on sterile branches characterised by pinnate branching in one plane and spirally disposed appressed leaves with a thick lamina of triangular, conical, or hexagonal form. He afterwards[3] extended the term to other Jurassic species and called attention to the striking resemblance of the fossil shoots to those of *Athrotaxis*. The photograph of *Athrotaxis cupressoides* shown in fig. 701 (p. 150) affords a very good idea of the habit of *Brachyphyllum*. Specimens in which the pinnate ramification is a conspicuous feature are more like shoots of *Thuya* or *Cupressus*, and on the smaller branches the leaves may assume a decussate arrangement. On older branches the leaves are often hexagonal and more or less convex, while on the branchlets they are more triangular or conical and are free at the apex, which in some forms is bent outwards from the axis (fig. 756). In nearly all cases specimens referred to *Brachyphyllum* are sterile and, except in examples where the preservation of the cones is too imperfect to afford any evidence of morphological characters, it is suggested that the generic name should be reserved for sterile branches and regarded as purely provisional. As Saporta[4] points out in his account of the genus, considerable confusion has been caused by attempts to assign species to several positions in the Coniferales on wholly insufficient evidence. Unger[5] figured a fertile specimen from the Rhaetic rocks of Franconia, which he referred to *Brachyphyllum speciosum* Münst., and this was re-figured by Schenk[6] as *Palissya aperta* though as others have pointed out the cones are very different from those of *Palissya*: as Nathorst[7] says, they have the characters of the genus *Elatides*, and the same is true of some cones figured by Saporta[8] and assigned by him to *Brachyphyllum*. The name *Elatides* is reserved for specimens characterised by a certain

[1] Gardner (86) p. 43, Pl. XI.   [2] Brongniart (28) A. p. 109.   [3] *Ibid.* (49) A. p. 69.
[4] Saporta (84) p. 310.   [5] Unger (49) Pl. v. figs. 3, 4.
[6] Schenk (67) A. Pl. XLII. figs. 1—13.   [7] Nathorst (97) p. 34.
[8] Saporta (84) Pl. 165, fig. 1; Pl. 167, figs. 2, 3; Pl. 171, figs. 7—9.

type of cone (fig. 742) and although typical examples of the genus bear leaves differing in their greater freedom from the axis from the more concrescent foliage of *Brachyphyllum*, it is not possible in all cases to draw a definite line between the two forms of shoot. Until a few years ago nothing was known as to the anatomical features of *Brachyphyllum* but the researches of Hollick and Jeffrey[1] have partially made good this deficiency: these authors investigated the structure of *Brachyphyllum macrocarpum* Newb. (= *B. crassum*) (fig. 758, G) from the Cretaceous beds of Kreischerville and demonstrated a close resemblance in some characters to recent Araucarias. They fully recognise that it would be unsafe to assume the presence of similar anatomical features in other species, though it is reasonable to expect the occurrence of such characters in many species not yet found as petrifications. Other examples of shoots of the *Brachyphyllum* habit furnishing information with regard to anatomy are *B. eathiense* Sew.[2] and Banc. and a Japanese species originally described as *Yezonia vulgaris* by Drs Stopes and Fujii[3] and since transferred to *Brachyphyllum*. As the great majority of specimens referred to *Brachyphyllum* give no anatomical information the generic name is usually applied to fossils exhibiting only external features; it is a form-genus. The introduction of anatomical characters, based on the examination of a very small number of examples, into a general definition might seriously mislead students with regard to the affinities of species known only as impressions. On the other hand as some species of *Brachyphyllum* exhibit anatomical features of diagnostic value the definition of the genus may be extended, in certain cases only to include the information furnished by such examples as those described on pages 322—328.

The inclusion of some species, *e.g.* the well-known type usually referred to *Thuites*, *T. expansus*, in *Brachyphyllum* is a change which may be regarded as retrograde, but an examination of specimens of that type shows the impossibility of recognising any constant verticillate disposition of the leaves such as would justify the adoption of *Cupressinocladus* or some other name implying

[1] Hollick and Jeffrey (09) B. p. 33.
[2] Seward and Bancroft (13) p. 869.
[3] Stopes and Fujii (10) p. 23; Jeffrey (10³).

affinity to recent Cupressineae. By slightly extending the use of *Brachyphyllum* we avoid the danger of giving a false impression of affinity and lighten the task of dealing with material which is of secondary botanical importance.

Brachyphyllum expansum (Sternberg).

In transferring this widely spread Jurassic species, founded by Sternberg on a specimen from the Stonesfield (fig. 754), Oxford-

FIG. 754. *Brachyphyllum expansum.* (Sedgwick Museum, Cambridge.)

FIG. 755. *Brachyphyllum expansum.* (Figured by Feistmantel as *Echinostrobus expansus.*)

shire, as *Thuites expansus*[1], from *Thuites* to *Brachyphyllum* the application of the latter name is extended to include Coniferous shoots in which the decussate arrangement of the leaves is more apparent than in the majority of species usually referred to

[1] Sternberg (23) A. p. 38, Pl. xxxviii. figs. 1, 2. For synonymy see Seward (04) B. p. 142.

Brachyphyllum. Schimper[1] included Sternberg's species in *Echino-strobus* and Saporta[2] adopted the designation *Palaeocyparis*. The small amount of evidence with regard to the structure of the cones does not afford an adequate reason for retaining the generic name *Thuites*.

Specimens from Jurassic rocks in India described by Feist-mantel[3] (fig. 755) as *Echinostrobus expansus,* superficially at least very similar to the European *Thuites expansus,* have recently been examined by Miss Holden[4]. The epidermal cells of the small decussate leaves are irregular in shape; the stomata are scattered but there is an astomatic area down the centre of the lamina. The stomata are sunk and have four accessory cells. Miss Holden points out that the epidermal features of this Indian type are different from those of *Brachyphyllum macrocarpum, B. Münsteri* and *B. affine*[5] in which rows of stomata alternate with strands of sclerenchyma: this difference is legitimately used as an argument in favour of retaining the generic name *Thuites* rather than em-ploying *Brachyphyllum*. It is, however, as a rule impossible to obtain any information with regard to the cuticular features, and from the external characters of impressions of foliage-shoots we cannot draw any satisfactory line between specimens referred to *Brachyphyllum* and *Thuites*. Miss Holden's work affords an illustration of the possibility of employing epidermal features as a means of separating shoots which in habit appear to belong to one generic type. So far as I know we have no data with regard to the epidermal structure of the European *Thuites expansus* and we cannot therefore say whether the Indian species are identical or not with those included in the same species from other regions.

In habit *Brachyphyllum expansum* agrees with *B. mamillare* and other types as also with recent species of *Thuya* and *Cupressus*: in some examples the branchlets are crowded and in others the ramification is much more open; the small appressed leaves are broadly triangular or longer and relatively narrower than in such

[1] Schimper (72) A. p. 333.   [2] Saporta (84) p. 600, Pl. 209.
[3] Feistmantel (76²) p. 60, Pls. IX., X.
[4] Holden, R. (15²) p. 221, Pl. XI. figs. 2, 5, 6. The specimens examined were kindly sent to the Cambridge Botany School by the Director of the Indian Geo-logical Survey.
[5] Hollick and Jeffrey (09) B.; Schenk (67) A.

species as *B. mamillare* or *B. crassum*: the apical portion of the
lamina is free and may be slightly falcate. In a few cases globular

FIG. 756.   *Brachyphyllum expansum*?.   (The original of Feistmantel's *Pachyphyllum*
    heterophyllum.   Calcutta Museum, Geol. Surv. India; nat. size.)

cones occur on the foliage-shoots characterised by spirally disposed
scales: in a specimen from the Stonesfield slate described in 1904

each cone-scale has a funnel-like cavity near one edge and the
upper side of the cavity is radially ridged[1]. No seeds have been
found in connexion with the cones. The male flowers are longer
and narrower and consist of numerous sporophylls attached at
right-angles and expanded distally into a peltate lamina. The
specimen reproduced in fig. 756 is the original of Feistmantel's
Pachyphyllum heterophyllum[2] from Indian Jurassic beds: on the
stouter axis there are spirally disposed triangular leaf-bases while
on the smaller branches the leaf-lamina is preserved and appears
to be thick, sub-falcate, and tetragonal. This specimen is in my
opinion indistinguishable from that shown in fig. 755, which
Feistmantel figures as *Echinostrobus expansus*[3] and both agree
superficially at least with European examples of *Brachyphyllum
expansum*.

Brachyphyllum mamillare Brongniart.

This specific name[4] has been applied to specimens from many
Jurassic localities and it might well be extended to others regarded
by authors as distinct species. An accurate specific determination
of the numerous *Brachyphyllum* shoots is indeed hopeless without
other characters than those afforded by impressions and casts. In
habit the species resembles *Athrotaxis cupressoides*: the branches
are given off at a fairly wide angle; the leaves are small, fleshy,
and more or less triangular with a median dorsal keel and usually
spirally disposed. There has been some confusion between this
species and Sternberg's *Thuites expansus*: the specimen from the
Yorkshire coast figured by Lindley and Hutton[5] under the latter
name, now in the Manchester Museum, is undoubtedly identical
with Brongniart's species. There is a considerable difference in the
degree of freedom of the upper part of the lamina from the axis;
in some specimens the leaf is almost entirely concrescent with the
axis and in others the leaves are more open and attached only by
the basal part of the lamina.

Feistmantel figures several specimens of *Brachyphyllum* from
Indian Jurassic localities under different names, many of which
appear to be indistinguishable superficially from *B. mamillare*.

[1] Seward (04) B. Pl. IX. fig. 4.
[2] Feistmantel (79) Pl. XI. fig. 4.   [3] *Ibid*. Pl. XI. fig. 2.
[4] Brongniart (28) A. p. 109.   [5] Lindley and Hutton (35) A. Pl. CLXVII.

Among these are some of the shoots referred by him to *Echinostrobus expansus*[1] and others described as *E. rajmahalensis* Feist. and *E. rhombicus*[2]. An examination of some of the figured specimens referred by Feistmantel to *Pachyphyllum* (= *Pagiophyllum*) *perigrinum* (Lind. and Hutt.) leads me to include them at least provisionally in *B. mamillare*. The generic distinction between the form-genera *Brachyphyllum* and *Pagiophyllum* is by no means always clearly marked.

Among many European examples of the *Brachyphyllum mamillare* form of Conifer, reference may be made to the illustrations by Saporta of the French Jurassic specimens referred to *Brachyphyllum Moreauanum* Brongn., *B. nepos* Sap. and a form with more slender branches, *B. gracile*[3].

Zeiller[4] records specimens of foliage-shoots with cones superficially resembling those of *Sequoia* from Lower Jurassic beds in Madagascar which he assigns to *Brachyphyllum* and compares with *B. nepos*.

Brachyphyllum spinosum Seward.

A Wealden species[5] founded on several well preserved specimens from the coast of Sussex characterised by the possession of short, thorn-like, lateral branches clothed with fleshy leaves with a longitudinally striated lamina of the usual *Brachyphyllum* type. Two or three of these spinous shoots occur at the same level on the parent-axis. The stouter branches are covered with spirally disposed polygonal leaf-bases, while on the more slender branches the broad and short leaves assume a more or less regular decussate disposition. In leaf-form and branching-habit this species agrees closely with several other examples of the genus, but the spinous shoots are a distinctive feature.

Brachyphyllum obesum Heer.

This species originally described from Lower Cretaceous strata in Portugal[6] is represented in the Potomac formation by specimens referred by Fontaine[7] to *Brachyphyllum crassicaule*, and there are

---

[1] Feistmantel (76[2]) Pl. IX. figs. 6—9; Pl. X. figs. 3, 4.
[2] *Ibid.* (79) Pl. XII. figs. 2, 10; (82) Pl. III. fig. 6.
[3] Saporta (84) Pls. 165—172.   [4] Zeiller (00) p. 3.
[5] Seward (95) A. p. 215, Pl. XVII.   [6] Heer (81) p. 20, Pl. XVII.
[7] Fontaine (89) B. p. 221, Pl. C. fig. 4; Pl. CIX. figs. 1—7.

many specimens recorded both from Jurassic and Cretaceous rocks
which differ in no important features from Heer's
type (fig. 757). An examination of branches of
the recent species *Cupressus Lawsoniana* shows
a considerable difference in the form of ramification depending on the development of numerous or few lateral shoots, and such differences
afford an argument against the use of distinctive
names such as *B. obesiforme* and others adopted by
Saporta[1] for Portuguese specimens. Apart from
the absence of thornlike branches this species is
hardly distinguishable from *B. spinosum*.

Fig. 757. *Brachy-
phyllum obesum.*
(After Heer; nat.
size.)

BRACHYOXYLON. Hollick and Jeffrey.

This generic name was proposed for pieces of wood from the
Middle Cretaceous beds in Staten Island originally regarded as that
of the plant which bore the foliage-shoots described from the same
locality by Hollick and Jeffrey as *Brachyophyllum macrocarpum*,
but as the result of further study it was recognised that lack of
proof of any connexion between wood and shoots necessitated a
new genus[2].

Brachyoxylon notabile Hollick and Jeffrey.

The tracheids of the xylem have separate pits usually in a
single row, but they are occasionally flattened and very rarely
there are two alternate rows of polygonal pits (fig. 758, A).
Normally there are no resiniferous cells in the xylem though these
occur in wounded specimens. The medullary rays are said to have
numerous pits on the radial walls. Jeffrey has described in detail
the wound-reactions of *Brachyoxylon*[3]: fig. 758, B represents part
of a transverse section showing a mass of resiniferous parenchyma
and a row of resin-canals stretching tangentially from the wounded
area. Wood exhibiting the same normal and traumatic features is
mentioned by Jeffrey from Martha's Vineyard and the Potomac
formation. It is pointed out that *Brachyoxylon* differs from typical
Araucarian wood in the frequent occurrence of circular and separate
bordered pits and in the power of developing traumatic resin-canals.

---

[1] Saporta (94) B. p. 176, Pl. xxxi.
[2] Hollick and Jeffrey (09) B. p. 54, Pls. xiii., xiv.   [3] Jeffrey (06).

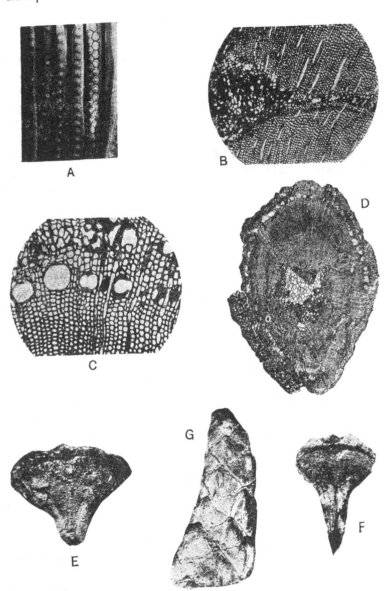

FIG. 758. A, B, *Brachyoxylon notabile*; A, tracheids of the secondary xylem;
B, traumatic resin-canals. C, D, *Araucariopitys americana*; C, traumatic
resin-canals; D, section of stem. E, F, *Protodammara speciosa*, cone-scales
(×7), see page 250. G, *Brachyphyllum crassum*. (After Jeffrey.)

In the combination of the Araucarian and the common type of tracheal pitting *Brachyoxylon* agrees with some other genera of Mesozoic woods, *e.g.* species of *Cedroxylon,* and in the formation of traumatic resin-canals it resembles *Abies*˙ and other genera of Abietineae as also *Sequoia.* Jeffrey's view is that *Brachyoxylon* is undoubtedly Araucarian though in its wound-reactions it differs from the present representatives of the Araucarineae: in this respect he considers the genus to hold the same relation to recent Araucarineae as *Sequoia* holds in respect of its power of developing resin-canals in response to injury to other allied genera in which no such reaction occurs. Admitting the Araucarian arrangement of pits on some though by no means on all tracheids, the sum of characters hardly warrants the inclusion of *Brachyoxylon* in the Araucarineae: as in several other Mesozoic genera there is in some degree a mixture of characters indicative of a generalised type and, while Jeffrey sees in this combination evidence of the derivation of Araucarian Conifers from an Abietineous ancestry, I venture to regard the spasmodic recurrence of the Araucarian type of pitting as a partial persistence of characters inherited from an ancient Araucarian stock.

Miss Holden[1] has described some wood from Cliffwood, New Jersey, which she refers to *Brachyoxylon,* differing from that described by Jeffrey in the presence of fibres in the secondary phloem, a feature associated generally with Cupressineae, Taxodineae, and the Podocarpineae. In the Cliffwood material the medullary rays are said to have smooth walls, a feature in which they differ from those of the Abietineae.

BRACHYPHYLLUM. *Brachyphyllum crassum* Lesquereux.

This name was given by Lesquereux[2] to a large branched vegetative shoot from the Dakota group, and for specimens, believed by Hollick to be identical with Lesquereux's species, from the Amboy clays Newberry proposed the name *macrocarpum*[3]: this specific name was not published and in the Amboy clay monograph[4] the designation *Brachyphyllum crassum* is adopted. The

[1] Holden, R. (14) p. 171.　　　　[2] Lesquereux (91) p. 32, Pl. II. fig. 5.
[3] Hollick in Newberry and Hollick (95) p. 51 (footnote).
[4] Newberry and Hollick (95) p. 51, Pl. VII. figs. 1—7.

same name was given by Tenison-Woods[1] to a form of *Brachyphyllum* from Queensland in 1883, the year in which Lesquereux
published the name *Thuites crassus*[2] for the Dakota specimens
afterwards transferred by him to *Brachyphyllum*. Neither author
gave an illustration of the type-specimen and the Australian type
is still unfigured, but Lesquereux's type is illustrated in the *Dakota
Flora*. It would therefore seem reasonable to retain the specific
name *crassum* rather than to adopt the designation *macrocarpum*
revived by Hollick and Jeffrey. This species was found in the
Middle Cretaceous beds in Staten Island in a condition which
enabled Hollick and Jeffrey to supply important information with
regard to anatomical characters. In their preliminary account[3] of
the foliage-shoots these authors included in *Brachyphyllum macrocarpum* some wood exhibiting well-defined characters suggestive
of Araucarian affinities, and Jeffrey[4] in another contribution speaks
of the wood as that of *Brachyphyllum*. In a subsequent description
of the Staten Island material the authors[5] state that they are no
longer in a position to affirm that the fragments of wood belong
to *Brachyphyllum*. This change of view is important as it was from
the characters of the wood that some of the arguments in favour of
an Araucarian affinity of the species *B. macrocarpum* were derived.
The structural features of the foliage-shoots alone, though in some
respects agreeing with those characteristic of recent Araucarineae,
are not known in sufficient completeness to settle definitely the
precise position of the species.

The foliage-shoots have triangular, appressed, leaves identical
in the form of the lamina with that in many Jurassic species of
Brachyphyllum (fig. 758, G), and in the method of branching as
also in the shape of the ultimate branchlets the specimens agree
with typical representatives of the genus. The branches have a
large pith containing nests of sclerous cells: external to the vascular
tissue is a ring of resin-canals and a deep-seated periderm, beyond
which are other canals belonging to the adnate leaves. Strands of
stereome occur immediately below the epidermis and these are
responsible for the longitudinal striations which often characterise

---

[1] Tenison-Woods (83).   [2] Lesquereux (83) p. 32.
[3] Hollick and Jeffrey (06) p. 200.   [4] Jeffrey (06).
[5] Hollick and Jeffrey (09) B. p. 55.

impressions of *Brachyphyllum* leaves. In the younger branches the
vascular tissue consists of separate bundles and a gap is formed on
the exit of the single leaf-trace: the trace divides in the outer
cortex into a number of fine strands 'which finally become lost in
a continuous band of transfusion-tissue' beneath the palisade
parenchyma[1]. No details are given with regard to the pitting of
the tracheids or the structure of the medullary rays, but the
authors state that the phloem showed no indication of the presence
of any thick-walled fibres. In a later paper Jeffrey[2] states that in
older stems of *Brachyphyllum crassum* the pits are flattened by
mutual contact though in younger branches this feature is often
not distinguishable. A single specimen is described as probably
a cone of *Brachyphyllum*[3]: this is, however, much smaller than
any cone previously recorded in connexion with *Brachyphyllum*
shoots and the anatomical data do not furnish any proof of its
morphological nature.

The species is recorded from the Magothy formation[4], Cliffwood,
and from the Raritan formation; Berry also describes a large
example from the Woodbine formation, Texas, as *B. macrocarpum*
var. *formosum*[5]. It is practically impossible to distinguish the
present species so far as external features are concerned from such
species as *B. obesum* Heer, *B. crassicaule* Font. and others[6].

Brachyphyllum eathiense Seward and Bancroft.

The type-specimen of this species was originally figured by
Miller as an 'imbricated stem' from Upper Jurassic rocks in the
North of Scotland[7]: it consists of a branched shoot bearing in
places some broadly triangular imbricate leaves with longitudinal
ridges on the surface of the lamina. The pith includes some
scattered thick-walled elements: no information of importance was
obtained as to the structure of the vascular tissue of the stele. The
short fleshy leaves have a well-protected epidermis succeeded by
palisade-tissue and groups of hypodermal fibres while the rest of
the mesophyll consists of parenchyma with secretory sacs and

[1] For figures, see Hollick and Jeffrey (09) B.   [2] Jeffrey (10³) p. 770.
[3] Hollick and Jeffrey (09) B. p. 37, Pl. IX. figs. 5, 6; Pl. XI. fig. 3; Pl. XIV. fig. 3.
[4] Berry (05) p. 44; (06) p. 168; (11³) p. 81.
[5] *Ibid.* (12³) p. 392, Pl. XXX.   [6] Seward (95) A. p. 218.
[7] Seward and Bancroft (13) p. 869, Pl. I. figs. 2—4.

portions of leaf-traces. The most striking feature is the occurrence of reticulately pitted, isodiametric tracheids, closely resembling those in recent Araucarian leaves and in *Brachyphyllum crassum*. Stomata were found on some of the leaves agreeing in the possession of four accessory cells (fig. 724, A, page 216) with those described by Jeffrey[1] and Thompson[2] in *B. crassum*. Though comparable with Araucarian leaves in the structure and distribution of the transfusion-tissue and in the branching leaf-traces there is hardly sufficient evidence to warrant any positive statement with regard to the relationship to recent genera of the American and Scottish species.

Brachyphyllum vulgare (Stopes and Fujii).

In their account of Upper Cretaceous plants from Hokkaido, Japan, Drs Stopes and Fujii[3] instituted a new genus *Yezonia* for some petrified shoots which they suggested should be placed in a special family of Gymnosperms. The specimens of foliage-shoots they described as *Yezonia vulgaris* and for a cone, which they consider may belong to the vegetative branches, the generic name *Yezostrobus* was proposed. The slender foliage-shoots bear appressed leaves, apparently spirally disposed, agreeing closely with those of recent Cupressineae in their form and relation to the axis; but in the absence of impressions their surface-features cannot be clearly determined. Anatomically the shoots agree very closely with *Brachyphyllum crassum*: the pith contains groups of sclerous cells; the leaf-traces branch repeatedly in the base of the leaf, and transfusion-tissue is abundant in the mesophyll. The secondary xylem shows uniseriate separate pits on the tracheids, and the medullary rays are 1—2 cells in depth. Jeffrey[4] drew attention to the striking resemblance between *Yezonia* and *Brachyphyllum crassum* and fully justified his substitution of *Brachyphyllum* for the new genus. Dr Stopes[5], while agreeing with this conclusion, points out that evidence furnished by fructifications can alone settle the question of generic identity; she states that the supposed cone attributed by Hollick and Jeffrey to *Brachyphyllum* differs widely from *Yezostrobus* which may be the cone of the Japanese species. The

[1] Jeffrey (10³) p. 768, Pl. LXV. fig. 6.   [2] Thompson (12³) Pl. VI. fig. 12.
[3] Stopes and Fujii (10) p. 23.   [4] Jeffrey (10³).
[5] Stopes (11³)

American cone may, however, be a vegetative bud, and nothing is known as to its seed-bearing appendages. *Yezostrobus* has not been proved to have any connexion with the foliage-shoots of *Brachyphyllum*. Disregarding the two cones, there can be no doubt as to the very close similarity between the American and Japanese shoots.

TAXODITES. Unger.

This generic name, adopted by Endlicher[1] from Unger, is employed for fossil species believed to be nearly related to the recent genera *Taxodium* and *Glyptostrobus*. Reference is made in the chapter on Coniferous woods to supposed examples of these genera. The separation of the two recent Conifers is based on features which cannot be applied to fossil impressions and even in the case of the existing types Beissner[2], following Bentham and Hooker, does not accept the Far Eastern species referred by Endlicher to *Glyptostrobus* as representatives of a distinct genus, but includes them in *Taxodium*. Heer describes some fragments of shoots from the Lower Cretaceous of Greenland as *Glyptostrobus groenlandicus*[3], but these are of little or no value as trustworthy records. Similarly his species *G. intermedius* from the Patoot beds[4] founded on dimorphic, sterile, shoots affords no substantial evidence of affinity to *Taxodium* or *Glyptostrobus*.

Taxodites europaeus Brongniart.

This species was first described by Brongniart[5] from Tertiary beds in Greece. The branches bear leaves varying in size and form, some being appressed and triangular while others are more elongate and freer from the axis: the oval or globular cones, with a maximum diameter of 15 mm., consist of 18—20 scales agreeing in their rounded crenulate edges and radially grooved surface with those of *Glyptostrobus*. Brongniart states that this species occurs also in Germany, Bohemia, and at Oeningen Heer[6] figures good examples of vegetative shoots and cones as *Glyptostrobus europaeus* from Oeningen; the scale-leaves are decurrent and the oval cones have semicircular scales with 6—8 grooves (fig. 759). This is described as one of the commonest fossils in the Swiss deposits and, as Heer

[1] Endlicher (47) p. 278.   [2] Beissner (91) p. 148.
[3] Heer (75) ii. Pls. XVI., XX., XXII.   [4] *Ibid.* (75) ii. Pl. LII.
[5] Brongniart (33).   [6] Heer (55) A. Pls. XIX., XX.

says, it bears a striking resemblance to the existing Chinese type
Glyptostrobus heterophyllus. Unger[1] describes well
preserved specimens from Greece and the species
is recorded, on the evidence of cones as well as
sterile shoots, from Leoben[2] and other localities in
Styria[3], also from Miocene beds in Bohemia[4].
Laurent[5] figures examples from Aquitanian beds
in the Puy-de-Dôme and Saporta and Marion[6]
refer to *Glyptostrobus europaeus* fragments of sterile
branches and an imperfect cone from Pliocene beds
in the Province of Ain and mention the occurrence
of the same type in Pliocene strata in the valley
of the Arno. Vegetative shoots are recorded from
Tertiary beds in Bosnia[7] and Nathorst[8] found the
species in Arctic Ellesmere Land. The sterile fragments figured by Goeppert and Menge[9] from the
Oligocene beds on the Baltic coast, though possibly correctly
determined, afford no proof of affinity to the genus *Glyptostrobus.* Some very good specimens from Eocene beds at Reading
are described by Gardner[10] as examples of this species but the
cones are immature and do not furnish convincing evidence of
close relationship to the recent genus. The same remark applies
to specimens figured by this author from Bournemouth. Specimens
from the latter locality, characterised by their long slender branches
with spirally disposed leaves having long decurrent bases and projecting apices, are referred to a distinct species *Taxodium eocaenicum*
and compared with the Floridan Conifer *Taxodium distichum* var.
imbricataria Mett. Gardner points out with reason that specimens
described by Heer[11] from Miocene beds in Greenland and Alaska
as *Glyptostrobus europaeus* and from rocks of the same age in

Fig. 759. *Taxo-
dites europaeus.*
(After Heer;
nat. size.)

---

[1] Unger (67) Pl. I. [2] Ettingshausen (88²) Pl. II.
[3] *Ibid.* (90) Pl. I.
[4] *Ibid.* (67²) Pls. X., XI.; Velenovský (81); Unger (52) Pl. XXXIV.
[5] Laurent (12) Pls. V., VI.
[6] Saporta and Marion (76) Pls. XXIII., XXXVII.
[7] Engelhardt (12) Pls. XXXII., XXXVI., XXXVII. [8] Nathorst (15²).
[9] Goeppert and Menge (83) A. Pl. XVI.
[10] Gardner (86) Pls. III., IV., XXIV.
[11] Heer (68) i. Pls. III., XLV.; (71) iii. Pls. I., III.; (77) i. Pls. XI., XII., XXVI.; (78)
Pls. IX., XIII.; (83) Pls. LXX., LXXV.

Spitzbergen as *G. Ungeri* may be fragments of Conifers more closely allied to *Sequoia.*

The fossils originally referred by Lesquereux[1] to *Glyptostrobus gracillimus* from the Dakota group were afterwards transferred by him to *Frenelites Reichii,* at a later date removed by Newberry to *Sequoia* and finally described by Jeffrey[2] as *Geinitzia gracillima.*

Neglecting records based on sterile specimens only it is clear that Conifers closely allied to *Taxodium* and especially to *Glyptostrobus heterophyllus* of China were abundant in the Tertiary floras of Europe.

Taxodites miocenicum (= *Taxodium distichum miocenicum* Heer).

Numerous specimens of branches and some male and female flowers have been described by Heer[3] from Miocene beds in Greenland, Spitzbergen, Grinnell Land, Northern Siberia, and Sachalin Island as *Taxodium distichum miocenicum.* Though in certain cases the material is too imperfect to determine with accuracy, some of the fossils bear a striking resemblance to *Taxodium distichum* both as regards vegetative features and cones. In this species Heer includes specimens originally named by Sternberg *Phyllites dubius* and afterwards transferred to *Taxodium.* The shoots bear distichous, linear leaves, reaching a length of 2 cm. and 2—3 mm. broad. The absence of a decurrent base is spoken of as a character distinguishing *Taxodium* from *Sequoia Langsdorfii.* Specimens from Grinnell Land now in the Dublin Museum described by Heer as *Taxodium distichum miocenicum* bear leaves contracted at the base but not decurrent. A specimen from Grinnell Land said to be a male inflorescence[4] consists of an axis bearing a few oval buds 3 mm. long which may possibly be groups of microsporophylls, but the figures are far from convincing. Nathorst[5] speaks of the occurrence of foliage-shoots, flowers, and seeds in the Tertiary *Taxodium* shales of Spitzbergen. Ettingshausen[6] figures shoots and flowers from Bilin as *Taxodium dubium* which are in all probability closely allied to the recent species. Goeppert and

---

[1] Newberry and Hollick (95) Pl. IX.   [2] Jeffrey (11).
[3] Heer (68) i. Pls. II., XII., XLV.; (71) Pls. III., IV., etc.; (77) Pls. XIII., XXV.; (78) Pls. VIII., IX.; (83) Pls. LXX., etc.
[4] Heer (68) i. Pl. II.; Schimper and Schenk (90) A. p. 294, fig. 203.
[5] Nathorst (11[3]) p. 223.   [6] Ettingshausen (67[2]) Pl. X.

Menge[1] refer some detached leaves from the Baltic amber to *Taxodium distichum* and a cone is referred to *Taxodites Beckianus* but without any clear evidence of affinity to the recent genus. Lingelsheim[2] records some wood from Tertiary rocks in Silesia which he refers to *Taxodium* and at the same locality he found masses of pollen some of which he assigns to that genus. Engelhardt and Kinkelin[3] describe cones of the *Taxodium* type as *Taxodium distichum* var. *pliocenicum* from the Frankfurt basin.

Heer's species is also recorded from several other Tertiary floras[4] and, despite the fragmentary nature of the material, there is good reason for regarding the evidence as an indication of the widespread occurrence of a Conifer in Tertiary Europe closely related to the Swamp Cypress of North America. The species is stated to be abundant in Pleistocene beds in North America from New Jersey to Maryland, Virginia and elsewhere on the eastern side of the continent[5]; it is represented by deciduous twigs, cones, seeds, stumps, and knees[6]: its occurrence points to the existence of Cypress swamps over a wide area, also to the migration of the existing species towards the south.

CUNNINGHAMIOSTROBUS. Stopes and Fujii.

Cunninghamiostrobus yubariensis Stopes and Fujii. This genus is founded on a single detached cone from Upper Cretaceous rocks at Hokkaido in Japan[7] which in size and form agrees with cones of *Cunninghamia*, and the anatomical features of the cone-scales support this comparison. The cone, 2×3 cm., is intermediate in size between those of *Cunninghamia sinensis* (*cf.* fig. 684, K) and *C. Konishii*; the scales being more like those of *C. sinensis*; they are 9—10 mm. across and characterised by the presence of a median pad of tissue projecting slightly from the upper surface presumably close to the attachment of the ovules, but no ovules or seeds were found and the open habit of the cone indicates that the seeds had been shed at the time of fossilisation: three pro-

[1] Goeppert and Menge (83) A. Pl. xvi. figs. 227—229.
[2] Lingelsheim (08) p. 34.
[3] Engelhardt and Kinkelin (08) Pl. xxiii. figs. 19—21.
[4] Squinabol (92) Pl. xvi.; Berry (09) p. 22, fig. 1.
[5] Berry (07²); (09²); (12⁴); (15).   [6] *Ibid.* (09²), figs. 1, 2.
[7] Stopes and Fujii (10) p. 45, Pl. v. figs. 27—34.

tuberances on a scale at the apex of the cone may represent aborted ovules though the nature of these is problematical. Near the base of a cone-scale there is a single transversely elongated vascular bundle which subdivides higher in the scale into a series of normally orientated vascular strands, and in one scale a much smaller bundle, probably an ovular trace, was found immediately above the main strand. The parenchymatous ground-tissue contains a few sclerous elements and several resin-canals, the larger ducts forming a series across the scale, and near the base a single large canal occurs below the broad vascular bundle as in *Cunninghamia*. Groups of transfusion-tracheids occur between the vascular strands.

The preservation of the tissues of the cone-axis is not good enough to throw any light on the question of affinity and it is from the morphology of the seedless scales that any conclusions must be drawn. The cone-scales show no indication of a division into the two organs characteristic of the Abietineae nor is there any evidence of a ligular outgrowth like that of an Araucarian scale. The resemblances in both form and anatomical characters to the sporophylls of *Cunninghamia* exhibited by the fossil cone appear to be such as to justify the employment of a generic name implying close relationship.

Cunninghamites Presl.

The employment of this name by many authors for sterile branches (*e.g.* fig. 805) superficially resembling foliage-shoots of *Cunninghamia sinensis* suggests an affinity which is not supported by any substantial evidence and while in some cases the fossils may belong to plants closely allied to the recent genus, there is no definite justification for assuming such alliance. The Lower Cretaceous species of *Cunninghamites* and similar forms are therefore relegated to Halle's genus *Elatocladus*.

MORICONIA.   Debey and Ettingshausen.

This generic name was applied by Debey and Ettingshausen[1] to some obscure impressions from the Cretaceous beds of Aix-la-Chapelle which they described as portions of a plant 'incertae sedis'; the specimens superficially resemble the pinnae of a fern with broadly linear pinnules, but the occurrence of curved lines

[1] Debey and Ettingshausen (55) B. p. 239, Pl. VII. figs. 23—27.

at right-angles to the long axis of the pinnules (fig. 760) suggested that some at least of the vein-like markings might be the boundaries of small scale-like leaves similar to those of *Libocedrus* and other Conifers. Saporta[1], in his reference to the genus in an account of the Sézanne flora, assigns *Moriconia* to the Cupressineae, a determination in accordance with the habit of the foliage-shoots, though in the absence of reproductive organs it is impossible to fix its position more precisely. The characteristic features are the pinnate branching, the flattened form of the branches, and the geometrically regular decussate short and broad leaves. The genus is recorded only from Lower and Middle Cretaceous rocks.

Moriconia cyclotoxon Debey and Ettingshausen.

This, the type-species, is recorded from Cretaceous rocks at Aix-la-Chapelle, from the West coast of Greenland and the Atlantic coastal plain. Heer[2] figured an imperfectly preserved specimen from Disco as *Pecopteris kudlistensis* in which an indication is given of the occurrence of the actual leaves, but some years later[3] he described well preserved examples as *Moriconia cyclotoxon*, and, as the result of an inspection of drawings supplied by Debey, identified them with the type-species. The same type is recorded from the Amboy clays (fig. 760)[4], Staten Island[5] and Block Island. As Hollick points out, a large impression included by Heer in *Moriconia* should rather be referred to *Brachyphyllum*: in the arrangement of the leaves and in the form of the short and blunt lateral foliage-shoots *Moriconia* agrees closely with some examples of *Brachyphyllum crassum*[6] in which the leaves appear to be regularly decussate. The leaves of *Moriconia* are wholly appressed and the upper edge of the lamina is rounded and almost truncate; a median line, possibly due to the presence of a dorsal keel, runs down the middle of the exposed broad surface of the shoots.

FIG. 760. *Moriconia cyclotoxon*. (After Newberry and Hollick; nat. size.)

[1] Saporta (68) A. p. 301.   [2] Heer (75) ii. p. 97, Pl. xxvi. fig. 18.
[3] *Ibid.* (82) i. p. 49, Pl. xxxiii.; (83) Pls. liii., liv.
[4] Newberry and Hollick (95) p. 55, Pl. x.
[5] Hollick (06) Pl. iii.; Berry (03) Pl. xlviii.   [6] Berry (06) Pl. ix. fig. 1.

CRYPTOMERITES. Bunbury.

The species *Cryptomerites divaricatus,* for which Bunbury[1] proposed this generic name, is.more probably Araucarian than a type allied to *Cryptomeria*: the choice of the term was suggested solely by vegetative characters and Bunbury recognised that these agreed with species of *Araucaria* as well as with *Cryptomeria.* The designation *Cryptomerites* should be restricted to fossils which there is good reason for believing to be allied to the recent genus.

Cryptomerites du Noyeri (Baily).

Baily[2] figured a sterile piece of foliage-shoot from Eocene leaf-beds in County Antrim as *Sequoia du Noyeri* which Gardner[3] subsequently stated to be identical with specimens obtained from the same locality bearing cones similar to those of *Cryptomeria.* Gardner described the Irish specimens and others from Mull as *Cryptomeria Sternbergii* (Goepp.), the specific name being adopted because he considered some examples figured by Ettingshausen[4] as *Araucarites Sternbergii* Goepp. to be identical with the Irish fossils, though most of the specimens described by authors, including the author of the species, as *Araucarites Sternbergii* are believed to be identical with *Araucarites Goepperti* Sternb. In these circumstances it seems desirable to employ the specific name *du Noyeri* used by Baily. Gardner's material consists of foliage-shoots agreeing in their spirally disposed leaves, 4—7 mm. in length and falcate in form, with branches of some species of *Araucaria, Dacrydium,* and *Cryptomeria japonica.* The occurrence of associated cones, in some cases attached to the vegetative shoots, affords fairly good evidence in support of comparison with *Cryptomeria.* The sub-globose cones, 15—20 mm. in diameter, consist of a comparatively small number of scales attached by a narrow base and gradually widening towards the distal edge which is deeply fringed. The general appearance of the cones, especially those from Glenarm in Antrim, is similar to those of *Cryptomeria* (*cf.* fig. 684, M) and taking into account the characters of the sterile branches the assumption of affinity to that genus appears to be well founded, though actual proof of close relationship is lacking. Gardner includes in *Cryptomeria Sternbergii*

[1] Bunbury (51) A. p. 190, Pl. XIII. fig. 4; Seward (00) B. p. 287.
[2] Baily (69) Pl. xv. fig. 4.   [3] Gardner (86) p. 85, Pls. x., xx., xxi.
[4] Ettingshausen (55) Pl. v.

specimens figured by Ettingshausen from Monte Promina as *Araucarites Sternbergii* and some of the impressions from Greenland referred by Heer[1] to *Sequoia Sternbergi*. The Miocene fragments figured by Heer afford no evidence of affinity other than that of leaf-form, and the use of the term *Cryptomerites* should therefore be avoided. If the Eocene plant is correctly regarded as closely allied to *Cryptomeria* it supplies another striking illustration of the change in the geographical distribution of Conifers since the early part of the Tertiary period.

CRYPTOMERIOPSIS. Stopes and Fujii.

Cryptomeriopsis antiqua Stopes and Fujii. The name *Cryptomeriopsis*[2] was proposed for some petrified twigs from Upper Cretaceous beds in Japan resembling in habit and structural features the recent Conifer *Cryptomeria japonica*. The xylem of the axis consists of tracheids with uniseriate separate, circular, bordered pits; there are no resin-canals and no xylem-parenchyma; the presence of the latter tissue is recorded by Suzuki in a second Japanese species *C. mesozoica*[3]. The medullary rays are usually one-cell deep in the type-species and there are a few (1—3 in *C. mesozoica*) oval pits in the field. In *C. antiqua* the phloem is said to consist of soft tissue only, but fibres occur in *C. mesozoica*. An undivided leaf-trace supplies each leaf. The four-sided leaves are characterised by the presence of three canals, a large central canal below the vascular bundle and two lateral ducts; the vascular bundle is accompanied by well-developed lateral groups of transfusion-tracheids. The leaves of *C. mesozoica* differ in a few details from those of the type-species. Prof. Jeffrey[4] maintains that *Cryptomeriopsis* is generically identical with *Geinitzia* as described by Hollick and Jeffrey from Staten Island and should be included in the Araucarineae. Dr Stopes[5] adheres to the view that the Japanese fossils are closely allied to *Cryptomeria* and afford no evidence of affinity to *Araucaria*: the structure of the xylem shows no Araucarian features in the pitting of the tracheids and, while accurate determination of systematic position must depend upon the evidence of reproductive shoots, the evidence of the vegetative shoots favours comparison with *Cryptomeria* rather than with Araucaria.

[1] Heer (75) iii. Pl. ii. figs. 1—4.
[2] Stopes and Fujii (10) p. 52, Pl. i. fig. 11; Pl. vi. figs. 35—41.
[3] Suzuki (10) p. 185.   [4] Jeffrey (10³) p. 771.   [5] Stopes (11³)

CHAPTER XLVII.

CALLITRINEAE.

It has already been pointed out that there is good reason for treating the three existing genera *Callitris, Widdringtonia,* and *Actinostrobus* as members of a distinct family. The genus *Tetraclinis,* as Saxton[1] has shown, while agreeing with the Callitrineae in certain features, exhibits a closer resemblance in its gametophyte to the Cupressineae and is regarded as a type connecting the two families Cupressineae and Callitrineae. So far as external characters are concerned, and these are the features from which the palaeobotanist is compelled to draw such conclusions as he can, *Tetraclinis* falls into line with the Callitrineae. The discontinuous distribution of the recent species of these four genera suggests antiquity and a former more extended range. Palaeobotanical literature contains numerous records of *Widdringtonia, Callitris, Frenela* or *Frenelites* based in many cases on sterile shoots and sometimes on cones and seeds more or less closely resembling those of recent forms. The generic name *Frenela* has now been discarded in favour of *Callitris*: it was proposed by Miquel in 1826 to avoid confusion between *Callitris* and *Calythrix,* the latter being the name of a Myrtaceous genus. An inspection of the published figures of supposed fossil representatives of the Callitrineae shows that the name *Widdringtonia* or *Widdringtonites* has sometimes been applied to fertile shoots with cones differing in the number of the valves from those of recent species and more closely resembling the cones of *Callitris, Tetraclinis,* or *Actinostrobus.* Moreover the number of valves in recent cones, though usually constant, is not invariably the same and in imperfectly preserved specimens it is often difficult to differentiate satisfactorily between the four genera. In the case of many sterile shoots preserved as impressions it is practically impossible to distinguish clearly between those of the Callitrineae and

---

[1] Saxton (13²); (13³).

slender branches of *Juniperus, Thuya,* and other Cupressineae. Even when cones are preserved there is some danger of confusion with fruits of certain Dicotyledons, *e.g. Lagerstroemia macrocarpa* (Lythraceae). In view of the difficulties of precise determination the most convenient course is to adopt the generic name *Callitrites* in a comprehensive sense without as a rule attempting to assign the fossil to one of the existing genera of the Callitrineae.

CALLITRITES. Endlicher.

Endlicher[1] employed the generic names *Widdringtonites, Callitrites, Frenelites,* and *Actinostrobites,* but the material seldom justifies such discrimination. The name *Actinostrobites* was proposed in the first instance for some cones described by Bowerbank[2] from the London Clay as *Cupressites globosus* and *C. elongatus* but Gardner[3], who examined the original specimens, is sceptical as to their connexion with the Callitrineae. Ettingshausen[4] described a small cone from Miocene beds in Carinthia as *Actinostrobus miocenica* on the ground that there appear to be traces of scales at the base of each of the six small linear valves of the cone. The specimen is too imperfect to be determined with any accuracy. It is impossible to express any considered opinion with regard to the validity of the numerous Tertiary records of *Callitris* and *Widdringtonia* without access to the actual material, though many of the illustrations lend strong support to the identification of the specimens as examples of some Callitrineous type. Despite the imperfection of many of the records there can be no doubt as to the former occurrence of representatives of the Callitrineae in Tertiary floras in Europe.

The pinnately branched sterile shoots referred to *Widdringtonites keuperianus* Heer[5] from the Trias of Switzerland and Germany bear a close resemblance to some forms of *Walchia* and there is no sound reason for assigning the species to the Callitrineae. Saporta[6] described fragments of branches from the Lower Lias of France as examples of Heer's type, but in this case also no

[1] Endlicher (47) p. 271.   [2] Bowerbank (40) p. 52, Pl. x.
[3] Gardner (86) p. 20.
[4] Ettingshausen (72) p. 164, Pl. II. figs. 9—12.
[5] Heer (65) A. p. 52, fig. 31; Schütze (01) Pl. x.; Schenk in Schimper and Schenk (90) A. p. 311.   [6] Saporta (84) Pl. 201, fig. 1.

cones were found. Similarly *Widdringtonites gracilis* Sap. and
W. creyensis Sap. from the Corallian and Kimeridgian of France[1]
respectively are founded solely on sterile shoots. The specimen
figured by Eichwald[2] from Jurassic rocks on the southern border
of the Caspian sea as *Widdringtonites denticulatus* has the habit of
an *Araucaria* and the supposed cone, which may be some foreign
body not actually attached, affords no evidence of affinity to the
Callitrineae. Zeiller[3] describes a small fragment from Liassic beds
in the Commune of Cherveux bearing small rhomboidal decussate
leaves similar to *Widdringtonites liassinus* (Kurr) as figured by
Salfeld and to *W. keuperianus*, but the material affords no definite
indication of relationship to the Callitrineae.

Callitrites Reichii (Ettingshausen).

This species, recorded from several Cretaceous localities in the
Eastern United States and elsewhere, is in many cases represented
only by slender sterile shoots and its position among the Coniferae
is by no means clearly established. It was founded by Ettingshausen[4] as *Frenelites Reichii* on some branched shoots from Cretaceous rocks in Saxony and afterwards described by Heer[5] from
the Patoot beds of West Greenland under the generic name
Widdringtonites though without satisfactory evidence in support
of relationship to *Widdringtonia*. This species is one of the commonest Conifers in the Amboy clays of New Jersey, but no cones
are figured by Newberry[6] in his monograph except two small
examples which it is suggested may be immature microstrobili.
Velenovský[7] figures sterile branches from the Perucer beds of
Bohemia and an ovate cone, 13 mm. long, with four valves, which
resembles a small cone of *Actinostrobus* and those described by
Berry as *Widdringtonites subtilis*. Some of the twigs bear terminal
elliptical bodies regarded as male flowers. The leaves of this species
are usually spiral and, with the exception of the apex, closely
appressed. *Callitrites Reichii* is also recorded by Krasser[8] from the

[1] Saporta (84) Pls. 201, 202.　　[2] Eichwald (68) p. 43, Pl. IV. fig. 9.
[3] Zeiller (11) Pl. II. fig. 6.　　[4] Ettingshausen (67) p. 246, Pl. I. fig. 10.
[5] Heer (82) i. p. 13, Pl. LII. figs. 4, 5.
[6] Newberry and Hollick (95) Pl. VIII.
[7] Velenovský (85) B. p. 27, Pls. VIII., X.; (87) figs. 14—16.
[8] Krasser (96) B. p. 126, Pls XIV., XVII.

Cenomanian of Moravia where it is represented by both sterile and fertile shoots; the cones are quadrivalvate. It occurs in the Middle Cretaceous of Staten Island though without any cones: Hollick and Jeffrey[1] regard the shoots referred by them to *Widdringtonia Reichii* as Araucarian on the ground that the bordered pits on the tracheal walls are usually contiguous. There is, however, no substantial reason for assigning these vegetative organs to the Araucarineae though the structure of the wood shows an Araucarian tendency. Berry[2] records the species from the Cenomanian Raritan formation of New Jersey and he expresses the opinion that the species is closely allied to some Potomac specimens described by Fontaine[3] as *Taxodium ramosum*, but in the absence of cones a definite determination of affinity is hardly possible.

Callitrites subtilis (Heer).

Founded by Heer[4] on slender twigs bearing spirally disposed, appressed, leaves from the Cretaceous beds of Atanekerdluk in Greenland and described by Newberry[5] from the Amboy clays. Hollick[6] and Berry[7] have also recorded the species from Cretaceous strata in different parts of the Eastern United States and the latter author figures examples from Upper Cretaceous beds in South Carolina[8]. The epidermal cells are regularly rectangular and the stomata are surrounded by 5—6 accessory cells. Berry figures conical cones, 7—9 mm. long by 4—5 mm. in diameter, composed of four thick scales differing somewhat in shape from the cones of recent species. The sterile shoots of this species bear a close resemblance to *C. Reichii* and the two species have often been confused; also to *Cyparissidium minimum* as figured by Velenovsky[9], *Juniperus macilenta* Heer[10] and *Widdringtonites fascicularis* Holl.[11]

---

[1] Hollick and Jeffrey (09) B p. 29, Pls. v., viii., xx.; Hollick (06) p. 44, Pl. iv. figs. 6—8.

[2] Berry (11³) p. 87, Pl. viii.

[3] Fontaine (89) B. p. 251, Pls. cxxiii., cxxiv., etc.; Berry (11⁴) p. 302.

[4] Heer (74) B. Pl. xxviii. fig. 1.

[5] Newberry and Hollick (95) p. 57, Pl. x. figs. 2—4.

[6] Hollick (06) p. 45, Pl. iv. figs. 2—5.   [7] Berry (12²).

[8] *Ibid.* (14) p. 25, Pl. xi. figs. 14—17.

[9] Velenovský (85) B. Pl. ix. figs. 6, 7; Pl. x. fig. 4.

[10] Heer (75) ii. Pl. xxviii. fig. 1c.

[11] Hollick (06) Pl. iv. fig. 1.

Callitrites curta (Bowerbank).

Bowerbank referred several pyritised cones from the London Clay of the Island of Sheppey to *Cupressites* and some of them he compared with species of *Callitris*. Gardner[1], as the result of an examination of Bowerbank's type-specimens, reduced the number of species and adopted the name *Callitris*. The specimens, in the British Museum, assigned to *Callitrites curta* are conical cones composed of 4, 5, or rarely 6 thick and woody valves which are sometimes unequal in size: the largest has a diameter of 2 cm. Fig. 761, A shows a cone of five valves and B, C, are two views of a section of a cone consisting of four valves. A similar type described by Gardner as *Callitris Ettingshauseni*[2], also from Sheppey, is represented by globular cones 12—15 mm. in diameter and

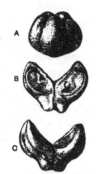

FIG. 761. *Callitrites curta.*
A, cone of five valves.
B, C, two views of a section of a four-valved cone.
(British Museum, drawn from specimens described by Gardner; nat. size.)

composed of 6—8 scales. These two species are probably correctly referred to the Callitrineae though the pyritised cones are the only portions of the plant preserved in the Sheppey clay. Gardner states that Ettingshausen[3] is incorrect in recording *Callitrites curta* from the Isle of Wight.

Callitrites Brongniarti (Endlicher).

This species, first described by Brongniart[4] as *Equisetum brachyodon* from the Paris Basin, is recorded from many European localities, in some cases represented only by sterile shoots but frequently also by cones and small winged seeds. Some well preserved specimens are figured by Unger[5] from the Tyrol (fig. 762, A, A′) under the name *Thuyites callitrina* characterised by regularly whorled leaves, apparently four at each node, with a relatively long and narrow appressed lamina and a small free apex and by valvate cones. Unger[6] subsequently described good specimens as

[1] Gardner (86) p. 21, Pl. IX. figs. 7, 21.   [2] *Ibid.* Pl. IX. figs. 1—6.
[3] Ettingshausen (79) p. 392; (80) p. 231.
[4] Brongniart (22) A. p. 329, Pl. V. fig. 3; Endlicher (47) p. 274.
[5] Unger (47) p. 22, Pls. VI., VII.   [6] *Ibid.* (67) p. 42, Pl. I. figs. 1, 2.

Callitris Brongniarti from Miocene beds in Euboea, but Saporta[1] considers these impressions to be more closely allied to *Widdringtonia* and renames them *Widdringtonia kumensis*. Good examples of quadrivalvate cones (fig. 762, B) are figured by Saporta[2] from the Eocene beds of Aix and Armissan in Provence, showing in some cases two outer broader valves and two internal laterally compressed valves. Ettingshausen[3] states that the species is very abundant at Häring in the Tyrol: that author describes some

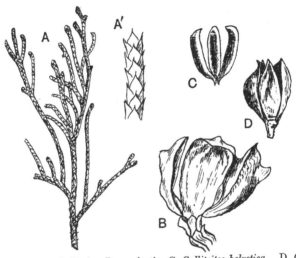

Fig. 762. A, A', B, *Callitrites Brongniarti*. C, *Callitrites helvetica*. D, *Callitrites europaea*. (A, A', after Unger; B, after Saporta; C, after Heer; D, after Engelhardt and Kinkelin.)

sterile shoots from Eocene beds in New South Wales as *Callitris prisca*[4] which he compares with *C. Brongniarti*. Well preserved shoots are described by Watelet[5] from the Paris Basin. Engelhardt[6] records the species from Oligocene beds in Bohemia but on the inadequate evidence of a winged seed; it is recorded also by Engelhardt and Kinkelin[7] from the Pliocene beds of the Frankfurt

[1] Saporta (68) p. 316.
[2] *Ibid.* (62) p. 209, Pl. II. fig. 6; Pl. III. fig. 1; (65²) p. 39, Pl. I. fig. 6.
[3] Ettingshausen (55) p. 34, Pl. V. figs. 7—35.
[4] *Ibid.* (86) p. 95, Pl. VIII. figs. 3, 4. [5] Watelet (66) A. Pl. XXXII.
[6] Engelhardt (85) p. 314, Pl. VIII. fig. 32.
[7] Engelhardt and Kinkelin (08) Pl. XXIII. fig. 5.

district though on slender grounds. It may be, as Masters suggested, that some at least of the Tertiary specimens included in *Callitrites Brongniarti* are more closely allied to the existing genus *Tetraclinis* than to *Callitris*.

Among other species that may be included in *Callitrites* are *C. brachyphylla* (Sap.) and *C. antiqua* (Sap.) from Provence[1], represented by shoots with spiral, sub-opposite or opposite appressed scale-leaves and by globular cones with four valves. As Solms-Laubach says[2], the cones agree closely with those of *Widdringtonia*, though it would be difficult to decide between that genus and *Tetraclinis*.

Some good specimens are figured by Heer[3] from the Oeningen beds as *Widdringtonia helvetica*, now transferred to *Callitrites*, consisting of branched filiform foliage-shoots with small appressed leaves and cones with four valves (fig. 762, C).

Fragments of branches with small appressed leaves in opposite pairs from the Oligocene amber beds of the Baltic coast are described by Goeppert and Menge[4] as three species of *Widdringtonites*, and in one case, *W. legitimus*, the species is founded on a cone 6 mm. long and 2·5 mm. broad which is not above suspicion as a record of a Callitrineous strobilus. *Frenela europaea* and *F. Ewaldana* described by Ludwig[5] from Tertiary beds near Frankfurt are founded on unconvincing specimens. Engelhardt and Kinkelin[6] describe pyramidal cones with 5—6 valves 1—1·5 cm. long (fig. 762, D), which they refer to *Frenelites europaeus*, from the Upper Pliocene beds of the Lower Main valley. Many other similar instances might be quoted, but on the other hand there is ample evidence of the presence in the earlier Tertiary floras in Europe of Conifers agreeing both in vegetative and reproductive shoots with existing species now confined to Africa and Australia.

FRENELOPSIS. Schenk.

Schenk[7] instituted this generic name for specimens originally described by Ettingshausen[8] from Wealden beds in Silesia as

---

[1] Saporta (62) Pl. II. fig. 7; (62²) Pl. III. fig. 3; (65²) Pl. I. fig. 4; (73) Pl. II. fig. 1.
[2] Solms-Laubach (91) A. p. 60.   [3] Heer (55) A. p. 48, Pl. XVI. figs. 2—18.
[4] Goeppert and Menge (83) A. p. 39.   [5] Ludwig (59) A. pp. 69, 136.
[6] Engelhardt and Kinkelin (08) Pl. XXIII. fig. 5.   [7] Schenk (71) p. 13, Pl. I.
[8] Ettingshausen (52) p. 26, Pl. I. figs. 6, 7.

Thuites Hoheneggeri on the ground that the external features of the vegetative shoots indicate an affinity to the recent genus *Frenela* (= *Callitris*) rather than to *Thuya* or *Cupressus*. The resemblance to *Callitris* was recognised by Ettingshausen. The most striking features of *Frenelopsis* are the comparatively long internodes of the jointed stems and branches (fig. 763, A), the occurrence of appressed leaves in opposite pairs or four in a verticil, concrescent with the whole internodal surface and projecting slightly above each nodal line as small broadly triangular scales, the presence of longitudinal lines of small dots on the internodal regions due to rows of stomata characterised by 4—5 accessory cells surrounding the depressed guard-cells[1] (fig. 763, D, E). The smaller branches closely resemble those of species of Cupressineae (fig. 763, C) or Callitrineae in leaf-form and branching, but older branches from which the leaves have partially or wholly disappeared often differ considerably from the younger foliage-shoots and by themselves afford little or no indication of their true nature. Further details are given in the description of representative species.

Frenelopsis is characteristic of Wealden or higher horizons in the Lower Cretaceous series; it occurs in Silesia, Bohemia, Portugal and the South of France and in some North American localities, particularly in the Potomac formation. Heer[2] records the species from Lower Cretaceous rocks in West Greenland but some of the original specimens which I had an opportunity of examining in the Stockholm Museum afforded no satisfactory evidence of their systematic position. Though assigned by Heer to the Gnetales, *Frenelopsis* is usually regarded as a Conifer agreeing with *Callitris* more closely than with any other existing genus. In their description of some fossil shoots referred by Newberry[3] to *Frenelopsis gracilis* Hollick and Jeffrey[4], who institute a new genus *Raritania* for this species, state that they have reason to believe that some American specimens correctly assigned to *Frenelopsis* are examples of Gnetalean plants. Nothing is known of any reproductive organs, but such information as we have with regard to the habit of the vegetative shoots and the structure of the stomata would seem to

---

[1] Zeiller (82) A. p. 231, Pl. xi.; Thompson (12³) Pl. v.
[2] Heer (75) ii. p. 73, Pl. xviii. figs. 5—8; (82) i. p. 7, Pl. ii. figs. 1—3.
[3] Newberry and Hollick (95) p. 59, Pl. xiii. figs. 1—3.
[4] Hollick and Jeffrey (09) B. p. 26.

be in favour of including this Lower Cretaceous genus among the
Coniferales and regarding it as probably allied to the Callitrineae.
But the data are insufficient to form the basis of any definite state-
ment as to the position of the genus.

{ *Frenelopsis Hoheneggeri* (Ettingshausen).
{ *Frenelopsis occidentalis* Heer[1].

Though the specimens referred to these two species may be
correctly separated their close agreement in habit points to a
single type so far at least as concerns the characters as a whole.
The specimen represented in fig. 763, A, B, originally described by
Zeiller from the province of Gard, illustrates the method of branching and the form of the leaves borne in whorls of four. Schenk, in
his account of Lower Cretaceous material of *F. Hoheneggeri* from
Wernsdorf in the Carpathians, states that the leaves are in decussate
pairs or sometimes in verticils of four. The epidermal cells have
straight walls and a thick cuticle; the stomata form longitudinal
rows on the internodes and are characterised by the presence of
4—5 accessory cells overarching the stomatal depression[2].

Specimens described from Bohemia by Velenovský[3] as *F.
bohemica* resemble *F. Hoheneggeri* both in habit and in the structure
of the stomata.

Frenelopsis ramosissima Fontaine.

This species is represented by numerous well preserved specimens in the Potomac formation[4]: some of the stems have a diameter
of 5 cm. and lateral branches are given off in whorls of 3—5; there
are three leaves at each node with broadly triangular apices and
concrescent decurrent bases as in *F. Hoheneggeri* (fig. 763, C). The
stomata are arranged in longitudinal rows and agree in the possession of a rosette of accessory cells (fig. 763, D) with *F. Hoheneggeri*:
several of the epidermal cells are provided with short spinous processes[5]. This species is represented by specimens showing clearly
the cupressoid habit of the smaller foliage-shoots (fig. 763, C).

[1] Heer (81) p. 21, Pl. XII. figs. 3—7; Saporta (94) B. pp. 139, 199. 214; Pls.
XXXVI., XXXVIII.
[2] Thompson (12³).   [3] Velenovský (88) figs. 1—3, 10.
[4] Fontaine (89) B. p. 215, Pls. 95—101; Berry (11) p. 422, Pls. LXXI., LXXII.
[5] Berry (10²).

In another Potomac species, *F. parceramosa* Font.[1], there
appears to be a single leaf at each node: this form resembles some
specimens from Lower Cretaceous rocks in Mexico which Nathorst[2]
made the type of a new genus *Pseudofrenelopsis,* but the features

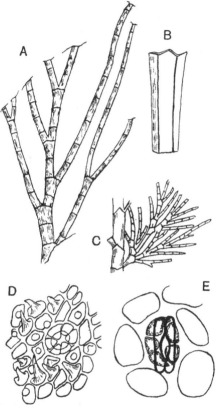

FIG. 763.   A, B, *Frenelopsis Hoheneggeri.*   C, D, *F. ramosissima.*   E, *F. occidentalis.*   (A, B, after Zeiller; C, D, after Berry; E, after Thompson.)

are hardly sufficiently well exhibited to throw much light on the
nature of the shoots. It is possible that some Wealden branches
from English strata described as *Becklesia anomala* Sew.[3] may
belong to some species of *Frenelopsis,* but the absence of any leaves
or nodal marks precludes their inclusion in this genus.

[1] Fontaine (89) B. p. 218, Pls. CXI., CXII., CLXVIII.
[2] Nathorst (93) p. 52, figs. 6—9.   [3] Seward (95) A. p. 179, Pl. XIV.

Further light will no doubt be thrown on the nature of *Frenelopsis* when the results of the investigations of Hollick and Jeffrey are published: it may be that Heer was correct in his attribution of the Portuguese specimens to the Gnetales though in the habit of the branching, especially in *F. ramosissima*, and in the structure of the stomata there is a closer resemblance to recent Callitrineae than to any other plants. The genus ranges from Wealden to Cenomanian rocks.

Sequoiineae.

In view of the restricted range of the two surviving species of *Sequoia* and the peculiarities of the genus, to which expression is given by the institution of the family-name Sequoiineae[1], the question of geological antiquity and past distribution assumes a special interest. Reference has already been made to fossil wood presenting features now found in *Sequoia*, but it is very doubtful if the anatomical characters of the recent species are sufficiently well defined to enable us to discriminate between the wood of *Sequoia* and certain other Conifers. Many of the impressions of vegetative shoots and cones described as *Sequoia* from Jurassic and especially Lower Cretaceous strata do not bear a close scrutiny. The widely spread species often referred to as *Sequoia Reichenbachii* affords no real evidence of affinity to the recent genus and the same remark applies to specimens included in Heer's genus *Sphenolepidium* and compared by authors with *Sequoia*. Some of the imperfectly preserved Jurassic cones agreeing superficially with those of *Sequoia* may well belong to species of Sequoiineae.

Though in the majority of instances Jurassic and Cretaceous records do not prove the former presence of *Sequoia* or a closely allied type, some of them afford justification for the belief that the American trees are survivals from at least the later floras of the Mesozoic era. On the other hand Tertiary strata in many parts of the world supply clear evidence of the wide distribution of *Sequoia* or some nearly related Conifers in Europe and elsewhere[2]. The inference suggested is that the recent species survive in California because of the greater possibilities of migration towards the more

[1] See page 151.

[2] Mr E. W. Berry (16) has recently published a sketch-map illustrating the world-wide distribution of fossils referred to *Sequoia*.

genial south on the American continent than in Europe where the retreat from Arctic regions ended in extinction.

Penhallow[1] records some petrified wood from Cretaceous strata in Alberta which he names *Sequoia albertensis* and regards as very similar to the wood of *Sequoia sempervirens*, but the evidence in favour of a reference to the existing genus is inconclusive. Resin-cells are scattered through the wood; the medullary rays have 1—2 bordered pits in the field, the broadly elliptical pore being generally diagonal to the cell-axis.

Tertiary wood from different localities in North America is referred to *Sequoia* on evidence that is far from conclusive. Prof. Jeffrey[2] described a particularly well preserved piece of stem from the Miocene auriferous gravels of the Sierra Nevada, near the home of *Sequoia gigantea*, as *Sequoia Penhallowi*; though I am informed that he is now inclined to refer the wood to the Abietineae. In his account attention is called to certain features, *e.g.* the pitting on the end-walls of the medullary-ray cells, the scarcity of xylem-parenchyma, and the presence of vertical and horizontal resin-canals, believed to be traumatic, which are certainly suggestive of abietineous affinity. Prof. Penhallow[3] described two species from Eocene beds in the North-West Territory as *Sequoia Langsdorfii* and *S. Burgessii*, both of which were previously described by Dawson but assigned by him to different positions. In the wood believed to belong to the plant which bore the well-known twigs recorded by many authors as *S. Langsdorfii* resin-cells are numerous and scattered and resin-canals are present only in a rudimentary form on the outer face of the summer-wood. The pitting of the medullary-ray cells is not described. A peculiar feature in *S. Burgessii*, if the wood is correctly referred to *Sequoia,* is the occurrence of two kinds of medullary rays, uniseriate and fusiform, the latter containing resin-canals. No resin-canals occur in the wood. Attention has been called (p. 171, fig. 712) to the abundance of petrified stems in the Lower Tertiary deposits in the Yellowstone Park: some of these are named by Mr Knowlton[4] *Sequoia magnifica*. A few of the trunks reach a diameter of 6—10 ft.

[1] Penhallow (08) p. 83, figs. 1—6.
[2] Jeffrey (04).　　　　　　　　　[3] Penhallow (03) pp. 41—46, figs. 2—8.
[4] Knowlton (99) p. 761, Pls. CIV., CV., CX., CXI., CXVII.

and a height of 30 ft. (fig. 764). The details are imperfectly preserved: a few of the tracheids show traces of single and double
rows of small bordered pits, but no pits are shown on the walls of
the medullary-ray cells. Resin-parenchyma is abundant and

FIG. 764. Petrified tree in the Yellowstone National Park (*Sequoia magnifica*
Knowlton). (From a photograph kindly supplied by Prof. Knowlton.)

scattered as in *Cupressinoxylon*: it is doubtful whether the wood
of *Sequoia* can be distinguished from that of some other genera
included in the genus *Cupressinoxylon*. Specimens of wood from
the Tertiary coal-field of Aichi-Gifu in the middle region of Hondo,

the main Island of Japan, recently described by Yasui[1] as *Sequoia hondoensis* has the following characters: narrow annual rings, tracheal pits usually uniseriate though often biseriate and opposite on the broader tracheids, rims of Sanio present, medullary-ray cells with oval bordered pits on the lateral walls but unpitted elsewhere, resin-cells scattered through the spring- and summer-wood, resin-canals present which are believed to be traumatic. The occurrence of this wood according to the author of the species 'completes in an interesting way the evidence for the existence' of *Sequoia* 'in Cenozoic times throughout temperate regions of the whole northern hemisphere.' While it is probable that the Sequoiineae were very widely spread in the Tertiary period it is open to question if the anatomical evidence is sufficiently clear to justify the reference of the Japanese wood to *Sequoia*. The chief reason for the adoption of that generic name is the occurrence of resin-canals similar to the traumatic ducts in the recent species.

The following descriptions include fossils which cannot be referred to *Sequoiites* and others which may reasonably be so named.

SEQUOIITES. Brongniart.

Sequoiites problematica (Fliche and Zeiller).

This species, originally described as *Sequoia problematica*[2], is founded on a small elliptical cone from Upper Jurassic rocks in the Boulogne district: in the form of the scales, which show a ridge extending from the edges of the distal surface to a central depression in the middle of the cone-scales, the fossil suggests affinity to the recent genus. Zeiller[3] also records a cone from Jurassic strata in Madagascar associated with branches of the *Brachyphyllum* type which he says presents all the characters of *Sequoia*. It must, however, be admitted that in both these cases close relationship to *Sequoia* has not been demonstrated.

Under the name *Sequoia minor* Velenovský[4] describes specimens from the Lower Cretaceous strata of Bohemia consisting of foliage-shoots with small imbricate linear-lanceolate leaves and a small terminal, spherical, cone the sporophylls of which have rhomboidal distal ends and a central umbo: but as in most fossils referred to *Sequoia* the evidence of generic affinity is inadequate.

---

[1] Yasui (17).   [2] Fliche and Zeiller (04).   [3] Zeiller (00).
[4] Velenovský (87) p. 638, figs. 11, 12.

Sequoiites giganteoides (Stopes).

This species, under the name *Sequoia giganteoides*, has recently been founded by Dr Stopes[1] on a small petrified fragment of a very

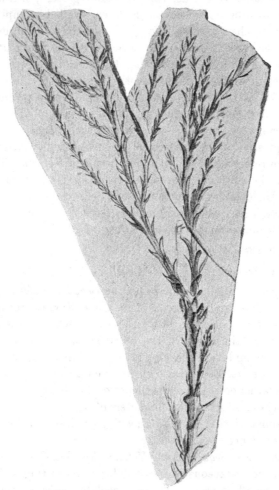

FIG. 765.   *Sequoiites concinna*.   Foliage-shoot from the Senonian of Greenland.
(Stockholm Museum; nat. size.)

slender foliage-shoot from the Lower Greensand of Luccomb Chine in the Isle of Wight. The pith contains stone-cells, and a single,

[1] Stopes (15) p. 70, Pl. II. text-fig. 16.

XLVII]   SEQUOIITES   351

undivided leaf-trace enters each decurrent leaf-base; there is a central large resin-canal in the leaves and a considerable development of transfusion-tracheids on each side of the secretory passage. Palisade-cells are a conspicuous feature and one or two layers of hypoderm fibres occur next the lower epidermis. The author of the species points out the close resemblance between the fossil and the leaves and shoot-axis of *Sequoia gigantea*.

Sequoiites concinna Heer.

Heer[1] described several specimens of foliage-shoots and cones from the Patoot beds in West Greenland as *Sequoia concinna*, the commonest Conifer in these rich Lower Cretaceous strata. The form of the sparsely branched shoots with their long and slender branchlets and straight or slightly curved, decurrent, acuminate, leaves (fig. 765) agree closely with those known as *Sphenolepidium Sternbergianum* from English and other Wealden rocks, as also with the shoots of *Sequoiites Couttsiae*. The oval cones, 23 × 20 mm., consist of a few scales with 5—6 angled thick distal ends on which there is a median transverse line and a central scar.

This species, represented by sterile shoots and cones, has recently been recorded by Berry[2] from Upper Cretaceous beds in Pike County, Arkansas.

Conites. *Conites Gardneri* (Carruthers).

Carruthers[3] described a cone and a piece of vegetative shoot from the Gault of Folkestone as *Sequoiites Gardneri* but neither specimen affords any satisfactory evidence of relationship with *Sequoia*. The shoot is of the *Pagiophyllum* type, and the cone, 2·5 × 1·5 cm., consists of spirally disposed scales with four-sided rhomboidal distal ends. There is no information with regard to the seeds: the data being wholly insufficient to serve as a criterion of affinity, the generic name *Conites* is substituted for *Sequoiites*. A specimen figured by Lange[4] from the Aachen Sands as *Carpolithes hemlocinus* Schloth. and compared by him to a *Sequoia* cone agrees closely with the English species.

---

1 Heer (83) p. 13, Pls. LI., LII., etc.
2 Berry (17) p. 172, Pl. VII. figs. 1—5.
3 Carruthers (69²) p. 7, Pl. I. figs. 7, 8.
4 Lange (90) Pl. XXXII. fig. 7.

Conites ovalis (Carruthers).

The type-specimen from the Gault of Folkestone[1] is an oval cone 6 cm. long and about 2·5 cm. in diameter; the scales are cuneate and the exposed ends transversely elongated and hexagonal (fig. 766). It bears a close resemblance to *Geinitzia gracillima*, but in the absence of any details with regard to anatomical features or seeds the non-committal name *Conites* is employed.

Sequoiites Holsti Conwentz ex Nathorst MS.

This species[2], from the Holma sandstone (Senonian) of Sweden, is founded on fragments of foliage-shoots covered with spirally disposed, appressed, broadly triangular leaves. The specimens are not well enough preserved to show in detail the anatomical features, but Conwentz considers such characters as he was able to recognise favourable to Nathorst's adoption of the generic name *Sequoiites*. The species is,

FIG. 766. *Conites ovalis.* (After Carruthers; nat. size.)

however, not above suspicion as a record of a Conifer closely allied to *Sequoia*.

Sequoiites Langsdorfii (Brongniart).

Brongniart[3] instituted this Tertiary species under the generic name *Taxites*, and Heer[4] in his description of foliage-shoots from Miocene beds in Switzerland adopted the designation *Sequoia*. In habit *S. Langsdorfii* is practically identical with *Sequoia sempervirens* and by many authors it is spoken of as the direct ancestor of the recent species. Under this species Schimper[5] includes a fairly long list of synonyms—species referred to *Taxites, Taxodium, Cupressites*, and other genera—which serves to emphasise the fact that impressions of sterile branches with distichous, linear, leaves cannot

[1] Carruthers (71) p. 3, with text-figure.
[2] Conwentz (92) p. 28, Pls. III., IV., VIII.
[3] Brongniart (28) A. p. 108.   [4] Heer (55) A. p. 54, Pls. XX., XXI.
[5] Schimper (72) A. p. 216.

in many cases be identified with *Sequoia* with absolute certainty. Differences are pointed out by authors in their description of species between the leaves of such recent forms as *Taxus baccata* and *Sequoia sempervirens*, but an examination of actual specimens reveals the inadequacy of such fine distinctions as are sometimes quoted. Our knowledge of the cones is confined to external characters and these afford a more substantial basis than the foliage-shoots on which to form an opinion with regard to the striking similarity between the Tertiary and existing species. *Sequoiites Langsdorfii* is recorded by Gardner[1] from the Eocene beds in Mull, but the identification rests on sterile branches bearing linear-lanceolate decurrent leaves 6—9 mm. long; the main axis of a branched specimen bears scale-like leaves appressed to the stem except at the distal end of the lamina and, as in the recent species, scale-leaves occur at the base of each lateral shoot. The species is recorded also from Styria[2], from Miocene beds in Greece[3] where it is represented by foliage-shoots and cones, from Italy[4], Germany, and other European localities. A very similar form, originally named by Brongniart[5] *Taxites Tournali*, is described by Gardner[6] from the Middle Bagshot beds of Bournemouth and with it he unites *S. Hardti* Heer founded on material from Bovey Tracey. *S. Tournali* is characterised by the association of distichous (fig. 794, A, B, p. 408) and smaller appressed leaves like those of *Sequoia gigantea*, an association also met with in *S. sempervirens*. *S. Tournali* is recorded by Saporta[7] who figures branches and cones from Eocene strata in Provence. Laurent[8] figures fragmentary specimens, which he refers to *S. Langsdorfii*, from the Aquitanian series in the Puy-de-Dôme.

Sequoiites Langsdorfii is very abundant in Arctic Miocene rocks: Nathorst[9] speaks of numerous branches in Tertiary clays in Ellesmere Land in a remarkable state of preservation so that they could be washed out and isolated like dried specimens in a herbarium. A peculiarity of the Ellesmere specimens is the occurrence of very

[1] Gardner (86) p. 41, Pl. x. fig. 1.   [2] Ettingshausen (57) Pl. I. fig. 3.
[3] Unger (47) Pl. II. figs. 17—23. Saporta (68) refers *S. Langsdorfii* as figured by Unger to *S. Tournali*.
[4] Squinabol (92) p. 26, Pls. xv., xvi.   [5] Brongniart (28) A. p. 108.
[6] Gardner (86) p. 40.   [7] Saporta (65²) Pl. II. fig. 1.
[8] Laurent (12) p. 65.   [9] Nathorst (11³) p. 225.

fine teeth on the edge of the lamina[1]. Similar teeth are stated by
Nathorst to have been seen in one or two examples of *Sequoia
sempervirens*, and it suggested that the papillae, which are a normal
feature of the recent species, were more strongly developed in the
Tertiary type. Heer[2] records the species from Miocene beds in
Greenland and states that it is one of the commonest Conifers in
Disco Island, from the Mackenzie River, Alaska, Spitzbergen, and
Sachalin Island. The fragments
reproduced in fig. 767 were collected in Disco Island and are now
in the Dublin Museum with other
fossils described by Heer; the long
linear leaves, A, are decurrent and
in some cases the lamina shows
fine transverse striations: the
smaller leaves shown in fig. 767, B,
are referred by Heer to a distinct
species *S. brevifolia*, but there is no
important difference between the
two forms. Palibin[3] figures sterile
shoots from the Sichota-Alin mountains. Penhallow[4] records the species from British Columbia and
other localities, and to the same
type he assigns some petrified wood

Fig. 767. *Sequoiites Langsdorfii*. (From specimens in the Dublin Museum described by Heer.)

from the Queen Charlotte Islands though without any real evidence
of connexion. Twigs and cones are described by Schmalhausen
from Tertiary strata in the New Siberian Islands[5], and the species
is said to be one of the most abundant and widely distributed
types in the Yellowstone National Park[6]. Remains of more than
one species of *Sequoia* are recorded from Florissant, Colorado,
which has recently been described as a Miocene Pompeii: the sedi-

[1] Nathorst (15²) p. 10, Pl. I. figs. 1—15.
[2] Heer (68) Pls. II., XX., LV., etc.; (71) Pls. XL., XLIII., etc.; (75) iii. Pl. II.;
(77) i. Pls. XII., XIII., XXV.; (78) v. Pl. I.; (82) i. Pl. LIII.
[3] Palibin (04) Pl. II.
[4] Penhallow (02) pp. 44, 68; (03) p. 41.
[5] Schmalhausen (90) Pl. I. figs. 2—11.
[6] Knowlton (99) p. 682.

ments of an ancient lake mixed with volcanic ash contain many plant and insect remains and Prof. Cockerell's careful investigations have led to the discovery of several new types[1]. Staub[2], who records the species from Aquitanian beds in Hungary, gives a list of references to other authors.

Making allowances for doubtful identifications based on sterile branches there remain a sufficient number of authentic records to demonstrate the wide range of this species and allied forms in Europe and the Arctic regions during the Eocene and Miocene periods. *S. Langsdorfii* is said to occur in beds of Lower Pliocene age in France[3] and a minute cone, only 2 by 1·9 mm. has recently been described by Mr and Mrs Reid from Pliocene deposits in Holland[4]. The Dutch specimen is referred to *Sequoia* with some hesitation and it is suggested it may be an immature cone of an undescribed species, which possibly marks the last appearance of the genus in Europe.

Sequoiites Couttsiae (Heer).

This species was founded by Heer[5] as *Sequoia Couttsiae* on material from Oligocene beds which form a basin-shaped depression in the granitic rocks of Dartmoor in Devonshire. The material consists of foliage-shoots (fig. 768, A, B), similar in habit to those of the recent species *Sequoia gigantea*, and globose or sub-globose cones with peltate scales and winged compressed seeds like those of *Sequoia sempervirens*. Several seeds are said to occur on each cone-scale. Beust[6] examined wood from Bovey Tracey in which he found tracheids with separate bordered pits and resiniferous xylem-parenchyma as in the recent species. Mr and Mrs Clement Reid[7] have recently investigated the Bovey Tracey material and their conclusion is that '*Sequoia Couttsiae* is a true *Sequoia* and close to the living *Sequoia sempervirens* and *S. gigantea*.' They give the following description of the cones: 'Broadly oval and abruptly narrowed into the stalk, or somewhat cordate; at the base are a few small recurved wedge-shaped barren scales, the lower ones having their stalks strongly reflexed, the middle ones with stalks

---

[1] Cockerell (06), (08), (08²), (08³).   [2] Staub (87) B. Pl. XIX. p. 249.
[3] Depape (13).   [4] Reid, C. and E. M. (15) p. 55, Pl. I. fig. 13.
[5] Heer (62) p. 1051, Pls. LIX.—LXI.   [6] Beust (85) Pl. III. figs. 1—8.
[7] Reid, C. and E. M. (10) p. 170, Pl. xv. figs. 23—27.

at right-angles to the axis; at the apex is a rosette of a few almost
sessile barren scales; the arrangement of the scales is distinctly
spiral. It is not easy to count the number of the scales, as none of
the cones we found are perfect. There would seem to be 20—24
fully developed scales, besides a few undeveloped round the apex
and base.' The scales vary in shape (fig. 768, C, D) and are cuneate
or umbrella-shaped, the rugose distal ends have lines radiating
from a central umbo; the winged seeds are pendant beneath the
thick involuted margin and on the upper surface of one scale five
seeds were found. Preparations of the cuticular membrane of the
leaves showed irregularly scattered stomata, each surrounded by

Fig. 768.   *Sequoiites Couttsiae.*   Twigs A, B, and cone-scales C, D, from Bovey
Tracey.   (Photographs by Mr and Mrs Clement Reid; ×3.)

a ring of four or occasionally five cells. Gardner[1] has also described
specimens from Bovey Tracey and Hampshire characterised by
imbricate keeled decurrent leaves with a free, divergent or falcate,
apex and in older branches by more obtuse appressed leaves. The
cones in size and form resemble those of *Sequoia sempervirens* while
the vegetative branches agree with *S. gigantea.* There are 3—5 seeds
on each scale. Several examples of *Sequoiites Couttsiae* are figured
by Heer[2] from Miocene beds in West Greenland and he speaks of the
species as the commonest Conifer in Disco Island. Gardner points
out that the northern form has larger cones and stouter foliage-
shoots than the British type and proposes for it a new specific name

---

[1] Gardner (86) p. 36, Pl. VI.

[2] Heer (68) p. 94, Pls. III., VIII., XLV.; (71) Pls. XL., XLIII., etc.; (83) Pl. LXVIII.

S. Whymperi. This more robust form occurs also in Spitzbergen, on the Mackenzie River, and elsewhere. Knowlton[1] records *S. Couttsiae* along with other species from beds probably of Miocene age in the Yellowstone Park: he assigns to this type specimens described by Lesquereux from Colorado as *Glyptostrobus Ungeri* and others from the Fort Union Group referred by Newberry to *Glyptostrobus europaeus.* The cone-bearing branches figured by Lesquereux[2] from the Western Territories as *Sequoia affinis* bear a close resemblance to *S. Couttsiae.* The latter species is recorded by Penhallow[3] from the Eocene beds on the Deer River in Canada (lat. 51° and 54° N.). The same or a closely allied type is recorded from Miocene beds in Alsace[4], and Saporta[5] describes very good examples of *S. Couttsiae* from the Eocene beds at Armissan in Provence. According to Gardner the material referred by Saporta to Heer's species includes at least two other species. Specimens described by Schmalhausen[6] from Eocene beds in South-West Russia as *S. Couttsiae,* though possibly correctly named, are not convincing. Palibin[7] records this species from Oligocene beds at Molotytchi in the Fatej district, Russia, and discusses the geological age of the strata from which Schmalhausen's plants were obtained.

Ettingshausen's specimens from Bilin in Bohemia assigned by him to *Taxodium dubium* may, as Gardner says[8], be examples of *S. Couttsiae.*

GEINITZIA. Endlicher.

The name *Geinitzia* was given by Endlicher[9] to a piece of sterile shoot from Lower Cretaceous strata in Saxony previously figured by Geinitz[10] as *Araucarites Reichenbachii* (fig. 769), and in the new genus was also included *Cryptomeria primaeva* Corda[11]. Both these species were referred by Endlicher to *Geinitzia cretacea.* Corda's species was founded on several foliage-shoots from Lower Cretaceous rocks in Bohemia with the habit of *Araucaria excelsa* and in one or two instances bearing what appear to be terminal buds described by Corda as small cones. In the first instance *Geinitzia*

1 Knowlton (99) B. p. 681.   2 Lesquereux (78) B. Pl. LXV.
3 Penhallow (02) p. 50   4 Bleicher and Fliche (92) p. 382.
5 Saporta (65²) Pl. II.   6 Schmalhausen (83²) Pls. XXXII., XXXVI.
7 Palibin (01) p. 499.   8 Gardner (86) p. 39.
9 Endlicher (47) p. 280.   10 Geinitz (42) Pl. XXIV. fig. 4.
11 Corda in Reuss (46) B. Pl. XLIII. figs. 1—11.

was applied to branches without any recognisable cones. In 1868 Heer[1] figured specimens from the Kome beds in Greenland which he believed to be identical with *Araucarites Reichenbachii* Gein. though the foliage-shoots bear shorter leaves than those on the type-specimen of Geinitz: Heer states that he was able to examine the type-specimen and assured himself of the specific identity of the German and Greenland specimens; he substituted the generic name

Fig. 769. *Geinitzia Reichenbachii.*
(After Geinitz.)

Fig. 770. *Geinitzia Reichenbachii.*
(After Heer; nat. size.)

Sequoia for *Araucarites* on the ground that some cones in the Tübingen Museum from Lower Cretaceous beds in Moravia, attached to branches apparently identical with *Araucarites Reichen-bachii*, presented a very close resemblance to those of recent Sequoias. The Moravian specimens, which he afterwards figured[2], are oval and the cone-scales have distally expanded distal ends (fig. 770) like those of *Sequoia*, but no evidence was obtained as to the number of seeds. Additional examples of vegetative shoots and cones were described by Heer[3] from Greenland as *Sequoia*

[1] Heer (68) Pl. xliii.   [2] *Ibid.* (69) Pl. i.
[3] *Ibid.* (75) ii. Pls. xii., xx., xxxiv., etc.

Reichenbachii and this species is recorded by authors from many Lower Cretaceous localities, but in no case is any conclusive evidence brought forward in support of the assumed generic identity with *Sequoia*. Specimens of *Sequoia Reichenbachii* with foliage and cones are figured by Velenovský[1] from Bohemia showing clearly the characteristic peltate cone-scales, and similar examples though with rather larger cones are described from Lower Cretaceous strata in North America[2]. On the other hand the name *Sequoia Reichenbachii* is applied in some cases to fragments of sterile branches unaccompanied with cones[3]: in one instance[4] evidence was obtained of the occurrence of separate circular bordered pits on the tracheids of some vegetative branches from the Cretaceous beds of Aix-la-Chapelle. It is impossible to say whether such shoots bore cones like those of *Geinitzia* or *Elatides*; some at least belong to *Elatides curvifolia*.

Two conclusions are suggested by an examination of the records so far quoted: the use of the generic name *Sequoia* is not based on any solid foundation and, secondly, it is unsafe to assume that fragments of sterile branches bearing falcate leaves similar to those on fertile shoots referred to *S. Reichenbachii* belong to that species. The common occurrence of Mesozoic specimens agreeing more or less closely with *Araucaria excelsa*, while demonstrating the abundance of that form of vegetative shoot, by no means proves the equally wide occurrence of one specific type. It has, for example, been shown by Nathorst[5] that the branches from Lower Cretaceous or Upper Jurassic rocks in Spitzbergen figured by Heer[6] as *Sequoia Reichenbachii* are examples of *Elatides curvifolia* (Dunk.). The genus *Elatides*[7] is characterised by cones differing in their flatter scales and more elongated form from those usually assigned to *Sequoia* though the foliage-shoots are of the same type. It is therefore advisable to adopt some provisional generic term for sterile shoots resembling in habit those of *Araucaria excelsa* and which in the absence of cones cannot be safely assigned to a genus founded on the cone-characters. The name *Pagiophyllum*[8] serves

---

[1] Velenovský (85) B. Pls. VIII., IX.
[2] Ward (99) B. Pls. 165, 166; Hollick (06) Pl. II. fig. 40; Pl. III. figs. 4. 5.
[3] Krasser (96) B. Pl. XVII. fig. 14; Schenk (71) Pl. XXIV. figs. 6, 7.
[4] Lange (90) p. 660.   [5] Nathorst (97) p. 35.
[6] Heer (75) ii. Pls. XXXVI., XXXVII.   [7] See page 270.   [8] See page 274.

this purpose and it should be applied to sterile branches of the
Araucaria type which cannot reasonably be referred to *Elatides*,
Geinitzia, or other genera connoting certain types of fertile shoot.
It has, however, been pointed out that in the first instance *Geinitzia*
was applied to sterile shoots, but later this designation came to be
associated with cones of elongate-oval form bearing peltate scales.
In 1852 Unger[1] applied *Geinitzia* to a specimen from Neustadt
consisting of a slender piece of foliage-shoot and an imperfectly
preserved cone similar to the cones of Heer's *Sequoia Reichenbachii*
but longer in form. Subsequently Heer[2] described under the name
Geinitzia formosa shoots and cones from Lower Cretaceous strata
at Quedlinburg: the cones are similar in form to that figured by
Unger and bear cone-scales with polygonal distal ends having a
central umbo and radially disposed lines on the exposed surface.
Schenk[3] also gives good drawings of *Geinitzia formosa*. A well preserved cone very like Heer's *G. formosa* was described by Newberry[4]
from the Amboy clays as *Sequoia gracillima*, the specific name
having been previously used by Lesquereux for sterile branches
from Dakota in conjunction with the generic name *Glyptostrobus*.
Newberry adopted Lesquereux's specific term because he found in
the Dakota beds cones like that from the Amboy clays associated
with the branches described by Lesquereux. Newberry's cone is
practically identical with that of Heer's *Geinitzia formosa*, but it is
noteworthy that the former is borne on a slender branch having
small appressed leaves in place of the more spreading falcate leaves
of Heer's species. This difference in the foliage is of secondary
importance in comparison with the close resemblance between the
cones. Subsequently Jeffrey[5] obtained good cones from the
Matawan formation apparently identical with *Sequoia gracillima*
(Lesq.) as figured by Newberry and he was able to investigate the
anatomical features. The pith of the cone-axis contains groups of
sclerous cells; the phloem differs from that of *Sequoia* in the absence
of fibres, while the secondary wood has no resin-cells—another
difference from *Sequoia*: the tracheal pits are circular and in no
case contiguous and there are no rims of Sanio. The latter feature

[1] Unger (52[2]).   [2] Heer (71[2]) p. 6, Pls. I., II.
[3] Schimper and Schenk (90) A. p. 299.
[4] Newberry and Hollick (95) p. 50, Pl. IX. figs. 1—3.   [5] Jeffrey (11).

is regarded by Jeffrey as an essential character of the Araucarineae and the absence of any Abietineous pitting in the medullary-ray cells is another Araucarian feature. The conclusion drawn by Jeffrey is that despite the absence of Araucarian pitting on the tracheids the anatomical details point to an Araucarian relationship, the wood of the cone-axis having the characters of Sinnott's genus *Paracedroxylon.* No information was obtained with regard to the seeds. Jeffrey's examination of the cone shows, as he says, that it does not agree structurally with the cones of recent Sequoias, but the reference to the Araucarineae rests on a slender basis.

In their account of the Kreischerville plants Hollick and Jeffrey[1] describe some sterile branches as *Geinitzia Reichenbachii* (fig. 806, D; page 437) which agree closely with specimens referred by authors to *Sequoia Reichenbachii* though they might equally well be identified with *Elatides curvifolia* (Dunk.). For such sterile twigs the name *Pagiophyllum* would be preferable. The pith of the Kreischerville shoots contains groups of sclerous cells; the leaf-bases show in transverse section three resin-canals and these are enclosed by the transfusion-tissue which accompanies the vascular bundle. In its distribution the transfusion-tissue differs from that in *Sequoia,* which is confined to the flanks of the vascular strand, and agrees with the corresponding tissue in Araucarian leaves. There is no xylem-parenchyma and the tracheids have 1—2 rows of bordered pits, in contact or sometimes separate and if in two rows alternate. The wood agrees with that described by Hollick and Jeffrey as *Brachyoxylon* and shows a decided Araucarian affinity. In the absence of cones attached to the shoots it is not possible to settle definitely the systematic position of the specimens. A fact in favour of identifying the branches with *Sequoia* (or more appropriately *Geinitzia*) *Reichenbachii* is the occurrence in the same beds of detached cone-scales very similar to those of *G. gracillima,* which are referred to two new genera, *Eugeinitzia* and *Pseudogeinitzia.*

EUGEINITZIA. Hollick and Jeffrey.

Eugeinitzia proxima Hollick and Jeffrey.

The scales on which this species is founded[2] closely resemble those of recent species of *Sequoia* and *Geinitzia gracillima.* The

---

[1] Hollick and Jeffrey (09) B. p. 38.

[2] *Ibid.* p. 43, Pl. x. fig. 10; Pl. xxv. figs. 1—3.

vascular bundles were found to be arranged round the margin of
the peltate portion of the scales and completely surrounded by
transfusion tissue, 'a feature of marked contrast to the scale-
bundles in *Sequoia* and at the same time one which indicates a
strong affinity with the Araucarineae.' The mature scales afforded
no indication of the number or place of attachment of the seeds,
but an immature cone lent support to the view that each scale bore
four ovules on the peduncle near the cone-axis. Hollick and
Jeffrey regard the scales as Araucarian and think it probable that
they were connected with the twigs named by them *Geinitzia
Reichenbachii*.

PSEUDOGEINITZIA. Hollick and Jeffrey.

Pseudogeinitzia sequoiiformis Hollick and Jeffrey.

A special generic name[1] is given to some four-sided scales on the
ground that they not only differ in their tetragonal form from the
hexagonal scales of *Eugeinitzia* but probably belonged to a smaller
cone. As in the former type the vascular bundles are enclosed by
transfusion-tracheids. The investigations of the American botanists
show that the sterile branches, *G. Reichenbachii*, exhibit certain
Araucarian tendencies and that the cone, *Geinitzia gracillima*, as
also the detached cone-scales, *Eugeinitzia* and *Pseudogeinitzia*,
cannot be included in *Sequoia*.

Until more is known of the morphological nature of the cones
described by Heer and other authors as *Sequoia Reichenbachii*,
S. ambigua, etc., their relationship to existing Conifers cannot be
settled, but meanwhile it would seem convenient to include both
the smaller oval cones and the longer forms represented by *G.
gracillima* in the same genus *Geinitzia*, applying the name to cones
having spirally arranged scales with peltate distal ends superficially
resembling those of *Sequoia*. The name *Sequoia*, much too freely used
by palaeobotanists, has in some cases[2] been applied to cone-bearing
branches that are almost certainly identical with *Sphenolepidium
Kurrianum* (Dunk.). On the other hand for sterile foliage-shoots
unconnected with cones the non-committal name *Pagiophyllum* is
suggested on the ground that foliage-shoots alone cannot be more
precisely determined.

[1] Hollick and Jeffrey (09) B. p. 45, Pl. x. fig. 14; Pl. xxv. fig. 4.
[2] Heer (69) p. 11, Pl. i. figs. 10—13 ('*Sequoia fastigiata*').

SPHENOLEPIDIUM. Heer.

Heer[1] instituted *Sphenolepidium* in place of *Sphenolepis*, proposed by Schenk[2] for Wealden Coniferous branches, because of the previous use of the latter name by Agassiz for a genus of fishes. Berry[3] has recently reverted to the original form *Sphenolepis* on the ground that its employment by zoologists is not a serious objection. Schenk's definition of his genus includes both vegetative organs and cones, but it is desirable that the name *Sphenolepidium* should be restricted to fertile specimens or at least to specimens which can with reasonable certainty be connected with cone-bearing examples. The habit of the foliage-shoots of the two best-known Wealden and Lower Cretaceous species, *S. Sternbergianum* and *S. Kurrianum*, is of the type which leads authors to employ such generic terms as *Sequoia*, *Athrotaxites* or *Athrotaxopsis*, *Widdringtonites*, *Glyptostrobus*, *Araucarites*, and *Cyparissidium*, but in the absence of cones it is impossible to feel confidence in any attempts to distribute such sterile specimens among genera which are characterised not only by a certain form of foliage-shoot but also by a particular type of cone.

The generic name *Sphenolepidium* should be retained only for specimens with small, more or less globose, cones possessing spirally disposed cone-scales, cuneate, relatively broad and fairly thick. The cones are much smaller and have relatively broader and thicker scales than those of *Elatides* though there is no essential difference in the vegetative characters of the two genera. No cones have been described throwing any light on the affinity of the genus and like many others it must be left for the present in the category of *Coniferae incertae sedis*. The leaves are spirally disposed on the comparatively slender branches and are either ovate, triangular, and free only in the acuminate region, or longer and more spreading and falcate; the latter type agrees with *Pagiophyllum* while some forms bearing *Sphenolepidium* cones are rather of the *Brachyphyllum* type. Many of the specimens recorded as *Sphenolepidium* afford no evidence as to the nature of the cones and should be assigned to *Pagiophyllum* or *Brachyphyllum*. The genus is characteristic of Wealden or Lower Cretaceous strata and is represented in several European districts and in North America.

---

[1] Heer (81) p. 19.   [2] Schenk (71) B. p. 243.   [3] Berry (114) p. 290.

Sphenolepidium Sternbergianum (Dunker).

This species was originally described by Dunker[1] from North
Germany as *Muscites Sternbergianus* and by later authors placed
in *Araucarites, Widdringtonites,* and other genera[2]. It is impossible
to determine the specific limits of this species[3] and *S. Kurrianum*
(fig. 771): the cones exhibit no well-defined distinguishing characters and the chief distinction
is the more spreading foliage of the Araucarian
or *Pagiophyllum* type of *S. Sternbergianum*. As
Berry suggests, this species—described from the
Potomac formation and elsewhere in North
America and from several European localities—
is probably represented in the Lower Cretaceous
flora of Greenland under such names as *Glyptostrobus groenlandicus* Heer and *Sequoia fastigiata.*
Some of the English and German fossils attributed to *S. Sternbergianum* are almost certainly
examples of *Elatides curvifolia.*

Sphenolepidium Kurrianum (Dunker).

Dunker[4] originally adopted the generic name
Thuites; later authors preferred *Brachyphyllum,
Widdringtonites, Araucarites* and other names.
Fontaine[5], who records this species from the
Potomac formation, includes in his genus *Athrotaxopsis* specimens which cannot be distinguished
by any features of morphological importance
from *Sphenolepidium Kurrianum.* The leaves are
ovate, more or less appressed, agreeing with
Brachyphyllum or in some examples intermediate
between the type of foliage assigned to *Pagiophyllum* and *Brachyphyllum.* The Wealden specimen reproduced in fig. 771 is placed in *Sphenolepidium* because of its association with branches,
identical in habit, bearing cones; if found as an isolated

Fig. 771. *Spheno-
lepidium Kurria-
num.* From the
Wealden of Sus-
sex. (British Mu-
seum, V. 2303;
$\frac{5}{6}$ nat. size.)

[1] Dunker (46) A. Pl. vii. fig. 10.
[2] For references, see Seward (95) A. p. 205; Berry (114) p. 293.
[3] Seward (112) p. 685.
[4] Dunker (46) A. p. 20.   [5] Fontaine (89) B. Pls. cxxxv., etc.

fossil it would be referred to *Brachyphyllum*. The figured specimen shows the variable form and size of the leaves and there is good reason to believe that the plants represented by the fragments included in one or other species of *Sphenolepidium* were characterised by a considerable range in the habit of the foliage-shoots, a fact which renders of little importance the separation into *S. Kurrianum* and *S. Sternbergianum* based on the form of the leaves in detached branches. The small cones borne terminally on slender branches resemble superficially the cones of *Athrotaxis*, but no facts are available as to the structure of the cone-scales and there is no evidence on which to found an opinion as to the position of the genus.

SCIADOPITINEAE.

Though several fossil plants have been compared with the existing species *Sciadopitys verticillata*, in no case is there any conclusive evidence of the occurrence of this type of Conifer. Schmalhausen founded the genus *Cyclopitys*[1] for impressions of shoots from Russia bearing whorled linear leaves which he believed to be closely allied to or generically identical with *Sciadopitys*. Zeiller[2], who brought forward strong arguments for assigning the strata regarded by Schmalhausen as Jurassic to the Permian period, considers *Cyclopitys* to be an Equisetaceous plant. Detached linear leaves similar to those of *Cyclopitys* are abundant in many Jurassic floras and, as Nathorst[3] says, they may be compared with several recent genera including *Sciadopitys*, but without anatomical data accurate determination is impossible. It is stated by Schenk[4] that the Cretaceous leaves described by Heer as *Pinus Crameri* agree in their epidermal features with the foliage of *Sciadopitys*, but in this as in other cases generic identity or even close relationship has not been demonstrated. Goeppert and Menge[5] describe some single leaves preserved in Baltic amber as *Sciadopitytes linearis* and *S. glaucescens*; they speak of the leaves as having a single vein on the upper face and two veins on the lower surface though it is not clear what morphological feature is represented by the 'veins.'

[1] Schmalhausen (79) A. p. 39.   [2] Zeiller (96) A. p. 477.
[3] Nathorst (97) p. 19.   [4] Schimper and Schenk (90) A. p. 293
[5] Goeppert and Menge (83) A. p. 36, Pl. XIII. figs. 117—123.

Schenk states that these Oligocene leaves are Dicotyledonous and
not the leaves of a Conifer.

Specimens of fossil wood have been described exhibiting certain
features, especially the pitting of the medullary-ray cells, similar to
those of *Sciadopitys*[1] but the occurrence of such features in other
recent genera precludes a definite reference to any one type.

SCIADOPITYTES. Goeppert.

This name has recently been revived by Halle[2] for two species of
Cretaceous leaves from Greenland, one of which, *Pinus Crameri*
Heer, though compared by Schenk with the leaves of *Sciadopitys*,
was not actually included in *Sciadopitytes*, while the other is a new
species, *Sciadopitytes Nathorsti*. Halle describes these leaves as
'Conifer-like in habit, with a dorsal groove which is protected by
elongated papillae and whose epidermal tissue differs from that of
the rest of the leaf through a non-seriate arrangement of the cells
and the occurrence of stomata.'

The outstanding feature of the leaves of *Sciadopitys* is the
double nature of the lamina and the morphological peculiarities
which have led to its recognition as a phylloclade; but, as Halle
admits, there is no evidence that the fossils are other than ordinary
simple leaves. The interesting characters described by Halle amply
justify the use of a generic name separating the leaves from those
known only as impressions, without any structural features preserved, and referred to *Pityophyllum*. It is, however, open to question whether the name *Sciadopitytes* does not imply more than the
facts support. The leaves named by Heer[3] *Pinus Crameri* are
about 12 mm. long and 2·5 mm. broad: the apex is bluntly rounded
and the base is slightly widened. Halle points out that there is
evidence that the leaves were cylindrical. The carbonised leaves
of this species form thick masses in the shale and excellent preparations of the cuticle can be obtained. Halle considerably
extends Schenk's account of the epidermal characters. The
apparent midrib is a groove and there is no indication of a true
median vein. Rather large stomata are crowded in the groove and

[1] See page 138.
[2] Halle (15) p. 508.
[3] Heer (68) i. Pl. xliv. figs. 7—18; (75) ii. Pl. xxiii. pp. 9—15; Halle (15)
p. 509, Pl. xiii. figs. 1—13.

are surrounded by somewhat tangentially elongated cells, the other cells between the stomata being much smaller. The cells on the sloping sides of the groove bear cylindrical papillae. On the whole the structure recalls that of the recent *Sciadopitys* though as Halle shows there are certain differences.

The second species, *Sciadopitytes Nathorsti*[1], was discovered by Nathorst in the Middle beds of Atanekerdluk in West Greenland. The leaves are at least 40 mm. long and about 1 mm. broad: the stomata are confined to the groove as in *S. Crameri* and numerous papillae are borne on the borders of the median depression. The stomata are not so crowded as in *S. Crameri* and differ less in size from the other epidermal cells. These two species, though exhibiting some similarity to *Sciadopitys*, can hardly be assumed to belong to plants more closely allied to the recent Japanese Conifer than to other existing forms. The occurrence of the characters described by Halle may be recognised by adding the name *Sciadopitytes* after the non-committal term *Pityophyllum*.

[1] Halle (15) p. 512, Pl. XII. figs. 16—29.

CHAPTER XLVIII.

ABIETINEAE.

THE relative antiquity of the different families of the Coniferales is a question which every student of the geological history of the group desires to answer. Reference has already been made to the different views that are held with regard to the phylogenetic relations of the Araucarineae and the Abietineae: conclusions on this subject are based partly on the morphological characters exhibited by recent types and in part on palaeobotanical data. The evidence afforded by petrified wood is briefly dealt with in Ch. XLIV: this shows that the features associated with modern Abietineae do not stretch as far back into the past as is the case with the type represented by the wood of the Araucarineae. The evidence derived from a study of impressions of foliage-shoots and cones as well as the meagre data supplied by petrified cones is less easy to interpret because of the greater imperfection of the records. The southern distribution of the Araucarineae predisposes the student in favour of a southern origin, while the essentially northern range of the Abietineae suggests that this family had its birth north of the equator. But conclusions based on such considerations require confirmation from other kinds of evidence. In the Jurassic-Wealden period the Araucarineae were well represented in the northern hemisphere and the impression gained from a survey of Jurassic records is that the Araucarineae shared with other types an almost world-wide distribution. It is much easier for a palaeobotanist to form an opinion as to the period of maximum development and vigour of a given set of plants than to discover a substantial foundation on which to rest a view as to the first appearance or the original home of the earliest representatives of the family-type. It is, perhaps, significant that the Araucarineae are represented in the Jurassic floras of Graham Land on the edge of Antarctica, Australia, and India. The Abietineae, on the other hand, do not bulk largely in Mesozoic floras before the closing stages of the Jurassic period and more especially in the earlier days of the Cretaceous era. The abundance of Abietineous cones in Lower Cretaceous strata, a period later than that in which the Araucarineae are abundantly preserved in plant-bearing deposits, at least points to a later maximum development of the Abietineae, and such data as we have seem to favour a northern rather than a southern origin. Winged seeds, hardly distinguishable from those of modern Pines (fig. 788, p. 396), from Rhaetic beds in the South of Sweden, foliage-shoots from beds of the same age exhibiting features now associated with the Abietineae, demand serious consideration in connexion with the antiquity of the family, though it can hardly be maintained that they furnish proof of the existence in Rhaetic and Liassic floras of true Abietineae. The occurrence of a winged pollen-grain (fig. 491, G; Vol. III. p. 298) in the partially decayed wood of *Antarcticoxylon* might be urged as a plea for a southern origin of the family, but an extended bladder-like exine is not a monopoly of the microspores of the Abietineae.

The following types selected in illustration of the fossil records of the Abietineae show how difficult it is in many cases to determine the precise position within the family to which cones or foliage-shoots should be assigned. Palaeobotanical literature contains many species referred to *Abies* or *Abietites*, *Cedrus*, and other genera, but it is usually impossible from the available data to carry identification so far. A few examples may be quoted: certain Lower Cretaceous cones bear a very close resemblance to those of *Cedrus*[1], but an examination of some of the less familiar cones of existing species of *Abies* and *Picea* shows that the reasons for connecting the fossils with *Cedrus* are not entirely satisfactory. The fossil wood described under *Cedroxylon* does not denote that the parent-plants were more closely allied to *Cedrus* than to some other genera of the same family. Boulay[2] has described some seeds from Miocene beds in France as *Cedrus vivariensis* which he unhesitatingly regards as generically identical with those of recent Cedars, and there is no reason to doubt the correctness of this conclusion. Cone-scales bearing two seeds from Miocene beds in Spitzbergen described by

[1] See page 385.
[2] Boulay (87) p. 235.

Heer[1] as *Pinus (Cedrus) Lopalini* may belong to a true cedar cone, but the evidence is hardly convincing. It is not too much to say that even Tertiary records of Conifers seldom enable us to discriminate between individual genera. In the absence of anatomical data the needle-like leaves scattered through Mesozoic and Tertiary strata cannot be identified with reasonable certainty. From Upper Pliocene beds in Germany Geyler and Kinkelin[2] described a cone as *Abies Loehri*, and this has more recently been identified as *Keteleeria* (fig. 786, C, p. 394) by Engelhardt and Kinkelin[3] on the strength of its external resemblance to *K. Davidiana*. The reference to *Abies*[4] of some leaves enclosed in the Baltic amber affords an example of the assistance afforded by characters recognisable in well preserved material, and it is probable that a fuller knowledge of the epidermal characters of recent Conifer leaves may supply a useful aid to more precise identification.

PITYITES. Gen. nov.

Endlicher[5] employed the name *Pinites* for leaves, male flowers, and cones considered to be closely allied either to recent species of *Pinus* or to some other genus of the Abietineae, such as *Abies*, *Larix*, or *Picea*. Many authors have adopted the generic name *Pinus* in cases where the evidence appears to them sufficiently strong to indicate identity with the existing genus, but it is only cones and foliage-shoots from Tertiary and Pleistocene beds that can as a rule be definitely assigned to such a position. It may, perhaps, be carrying consistency too far to restrict Endlicher's designation to such specimens as there is good reason for connecting with the recent genus *Pinus*; but the more restricted use of *Pinites* has the merit of being less likely to mislead the student and, chiefly on that account, I propose to adopt the genus *Pityites* for Abietineous fossils which cannot with confidence be referred to a more precise position. In practice this designation will not often be employed as in most cases cones and vegetative organs occur as separate fossils and are most conveniently described under the

---

[1] Heer (78) i. Pl. IX. figs. 6—8.

[2] Geyler and Kinkelin (90) p. 16, Pl. I. figs. 13—15.

[3] Engelhardt and Kinkelin (08) p. 216, Pl. XXVI. fig. 7.

[4] Goeppert and Menge (83) A. Pl. XIII. figs. 107—110.

[5] Endlicher (47) p. 283.

terms suggested by Nathorst and mentioned below. *Pityites* is, however, appropriate for such specimens as those represented in figs. 772, 773 which show a direct connexion between cones and foliage-shoots.

Goeppert adopted *Pinites* for fossil wood in a wide sense, but it has long been the custom to describe petrified wood agreeing structurally with recent Pines and other members of the Abietineae under Kraus's term *Pityoxylon*. Nathorst[1], with a view to greater convenience, proposed certain subgeneric names as qualifying epithets indicating the nature of the fossils but not implying a direct connexion with *Pinus*: he adopted the names *Pityanthus* for male flowers suggesting alliance with those of some Abietineous genus, *Pityostrobus* for cones, *Pityolepis* for cone-scales, *Pityospermum* for seeds, *Pityocladus* for vegetative shoots, and *Pityophyllum* for detached leaves. To these the name *Pityosporites*[2] has recently been added.

The generic or rather subgeneric term *Pityophyllum* is apt to mislead the student if used in conjunction with *Pinites*: the leaves so named, as Nathorst admits, are in many instances almost certainly derived from plants which do not belong to the Abietineae. Under *Pityophyllum* are included both needle-like leaves which are probably Abietineous with others having a broader lamina (fig. 776) and much more likely to be connected with such genera as *Cephalotaxus*, *Torreya*, or *Podocarpus*.

The term *Pityosporites*[3] is proposed for microspores provided with wings similar to those of *Pinus* and other members of the Abietineae, though in this case also relationship with another family, namely the Podocarpineae, is not excluded. These terms whether used as subgeneric titles or as generic designations serve a useful purpose for *disjuncta membra*, while the name *Pityites* is employed for specimens of a more complete kind. The name *Abietites* has often been used for vegetative shoots and cones[4] which there is no adequate reason for assigning to a position nearer to *Abies* than to other genera of the same family: it is desirable to restrict the term to fossils which afford evidence of

---

[1] Nathorst (97) p. 62; (99) p. 16.   [2] Seward (14) p. 23.   [3] *Ibid.*
[4] *E.g.* Geinitz (80) p. 12; Fontaine in Ward (05) B. Pl. LXVIII. figs. 14—17; Thomas (11) Pl. IV. fig. 16; Pl. V. figs. 1, 2.

affinity to the recent genus. Similarly, names such as *Laricites Cedrites* and others implying a more precise determination than is suggested by *Pityites* may conveniently be used either as sub-generic or generic terms.

In the account of recent Conifers allusion is made to the views held by students of fossil plants with regard to the relative position of the Abietineae and the Araucarineae in a chronological sequence. The types selected for description are intended to serve as guides to those who wish to draw conclusions from the geological records, but so long as we have to trust chiefly to impressions without the more certain guidance of anatomical data the inferences drawn cannot be regarded as other than provisional. The evidence of fossil seeds is difficult to interpret, as its value depends on the amount of importance to be attached to the occurrence of speci-mens closely resembling in the form of the wing the seeds of recent Pines and other Conifers. The winged seeds of *Agathis* differ in the shape of the membranous appendage from those of Abietineous species, and the oldest winged seeds attributed to the Abietineae, from Rhaetic rocks, exhibit a closer agreement with the Abietineous type. On the other hand it is questionable whether the form of a wing constitutes a safe criterion of affinity. A similar difficulty is presented by 'winged' pollen-grains: a bladder-like extension of the exine though usually associated with the Abietineae is a character which is not confined to that family. Foliage-shoots like those of recent Abietineae are recorded from Rhaetic rocks and later Mesozoic strata, but we have no means of determining in the case of the oldest examples whether their superficial resemblance to branches of *Cedrus* and other genera has a phylogenetic signifi-cance. The generic name *Pinites* is applied by Renault[1] to a slender branch from Permian rocks in France bearing spirally disposed filiform leaves 3 cm. long apparently borne singly and directly on the main axis, not on short shoots. It is elsewhere[2] suggested that this specimen, *Pinites permiensis*, may belong to a plant allied to *Dicranophyllum*: there is certainly no adequate reason for the employment of the generic term *Pinites*. Similarly an impression figured by Stur[3] from the culm of Altendorf as *Pinites antecedens*,

---

[1] Renault (93) A. Pl. xxxii. fig. 1; (96) A. p. 377.
[2] Page 101.   [3] Stur (75) A. Pl. xiv. fig. 4.

which I was able to examine in the Vienna collection, is too fragmentary to be determined. The occurrence of linear leaves in fascicles is in itself no real evidence of Abietineous affinity: the clustered leaves of *Czekanowskia* and *Phoenicopsis*, especially the former, though essentially similar in habit to the foliage-shoots of some Abietineae are generally believed to belong to plants of another class. The evidence furnished by petrified wood has already been considered: the important point is that there is no satisfactory case of the occurrence of fossil wood of Palaeozoic age[1] having typical Abietineous features, a fact of importance in relation to the widely spread Palaeozoic woods agreeing in essentials with the Araucarian type.

Pityites Solmsi Seward.

This name was proposed for some cones attached to foliage-shoots as well as detached cones and vegetative branches from Wealden rocks on the coast of Sussex[2]: the type-specimens form part of the rich Rufford collection in the British Museum. The branches are covered with the elongated persistent bases of scale-leaves and in the axils of these are borne numerous long needles (fig. 772). The cones are oblong and bear broad, rounded, scales like those of *Pinus Strobus*, *P. excelsa* (fig. 773; *cf.* fig. 704), *Picea* and *Abies*; they agree closely with *Pityostrobus Carruthersi* (Gard.) as also with *P. Andraei* (Coem.)[3] from Lower Cretaceous rocks in Belgium and with the smaller cones from the Potomac formation described by Fontaine[4] as *Abietites ellipticus*. The preservation is not sufficiently good to show the number of leaves in each foliage-spur: the needles may have been borne in dense clusters as in *Cedrus*. In general habit the species resembles *Cedrus* and *Larix* though the greater length of the needles is more in accordance with recent species of *Pinus*. Shoots similar to those of this species are represented by *Prepinus statensis* Jeff.[5] from the Cretaceous beds of Kreischerville. Dr Stopes[6], following the example of Berry, refers this species to *Abietites*.

[1] See page 220, also Thomson and Allin (12).
[2] Seward (95) A. p. 196, Pls. XVIII., XIX.   [3] Gardner (86²).
[4] Fontaine (89) B. Pl. CXXXIII. figs. 2—4.
[5] Hollick and Jeffrey (09) B. p. 19.
[6] Stopes (15) p. 157.

Pityites (Pinites) eirensis sp. nov.

In his account of petrified material from Franz Josef Land
Solms-Laubach[1] describes sections of a *Pinus*-like leaf from Bell

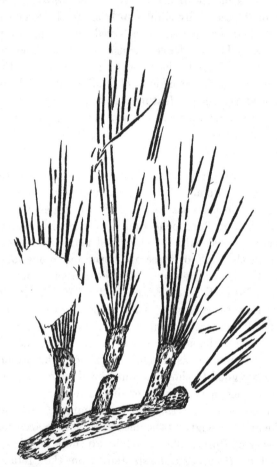

FIG. 772. *Pityites Solmsi.* (British Museum, V. 2169; nat. size.)

Island (Eira harbour) probably of the same geological age as the
plant-beds of Cape Stephen, which is believed to be Upper Jurassic
or Lower Cretaceous. Through the kindness of the Director of the

---

[1] Solms-Laubach (04) p. 12, Pl. I. fig. 14; Pl. II. fig. 3. For evidence as to
geological age, see Newton and Teall (97), (98); Nathorst (99).

Geological Survey I have been able to examine the sections in the
Jermyn Street Museum. Graf Solms-Laubach describes the leaves
as oval in section, the upper face strongly convex and the lower
almost flat as in two-needled Pines, but as shown in fig. 774, A the
leaves may be approximately cylindrical (1 mm. in diameter), like
those of *Pinus monophylla* or the leaves of *Cedrus*. There is a single
vein accompanied by some radially disposed transfusion-tracheids,
the whole being enclosed in a single layer of rather thick-walled

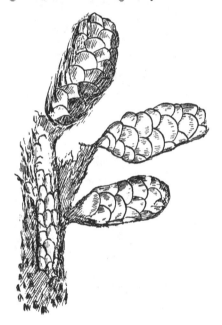

FIG. 773. *Pityites Solmsi.* (British Museum, V. 2146; nat. size.)

cells. There is no distinct division of the bundle into two halves
but there are indications of the presence of a broad median medullary ray. The mesophyll-cells have prominent infoldings precisely
as in recent Pines, *Cedrus*, and some other Abietineae (fig. 774, B;
cf. fig. 694): the epidermis has a thick cuticle and below it are
1—2 layers of small thick-walled elements. Solms-Laubach speaks
of two resin-canals, one at each side of the lamina, but I was unable
to distinguish any undoubted canals in the leaf shown in fig. 774.
The occasional absence of canals in Abietineous leaves normally

possessing them is mentioned in Chapter XLIII. The leaf for which
the specific name *eirensis* (from Eira harbour) is proposed affords

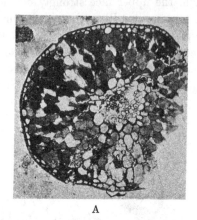

A

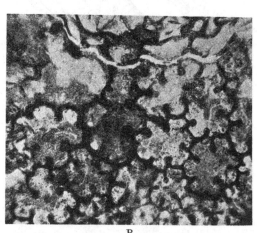

B

FIG. 774. *Pityites (Pinites) eirensis.* A, section of leaf. B, mesophyll enlarged.
(Museum of the Geological Survey.)

an interesting example of an Abietineous type, in all probability of
Upper Jurassic age, exhibiting a remarkable resemblance to certain
recent species especially *Pinus monophylla.*

PITYOCLADUS. Nathorst.

Under this name, used as a subgenus of *Pinites*, Nathorst[1] includes branches bearing short shoots similar in habit to those of *Cedrus* and *Larix*. Branches of this type, bearing leaves and cones, are illustrated by *Pityites Solmsi* (figs. 772, 773), but in some cases such vegetative shoots occur as detached fossils and it is to them that Nathorst's term may conveniently be applied. The striking resemblance of the fossil specimens to shoots of *Cedrus* and *Larix* and the frequent association or attachment of needle-like leaves afford strong grounds for assigning the branch-fragments to the Abietineae.

Pityocladus Nathorsti Seward.

In his description of *Schizolepis Follini* Nath.[2] from Rhaetic rocks in Scania, Nathorst includes not only cones with lobed scales characteristic of *Schizolepis* but leaves and branches. Solms-Laubach[3] expressed the opinion that we know nothing of the foliage of *Schizolepis* 'for there is nothing to make it even probable that the numerous needles which lie one above another in the beds at Palsjö, any more than the branches beset with needle-bearing shoots which Schenk has referred to this genus, have any connexion with *Schizolepis*.' In a later account of *Schizolepis*, Nathorst[4] suggests the advisability of separating the leaves and branches from the *Schizolepis* cones, though as he says the association of the two sets of organs in more than one locality may be significant. It is, therefore, preferable to assign the vegetative organs to *Pityocladus,* at the same time keeping in mind the possibility of an original connexion with the cones described under the generic name *Schizolepis*. In order to avoid confusion I have removed the branches and associated leaves from *Schizolepis Follini* to a distinct species *Pityocladus Nathorsti*. The specimens figured by Nathorst consist of (i) a fairly stout axis bearing a smaller lateral shoot like that on which the leaf-clusters of *Cedrus* and *Larix* are borne; (ii) separate short shoots characterised by zones of small scars alternating with smooth areas; (iii) numerous crowded linear leaves.

[1] Nathorst (97) p. 62.   [2] *Ibid.* (78) B. p. 28.
[3] Solms-Laubach (91) A. p. 70.   [4] Nathorst (97) p. 38.

Pityocladus longifolius (Nathorst). [And *Pityophyllum longifolium*
Nathorst.]

This species was first described by Nathorst[1] from Rhaetic beds
in Scania as *Taxites longifolius* and afterwards as ? *Cycadites*.

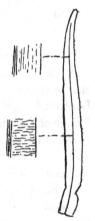

Fig. 776. *Pityophyl-
lum longifolium.*
(Nat. size.)

Fig. 775. *Pityocladus longifolius.* Leaves attached
to a short shoot covered with scales. (Nat. size:
a piece of lamina enlarged to show the fine transverse striations. From a specimen from Scania
in the Stockholm Museum.)

Möller[2] adopted Nathorst's generic name *Pityophyllum* for detached leaves and that designation should be used for specimens

[1] Nathorst (78²) B. p. 50.   [2] Möller (03) p. 40, Pl. vi.

which afford no evidence as to the nature of the axis. The specimen from the Stockholm Collection represented in fig. 775, from the Rhaetic rocks of Scania, is especially interesting as affording one of the few examples of leaves of the type known as *Pityophyllum longifolium* attached to an axis covered with short scales. The lamina varies from 1 to 5 mm. in breadth and may be broader: as seen in the enlarged portion the lamina is transversely wrinkled, an appearance characteristic of most forms of the broader *Pityophyllum* leaves and probably produced by contraction on drying. There is little difference between this species and the leaves figured by Nathorst from Scania as *Schizolepis Follini*. An accurate specific delimitation of *Pityophyllum* leaves is hopeless. Detached leaves (fig. 776) similar to those shown in fig. 775 are recorded by Möller from Lower Jurassic beds in Bornholm, by other authors from Jurassic strata in Turkestan[1], South Russia[2], Oregon[3], Spitzbergen[4], and elsewhere.

Pityocladus Schenki Seward.

Schenk[5] also refers to *Schizolepis* several specimens of branches and leaves as well as cones which he includes in *Schizolepis Braunii* Schenk. The larger branches bear leaf-cushions and short lateral shoots with scale-leaves at the base, and in some examples tufts of needles occur on the short shoots. A specimen described by Braun as *Isoetes pumilus* is identified by Schenk in his monograph of the Rhaetic Flora[6] as a leaf-bearing shoot like that of a recent Pine. The Swedish specimen represented in fig. 775, also of Rhaetic age, is similar to those included by Schenk in *Schizolepis Braunii*.

Pityocladus kobukensis Seward.

This species originally described from the Jurassic beds on the Kubuk River in Chinese Dzungaria as *Pinites*[7] is founded on branches bearing short shoots almost identical with *P. Schenki*. Short shoots not more than 1 cm. long are borne spirally on a thicker axis and covered with small leaf-scars (fig. 777) exactly as in the corresponding shoots of *Cedrus* or *Larix*. With the

[1] Seward (07[4]) B. p. 32. [2] Thomas (11) p. 78, Pl. VII. figs. 58—61.
[3] Fontaine in Ward B. (05) Pl. XXXV. [4] Nathorst (97).
[5] Schenk (67) A. Pl. XLIV. figs. 1—4. [6] *Ibid.* Pl. XLIV. fig. 2.
[7] Seward (11) p. 54, Pl. IV. figs. 47—51; Pl. V. fig. 65.

branches are associated numerous needles, 1 mm. broad and at
least 5 cm. long, sometimes covering the
whole surface of the rock. The specimens
agree closely with the foliage-shoots of
Pityites Solmsi: similar examples are
described by Ettingshausen[1] from Liassic strata as *Halochloris baruthina*
Ett. A branch with short shoots from
Jurassic rocks in Amurland described
as *Pinites* sp. cf. *P. kobukensis*[2] may be
specifically identical with the Dzungaria
fossils: an example of the same type
lent to me by Dr Krystofovič from
Jurassic beds of Amurland shows a
forked lateral foliage-shoot. This author
has recently described a specimen from
Jurassic rocks in Transbaikalia as *Pinites* (*Pityophyllum*) cf. *P. kobukensis*[3].

FIG. 777. *Pityocladus kobukensis*; *b*, branch-scar. (After
Seward; nat. size.)

Similar though smaller specimens
of Abietineous short shoots are described by Nathorst[4] as *Pinites* (*Pityo-
cladus*) spp. *a* and *b* from Upper Jurassic beds in Spitzbergen, and
compared by him with *Pityites Solmsi*.

PITYOPHYLLUM. Nathorst.

This name[5] is applied to detached leaves of needle-like form like
those of recent Pines or to long linear leaves broader and flatter
than the needles of *Pinus*. Some of the specimens referred to
this genus are very similar to the leaves of *Keteleeria*. In a few
cases (fig. 775) the leaves are still attached to a short shoot but
usually they occur as detached specimens (fig. 776). The genus is
met with in Rhaetic strata but is specially abundant in Jurassic
floras and persists through Cretaceous and Tertiary rocks. The
leaves generally described under this generic term are broader and
flatter than such leaves as those of *Pityites Solmsi*[6] and recent

[1] Ettingshausen (52) B. Pl. II. fig. 4.
[2] Seward (12³) Pl. III.   [3] Krystofovič (15) Pl. VI. fig. 9.
[4] Nathorst (97) Pl. III. figs. 28—30, Pl. IV. figs. 13, 14, 23.
[5] *Ibid.* (97) p. 62.   [6] See page 374; also Nathorst (97) Pl. V. figs. 1—10

Pines, and the presence of a fine transverse wrinkling on the lamina is a characteristic feature. *Pityophyllum*, if employed for both the narrower and broader forms, includes specimens which in all probability belong to Conifers of more than one family: some are certainly Abietineous but the flatter and broader forms bear a closer resemblance to leaves of some species of *Podocarpus, Cephalotaxus* or *Torreya*. Nathorst, who instituted the generic name *Pityophyllum*, recognises that many of the specimens so named have no real botanical value. Detached leaves of the type included in this comprehensive genus are of little interest; but it is noteworthy that such species as *P. Nordenskiöldi* (Heer) and similar forms are characteristic fossils in Jurassic and Cretaceous strata.

Pityophyllum Lindströmi Nathorst.

Under this name Nathorst[1] includes leaves described by Heer from Cretaceous strata in Greenland as *Pinus Quenstedti* and *P. Peterseni*, also specimens from Upper Jurassic beds in Spitzbergen. The leaves reach a length of at least 8 cm. and are 1—2 mm. broad; the lamina tapers gradually towards the base and is more abruptly narrowed in the apical region; there is a prominent midrib on one side and sometimes indications of two finer marginal 'veins,' also other longitudinal striations which may mark the position of rows of stomata. These leaves are broader than those of *Pityites Solmsi* and narrower than very similar specimens described by Heer, Nathorst, and other authors as *Pinus* or *Pityophyllum Staratschini*[2] from Cretaceous and Jurassic rocks. *Pityophyllum Nordenskiöldi*[3] (Heer) from rocks of the same age is another similar form having a tendency to a slightly sickle-shaped and transversely wrinkled lamina (*cf.* fig. 776). *Pityophyllum* is abundantly represented in Jurassic Floras[4]: the specimens are, however, of very little interest to the botanist as it is impossible to assign them to a family position in the Coniferales.

PITYOSTROBUS. Nathorst.

This name is used in preference to Feistmantel's genus *Pinostrobus*, recently resuscitated by Dr Marie Stopes, on the ground

---

[1] Nathorst (97) pp. 40, 67, Pls. v., vii.; (99) p. 20.

[2] *Ibid.* (97) pp. 41, 68, Pls. v., vi.   [3] Heer (78) ii. Pl. ii.; Nathorst (97) p. 18.

[4] For references see Möller (03) p. 39; Seward (11) p. 53; Krystofovič (10) Pl. iii. fig. 10; Thomas (11) p. 78.

that Nathorst's term is more appropriate for specimens which do
not afford evidence of closer affinity to *Pinus* than to other genera
of the Abietineae. In cases where the specimens may reasonably
be regarded as more nearly allied to *Pinus* than to any other genus
the designation *Pinites* may be added.

There can be no question of the abundance of Abietineous
Conifers in Tertiary floras and it is equally true that cones of the
Pityostrobus type are widely spread in Lower Cretaceous strata
especially in Europe. The evidence furnished by cones clearly
points to the existence in Upper Jurassic floras of Conifers closely
resembling in the general form of their strobili recent members
of the Abietineae. The wide distribution of cone-scales and cones
of the Araucarian type in Middle Jurassic floras is in striking contrast to the scarcity of cones of the Abietineous form in rocks older
than the uppermost Jurassic and Lower Cretaceous series.

Pityostrobus dejectus (Carruthers).

Carruthers[1] speaks of this Kimeridge cone from Dorsetshire as
the oldest example of a Pine-cone. It is represented by a single
imperfectly preserved specimen, 2 × 2 cm., of globular form with
partially destroyed broad and thin cone-scales: though it superficially resembles some recent Abietineous cones there is scarcely
enough evidence to warrant its inclusion in the Abietineae. The
cone was first described as *Pinites depressus* but owing to the
previous use of that specific name by Coemans it was re-named
P. dejectus[2].

Pityostrobus strobiformis (Fliche and Zeiller).

A species, described as *Pinites strobiformis*[3], from Portlandian
rocks near Boulogne founded on a single incomplete cone similar
in form and in the possession of apparently flat, imbricate, scales
to *Pinus excelsa* (*cf.* fig. 704, p. 154). The surface-features are not
shown on the weathered specimen and there is no definite information with regard to the number or position of the seeds, but as the
authors of the species state the narrow elongate and slightly curved
form of the cone, which was probably about 17 cm. long, affords a
valid reason for comparison with recent Pines.

[1] Carruthers (69²) p. 2, Pl. II. fig. 10.   [2] *Ibid.* (71) p. 2.
[3] Fliche and Zeiller (04) p. 802, Pl. XIX. fig. 6.

Pityostrobus Sauvagei (Fliche and Zeiller).

The type-specimen of this species[1], from the Portlandian of Boulogne, is an ovoid cone 4·5 cm. long characterised by distally expanded scales and resembling the small cones of *Pinus Laricio*. In the absence of further data precise identification is not possible though the fossil is probably correctly regarded as an Abietineous cone of the *Pinus Pinaster* type. Having regard to the fact that in this and the preceding species the determination is based solely on external form no very definite statement is admissible as to systematic position, but such evidence as there is favours the view that in these two cones we have Jurassic representatives of two sections of the genus *Pinus*.

Pityostrobus Dunkeri (Carruthers).

Several detached cones bearing imbricate scales, broad and flat like those of *Picea*, some species of *Abies* in which the ovuliferous scales are longer than the bract-scales, and certain species of *Pinus* characterised by flat scales instead of the woody scales of the *Pinus silvestris* and *P. Pinaster* type have been described from British Wealden strata as also from other countries. Gardner[2] instituted the following species: *Pinites Carruthersi, P. valdensis, P. cylindroides*[3], *P. pottoniensis*[3], but an examination of the type-specimens shows that the distinctive features are not sufficiently well marked to warrant so many specific names. The Lower Greensand specimen from Potton, *P. cylindroides*, is water-worn and the shape of the imperfect scales is not the original form; it may possibly be identical with *P. valdensis, P. Carruthersi*, and *P. pottoniensis*, and there are no important features in which these forms differ from the longer cones of *Pityostrobus Dunkeri*. The cones from Brook in the Isle of Wight named by Carruthers[4] *Pinites Dunkeri* were originally described by Mantell as *Abietites Dunkeri*[5]; they reach a length of over 33 cm. and have a breadth of 3 cm., they are elongate-oval and relatively narrow and the long scales are attached to a slender axis (fig. 778). The seeds, apparently two on each scale, are oval and compressed. Cones of similar form and length are described by

---

[1] Fliche and Zeiller (04) p. 804, Pl. XIX. fig. 7.
[2] Gardner (86²).
[3] See also Seward (95) A. p. 193; Stopes (15) pp. 138, 140.
[4] Carruthers (66²) Pl. XXI. figs. 1, 2.   [5] Seward (95) A. p. 194.

Fig. 778. *Pityostrobus Dunkeri*. (British Museum; nat. size.)

Velenovský[1] from Lower Cretaceous rocks in Bohemia as *Pinus longissima*, a species recently recorded by Dr Stopes[2] from the Lower Greensand of England.

Though in the absence of foliage-shoots cones of this type cannot be assigned with certainty to any one recent genus, their great length suggests comparison with those of *Pinus Lambertiana* and *P. excelsa* rather than with cones of recent species of *Picea*.

Pityostrobus Leckenbyi (Carruthers).

This species was first described by Carruthers[3] from a specimen in the Leckenby Collection, Cambridge, from the Lower Greensand of the Isle of Wight. It is 10 cm. long and 5 cm. in diameter; the

Fig. 779.   *Pityostrobus Leckenbyi.*   From the Lower Greensand of the Isle of Wight.   (After Carruthers; ½ nat. size.)

scales agree in external form with those of *Cedrus* and Dr Stopes has recently proposed the generic name *Cedrostrobus*[4] in order to emphasise this resemblance. Prof. Fliche[5] described a cone from the Argonne as *Cedrus oblonga* which he believed to be identical specifically with *Abies oblonga* of Lindley and Hutton, but Dr Stopes gives Fliche's name as a synonym of *Cedrostrobus Leckenbyi*. A cone of similar form is also described by Coemans from Belgium as *Pinus Corneti*[6] and compared by him with *Cedrus*.

[1] Velenovský (85) B. Pl. I. figs. 14—17.
[2] Stopes (15) p. 141, text-fig. 38.
[3] Carruthers (69²) Pl. I. figs. 1—5.   [4] Stopes (15) p. 143, text-fig. 39.
[5] Fliche (96) p. 200, Pl. VIII.   [6] Coemans (66) p. 11, Pl. V. fig. 3.

Though superficially very like a cone of *Cedrus* (fig. 779), *Pityo-strobus Leckenbyi* also strongly resembles some species of *Picea* and *Abies* in which the bract-scales do not project beyond the semini-ferous scales. I have adopted the non-committal term *Pityostrobus* as it is by no means certain that Carruthers' type is more closely allied to *Cedrus* than to *Abies*.

Dr Stopes includes in *Cedrostrobus* a second species, *Cedrostrobus Mantelli*[1], from the Lower Greensand of Kent which Carruthers originally named *Pinites*: she compares with it a Potomac cone described by Berry[2] as *Cedrus Leei*. But these species do not afford any proof of close relationship to the recent genus *Cedrus*. It is probable that some of the numerous cones found in Lower Cre-taceous rocks belong to trees having the characters of *Cedrus*, though in the absence of more decisive evidence than has so far been furnished it would seem preferable to retain the wider desig-nation *Pityostrobus*.

Pityostrobus Benstedi (Mantell).

The small oval cone on which Mantell[3] founded the species *Abies Benstedi* is from the Lower Greensand of Kent. It was subsequently

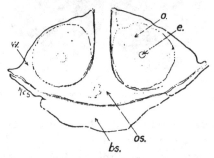

Fig. 780. *Pityostrobus Benstedi*. Tangential section showing, *os*, ovuliferous scale; *bs*, bract-scale; *o*, the two ovules on the ovuliferous scale; *e*, endosperm; *w*, wing. (After Stopes; × 8.)

described by Carruthers[4] as *Pinites* and regarded by him as prob-ably more nearly allied to *Cedrus* than to *Pinus*. Dr Stopes[5] has recently made a further examination of the structure of this type

[1] Stopes (15) p. 145, text-fig. 40.   [2] Berry (11) Pl. LXX. fig. 4.
[3] Mantell (46) p. 52, Pl. II. fig. 2.   [4] Carruthers (66²) p. 541.
[5] Stopes (15) p. 130, text-figs. 32, 33; Pls. X., XI.

and believes it to be more closely related to *Abies*. The cone-scales, as seen in fig. 780, show their double nature, and on the semini-ferous scale are two ovules provided with wings, *w*. The ovules are immature and there is a small space in the middle of the endo-sperm (fig. 780, *e*).

Pityostrobus (*Pinites*) *sussexiensis* (Mantell).

This Aptian (Lower Greensand) species, originally named by Mantell *Zamia sussexiensis*[1] and afterwards referred by Carruthers[2] to *Pinites*, has recently been more fully described by Dr Stopes[3] under *Pinostrobus*. The cone, 14 cm. long and nearly 5 cm. in diameter, bears overlapping scales with a thickened, curved, distal margin 2 cm. broad and, on the exposed surface of the specimen, 1·3 cm. deep. The seeds, two on each scale, have a corrugated stone-layer in the testa and bear massive, broad wings. A section through the middle of a scale shows an irregularly scattered double set of variously orientated vascular bundles and resin-canals. The species closely resembles *Pinus excelsa* and *P. Strobus* and is con-sidered by Dr Stopes to occupy a position between these two types.

The inference to be drawn from this and several other cones from Lower Cretaceous strata is that Abietineous cones having more or less flat scales as seen on the surface were more abundant in Europe in the early Cretaceous forests than those in which the distal ends of the scales are rhomboidal as in *Pinus Pinaster*.

Pityostrobus oblongus (Lindley and Hutton).

The type-specimen, a water-worn cone from Dorsetshire, pre-sumably from Lower Greensand rocks, was described by Lindley and Hutton as *Abies oblonga*[4]: it was assigned by some authors to *Pinites*. Williamson[5] gave an account of a cone from Sidmouth in Devonshire, which he referred to *Pinites oblongus*, though Dr Stopes[6] expresses a doubt as to the identity of his specimen with that described by Lindley. Schimper[7] employs the name *Cedrus* and Goeppert assigns the species to *Abietites*[8]. It is impossible to determine the position of the specimen represented in fig. 781 among the Abietineae.

[1] Mantell (43) p. 34.   [2] Carruthers (66²) p. 541, Pl. xx. figs. 5, 6.
[3] Stopes (15) p. 123, Pls. x., xi.   [4] Lindley and Hutton (35) A. Pl. 137.
[5] Williamson (86).   [6] Stopes (15) p. 135.
[7] Schimper (72) A. p. 299.   [8] Goeppert (50) p. 207.

The type-specimen bears a resemblance to *P. Leckenbyi* (fig. 779); the scales are broad and thin at the distal end and the axis is relatively slender. The French specimens from Lower Cretaceous rocks referred by Fliche to this species as *Cedrus oblonga* are considered by Dr Stopes to be specifically identical with *Pityostrobus Leckenbyi*.

FIG. 781. *Pityostrobus oblongus.* (After Lindley and Hutton, from Stopes; nat. size.)

Pityostrobus hexagonus (Carruthers).

A large cone 15 cm. long and 4 cm. in diameter composed of stout woody scales with hexagonal apophyses was described by Carruthers as *Pinites hexagonus* from the Gault of the South of England[1]; it agrees externally with recent cones of the *Pinaster* type but the distal ends of the scales are almost flat and nothing is known of the internal structure. The species may be compared with the Lower Cretaceous species *P. Quenstedti* Heer[2].

Pityostrobus (Pinites) Andraei (Coemans).

The cones of this species (fig. 782), the commonest type in the Lower Cretaceous rocks of Hainault[3], are 10—14 cm. long and

[1] Carruthers (71) p. 2, Pl. xv.   [2] Heer (71²).
[3] Coemans (66) p. 12, Pl. iv. fig. 4; Pl. v. fig. 1.

2·25 cm. in diameter. The cone-scales are compared with those of
Pinus excelsa, but the distal ends are stouter
than in the recent species and more like those
of P. Pinaster. Heer[1] compares P. Andraei
with his Pinus Quenstedti from Moravia in
which the scales have thick apophyses with a
central umbo. The needles of the Moravian
species are 20 cm. long and appear to be either
3 or 5 in a fascicle.

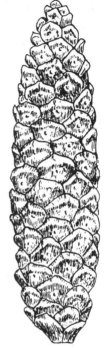

It is impossible within the limits of a general
text-book to discuss the bearings of the numerous Tertiary records of Abietineous cones,
many of them undoubtedly borne by species
of Pinus. A few examples only are mentioned
primarily in order to draw the attention of
students to the importance of making a critical
examination of Tertiary and Pleistocene Conifers. The neglect of Tertiary plants is largely
due to the unscientific treatment by authors of
detached leaves of Angiosperms which in many
instances are referred to recent genera on wholly
inadequate grounds, but the more trustworthy
nature of the material on which species of
Abietineous cones are founded deserves careful
consideration and would probably yield results
of considerable importance.

Fig. 782. Pityostrobus
(Pinites) Andraei.
(After   Coemans:
nat. size.)

Pityostrobus (Pinites) macrocephalus (Lindley and Hutton).

This species, founded on a cone 12 cm. long and 6 cm. in
diameter, was in the first instance described by Lindley and
Hutton[2] from an account furnished by Prof. Henslow and named
Zamia macrocephala; it was found near Dover and believed to be
derived from the 'Greensand formation.' A second specimen from
Faversham in Kent was described by the same authors as Zamia
ovata[3]. Endlicher[4] assigned the cones to Zamiostrobus and Miquel[5]

[1] Heer (69) p. 13, Pl. II. fig. 11.   [2] Lindley and Hutton (35) A. Pl. cxxv.
[3] Ibid. (37) A. Pl. ccxxvi A.   [4] Endlicher (40) p. 72.   [5] Miquel (42) p. 75.

proposed the name *Z. Henslowi*: their Abietineous nature was first
recognised by Corda[1], and Carruthers[2] subsequently gave some
account of the internal structure and employed the generic name
Pinites. The discovery of additional specimens *in situ* enabled
Carruthers to assign *P. macrocephalus* to Eocene beds at the
junction of the Woolwich and Thanet beds with the London Clay
Both Carruthers and Gardner[3] retain both specific names, but an
examination of the specimens convinces me that there are no
differences worthy of specific recognition. The following brief
account is based on an examination of sections in the British
Museum and in part on notes supplied by Mr Dutt of Queens'
College, Cambridge, who is preparing a fuller account of the
material[4]. The cones are ovoid-cylindrical and obtuse; the
weathered surface (fig. 783) shows slightly convex polygonal areas
without any trace of a central umbo. The axis is slender in comparison with that of most recent species of *Pinus*; the stele includes
a fairly large pith of thick-walled cells surrounded by a vascular
cylinder in which foliar gaps are formed by the exit of the double
sporophyll-traces. It is noteworthy that no resin-canals occur in
the xylem. A ring of large resin-canals lined with thin-walled
epithelial cells occurs outside the phloem. The cone-scales are
given off almost at right-angles and then bend sharply upwards
and become slightly broader near the surface of the cone (fig.
784, B). In one section a portion of a subtending bract-scale was
recognised. The seminiferous scales are composed of thick-walled
cells and contain idioblasts like those in Araucarian leaves, also
resin-canals: two ovules occur in a depression near the base of the
scales. The sporophyll-trace divides in the scale into several bundles,
and in places there are indications of a second series of inversely
orientated strands. The comparatively large ovules, nearly 1 cm.
long, are attached by a short stalk, and in places the remains of a
wing can be seen. Although the integument is thick and lignified
and the micropyle closed there are no embryos and no indication
of archegonia in the partially preserved nucellar tissue. In the

[1] Corda in Reuss (46) B.
[2] Carruthers (66[2]) pp. 536, 540, Pl. XXI.
[3] Gardner (86) pp. 63, 65, Pl. XIV.
[4] Dutt (16).

ovule shown in fig. 784, A, the contracted nucellus, n, forms a cylindrical column which presents a misleading resemblance to the

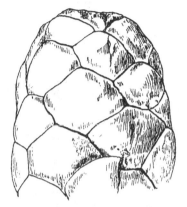

FIG. 783. *Pityostrobus (Pinites) macrocephalus.* (After Gardner; nat. size.)

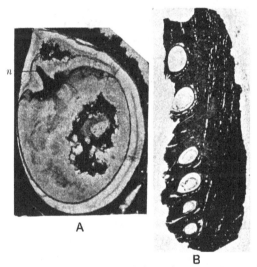

A

B

FIG. 784. *Pityostrobus (Pinites) macrocephalus.* A, section of an ovule; n, nucellus. B, longitudinal section of part of the cone. (From sections in the British Museum; B, slightly reduced.)

prothallus tent-pole of *Ginkgo* ovules: at its blunt apex are two winged pollen-grains. Prothallus-tissue is also represented.

Carruthers compared the species with *Pinus Pinaster* but the surface-features are more like those of the cones of *Pinus excelsa* (*cf.* fig. 704, p. 154) and similar types.

Pityostrobus (*Pinites*) *Plutonis* (Baily).

This species was founded by Baily[1] on part of a cone from the plant-beds in the basalts of Antrim and described, in greater detail

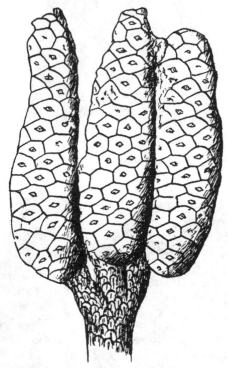

FIG. 785.   *Pityostrobus* (*Pinites*) *Plutonis*.   (After Gardner; nat. size.)

and from better material from the same locality, by Gardner[2]. The cones are 7—9 cm. long and 2—3 cm. in diameter, characterised by woody scales with sub-hexagonal apophyses with a central umbo and a rounded upper margin; there are two seeds with long and narrow wings on each scale. In one case (fig. 785) three cones are attached in an erect position to a branch covered with persistent leaf-bases. The foliage-leaves were borne in pairs and reached a length of 10—15 cm. A similar type with shorter needles from the

[1] Baily (69) Pl. xv. fig. 1.   [2] Gardner (86) p. 69, Pls. xv.—xviii.

same locality is named by Gardner *Pinus Bailyi*[1]. These Irish specimens agree in the cones and foliage-spurs with such recent species as *Pinus halepensis* and *P. Pinaster*, but the apparently erect position of the cones of the fossil type is a distinctive feature.

Tertiary cones similar to *Pityostrobus Plutonis* are illustrated by *Pinus robustifolia* Sap.[2] from Provence, *P. Kotschyeana* (Ung.), originally described by Unger and recorded by Tuzson[3] from Hungary, *P. transsylvanica* Pax[4], a North American form, which the author of the species compares with *Pinus Balfouriana*, and *P. prae-montana* described by Mogan[5] from Lower Austria.

Pityostrobus (Pinites) palaeostrobus (Ettingshausen).

This type originally described by Ettingshausen[6] from Häring in the Tyrol is recorded from many Tertiary localities. The cones are ovate sub-cylindrical with scales of the *Pinus Strobus* form and the needles are borne in fascicles of five. Heer[7] refers to this species some thin and long needles from the Miocene of Greenland, but it is not clear that the needles are in fives. The species is recorded from Hungary[8], Germany, France, and elsewhere. A cone of similar form is described by Unger[9] from the Oligocene of Kumi as *Pinus megalopis* and it is associated with quinary fascicles.

Pityostrobus MacClurii (Heer).

This species described by Heer[10] as *Pinus (Abies) MacClurii* from Miocene beds in Banks Land, lat. 74° 27′ N. is represented by a specimen in the Dublin Museum. The narrow oval cone, 6 × 1·5 cm., consists of imbricate scales with the upper margin rounded or irregularly truncate: some of the scales show indications of a pair of seeds. Though similar to cones of *Picea*, the fossil cannot be definitely assigned to any recent genus. Similarly, Miocene specimens of cones, scales, and leaves from Spitzbergen referred by Heer[11] to *Pinus Abies* L. do not afford satisfactory evidence of their generic position.

---

[1] Gardner (86) p. 73.
[2] Saporta (73) p. 94, Pl. II.
[3] Tuzson (09[2]) p. 240, Pls. XIV., XV.
[4] Pax (07) p. 310.
[5] Mogan (03) figs. 1—3.
[6] Ettingshausen (55).
[7] Heer (83) Pl. LXX. fig. 8; Pl. LXXXVII. figs. 5, 6.
[8] Staub (85).
[9] Unger (67) Pl. XVI.
[10] Heer (68) i. p. 134, Pl. XX. figs. 16—18.
[11] *Ibid.* (71) iii. p. 41, Pl. V. figs. 35—49.

While most of the Tertiary species of *Pityostrobus* agree closely with recent types some exhibit more or less striking peculiarities. A species described by Engelhardt and Kinkelin as *Pinus Timleri*[1] from Pliocene beds near Frankfurt is founded on pieces of large cones characterised by cone-scales with a conical distal end having 3 to 5 flat surfaces (fig. 786, A). The authors compare it with *Pinus Gerardiana* from Afghanistan.

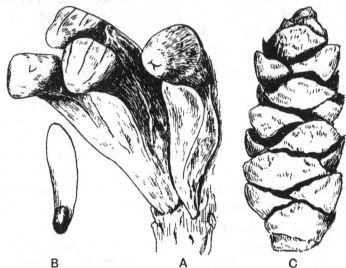

B   A   C

FIG. 786.   A, B, part of a cone and a seed of *Pinus Timleri*.   C, *Keteleeria Loehri*.   See page 370.   (After Engelhardt and Kinkelin; nat. size.)

The material obtained from Pleistocene beds is often well enough preserved to afford trustworthy data with regard to the later geological history of different genera.   Clement Reid[2] identified a small cone from the Cromer Forest bed as *Pinus silvestris*, and from this horizon in Sweden Nathorst[3] records the same species; it is recorded also from Pleistocene deposits associated with *Elephas primigenius* in France[4] and similar evidence has been obtained from Switzerland, Germany, Denmark, and other regions.   At a later date the former range of *Pinus silvestris* and other types is illustrated by the

[1] Engelhardt and Kinkelin (08) p. 205, Pl. xxv. figs. 1—4.
[2] Reid, C. and E. M. (15) Pl. i. figs. 7, 8.
[3] Nathorst (10) p. 1354.   [4] Fliche (00).

evidence of submerged forests and, as we ascend the scale, the records become more legible and the prehistoric merges into the historic era.

A cone apparently identical with the Spruce Fir (*Picea excelsa*) found in the Pre-glacial beds on the Norfolk[1] coast (fig. 787) is a relic of the flora which existed in England when the Rhine after receiving 'many large tributaries—now separate rivers—seems to have flowed across the present bed of the North Sea.' The same species is recorded from Pliocene beds on the Dutch-Prussian frontier, also from the valleys of the Main and Neckar, the specimens from the latter locality being referred by Glück[2] to *Picea excelsa* var. *alpestris*. Sernander[3] has discussed the past history of *Picea* in Scandinavia and quotes records of the occurrence of the genus in other parts of Europe. Similar instances of the wider range of Abietineous genera are given by Berry[4] and other authors who have described Pleistocene plants in North America. From the facts at present available it would seem that *Pinus* and allied genera were more abundantly represented in the Tertiary and Post-Tertiary floras in Europe than in American strata of the same age.

FIG. 787. *Picea excelsa*. From Pre-glacial beds at Mundesley, Norfolk. (After Reid; nat. size.)

PITYANTHUS. Nathorst.

Pityanthus granulatus (Heer). This species, described by Heer[5] from the Patoot (Cretaceous) beds in Greenland as *Ophioglossum granulatum* and afterwards described by Newberry[6] from the Amboy clays, has recently been identified by Dr Stopes[7] as a long microstrobilus of some Abietineous Conifer, probably a *Pinus*.

[1] Reid, C. and E. M. (08) Pl. xv. fig. 147.   [2] Glück (02).
[3] Sernander (93).   See also Andersson (10) and W. B. Wright (14) for excellent summaries of Pleistocene history.
[4] Berry (07); (10³); Penhallow (04).   [5] Heer (83) Pl. LVII. figs. 8, 9.
[6] Newberry and Hollick (95) Pl. IX. figs. 11—13.   [7] Stopes (11⁴) text-figs. 1, 2.

This author examined the American specimen, which she regards
as a fertile shoot of a three-needled Pine: the strobilus is 35 mm.
long and from it winged pollen-grains were isolated. In its unusual
length the strobilus resembles the male flowers of *Pinus australis*
from Florida.

PITYOSPERMUM. Nathorst.

The few specimens chosen for description afford examples of
some of the oldest records of fossils, agreeing in the form of the wing
with recent Abietineous seeds and, as far as I know, none have been
discovered in strata below the Rhaetic. From Tertiary rocks
numerous winged seeds are recorded, but these are of no special
interest and they are usually accompanied with foliage-shoots,
cones, or other fossils which afford more trustworthy data as to
relationship.

Pityospermum Lundgreni Nathorst.

Nathorst described several winged seeds from the Rhaetic beds
of Scania as *Pinus Lundgreni*[1]; they are 9—11 mm. long and 4 mm.
broad, the actual seeds being 3—4 mm. in
length. Two examples from Stabbarp in the
Stockholm Museum are represented in fig.
788. To the same species Nathorst referred
some imperfect cylindrical cones bearing
thin imbricate scales and reaching a length
of 3—5 cm. and a diameter of 1·2—2 cm.;
he also suggested the possibility that some
short shoots and long needle-like leaves described as *Schizolepis Follini* Nath. may

FIG. 788. *Pityospermum
Lundgreni.* From Stab-
barp in Scania; Rhaetic.
(Stockholm   Museum;
nat. size.)

belong to the plant which bore the cones and seeds. In a later
account of *Schizolepis*[2] he expressed the opinion that in the
absence of any proof of actual connexion the leaves and short
shoots should be separated from *Schizolepis* and included in
Pinites. These leaf-fascicles are described under the name *Pityophyllum* and the seeds, which occur as separate fossils, are alone
included in *Pityospermum Lundgreni*. The striking resemblance
of the seeds to those from Franz Josef Land (fig. 789) and recent

---

[1] Nathorst (78) B, p. 31, Pl. xiv. figs. 9 a, 13—17; Pl. xv. figs. 1—2.
[2] *Ibid.* (97) p. 38.

Abietineous seeds is a valid reason for suggesting the inclusion of the Rhaetic specimens in the Abietineae, though it would be going too far to conclude that the seeds were borne on cones generically identical with or even closely related to those of any existing representative of the family. A Pliocene seed figured by Engelhardt and Kinkelin[1] as *Pinus Timleri* (fig. 786, B) bears a close resemblance in the form of the wing to some of the Rhaetic specimens.

Pityospermum Nilssoni Nathorst.

This species, also from the Rhaetic flora of Scania[2], is characterised by the much longer wing (2·7 cm.) which in size and form differs much more widely than *Pityospermum Lundgreni, P. Nanseni*, and other Jurassic types from the wings of any recent seeds.

Pityospermum Nanseni Nathorst.

The seed shown in fig. 789, A, 11 mm. long, is drawn from a specimen in the Museum of the Geological Survey (Jermyn Street) collected by Dr Koettlitz in Franz Josef Land and of Upper Jurassic or Wealden age: this and other seeds are figured by Newton and Teall[3]. The name *Pityospermum Nanseni* was applied by Nathorst[4] to similar specimens obtained by Dr Nansen from the same region. Other winged seeds from Franz Josef Land closely resemble Heer's species *Pinus Maakiana*[5] from Jurassic rocks in Siberia. A seed, 1·2 cm. long, from Wealden beds in the South of England is reproduced in fig. 789, B[6]: this is possibly a distinct species, but the specific determination of separate seeds of this form is of little value unless the differences are well marked. The important point is the striking resemblance between such seeds as those shown in figs. 788, 789

A   B

FIG. 789. A, *Pityospermum Nanseni.* B, *Pityospermum* sp. (A, drawn from a specimen in the Museum of the Geological Survey figured by Newton and Teall; B, from a specimen, V 2323, in the British Museum from Wealden rocks.)

[1] Engelhardt and Kinkelin (08) Pl. xxv. fig. 4.
[2] Nathorst (78) B. p. 32, Pl. xv. figs. 17—19.
[3] Newton and Teall (97) Pl. xxxviii.
[4] Nathorst (99) p. 18, Pl. ii. figs. 12. 13.
[5] Heer (77) ii. Pl. xvi. fig. 1.   [6] Seward (95) A. p. 198.

and seeds of recent Pines and other Abietineae. It is not possible
to determine the precise generic affinity of seeds of this type, but
their practical identity with recent Abietineous seeds warrants
their reference to that family.

PITYOSPORITES. Seward.

This generic name has been adopted[1] for spores, provided with
bladder-like extensions of the exine, agreeing in size and form with
those of recent Abietineous genera. Winged pollen occur also in
the Podocarpineae, but the fossil examples so far recorded are much
more like the microspores of Abietineous genera than those of
Podocarpus, Dacrydium and *Microcachrys.*

Pityosporites antarcticus Seward.

In the course of examining sections of wood collected by
Mr Priestley on the Priestley Glacier (approximately lat. 74° S.) I
noticed two small microspores in the siliceous matrix of the partially
decayed stem[2]: one is shown in fig. 491, G (Vol. III. p. 298); the
longest axis is 80μ and the central part bears two bladders characterised by a fine surface-reticulation similar to that on recent spores.
A microspore of *Pinus silvestris* has a length of 75μ. It is very
unlikely that the spores have any connexion with the stem in which
they are preserved; they bear a much closer resemblance to the
microspores of Abietineous genera than to the spores of the Podocarpineae: the probability is that the Antarctic specimens belong
to some Abietineous Conifer though this cannot be definitely stated.
It is probable that the upper part of the Beacon Sandstone, from
which the boulder containing the fossil is believed to have been
derived, is not older than Lower Mesozoic, *e.g.* Rhaetic.

Pityosporites sp.

Among the spores found by Nathorst[3] in Liassic clay from Hör
in Scania were several winged microspores, one of which is reproduced in fig. 790, C from a photograph kindly supplied by Prof.
Nathorst. The length of the spore is about 100μ and in the
shape of the bladders it agrees closely with the microspores of
Picea excelsa[4].

[1] Seward (14) p. 23.   [2] *Ibid.* p. 23, Pl. VIII. fig. 45.
[3] Nathorst (08) p. 13, Pl. II.
[4] Kirchner, Loew, and Schröter (06) p. 151, fig. 68.

Pityosporites sp.

In his account of petrified plant-remains from Franz Josef Land, probably of Wealden or approximately Wealden age, Graf Solms-Laubach[1] mentions the occurrence of well preserved pollen with bladders and figures a piece of a cone with flat scales similar

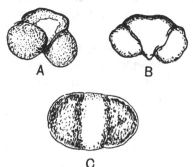

C

FIG. 790. A, B, *Pityosporites* sp. from Franz Josef Land. C, *Pityosporites* sp. from Scania. (A, B, from specimens in the Museum of the Geological Survey; C, after Nathorst.)

to that of a *Picea*. The drawings reproduced in fig. 790, A, B were made from specimens found in sections of the material examined by Solms-Laubach in the Geological Survey collection. The longest diameter is 70—90μ; in form and size the microspores resemble those of recent Pines. The apparently broad wall shown in fig. 790, A is the result of the obliquity of the section.

ENTOMOLEPIS. Saporta.

Entomolepis cynarocephala Saporta. Saporta[2] instituted this generic name for some cones from the Oligocene plant-beds of Provence 8—10 cm. long, ovate-elliptical, and composed of spirally arranged coriaceous scales not thickened at the apex but prolonged beyond the imbricate broad portion into a long recurved, acuminate and fimbriate, spinous process. No seeds have been found and there is no evidence as to internal structure. Saporta considers the cones to belong to some extinct type and, as Zeiller[3] says, they are probably Abietineous.

---

[1] Solms-Laubach (04) p. 11.

[2] ἔντομος, cut up; λεπίς, scale. Saporta (65²) p. 55, Pl. II. fig. 3.

[3] Zeiller (00) B. p. 278.

CROSSOTOLEPIS.   Fliche.

Crossotolepis Perroti Fliche.   This generic name[1] was proposed
for an imperfectly preserved cone, from Oligocene beds near
Embrun in the French Alps, of elongate-cylindrical form, 13·3 cm.
long and 3·5 cm. in diameter, characterised by the fimbriate edge
of the imbricate, highly inclined, scales, which bear two seeds.
The cone agrees closely with several recent Abietineous types
especially with *Picea Menziesii* and other species of *Picea*, but is
distinguished by the deeply fimbriate upper margin of the thin
cone-scales; it is difficult to determine how far this feature is the
result of secondary causes: Fliche believes it to be an original
character comparable with that which led Saporta to found the
genus *Entomolepis* for an Oligocene cone from Armissan in Provence.
It is not certain whether the seeds are winged.   Fliche is no doubt
correct in his conclusion that *Crossotolepis* is an Abietineous cone
closely allied to *Picea* and probably related to *Entomolepis*.   He
refers the two Oligocene genera to the Abietineae; they differ
from any recent forms in the greater dissection of the distal edges
of the seed-bearing scales, which in this respect are comparable
with the more feebly lobed scales of the cones of *Picea Engelmanni*
and other species.   Our knowledge of both genera is, however,
meagre and all that can be said is that the type-specimens afford
some evidence of the former occurrence of some Tertiary Abietineous
Conifers distinguished by the distally dissected scales from any
recent types.

PREPINUS.   Jeffrey.

This genus was instituted[2] for lignitic specimens of short shoots
and leaves from Middle Cretaceous beds on Staten Island, N.Y.
characterised by the large and indefinite number of leaves borne
on a single short shoot, the presence of a basal sheath of scale-
leaves, and by certain anatomical features, particularly the mesarch
structure of the single leaf-bundle, the occurrence of a complex
system of transfusion tissue, and other features.

---

[1] κροσσωτόs, fringed, tasselled.   Fliche (99) p. 474, Pl. xii.
[2] Jeffrey (08²) Pls. xiii., xiv.; Hollick and Jeffrey (09) B. p. 19, Pls. ix.,
xxii.—xxiv.

Prepinus statensis Jeffrey.

The short shoots (fig. 791, B), rather less than 1 cm. long, consist of a relatively broad axis bearing on the upper part numerous spirally disposed truncate portions of leaves, in some cases

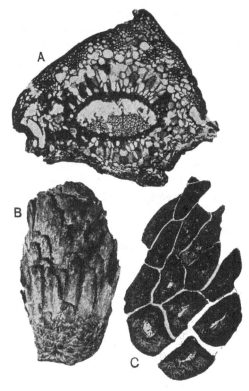

Fig. 791. *Prepinus statensis.* A. Transverse section of a detached leaf believed to belong to *P. statensis.* B. Short shoot showing the basal portions of needles and, below, the scars of scale-leaves. C. Transverse section through part of a leaf-fascicle. (After Jeffrey; A, × 30; B, × 7; C, × 9½.)

more than 20 on a single shoot. Scars of scale-leaves are represented by rhombic areas near the base. There is no evidence as to the nature of the branches on which the foliage-spurs were produced. The leaves are polygonal in section (fig. 791, A, C): there are two marginal resin-canals and a single median vascular bundle as in some recent Pines. There is a considerable development of sclerous tissue in the ground-tissue of the leaf and a complete absence of mesophyll with infolded walls like that of recent and some fossil Pines (*cf.* fig. 774). The phloem is represented by a crescentic space in the leaf shown in fig. 791, A. The dark zone surrounding the bundle consists of thick-walled and relatively long transfusion-tracheids and external to these is a broader sheath of short transfusion-tracheids, but there is no endodermal layer and no admixture of parenchyma with the tracheids. The xylem is composed partly of centripetal and in part of centrifugal elements : the centrifugal xylem forms an uninterrupted arc next the phloem, and between this and the transfusion-tissue on the lower side of the bundle seen in fig. 791, A the centripetal xylem is represented by radial rows of tracheids separated by spaces. Spiral proto-xylem elements occur between the two groups of metaxylem. It is interesting to find similar transfusion-tissue in some leaves of true Pines described by Jeffrey from the same beds, but their bundles are double and composed of centrifugal xylem only as in modern species.

The pith of the axis of a *Prepinus* shoot contains nests of sclerous cells: the leaf-traces pass through the cortex as single bundles, and the single ring of wood contains a row of resin-canals blocked by tyloses. The tracheids have uniseriate bordered pits which are often contiguous and separated by Sanio's rims.

Jeffrey compares this species with *Pinites* (=*Pityites*) *Solmsi* Sew.[1] (fig. 772) from English Wealden beds and with shoots described by Fontaine[2] from the Potomac series under Heer's generic name *Leptostrobus*.

Prepinus viticitensis Jeffrey.

This species[3] was founded on specimens of short shoots from the Lower Cretaceous clays of Gay Head, Martha's Vineyard, Massachusetts, which are considered to be closely allied to *Pityites Solmsi*. The wood of the axis, representing a single year's growth, contains two series of resin-canals and, as in the type-species, the canals of the leaves are in continuity with those in the cortex of the shoot-axis, whereas in recent Pines the leaf-canals end blindly.

---

[1] Seward (95) A. p. 196, Pls. XVIII., XIX. See page 373.
[2] Fontaine (89) B. p. 227, Pls. CI.—CIV. etc.   [3] Jeffrey (10).

As the choice of the name *Prepinus* implies, Jeffrey regards the type of foliage-shoot represented by these species as the direct ancestor of the leaf-spurs of recent Pines. The short shoots of *Prepinus* are smaller than those of existing species of *Pinus*, but in the numerous and spirally arranged leaves they resemble those of *Cedrus* and *Larix*. Anatomically the fossil leaves differ widely from any Abietineous types, and were it not for the occurrence of true Pine needles in association with *Prepinus*, which to some extent bridge the gap between *Prepinus* and *Pinus*, one might be sceptical with regard to the close affinity of *Prepinus* to recent Pines. Jeffrey compares the structure of the leaf of *P. statensis* with that of some leaves of *Cordaites*, but the agreement is probably not so close as Dr Stopes' description[1], quoted by Jeffrey, suggests. Jeffrey[2] regards the short shoots of *Pinus* and other Abietineae as a primitive attribute of the Coniferous stock and as one of several reasons for believing the Abietineae to be the oldest tribe of Conifers. Prof. Thomson[3] has recently discussed the value of the evidence based on the short shoots of *Pinus* and *Prepinus* and comes to the conclusion that the foliage-spurs of *Pinus* are specialised shoots and do not belong to the category of primitive forms. In *Cedrus*, *Larix*, and *Pseudolarix* the leaves are spirally disposed on the short shoots, while in *Pinus* they are fewer and cyclic. The frequent occurrence of more than the normal number of leaves on the foliage-spurs of *Pinus* has already been mentioned: in healthy plants supernumerary foliage-leaves are not uncommon and an increase in the number of needles is also induced by wounding. The spirally arranged scale-leaves below the whorled leaves on a short shoot of *Pinus* are homologous with the scale-leaves on ordinary branches, and on seedling Pines they are replaced by the primordial leaves: transitional forms occur between these three forms of leaf. The persistent short shoots of *Cedrus*, *Larix*, and *Pseudolarix*, as also of *Ginkgo*, are regarded as the more primitive condition as compared with the deciduous nature of the cyclic foliage-shoots of *Pinus*. Thomson notes that short shoots of *Pinus* may proliferate like those of *Cedrus* and *Larix*. He concludes that ancestrally 'the leaves of the Pines were spirally arranged on ordinary branches and that the spur is derived from

---

[1] Stopes (03).   [2] Jeffrey (10²) p. 331.   [3] Thomson (14).

this condition.' The fossil shoots from the Potomac series described as *Leptostrobus longifolius* differ in the larger number of the needles from modern Pines and resemble abnormal short shoots of *Pinus excelsa*[1] produced by wounding. The short foliage-shoot of *Prepinus* furnishes a more completely known example of a branch bearing spirally disposed leaves. In view of the palaeontological evidence and of the facts obtained from a study of recent Pines it would seem that, as Thomson holds, the present form of the Pine spur is the result of specialisation and not a primitive feature.

[1] Thomson (14), Pl. XXII. fig. 10.

CHAPTER XLIX.

PODOCARPINEAE.

THE data on which to base any conclusions as to the antiquity or former distribution of the genus *Podocarpus* or of Conifers believed to be closely allied to recent Podocarps are unfortunately derived from records which in the majority of cases are far from satisfactory and consist mainly of detached leaves. Velenovský[1] refers some linear leaves from the Perucer series of Bohemia to *Podocarpus* (*P. cretacea*), but they exhibit no distinctive characters. Some of the numerous leaves described as species of *Podocarpus* are in all probability correctly regarded as Tertiary representatives of the recent genus, but it is often impossible to state with any confidence that detached leaves should be referred to *Podocarpus*, or preferably to *Podocarpites*, rather than to a more comprehensive genus such as *Taxites* or *Elatocladus*. In spite of the fragmentary nature of the evidence and the fact that no undoubted example of a *Podocarpus* fertile shoot has been discovered, an examination of the published records leads to the conclusion that in Tertiary floras, particularly in those of Eocene age, species closely allied to existing Podocarps were abundant in Europe, a conclusion that is especially interesting from the point of view of the present geographical distribution of the Podocarpineae. A brief account of some of the better known examples of Tertiary species of *Podocarpus*, which are transferred to the genus *Podocarpites* in accordance with the practice usually adopted in the case of fossil species, may serve to illustrate the nature of the material and the wide range of the specimens.

Dr Guppy[2] in his very suggestive remarks on the present distribution and means of dispersal of *Podocarpus* writes: If we assign a home in the high latitudes of the northern hemisphere to a genus that was well represented in Europe in the Tertiary period, a movement of migration southward would explain most of the difficulties in the present distribution. The great vertical range

---

[1] Velenovský (85) B. Pl. XII. figs. 5—11.   [2] Guppy (06) p. 302.

of some of the species leads us to attribute a corresponding power of adaptation to the genus in respect of widely different climates.... With such a capacity for adaptation, migration of the genus would be rendered easy over the globe.' The geological history of the Podocarpineae is unfortunately very fragmentary but such data as are available lend support to the view that there was 'a centre of diffusion in the extreme north,' the present distribution of the

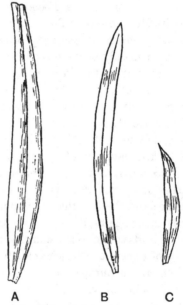

A   B   C

Fig. 792. *Podocarpites eocaenica.*
(A, after Heer; B, C, after Gardner; nat. size.)

family being as in the case of the Araucarineae the result of migration from other parts of the world where the plants are now represented only in the floras of a bygone age.

PODOCARPITES. Andrae.

Podocarpites eocaenica (Unger).

This species, from Eocene beds in South Styria, was founded by Unger[1] on detached sub-falcate leaves 3—12 cm. long (fig. 792, A) with a short petiole or a more or less sessile lamina.

[1] Unger (51) p. 158, Pl. xxiii. figs. 11—16.

Gardner[1] has described examples of this species from the Middle
Bagshot beds of Bournemouth, the Lower Bagshot of Alum Bay
in the Isle of Wight, and from other British localities. A good
impression from Bournemouth is reproduced in fig. 792, B: the
leaf is 7 cm. long and 3 mm. broad, the apex is sharply pointed
and there is no petiole. Fig. 792, C shows a leaf of slightly different
form which may belong to the species. No reproductive organs
have been found. Ettingshausen[2] records specimens of this type
from Eocene beds at Häring in the Austrian Tyrol where the
species is said to be abundant, from Leoben[3] in Styria, Bilin[4] in
Bohemia, Sagor in Carinthia[5], and elsewhere. Heer[6] states that
the species is common in Swiss Eocene deposits and Engelhardt[7]
describes examples from Oligocene rocks in Bohemia. A similar
or perhaps specifically identical type was figured by Lindley[8] from
Eocene plant-beds at Aix in Provence as *Podocarpus macrophylla*,
but Saporta[9], who figured additional specimens from the same
locality, proposed the name *Podocarpus Lindleyana* on the ground
that Lindley's designation implies identity with a recent species.
A leaf referred to *P. eocaenica* is figured by Massalongo[10] from
Tertiary beds in Italy.

Some vegetative shoots originally described by De la Harpe[11]
from Alum Bay as *Cupressites elegans* are referred by Gardner[12] to
Podocarpus and this determination derives support from the variety
in the foliage illustrated by his specimens: in some branches the
linear leaves are two-ranked while in others the leaves are spirally
disposed and three-sided, a diversity met with in recent species.
The imperfectly preserved fragment reproduced in fig. 793 from a
careful drawing by Miss Woodward of the actual specimen is
figured by Gardner as a fertile branch bearing a single seed with

---

[1] Gardner (86) p. 48, Pl. II. figs. 6—15.
[2] Ettingshausen (55) p. 37, Pl. IX. figs. 14, 15.
[3] *Ibid.* (88²) p. 277.
[4] *Ibid.* (67²) p. 118, Pl. XIII. figs. 1, 2.
[5] *Ibid.* (85) p. 6, Pl. XXVIII. fig. 12.
[6] Heer (55) A. p. 53, Pl. XX. fig. 3.
[7] Engelhardt (85) p. 315, Pl. VIII. figs. 37, 38.
[8] Lindley in Murchison and Lyell (29) p. 298, fig. A.
[9] Saporta (62) p. 216, Pl. II. fig. 7.
[10] Massalongo (59) p. 166, Pl. V. fig. 36.
[11] De la Harpe in Bristow (62) p. 111, Pl. V. fig. 3.
[12] Gardner (86) Pl. VIII.

a fleshy base as in certain existing species, but the details are too
indistinct to afford any proof of affinity to *Podocarpus*. Gardner also describes a globose wrinkled
seed, 16 mm. in diameter, as ?*Podocarpus argillae-
londinensis*[1] from the London Clay which bears
a close resemblance to the seeds of *Podocarpus
elata*. The specimens from Eocene beds in the
Island of Mull described by Gardner[2] as *Podocarpus borealis*, consisting of small falcate leaves
and seed-like bodies, are too imperfect to be
determined with accuracy. Fig. 794, C, C' represents a type from Bournemouth described as
Podocarpus incerta which differs from other species

FIG. 793.   *Podocar-
pites elegans.* Sup-
posed fertile shoot.
(Specimen in the
British   Museum
figured by Gard
ner.)

in the absence of a definite midrib; the linear-lanceolate coriaceous
leaves, reaching a length of 3 cm., are decurrent and appear to have
several parallel veins, a feature characteristic of the section *Nageia*:

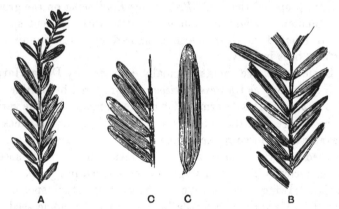

A   C   C   B

FIG. 794.   A, B, *Sequoiites Tournali* (see p. 353).   C, C', *Podocarpites incerta.* (From
specimens in the British Museum described by J. Starkie Gardner; A, B, V. 524;
C, V. 522; nat. size.)

it is, however, doubtful whether this species should be included in
a genus implying affinity to *Podocarpus*.   Some leaves figured by
Schmalhausen[3] from Oligocene strata in Russia as species of
Podocarpus are too imperfect to afford any trustworthy evidence
as to the occurrence of the genus *Podocarpites*.   Ettingshausen[4]

[1] Gardner (86) p. 52, Pl. IX. figs. 35, 36.   [2] *Ibid.* (87) A. Pl. XIII. figs. 3—11.
[3] Schmalhausen (83²) Pl. XXXII.   [4] Ettingshausen (86) Pl. VIII. figs. 25—27.

describes as *Podocarpus prae-cupressina* foliage-shoots and a very imperfectly preserved seed from Eocene rocks in New South Wales: the same author records *Podocarpus Parkeri*[1], a doubtful species, from New Zealand.

Podocarpites Campbelli Gardner.

Shoots bearing leaves similar to those of *P. eocaenica* are described by Gardner from the Eocene plant-beds in the basalts

Fig. 795. *Podocarpites Campbelli.* (After Gardner; nat. size.)

of Mull as *Podocarpus Campbelli*[2] (fig. 795).   The linear acuminate straight or slightly curved leaves with a contracted decurrent base are about 7 cm. long and possess a well-defined midrib.   Gardner states that the late Prof. Oliver regarded the fossils as the branches of some Podocarp; they are compared with *Podocarpus falcata* of the Cape and Tropical Africa and with the South African species *P. Thunbergii.*

[1] Ettingshausen (87) Pl. I. figs. 12—14.
[2] Gardner (86) p. 97, Pl. xxvi.

Fossils believed to be related to DACRYDIUM.

The records of the rocks afford very little information with
regard to the past history of Conifers allied to the recent genus
Dacrydium. The marked dimorphism of the foliage-shoots (fig. 708,
p. 160), their close resemblance to branches of some other Conifers,
as also to Lycopodiaceous plants and some of the larger Mosses,
are serious difficulties in the way of recognising representatives of
this genus among impressions of vegetative branches. It is interesting to find that the most promising piece of evidence of the
occurrence of a fossil type (*Stachyotaxus*) allied to *Dacrydium* is
furnished by a Rhaetic flora, a fact pointing to a high antiquity
of the plan of reproductive shoot characteristic of existing species.

Schenk[1] compares with *Dacrydium* some obscure and small
fragments from the Coal Measures of China which he described as
Conchophyllum Richthofeni, but there are no substantial grounds
for such comparison. The specimens consist of pieces of slender
axes bearing spirally disposed bracts or small leaves showing at the
base of the ovate-oblong lamina a slight depression from which a
seed may have fallen. The Lower Cretaceous foliage-shoots from
Bohemia described by Velenovsky[2] as *Dacrydium densifolium* have
no claim to be accepted as branches of a Podocarpineous Conifer.
Ettingshausen[3] figures from Eocene beds in Australia and New
Zealand sterile twigs assigned respectively to *Dacrydium cupressi-
noides* and *D. prae-cupressinum*: in neither case is there any evidence as to the nature of the reproductive organs, and the form
of the foliage-shoots might with equal probability be interpreted
as evidence of other Conifers or of some Lycopodiaceous plant.

STACHYOTAXUS. Nathorst.

Stachyotaxus elegans Nathorst.

The genus *Stachyotaxus*[4] was instituted for some Rhaetic
specimens from Scania originally named by Agardh *Sargassum
septentrionale* and *Caulerpa septentrionalis*; the former was re-
named by Nathorst *Carpolithes septentrionalis* and the latter

[1] Schenk (83) A. p. 223, Pl. XLII. figs. 21—26.
[2] Velenovský (85) B. Pl. XII. figs. 1—2.
[3] Ettingshausen (86) Pl. VIII. figs. 23, 24; (88) Pl. I. fig. 19.
[4] Nathorst (86) p. 98.

Cyparissidium septentrionale. The discovery of additional material led Nathorst to transfer some of the specimens to a new genus *Stachyotaxus.* The foliage-shoots of *Stachyotaxus elegans*[1] are dimorphic; some of the leaves are appressed and imbricate as in

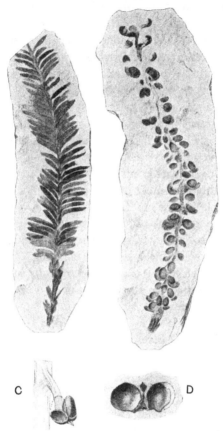

FIG. 796. *Stachyotaxus elegans.* A, B, ⅔ nat. size; C, slightly enlarged; D, × 3½. (After Nathorst.)

Cyparissidium and some other Conifers while others are linear and distichous (fig. 796, A), sessile and decurrent, with a lamina reaching a length of about 1 cm. The epidermal cells have straight walls and the stomata occur in two rows on the lower surface. The megastrobili (fig. 796, B) have the form of spikes about

[1] Nathorst (08[2]) p. 11, Pls. II., III.

5—6 cm. long; the fairly stout axis bears sporophylls, approximately at right-angles, consisting of a short and relatively thick
stalk expanded into a triangular scale bearing two ovate seeds,
3—3·5 mm. long, each being enclosed basally in a cupule (fig. 796,
C, D): the distal end of each sporophyll forms an upturned acuminate apex. The cuticle of the seed-coat shows that the latter
consisted of thick-walled cells, and within the testa Nathorst found
the remains of a much more delicate membrane, possibly representing the nucellus. Nathorst compares the sporophylls with
those of recent Dacrydiums, though in *Stachyotaxus* there are
normally two seeds on each sporophyll and not one as in the recent
genus, a difference possibly of no great importance. In habit
the fertile shoots of the fossil type are comparable with those of
Podocarpus spicata. Miss Gibbs[1] in her account of recent Podocarpineae expresses agreement with Nathorst's view that *Stachyotaxus* is probably a member of that family. Nathorst describes a
second Swedish species but from a slightly lower horizon in the
Rhaetic series. This species, *Stachyotaxus elegans,* is characterised
by longer and stouter megastrobili reaching at least a length of
12 cm. and by longer linear leaves 10 cm. long.

Hartz[2] refers to *Stachyotaxus septentrionalis* some sterile shoots
from Lower Jurassic, or Rhaetic, beds in East Greenland, and
Halle[3] draws attention to the superficial resemblance to the
Swedish type of some vegetative twigs from Graham Land which
he refers to the genus *Elatocladus.*

Strobilites. *Strobilites Milleri* Seward and Bancroft.

The specimen on which this species is founded[4] was obtained
by Hugh Miller from Upper Jurassic beds on the North-East coast
of Scotland and inaccurately figured in the *Testimony of the Rocks*[5].
Fig. 797 is from a careful drawing by Mr T. A. Brock of the original
specimen in the Edinburgh Museum. A slender axis bears numerous spirally disposed oval bodies (6 × 5 mm.) which are no doubt
seeds: each shows a differentiation into an inner portion surrounded

[1] Gibbs (12) p. 539.
[2] Hartz (96) Pl. xix. figs. 2, 3.
[3] Halle (13²) p. 83.
[4] Seward and Bancroft (13) p. 882, Pl. i. fig. 13.
[5] Miller (57) B. p. 493.

by a flat border and it may be that the latter is the impression of a sarcotesta. Another possible interpretation is that the oval bodies are seeds in intimate association with fertile bracts. The strobilus bears a close resemblance to *Stachyotaxus elegans* Nath.[1] from the Rhaetic of Sweden compared by the author of the species with an ovuliferous shoot of *Podocarpus spicata* and *Dacrydium Franklini* and believed to be allied to the recent genus *Dacrydium*, a view upheld by Miss Gibbs[2] in her account of recent Podocarps. It is not improbable that *Strobilites Milleri* is more closely allied to the Podocarpineae than to any other family of Conifers.

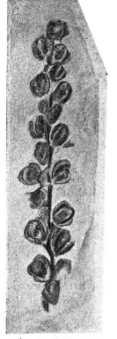

Fig. 797. *Strobilites Milleri.* (After Seward; Edinburgh Museum; nat. size.)

Saxegothopsis Dusén.

In his account of a Tertiary flora, possibly Oligocene, from localities on the Magellan straits Dusén[3] describes a single leaf as *Saxegothopsis fuegianus* on the ground that it resembles the leaves of *Saxegothaea conspicua*. The lamina, rather less than 2 cm. long, is linear-lanceolate with a spinous apex and a short stalk; no veins are shown in the drawing. There is no substantial reason for regarding this solitary fossil as a fragment of a Conifer allied to *Saxegothaea*. Dusén admits the lack of satisfactory evidence indicating generic identity, but the specimen hardly merits the distinction of being made the type of a new genus.

PHYLLOCLADINEAE.

PHYLLOCLADITES. Heer.

This generic name was given by Heer[4] to a fossil, subsequently transferred to a new genus *Drepanolepis*[5] (fig. 798, C), which affords

---

[1] Nathorst (08²) Pl. ii. figs. 1—27.   [2] Gibbs (12) p. 539.
[3] Dusén (99) p. 105, Pl. xi. fig. 10.
[4] Heer (75) ii. p. 124, Pl. xxxv. figs. 17—21.   [5] Nathorst (97) p. 43.

no real evidence of a relationship to the recent genus *Phyllocladus*, but Heer's term may be retained for a species described by Ettingshausen as *Phyllocladus asplenioides*[1].

Phyllocladites asplenioides (Ettingshausen).

This Tertiary species from New South Wales presents a close resemblance to *Phyllocladus* (fig. 675, p. 107) and is probably an Eocene representative of the genus. The specimens consist of fairly stout axes bearing cuneate and irregularly lobed leaf-like organs, in some cases apparently subtended by small scales—a circumstance which justifies Ettingshausen's suggestion that the lateral members are phylloclades. In one case a seed occurs at the base of a phylloclade. The phylloclades are practically identical with some forms of *Thinnfeldia*, particularly with American examples referred by Berry to his genus *Protophyllocladus*; the laminae of the Australian species agree both in form and venation with those of the Cretaceous American impressions, but 'in the latter there is no good reason for interpreting the leaf-like organs as flattened branches.

Ettingshausen's species is the only fossil that has come under my notice that has any substantial claim to be considered a satisfactory record of the recent genus *Phyllocladus*. In the account of *Thinnfeldia* in Volume II[2] reference is made to the resemblance of some impressions included in that genus to the phylloclades of *Phyllocladus*, a resemblance which led Ettingshausen to assign the type-species of *Thinnfeldia* to the Coniferae. Berry[3] considers that Ettingshausen's comparison with *Phyllocladus*, though not applicable to Jurassic and other of the older species of *Thinnfeldia*, is valid in respect of certain Middle and Upper Cretaceous forms for which he instituted the genus *Protophyllocladus*. Attention has previously been called[4] to the inadequacy of the evidence in support of the conclusion implied by the adoption of the name *Protophyllocladus*. The specimens for which this name was instituted consist of comparatively large coriaceous leaf-like impressions, linear or ovate-lanceolate with an entire, undulate, or crenulate margin, provided with a short petiole prolonged as a stout midrib from which numerous simple veins are given off at an acute angle. In

---

[1] Ettingshausen (86) p. 94, Pl. VIII. figs. 28—31.
[2] Page 543.   [3] Berry (03) B.   [4] Seward (04) B. p. 31.

venation and to a large extent in shape the fossils conform to the
characters of *Thinnfeldia*.

Protophyllocladus subintegrifolius (Lesquereux).

This species was originally described by Lesquereux[1] from
Dakota beds in Nebraska as *Phyllocladus subintegrifolius*. Heer[2]
described similar or possibly identical specimens from the Atane
beds of Greenland as *Thinnfeldia Lesquereuxiana* and included
Lesquereux's name as a synonym: these Greenland specimens
reach a length of 8 cm. and a breadth of 2 cm. Heer classes the
species among plants of uncertain position and compares the
impressions with the phylloclades of *Phyllanthus*. The species is
recorded by Hollick[3] from Martha's Vineyard and other localities,
by Berry[4] from the Raritan flora, and as *Thinnfeldia Lesquereuxiana* by Newberry and Hollick[5] from the Amboy clays and
other Cretaceous floras[6]: most of the examples are detached leaves
(or ? phylloclades), linear, spathulate, or ovate with an entire or
toothed margin while a few are branched (fig. 798, A, B). In no
single case is there any evidence in favour of regarding the specimens as phylloclades rather than leaves. Until additional facts
are obtained it would seem preferable either to retain the generic
name *Thinnfeldia* used by several authors or to adopt some title
which does not suggest a relationship to any recent genus. Zeiller[7]
has described a specimen from the Great Oolite of Marquise
(N. France) as *Protophyllocladus* sp.: this is the first European
record for Berry's genus.

A similar species is represented by *Protophyllocladus polymorphus*
(Lesq.) first described by Lesquereux[8] from Vancouver Island as
Salisburia polymorpha and afterwards transferred by Knowlton[9]
to *Thinnfeldia*. Another closely allied type is *Protophyllocladus
lobatus* Berry[10] from Upper Cretaceous rocks in South Carolina.
Some of these supposed phylloclades closely resemble flattened

[1] Lesquereux (74) p. 54, Pl. I. fig. 12.
[2] Heer (82) i. p. 37, Pl. XLIV. figs. 9, 10; Pl. XLVI. figs. 1—12.
[3] Hollick (06) p. 36, Pl. v. figs. 1—6.   [4] Berry (11³) p. 98, Pl. IX.
[5] Newberry and Hollick (95) p. 59, Pl. XI.
[6] Lesquereux (91) Pl. II. figs. 1—3.   [7] Zeiller (12) p. 13.
[8] Lesquereux (78) B. p. 84, Pl. LX. figs. 40, 41.
[9] Knowlton (93) p. 47. Pl. v. figs. 1—4; Berry (03) B. p. 442.
[10] Berry (03) B.; (14) p. 17, Pl. II. figs. 9—13.

leaf-like branches from Kreischerville for which Hollick and Jeffrey instituted the genus *Androvettia*[1] (fig. 806, A—C): some examples of that genus are clearly distinct as there are small leaves borne

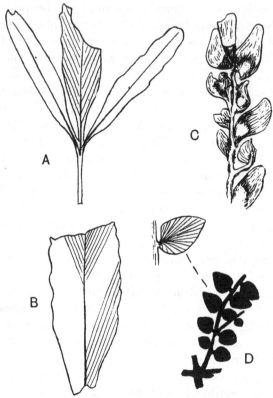

FIG. 798. A, B. *Protophyllocladus subintegrifolius.* C. *Drepanolepis rotundifolia.* D. *Phyllocladopsis heterophylla.* (A, B. after Newberry and Hollick; C, after Nathorst; D, after Fontaine. Nat. size.)

on the flat phylloclade, but other specimens are hardly distinguish-able from *Protophyllocladus.* Though it would be foolish to deny that the marked resemblance as regards form and venation between certain species of *Protophyllocladus* or *Thinnfeldia* and *Phyllocladus* may be significant, it is true that no satisfactory evidence has been produced in support of actual affinity to the recent genus.

[1] See page 436.

Palaeocladus Ettingshausen.

This name was applied to a single species, *Palaeocladus cunei-formis,* from New South Wales[1] founded on a specimen described as a linear cuneiform phylloclade with a median rib from which a few simple veins are given off at an acute angle, each passing up the middle of a lateral tooth; the teeth or serrations are regarded as lateral phylloclades concrescent with one another and with the main flattened axis. A few small scales occur at the base of the compound phylloclade. Ettingshausen's ingenious interpretation, though possibly correct, rests on too slender a basis to justify the assumption of any affinity to *Phyllocladus.*

Phyllocladopsis Fontaine.

This generic term was proposed for some branched foliage-shoots from the Potomac formation[2] characterised by small broadly ovate leaves with spreading veins. The type-species *Phyllocladopsis heterophylla* (fig. 798, D), as Fontaine and Berry state, bears a close resemblance to some forms of *Nageiopsis* and there can be little doubt as to the foliar nature of the appendages, which afford no evidence of morphological affinity to the phylloclades of a *Phyllocladus.* The species must be left for the present as a plant of uncertain position: it would seem more appropriate to adopt the generic name *Nageiopsis* than to make use of a designation suggesting a relationship which has not been established.

TAXINEAE.

TAXITES. Brongniart.

This generic name, first proposed by Brongniart[3] for some Tertiary and one Jurassic species, has been widely used for vegetative shoots bearing spirally disposed and distichously placed linear leaves (*e.g.* fig. 802) resembling in habit those of *Taxus, Sequoia sempervirens, Cephalotaxus* and some other recent Conifers. In the absence of anatomical characters or reproductive organs it is impossible to determine the precise position of shoots of this common form and the designation *Taxites,* as generally employed,

---

[1] Ettingshausen (86) p. 93, Pl. VIII. figs. 33, 33 *a*.
[2] Fontaine (89) B. p. 204. Pl. LXXXIV. fig. 5, Pl. CLXVII. fig. 4; Berry (03) B.
[3] Brongniart (28) A. p. 108.

cannot therefore be interpreted as indicative of affinity to *Taxus*.
Taxites has also been applied to fossil wood believed to belong to
Conifers allied to *Taxus* or other members of the Taxeae, but this
name has been superseded by *Taxoxylon*. *Taxites* is misleading
in its implication of relationship as the specimens so named afford
no proof of their systematic position within the Coniferales. Halle[1]
has proposed to employ the generic name *Elatocladus* for sterile
shoots of the *Taxites* type, the latter name being restricted to
fossils which there are good grounds for assigning to the Taxeae.
Among other genera to which authors have referred vegetative
shoots superficially similar to Taxus is *Palissya*[2], a genus founded
in part on a definite type of reproductive shoot very different from
that of *Taxus*. The investigation of the epidermal structure of
some specimens of Jurassic age placed by Feistmantel in *Palissya*
has led Miss Holden to institute a new genus *Retinosporites*[3] based
on characters recognisable only in the case of fossils which retain
their cuticular features. Another example of foliage-shoots of the
Taxites habit, which are shown by their fertile shoots to have no
connexion with the Taxeae, is *Stachyotaxus*[4]. *Taxites* should be
retained for fertile branches or reproductive organs which there is
some good reason for believing to be more closely related to *Taxus*
than to any other genus. So far as I am aware, none of the
specimens described as species of *Taxites* supply sufficient justification for being so designated. Fragments of sterile shoots from
the Miocene beds of Western Greenland described by Heer[5] as
Taxites Olriki and now in the Dublin Museum afford no proof of
close affinity to *Taxus*: the leaves are rather larger than those of
Sequoiites Langsdorfii and not decurrent.

The generic names *Cephalotaxites* and *Cephalotaxopsis* applied
by Heer[6] and Fontaine[7] respectively to foliage-shoots resembling
those of recent species of *Cephalotaxus* suggest an affinity that is
not supported by data based on reproductive shoots. While such
a designation as *Cephalotaxopsis* may be used without a definite
implication of relationship to the recent genus, *Cephalotaxites*,
like *Torreyites* and *Taxites*, is more appropriately restricted to

[1] Halle (13²) p. 83.   [2] See page 426.   [3] See page 432.   [4] See page 410.
[5] Heer (68) p. 95, Pl. I. figs. 21—24; Pl. XLV. fig. 1.
[6] *Ibid.* (83), p. 10.   [7] Fontaine (89) B. p. 235.

specimens which there is good reason for connecting with *Torreya* and *Taxus* respectively.

Among the numerous specimens of wood from Mesozoic and Tertiary beds referred by authors to *Taxoxylon*[1] there are very few which show true spiral bands on the secondary-xylem elements: there are no undoubted examples of the Taxineous type of stem from pre-Tertiary strata.

The foliage-shoots described as species of *Taxites* are as a rule valueless as records of *Taxus*, *Cephalotaxus*, or *Torreya*. Some small seeds very like those of *Taxus baccata* are described by Ludwig from the Oligocene lignites of West Germany as *Taxus margaritifera*[2] and similar examples were described by Heer[3] as *Carpolithes nitens* from the lignites of Bovey Tracey: the Bovey seeds have recently been named by Mr and Mrs Reid *Taxus* (?) *nitens*[4]. The latter authors refer some globose mucronate seeds from the preglacial deposits of Norfolk to *Taxus baccata*[5], and Clement Reid[6] records the same species from preglacial beds in Suffolk, from interglacial beds in Suffolk and from the peat below sea-level in the Thames valley. The genus *Taxus* ranges through parts of Europe, Asia, North America, Algeria, and occurs sporadically on the mountains of Sumatra, Celebes and the Philippines; it is noteworthy that there is little difference between the several species which are probably mere geographical forms[7]. The Yew is still wild in parts of Sussex, Hampshire, and Wiltshire, in a few localities in Scotland, but still rarer in Ireland. It is clear from the Pleistocene records that *Taxus* was formerly much more widely spread. Dr Conwentz[8] has shown that many places in Germany and the British Isles derive their names from the Yew, and the same author found that several prehistoric wooden articles in the Dublin Museum are made from Yew wood.

TORREYITES.

The evidence on which several fossil Coniferous branches from Cretaceous and Tertiary rocks have been referred to *Torreya* is in

---

[1] See page 202.   [2] Ludwig (61) p. 73, Pl. LX. fig. 19.
[3] Heer (62) p. 1078, Pl. LXX. figs. 15—23.
[4] C. and E. M. Reid (10) p. 172.
[5] *Ibid.* p. 171, Pl. XVI. figs. 42, 43; (08) Pl. XV. fig. 145.
[6] Reid (99) B. p. 151.   [7] Elwes and Henry (06) p. 99.   [8] Conwentz (01).

most cases unconvincing and with one exception no facts as to epidermal characters are available. Leaves of recent species of *Torreya* (fig. 694, B, p. 141) are characterised by two well marked stomatal grooves on the lower surface, and another feature is the absence of a prominent midrib: the leaves of *Cephalotaxus*, similar in form and size to those of *Torreya*, differ in the flat ungrooved lower surface and the prominence of the midrib on the upper surface of the lamina. We have no information with regard to any fossil seeds of the *Torreya* type, a type to which reference is made in the account of fossil Palaeozoic seeds. The present distribution of *Torreya* suggests that it was formerly more widely spread, but the data at present available do not admit of any very satisfactory statement of its past history.

Torreyites carolianus (Berry).

Berry described this species as *Tumion carolianum*[1], using the unfamiliar generic name which has been substituted by purists in nomenclature for *Torreya*. The material from Middle Cretaceous rocks in North Carolina consists of twigs with spirally disposed flat linear-lanceolate leaves 2·5—3 cm. long and with a maximum breadth of 3 mm., gradually tapering towards the slender apex and slightly contracted at the decurrent base. There is no distinct midrib, but in the proximal part of the lamina a more opaque band indicates the position of the vascular tissue: on either side of the middle line is a band in which the stomata are scattered; the long axis of the guard-cells tends to be at right-angles to the length of the leaf as in recent species and the fossil stomata generally resemble those of existing types. Some less satisfactory specimens from Upper Cretaceous beds in Georgia are doubtfully referred by Berry[2] to this species. In view of the characters of the vegetative fragments from Carolina it seems reasonable to adopt the generic name *Torreyites*. Berry has published a map showing the distribution of Cretaceous representatives of *Torreya*, but it is questionable whether the nature of the records constitutes a solid foundation.

Heer[3] has described two species, *Torreya Dicksoniana* and

---

[1] Berry (08²).   [2] *Ibid.* (14) pp. 107, 123.
[3] Heer (75) ii. p. 70, Pl. xviii. figs. 1—4; p. 71, Pl. xvii. figs. 1, 2; (82) i. Pl. ii. fig. 11.

T. parvifolia, from the Lower Cretaceous beds of Greenland: the leaves of the former exhibit a close agreement with those of the recent species though the evidence in support of generic identity is far from decisive. The second species differs in the smaller leaves: an examination of one of the figured specimens in the Stockholm Museum led me to the conclusion that it may be identical with an Upper Jurassic form from Scotland described as *Taxites Jeffreyi*[1]: there are no adequate grounds for the use of the name *Torreya*. Fontaine's Potomac species *Torreya virginica*[2] is founded on a piece of shoot bearing long linear leaves with no obvious midrib but with two strong lines between the middle and the edges of the lamina which suggest stomatal grooves. This specimen is of less value than the type-specimen of *Torreyites carolinianus* because of the lack of information with regard to the stomata. A second species of very little botanical value is described from the same formation as *Torreya falcata*[3]. Yokoyama's Upper Jurassic or Wealden species *Torreya venusta*[4] from Japan has no claim to be included among records of *Torreya*.

The Miocene species from Greenland, *Torreya borealis*, described by Heer[5] is founded on sterile twigs with broad linear leaves which afford no definite indication of relationship to the recent genus. Some fragments from Pliocene beds near Lyon described by Saporta and Marion[6] as *Torreya nucifera* var. *brevifolia* show two stomatal grooves and may be correctly identified. These authors regard the specimens described by Ettingshausen from Bilin in Bohemia as *Sequoia Langsdorfii* as fragments of a *Torreya* and re-name the species *T. bilinica*.

Some detached leaves and seeds from Upper Pliocene beds in the Main Valley (Frankfurt) are referred by Engelhardt and Kinkelin[7] to *Torreya nucifera fossilis*: the seeds bear a close resemblance to those of the existing species. While there is fairly good evidence from Tertiary localities of the comparatively recent occurrence of *Torreya* in Europe the records cannot be regarded as conclusive.

[1] Seward (11[2]) p. 688.
[2] Fontaine (89) B. p. 234, Pl. cix. fig. 8.
[3] *Ibid.* p. 235, Pl. cxiii. fig. 4.
[4] Yokoyama (89) p. 230, Pl. xxii. figs. 11, 12.
[5] Heer (83) p. 56, Pl. lxx, fig. 7a.
[6] Saporta and Marion (76) p. 87, Pl. xxii. figs. 6, 7.
[7] Engelhardt and Kinkelin (08) p. 191, Pl. xxiii. figs. 6—8.

VESQUIA. Bertrand.

Vesquia tournaisii Bertrand. The name *Vesquia*, after the French botanist Julien Vesque, was given by Bertrand[1] to seeds from Lower Cretaceous strata at Tournai which he described as intermediate in certain respects between *Taxus* and *Torreya*; they are two or three times as large as the seeds of *Taxus* and about one-third the size of those of *Torreya*. The ligneous shell is ribbed and on each side at the base is a large orifice marking the position of a vascular bundle which is continued through the length of the marginal ribs. The seeds are elliptical in transverse section and prolonged apically into a micropylar beak agreeing structurally with the micropyle of *Taxus* and *Torreya*. Bertrand also found anatomical features in the shell similar to those of the recent genera. In the absence of illustrations it is difficult to follow the description in detail, but the facts appear to favour Bertrand's conclusions with regard to the affinities of the fossil species.

CEPHALOTAXOPSIS. Fontaine.

Fontaine[2] gave this name to specimens of vegetative shoots, abundant in the Patuxent formation in the Potomac group, closely resembling in habit recent species of *Cephalotaxus*: the characters of the genus have been revised by Berry[3] who adds some particulars as to the structure of the epidermis. Fontaine's four species are reduced by Berry to two. No seeds have been found attached to the branches, but the American authors consider that some associated seeds may belong to the genus.

Cephalotaxopsis magnifolia Fontaine.

With this species Berry[4] includes *C. ramosa* Font. The branches are fairly robust and in some cases bear lateral shoots in whorls or pseudo-whorls; the leaves are distichous, linear-lanceolate, rather abruptly rounded at the base and tapering gradually to a mucronate apex, with an average length of 4—5 cm. and a breadth of 3—4 mm. Groups of bud-scale scars occasionally occur at the base of an ultimate shoot. The thick lamina may be transversely wrinkled as in *Pityophyllum*. There is a distinct midrib and a short distance

---

[1] Bertrand (83).   [2] Fontaine (89) B. p. 235.   [3] Berry (11) p. 374.
[4] *Ibid.* p. 377, Pl. LX. fig. 1; Fontaine *loc. cit.* Pls. CIV.—CVIII.

on either side of it is a stomatal groove in which stomata are irregularly scattered; the orientation of the guard-cells though not constant tends to be parallel to the long axis of the leaf. The epidermal cells are thick-walled and quadrangular or hexagonal and arranged in regular rows. Berry states that the stomatal grooves are a prominent feature and that there is some evidence of the occurrence of woolly hairs, characters suggestive of *Torreya* rather than *Cephalotaxus*. Branches from the Potomac beds with shorter leaves are referred to *Cephalotaxopsis brevifolia* in which is included *C. microphylla* Font. Berry points out that the photograph of *C. brevifolia* which he gives serves to 'emphasize the idealisation and inaccuracy of the former figures of this plant.'

Such evidence as is available favours the comparison of these species with recent Taxineae, but the structure of the leaves of *C. magnifolia*, so far as it is indicated in the epidermal preparations described by Berry, would seem to be in favour of a closer affinity to *Torreya* than to *Cephalotaxus*. The presence of depressed stomatal regions is a characteristic feature of *Torreya* and not of *Cephalotaxus*.

Heer[1] described a specimen from the Lower Cretaceous beds of Greenland as *Cephalotaxites insignis* consisting of a small piece of foliage-shoot with, apparently attached to it, an oval seed (18 × 13 mm.) in the form of a mould. Without examining the type-specimen it would be rash to accept the determination as correct. Berry[2] has described some seeds from Mid-Cretaceous rocks in Carolina as *Cephalotaxospermum carolinianum*, approximately 10 by 8 mm. in size, ovoid acuminate, and resembling the seeds of *Cephalotaxus*, but no anatomical features are preserved. Saporta's species *Cephalotaxus europaea*[3] founded on sterile shoots from the Aquitanian of Manosque in the South of France affords no convincing evidence of generic identity with *Cephalotaxus*.

Seeds from Upper Pliocene beds in the Main valley are referred by Engelhardt and Kinkelin[4] to three species of *Cephalotaxus*, the type-species being *C. francofurtana*. They bear a close resemblance to the recent seeds but this is hardly sufficient to prove their generic identity.

[1] Heer (83) p. 10. Pl. LIII. fig. 12.   [2] Berry (104).
[3] Saporta (93) p. 42, Pl. v. fig. 4.
[4] Engelhardt and Kinkelin (08) p. 194, Pl. XXIII. fig. 11.

CHAPTER L.

CONIFERALES INCERTAE SEDIS.

TRIOOLEPIS. Zeiller.

Trioolepis Leclerei Zeiller. This generic name was proposed by Zeiller for a cone from the Rhaetic flora of Tonkin[1] which he at first placed in the comprehensive genus *Conites*. In general appearance the specimen resembles a cone of a *Picea*; it is elongate-oval and incomplete, more than 10 cm. long and about 3 cm. in diameter. The impression shows numerous spirally disposed, imbricate, scales apparently thin, oval-linear in form, 12—15 mm. long and 6—7 mm. broad, suddenly contracted to an obtuse apex; the surface is marked by more or less distinct longitudinal folds and close to the base are slight depressions indicating the former presence of seeds 5 mm. long and 1·5 mm. broad. On some of the scales there is a faint curved trilobed line in the upper third of the ventral face which, it is suggested, may possibly mark the limit of an ovuliferous scale fused to a subtending bract-scale, but there is no substantial ground for any conclusion as to the morphological nature of the cone-scales. Zeiller states that there is no decisive evidence with regard to the systematic position of this fossil: the presence of three seeds recalls *Cunninghamia* though this in itself is probably of comparatively small importance.

MASCULOSTROBUS. Seward.

This designation was proposed for fossils which are in all probability male strobili of Gymnosperms[2].

Masculostrobus Zeilleri Seward.

This, the type-species, was discovered by the late Dr Gunn in Kimeridgian strata on the North-East of Scotland: it consists of a slender axis 13 cm. long bearing numerous small branches with

---

[1] Zeiller (03) B. p. 208, Pl. L. fig. 15; Pl. F, fig. 2.   [2] Seward (11²) p. 686.

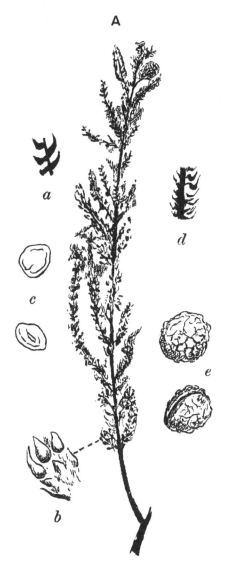

FIG. 799. *Masculostrobus*. A. *M. Zeilleri*. nat. size; *a*, *b*, portions enlarged; *c*, microspores from A; *d*, *Masculostrobus sp.*: *e*, spores from *d*. (British Museum.)

spirally disposed sporophylls. A few oval spores (fig. 799, c), 20—30 μ in diameter, were obtained from some of the sporophylls. The specimen is probably an inflorescence of male flowers of a Conifer; it bears some resemblance to Zeiller's *Pseudoasterophyllites Vidali*[1] from the Kimeridgian of Spain, but in that type the leaves are in whorls and the shoot appears to be sterile.

The smaller example shown in fig. 799, d, e, may be part of a microstrobilus of *Elatides curvifolia*, but in the absence of any connexion with that species it should be retained in *Masculostrobus*.

Möller and Halle[2] in their account of a flora from South-East Scania, probably of Wealden age, describe some fragments of microstrobili which they assign to *Masculostrobus*. The Cretaceous specimens from Kreischerville described by Hollick and Jeffrey as *Strobilites microsporophorus*[3] may be transferred to *Masculostrobus*. This species, represented by portions of small strobili a few millimetres long, is characterised by sporophylls in which the leaf-traces are enclosed by transfusion-tracheids, a feature suggesting comparison with modern Araucarineae, and the spores have two bladder-like wings. It may well be, as the authors of this type suggest, that the extension of the exine of the microspores, now restricted to the Abietineae and Podocarpineae, may be an ancient character and formerly more widely spread among the Coniferales.

PALISSYA. Endlicher.

Endlicher[4] instituted this genus for a type of Conifer previously described by Braun[5] from the Rhaetic flora of Franconia as *Cunninghamites sphenolepis*, characterised by distichous, linear, leaves and a megastrobilus composed of loosely imbricate scales. The name *Palissya* was selected to commemorate Bernard Palissy. The type-species, *P. Braunii*, was first figured by Goeppert[6] and described in more detail by Schenk[7] who, as Nathorst points out, included two distinct types of reproductive shoot under the name *Palissya*: one of them is retained in that genus while the other agrees with cones referred to *Elatides*[8]. Endlicher's designation

[1] Zeiller and Vidal (02) p. 7, Pl. II.
[2] Möller and Halle (13) p. 36, Pl. VI figs. 9—18
[3] Hollick and Jeffrey (09) B. p. 66.
[4] Endlicher (47) p. 306.   [5] Braun (43).   [6] Goeppert (50) Pl. XLVIII.
[7] Schenk (67) A. p. 175, Pl. XLI. figs. 2—14.   [8] *Ibid.* fig. 7.

has been employed by several authors for Mesozoic shoots with linear leaves unaccompanied by reproductive organs and having therefore no claim to be assigned to a genus characterised by a well defined type of strobilus. Further reference to the mis-application of *Palissya* is made in the account of some Indian specimens recently assigned to *Retinosporites*. In the absence of any evidence of the occurrence of strobili, shoots similar in habit to *Palissya* should be described under Halle's generic name *Elatocladus*. An impression of a shoot from Triassic beds in Bucks County, Pennsylvania, recently described by Wherry[1] as *Palissya longifolia*, would be more appropriately referred to *Elato-cladus* as it affords no indication of the nature of the fertile branches.

Palissya sphenolepis (Braun).

Nathorst[2] in his recent and able account of this species reverts to the older specific name on the ground that Endlicher's name *P. Braunii* is not in accordance with the laws of priority. It is possible though not certain that some specimens described under different names by Presl prior to the publication of Braun's account of *Cunninghamites sphenolepis* may belong to this species. Schenk considerably extended Braun's description and was the first to publish figures of ripe cones showing certain morphological features of the seed-bearing scales which authors have differently interpreted. *Palissya sphenolepis* is a Rhaetic species recorded from Franconia and Scania, possibly more closely allied to the genus *Cunninghamia* than to any other existing Conifer; but, as Nathorst suggests, it may belong to an extinct section of Gymnosperms. The foliage-shoots bear spirally disposed, two-ranked, leaves of the *Taxites* form; the lamina is narrow and linear with a median vein, decurrent on the axis as a persistent leaf-cushion. The epidermal cells have straight walls and the stomata, confined to the lower surface, occur in two rows. *Palissya* cannot be identified with any degree of certainty in the absence of well-preserved strobili. The mega-strobili are cylindrical and relatively narrow; in an immature con-dition they closely resemble those of *Elatides*, the surface being formed of the lanceolate, imbricate, distal ends of crowded cone-scales. It is the older strobili with elongated internodes that

---

[1] Wherry (16).   [2] Nathorst (08²).

constitute the most striking feature of the genus: the cone-scales (sporophylls) are entire, elongate-lanceolate (fig. 800) with an acuminate apex, and each bears 5—6 pairs of seeds characterised by a cup-like basal investment or cupule (fig. 800, C). The sporophylls have a strongly developed keel on the lower surface and a less distinct median rib between the two rows of seeds on the upper face (fig. 800, A, B). There is no evidence to support the view that the cone-scales are double[1]. Schenk[2] described the cone-scales as bearing 10—12 seeds on the edge and Saporta[3] believed the seeds to be lobes of a seminiferous scale, each lobe supporting one seed. Nathorst's investigation of Scanian material has thrown a welcome light on the nature of the mega-strobili as interpreted by previous authors. The bodies described by Schenk as seeds are projecting spherical casts of cup-like organs which originally embraced the lower portions of the seeds. The morphological nature of the cupule cannot be determined, but as Nathorst suggests it may correspond to the epimatium[4] which partially encloses the seeds of *Dacrydium* and other recent

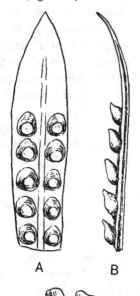

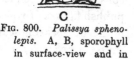

FIG. 800. *Palissya spheno-lepis.* A, B, sporophyll in surface-view and in section. C, cupules of two seeds. (After Nathorst.)

Conifers. Nathorst compares the cone-scales of *Palissya* with those of the genus *Cunninghamia* (fig. 684, K, p. 116) in which each sporophyll bears three seeds on the adaxial side of a membranous outgrowth stretched across the scale. The resemblance would seem to be closer than Nathorst suspects as the membrane in *Cunninghamia* does not arise as a continuous strip of tissue but as three separate ligule-like pieces, one on the abaxial side of each seed.

[1] Solms-Laubach (91) A. p. 73.
[2] Schenk in Schimper and Schenk (90) A. p. 336.
[3] Saporta (84) p. 513.
[4] See page 118.

ELATOCLADUS. Halle.

This genus was founded with a view to reduce the 'present state of intolerable confusion in the classification of the Conifers[1].' Halle expresses the opinion, with which most botanists cannot but agree, that sterile shoots of Conifers should not be described under generic names which imply the possession of a certain type of cone. The occurrence of foliage-shoots of similar or even identical habit in certain recent genera possessing distinct types of reproductive shoots serves to emphasise the unscientific character of the too common practice of assigning fossils to genera distinguished by a particular form of cone even though the specimens in question afford no evidence of the nature of the fertile branches. The generic name *Taxites* has been widely employed for dorsiventral branch-fragments bearing linear leaves with a pseudo-distichous arrangement as in the recent *Taxus*, but notwithstanding the employment of *Taxites* by authors who do not intend to convey the impression of relationship to *Taxus*, it is preferable to reserve *Taxites* for specimens which there are reasonable grounds for believing to be related to the Taxeae. Halle insists that it is undesirable to use one name for dorsiventral shoots and another for shoots with spirally disposed leaves, as fossil forms are known, *e.g. Stachyotaxus elegans*, Nathorst, in which both types occur on the same plant. Similar cases of dimorphism are well illustrated by *Dacrydium* and other recent Conifers (*cf.* fig. 708, p. 160). *Entocladus* is proposed for sterile coniferous branches of the radial or dorsiventral type, 'which do not show any characters that permit them to be included in one of the genera instituted for more peculiar forms.'

The name serves a useful purpose for sterile shoots which it has been the custom to include in *Taxites* and for types such as *Elatocladus heterophylla* Halle, which bear both distichous, linear leaves and crowded scale-like leaves similar to those of *Brachyphyllum* and some forms referred to *Pagiophyllum*. It is, however, desirable to retain *Brachyphyllum* and *Pagiophyllum* for sterile shoots exhibiting no well marked dimorphism and bearing fleshy appressed leaves and four-sided falcate leaves respectively. Used

---

[1] Halle (13²) p. 82. ἐλάτη, Pine or Fir; κλάδος, shoot.

in this narrower sense *Elatocladus* is more likely to serve the object
which the author had in view. It is noteworthy that in some
specimens of *Elatocladus* (*Taxites* spp.) the leaves have a transversely
wrinkled lamina, a feature usually associated with the detached
linear leaves assigned to the genus *Pityophyllum*.

Elatocladus heterophylla Halle.

The shoots of this species from the Jurassic flora of Graham
Land[1] are freely branched and the ultimate branches show a

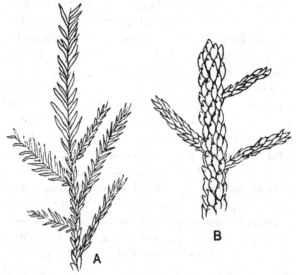

FIG. 801.   *Elatocladus heterophylla*.   (After Halle; nat. size.)

tendency towards a distichous arrangement. The radially symmetrical branches bear short and thick, acute, more or less appressed
leaves or the lamina may be slightly falcate, while the dorsiventral
shoots have narrow linear acute leaves in two ranks (fig. 801).
A faintly marked midrib is present in both forms of leaf. Halle
states that the shorter leaves differ from those of *Brachyphyllum*
in being rather longer and less closely appressed to the axis, though
this is a difference of secondary importance. The main interest
of the species is its dimorphism.

In *Elatocladus* is also included the Indian species, recorded by

[1] Halle (13²) p. 84, Pl. VIII. text-fig. 18.

Halle[1] from Graham Land, originally described by Oldham and Morris as *Cunninghamites confertus* and subsequently removed by Feistmantel to *Palissya*: the leaves are distichous with a sessile and decurrent lamina attached at a wide angle. The apex of the leaves is obtuse and a midrib is present. In *Elatocladus conferta* Halle includes the Australian form *Palissya australis* as figured by Stirling[2].

Elatocladus zamioides (Leckenby ex Bean MS.).

This type, from the Middle Jurassic of Yorkshire, was described by Leckenby[3] as *Cycadites zamioides* and subsequently transferred

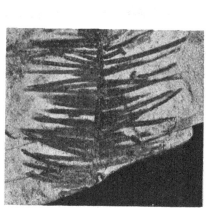

FIG. 802. *Elatocladus plana.* (Specimens figured by Feistmantel as *Taxites planus*; nat. size.)

to *Taxites*[4]. The type-specimen, in the Sedgwick Museum, Cambridge, consists of a slender axis bearing two-ranked spirally

[1] Halle (13[2]) p. 86, Pl. VIII.   [2] Stirling (00) Pl. III. figs. 8, 9.
[3] Leckenby (64) A. p. 77, Pl. VIII. fig. 1.
[4] Seward (00) B. p. 300, Pl. X. fig. 5.

attached linear leaves slightly more than 2 cm. long and 1·8 mm. broad, basally contracted, with an acute apex. It closely resembles shoots of recent Taxeae and *Sequoia sempervirens,* also some Potomac species included by Fontaine[1] in *Cephalotaxopsis* though there is no evidence of relationship to the recent *Cephalotaxus.* Shoots of similar habit are figured by Feistmantel[2] from Indian Jurassic beds as *Taxites planus* but the lamina is not contracted at the base. Two of Feistmantel's figured specimens are reproduced in fig. 802. These afford good examples of fossil branches which it has been the custom to refer to *Taxites,* but without information with regard to the epidermal characters it is impossible to determine their affinities. The form of the leaf-bases agrees with that shown in fig. 803 and it is probable that *Taxites planus* may be another example of Miss Holden's genus *Retinosporites,* though in the absence of anatomical data *Elatocladus* is the more appropriate designation.

RETINOSPORITES.   Holden.

Feistmantel used the name *Palissya* for some Indian Jurassic vegetative coniferous shoots which afford no evidence of affinity to that genus as represented by *P. Braunii.* Some of his fossils may be identical with the British species *Taxites zamioides,* now assigned to *Elatocladus,* while the examples described by Feistmantel as *Palissya* sp. and *Palissya indica* have been transferred to a new genus *Retinosporites.* The Indian impressions afford no evidence of a midrib; the upper epidermis consists of cells with straight walls and there are no stomata, while on the lower face of the lamina stomata are irregularly scattered, the long axis of the guard-cells being more or less parallel to the margin of the leaf. The absence of a midrib, at least so far as regards impressions and cuticular preparations, led Miss Holden[3] to separate the Indian specimens from *Palissya* and *Taxites* as vegetative shoots included by authors in genera having leaves with a distinct median vein and in which the stomata are in rows on the lower surface. The generic name *Retinosporites,* spelt by Miss Holden *Retinosporitis,* is proposed on the ground that the only flat-leaved Conifers among

---

[1] Fontaine (89) B. Pls. CVI.—CVIII.   [2] Feistmantel (79) Pls. XIII.—XV.
[3] Holden, R. (15²).

those examined showing similar epidermal characters were certain seedlings of the *Retinospora* type. Miss Holden recognises that *Retinosporites* suggests relationship to such recent Cupressineous species as are included under the genus *Retinospora*, but she states that no such implication is intended. The new designation, though not very happily chosen, may be retained for shoots with linear leaves (fig. 803) without a midrib and having the stomatal features described in Miss Holden's account of *R. indica*.

Retinosporites indica (Oldham and Morris).

An Indian species originally described by Oldham and Morris as *Taxites indicus* and transferred by Feistmantel to *Palissya*. The leaves are linear and decurrent (fig. 803, A) and without a midrib. The epidermal cells have straight walls and the stomata, though occasionally present on the upper surface, are scattered on the lower epidermis as in the *Retinospora* foliage of *Thuya* or *Juniperus*, without any indication of a median astomatic region such as one would expect in leaves possessing a midrib. There are generally six accessory cells and the guard-cells are sunk below the level of the epidermis (fig. 803, B).

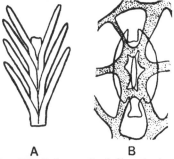

FIG. 803. *Retinosporites indica*. A, piece of shoot; B, stoma. (After R. Holden.)

Sterile foliage-shoots formerly referred to the genus
Cunninghamites.

Presl[1] gave the name *Cunninghamites* to some sterile shoots from Rhaetic and Lower Cretaceous strata on the ground of their resemblance to branches of *Cunninghamia* and both his specimens and the large number, particularly from Cretaceous beds, referred by authors to Presl's genus afford no real evidence of affinity to the recent Conifer. The type-species *Cunninghamites oxycedrus*, from Lower Cretaceous rocks in Saxony, is probably identical with Corda's species *Cunninghamia elegans* (fig. 804) from Lower Cretaceous beds in Bohemia. Presl also included in *Cunninghamites*

[1] Sternberg (38) A. Pl. XLVIII. fig. 3; Goeppert (50) Pl. XLVII.

branches from the Keuper of Germany which he named *C. dubius*: this species is identified by Saporta[1] with *Palissya Braunii*, but the latter name is now restricted to shoots bearing a particular form of strobilus. Nathorst[2], on the other hand, suggests that *C. dubius* may belong to *Elatides*, and the same author instituted a new generic name *Camptophyllum*[3] for fragmentary foliage-shoots

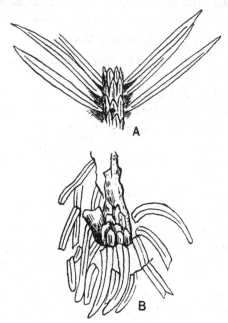

FIG. 804.   A, *Elatocladus elegans?*   B, *Elatocladus Schimperi*.   (A, after Velenovský; B, after Nathorst.   Nat. size.)

from the Rhaetic of Scania which he named *C. Schimperi* (fig. 804, B): these bear a close resemblance to *C. elegans*; the linear-lanceolate leaves are 15—20 mm. long and the lamina is recurved, probably as the result of drying. It is impossible to determine the position of this type and it may conveniently be transferred to *Elatocladus*. Some of Nathorst's specimens are also figured by Schenk[4]. The name *Cunninghamites* is given by Oldham and

[1] Saporta (84) p. 511.   [2] Nathorst (08²) p. 10.
[3] *Ibid.* (78) B. Pl. XVI. figs. 13—16.
[4] Schenk in Schimper and Schenk (90) A. p. 351, fig. 236.

Morris[1] to Jurassic Indian shoots which Feistmantel[2] afterwards transferred to *Palissya*: these have recently been included by Halle[3] in *Elatocladus* as *E. conferta* and recorded by him from the rich Antarctic flora of Graham Land.

In no case have we any definite information with regard to the cones borne by the *Cunninghamites* type of shoot. Velenovský[4] figures some Lower Cretaceous cone-scales from Bohemia, which he refers to *Cunninghamia*, resembling in shape the scales named by Hollick and Jeffrey *Protodammara*, but the resemblance may be only superficial.

The vegetative branches assigned by authors to *Cunninghamites* have linear-lanceolate leaves usually showing a distinct midrib and often other parallel lines on the lamina which are probably due to hypodermal fibres. The leaves may reach a length of 6 cm. and are 1—4 mm. broad; the edge is entire and finely serrate as in *Cunninghamia sinensis*. A characteristic feature is the occurrence of persistent decurrent leaf-bases on the branches which in some specimens that have lost the free portion of the lamina present a close similarity to *Brachyphyllum*. Some of the examples of *Cunninghamites* may well be shoots of a plant allied to *Araucaria Bidwillii* or *A. brasiliensis*. There is no doubt that under *Cunninghamites* are included branches of many different Conifers.

Elatocladus elegans (Corda).

Originally described by Corda[5] from Lower Cretaceous rocks in Bohemia as *Cunninghamia elegans*, this species is recorded from many Cretaceous localities; from the Patoot beds in Greenland[6], Moravia[7], Westphalia[8], Upper Cretaceous beds in Bulgaria[9], the Amboy clays[10], Cliffwood, Martha's Vineyard[11], Georgia, Carolina[12], and other places in North America[13]. The specimen shown in fig. 805 from Moravia shows a midrib in a few leaves, while in the

[1] Oldham and Morris (63) Pl. xxxii. fig. 10.
[2] Feistmantel (76²) p. 55.
[3] Halle (13²) p. 86, Pl. viii. [4] Velenovský (87).
[5] Corda in Reuss (46) B. Pl. xlix.
[6] Heer (83) Pl. liii. fig. 1. [7] *Ibid.* (69) Pl. i.
[8] Hosius and von der Marck (80) B. Pl. xxxvii.
[9] Zeiller (05²) Pl. vii. fig. 14. [10] Newberry and Hollick (95) Pl. v.
[11] Hollick (06) Pl. iii. fig. 1. [12] Berry (10⁴) Pl. xx.; (14) p. 106.
[13] See Hollick (06) for other references.

Bohemia specimen represented in fig. 804 the midrib is more obvious and the leaf-bases have a more regular form. The branching is sparse and not pinnate. Velenovský[1] assigns some branches to *C. stenophylla* but these may be younger forms of *C. elegans*. Similarly *C. squamosus*[2], as figured by Heer and other authors, affords no satisfactory evidence of specific difference from *C. elegans*. Impressions from the Atane beds of Greenland described by Heer[3] as *C. borealis* have been compared by Schenk with *Sequoia* and also referred by him to *Torreya*: there is no possibility of deciding the precise systematic position of these and similar specimens. Ettingshausen[4] has described as *Cunninghamites miocenica* fragments of shoots from Sagor in Carinthia bearing linear leaves with a finely serrate edge.

ANDROVETTIA. Hollick and Jeffrey.

This genus was instituted[5] for Cretaceous fossils from Staten Island superficially resembling Fern leaves with a pinnate venation and an irregularly lobed or incised margin. The leaf-like fragments are, however, stem-structures bearing minute scale-like leaves attached to the edges and surface. In habit these phylloclades agree with *Phyllocladus*, but on anatomical grounds the authors of the genus regard it as Araucarian though the evidence is far from convincing.

Androvettia statenensis Hollick and Jeffrey.

Some of the specimens show no indication of their phylloclade-nature and, as impressions, would be identified as Fern pinnules or referred to *Thinnfeldia*. Others, after bleaching in chlorine-water, showed a fairly stout vascular axis giving off simple or forked branches at an acute angle; small decurrent leaves free only at the apex occur on the margins of the shoots (fig. 806, A, B). In a few cases the phylloclades bear short axillary branches with immature cones, possibly microstrobili.

There are three vascular cylinders in the section reproduced in fig. 806, C, and in the narrow wings of the 'lamina' there are the

[1] Velenovský (85) B. p. 15.
[2] Heer (71[2]) Pl. I. figs. 5—7; Schimper and Schenk (90) A. p. 282; Berry (03) p. 64.
[3] Heer (82) B. Pl. XXIX. fig. 12.   [4] Ettingshausen (72) Pl. I. fig. 30.
[5] Hollick and Jeffrey (09) B. p. 22, Pls. III., VII., VIII., XXVIII., XXIX.

traces of two or three leaves. Several stomata occur on the surface of the phylloclade, each surrounded by 4—5 accessory cells. Sclerotic cells are present in the pith. The secondary xylem is of the coniferous type and the uniseriate bordered pits on the tracheids may be either separate and circular or flattened by contact. No resin-cells, such as occur in the wood of *Phyllocladus*, were recognised. The medullary rays are not described. The data are

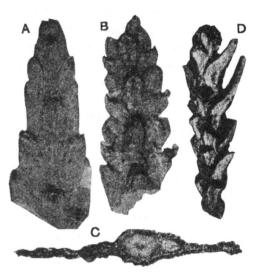

FIG. 806. A—C, *Androvettia statenensis.* D, *Geinitzia Reichenbachii.* (A, B, × 6; C, × 7; D, × 7; after Hollick and Jeffrey.)

FIG. 805. *Elatocladus elegans.* (Nat. size; after Heer.)

hardly sufficient to justify the inclusion of this type in the Araucarineae: the occasional flattening of the tracheal pits and the absence of resin-cells are not fatal to an alliance to *Phyllocladus*. The precise position of the genus within the Coniferales must for the present be left in doubt.

Androvettia elegans Berry.

This species, from the Upper Cretaceous of Georgia, is represented by dorsiventral fern-like vegetative shoots which, as Berry[1]

[1] Berry (14) p. 103, Pl. xviii.

points out, bear a close resemblance to species of *Moriconia* (*cf.* fig. 760). The branches are distichously arranged and the opposite lateral leaves are stout, falcate, and decurrent while those on the upper and lower faces are represented by scales on the middle line of the phylloclades.

The same author describes specimens from beds in North Carolina[1] referred to the lower half of the Upper Cretaceous as *Androvettia carolinensis*.

DACTYOLEPIS. Hollick and Jeffrey.

Dactyolepis cryptomerioides Hollick and Jeffrey. The generic name[2] was instituted for some detached, cuneate, cone-scales from the Cretaceous beds at Kreischerville in Staten Island, approximately 4 mm. long, composed of an upper and a lower segment. The upper portion is divided distally into as many as seven irregular short finger-like processes and the lower part is entire. Each of the processes possesses a single vascular bundle 'completely surrounded by a cordon of transfusion-tissue, thus betraying its Araucarineous relationship.' The scales which are without seeds are compared with those of *Voltzia*. There is, however, no proof that *Voltzia* had double scales. The view that *Dactyolepis* is Araucarian may fairly be said to rest on an insufficient basis.

RARITANIA. Hollick and Jeffrey.

The name *Raritania*[3], after the Raritan formation, was given to some Cretaceous fossils from Kreischerville identical with New Jersey specimens described by Newberry as *Frenelopsis gracilis*[4] on the ground that they belong to a type distinct from *Frenelopsis* as generally understood.

Raritania gracilis (Newberry). The specimens so named consist of slender, dichotomously branched, axes bearing minute leaves resembling *Psilotum triquetrum* and in the form of the branching the leaves of *Baiera Lindleyana* (Schimp.). The distinguishing feature is the occurrence of the prickle-like leaves (fig. 807, B) invisible to the unaided eye (fig. 807, A). A small

---

[1] Berry (10[4]) p. 183, Pl. XIX.
[2] Hollick and Jeffrey (09) B. p. 52, Pl. X. figs. 12, 13.
[3] *Ibid.* (09) B. p. 26, Pls. VI., IX., X., XX.
[4] Newberry and Hollick (95) p. 59, Pl. XII. figs. 1—3 *a*.

imperfectly preserved cone was found on a peduncle having leaves
similar to those on the vegetative twigs. Some fragmentary
lignitic branches (fig. 807, C) associated with the impressions
showed the anatomical characters of a Conifer; but Hollick and
Jeffrey, though believing that the fragments 'almost certainly'

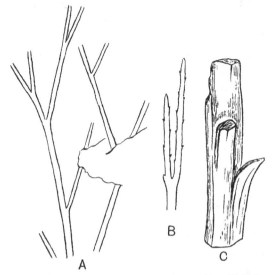

Fig. 807.   A, B, *Raritania gracilis*; C, *Raritania*?.   (After Hollick and Jeffrey.
A, B, × 6; C, × 10.)

belong to *Raritania,* admit that there is no proof of their identity
with the dichotomously branched impressions. The secondary
xylem of the cylindrical stele of the twigs, one of which is shown
in fig. 807, C, agrees with that of *Brachyphyllum macrocarpum*
Newb. The uniseriate bordered pits are occasionally flattened;
the cortex, confluent with the decurrent leaves, contains sclerotic
cells and each leaf has a resin-canal subtending the leaf-trace.
The genus is referred to the Araucarineae on the evidence of the
occurrence of flattened pits at the ends of some of the tracheids,
a conclusion difficult to accept without considerable reservation.

SCHIZOLEPIS.   Braun.

Braun[1] instituted this genus for a strobilus from Rhaetic
rocks in Germany which he called *Schizolepis liaso-keuperinus,*

———
[1] Braun (47) p. 86.

characterised by its deeply split cone-scales. Schenk[1] subsequently substituted the name *S. Braunii* and included under that designation Braun's *Isoetites pumilus*, a species founded on a foliage-shoot, also some other similar vegetative branches believed to belong to the plant which bore the cones. As here used, the term *Schizolepis* is restricted to cones and cone-scales since there is no definite evidence as to the nature of the foliage-shoots connected with the strobili. *Schizolepis* cannot be referred on any satisfactory grounds to a definite position among the Coniferales: it is possibly an extinct type allied to recent Abietineae, but until more is known with regard to the morphology of the cone-scales the systematic position must be left an open question.   The genus is represented by strobili from Rhaetic beds in Franconia, Scania, and Poland; detached scales from Middle Jurassic floras are also included in *Schizolepis* (fig. 808), and Nathorst has described incomplete strobili from Upper Jurassic or Wealden strata in Spitzbergen.   Attention has been called to a resemblance between *Schizolepis* scales and the fertile leaves of *Tmesipteris*[2], but there is no reason for regarding this as indicative of relationship.   More than one author has compared the bilobed cone-scales of *Schizolepis* with the 3—5-lobed scales of *Voltzia* and *Cheirolepis* though this comparison rests on a feature which in itself is no proof of affinity. A comparison may also be suggested with the reflexed cone-scales of *Picea Breweriana*.

Schizolepis Braunii Schenk.

It has already been pointed out that under this name Schenk[3] included both cones and vegetative shoots though he recognised the lack of any decisive evidence of common parentage.   While agreeing with Nathorst that the association with *Schizolepis* strobili of similar vegetative shoots both in Germany and Sweden may be more than accidental, in the present state of our knowledge it is preferable to refer the leaves and branches to *Pityophyllum* or *Pityocladus*.   In the younger strobili the bilobed scales are more or less pressed against the axis and in older examples they are more spreading: each scale has two lanceolate lobes and is attached

---

[1] Schenk (67) A. p. 179.
[2] Nathorst (97) p. 61.
[3] Schenk (67) A. p. 179, Pl. XLIV.; Schimper and Schenk (90) A. p. 306.

by a narrow stalk-like basal portion (fig. 808, B). Schenk states that there are two anatropous seeds to each scale, but it is not clear if the actual seeds are present.

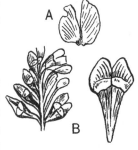

Schizolepis Follini Nathorst.

In this species from the Rhaetic flora of Scania[1] the bilobed scales are sessile and broader than in *S. Braunii*. In some specimens there appear to be two seeds near the base of a scale. Saporta[2] regards the cone-scales of the type-species and *S. Follini* as double structures, the transverse line shown in Schenk's figures below the forking of the scale being the limit of the bract-scale, while the thinner distal lobed part represents the seminiferous scale; an interpretation which rests on very slender evidence. To this species Raciborski[3] refers a specimen from rocks possibly of Rhaetic age in Poland.

FIG. 808. *Schizolepis.* A, *Schizolepis Moelleri.* B, *Schizolepis Braunii.* C, *Schizolepis(?) retroflexa.* (A, after Seward; B, after Schenk; C, slightly enlarged, after Nathorst.)

Schizolepis Moelleri Seward.

Under this name[4] some detached bilobed scales without seeds were described from Jurassic rocks in Turkestan (fig. 808, A) and the South of Russia[5], and Krystofovič has recently discovered a cone of the same species in Jurassic strata in Transbaikalia[6].

Schizolepis cylindrica Nathorst.

Founded on a long and narrow, incomplete, strobilus from Upper Jurassic rocks in Spitzbergen[7] bearing deeply lobed scales. A second species, *S. retroflexa*[8] (fig. 808, C), assigned with some

[1] Nathorst (78) B. p. 28, Pls. xiv., xv.
[2] Saporta (84) p. 502.
[3] Raciborski (92) p. 354, Pl. ii. figs. 1, 20 *d.*
[4] Seward (07²) p. 36, Pl. vii. figs. 64—66.
[5] Thomas (11) p. 79, Pl. v. fig. 4.
[6] Krystofovič (15) p. 95, Pl. vi. figs. 10. 11.
[7] Nathorst (97) p. 39, Pl. ii. figs. 1, 2.
[8] *Ibid.* p. 60, Pl. iii. figs. 31, 32; Pl. vi. figs. 11, 12.

CONIFERALES INCERTAE SEDIS

doubt to *Schizolepis*, is characterised by pendulous stalked scales: in neither of these Spitzbergen forms are there any seeds on the strobili. Nathorst quotes a species from Portugal described by Saporta[1] as *Palaeolepis bicornuta* as being possibly allied to *Schizolepis*, but there are not sufficient grounds for assuming any close affinity. He also draws attention to the resemblance of the bilobed leaves or scales of *Schizolepis*(?) *retroflexa* to the fertile leaves of *Tmesipteris*; the relationship of the Spitzbergen fossils must be left uncertain pending more evidence.

DREPANOLEPIS. Nathorst.

Heer described a specimen of Jurassic age from Spitzbergen as *Phyllocladites rotundifolia*[2] which he considered to be closely allied to *Phyllocladus*: an examination of the type-specimen led Nathorst[3] to institute a new generic name *Drepanolepis*. As Nathorst's revised description and more accurate drawing show, there are no substantial grounds for assuming any relationship between the fossil and *Phyllocladus*. *Drepanolepis rotundifolia* consists of a fairly stout axis bearing spirally disposed thick, falcate scales each of which bore a seed, or possibly a sporangium, near the base (fig. 798, C). A similar type is described by Nathorst as *Drepanolepis angustior*[4] characterised by the narrower form of the scales and a broader axis. Both species may be described as strobili of open habit with single seeded sporophylls: it is impossible to determine the systematic position of the genus, though as Nathorst says it is probably a type of fertile Gymnospermous shoot. There is no reason for comparing the specimens with *Phyllocladus*.

SCHIZOLEPIDELLA. Halle.

Schizolepidella gracilis Halle. The specimens on which this genus is founded are from the Hope Bay flora in Graham Land[5], probably of Middle Jurassic age; they consist of slender sterile shoots reaching a maximum length of 12 cm. and 2 mm. broad, rarely branched and bearing small leaves, 2 × 1·5 mm., apparently

[1] Saporta (94) B. Pl. xxxiii. fig. 4.
[2] Heer (75) ii. p. 124, Pl. xxxv. figs. 17, 18.
[3] Nathorst (97) p. 43, Pl. vi. figs. 24, 25.
[4] *Ibid.* p. 71, Pl. iii. figs. 33—37. [5] Halle (13²) p. 90, Pl. ix. figs. 18—21.

spirally disposed (fig. 809). The lamina is rounded, ovate or obovate, and always bilobed at the broad apex: no veins were detected. As Halle says, it is impossible to determine the affinities of the fragments but he thinks they may belong to pendulous branches of a Conifer. Attention is called to a resemblance to some Hepaticae, and a possible relationship to *Lycopodium* or the Psilotales may also be suggested. The choice of the generic name is not intended to imply anything more than a superficial similarity between the leaves and the bilobed strobilar appendages of *Schizolepis*[1].

A   B

FIG. 809. *Schizolepidella gracilis.* (After Halle; A, enlarged; B, nat. size.)

CYPARISSIDIUM. Heer.

This name was given[2] to foliage-shoots and cones, from the Urgonian rocks of Greenland, originally described as *Widdringtonites gracilis*[3]. The smaller sterile branches are indistinguishable from specimens referred by authors to *Widdringtonites* while the larger examples might be included in *Brachyphyllum*. The leaves are small, appressed, and imbricate, similar to those of some recent Cupressineae and Callitrineae but spirally disposed and not verticillate (fig. 810); the shoots agree also with *Microcachrys* and other recent Conifers. The cones are composed of a small number of flat scales (fig. 810, B) too imperfectly preserved to afford any definite evidence as to the affinities of the genus. Heer states that a detached cone-scale shows the impression of a single seed, but the material is insufficient to form the basis of a comparison with the Araucarineae; he points out a resemblance to *Cunninghamia* and mentions the striated surface of the fossil cone-scales as a distinctive feature, though that may be due, in part at least, to the state of preservation. The flat form of the cone-scales is a character in which *Cyparissidium* differs from genera such as *Sequoiites*, and from the Callitrineae the cones are distinguished by the spiral arrangement of the scales.

---

[1] See page 439.
[2] Heer (75) ii. Pl. xvii. fig. 5 *b, c*; Pls. xix., xx., xxi.; (82) pp. 16, 50, Pls. i., vii., xxviii.
[3] *Ibid.* (68) p. 83.

Cyparissidium is characteristic of Lower Cretaceous strata though Nathorst[1] has recorded a species, *C. Nilssonianum*, from Rhaetic rocks in Scania with cones having scales more pointed and lanceolate than those of Heer's species. A second Rhaetic species, *C. septentrionale*, has been transferred by Nathorst to the genus *Stachyotaxus*[2].

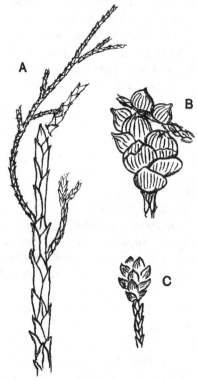

FIG. 810. *Cyparissidium.* A, B, *Cyparissidium gracile.* C, *C. minimum.*
(A, B, after Heer; C, after Velenovský; nat. size.)

Cyparissidium gracile Heer.

The type-species (fig. 810, A, B) is one of the most abundant plants in the Kome beds at Pattorfik in West Greenland and it occurs also in the Atane beds. The cones are 26 mm. long and 11—20 mm. broad with scales having a breadth of 12 mm., a

[1] Nathorst (86) p. 103, Pl. XXII.	[2] See page 410.

rounded distal border, a mucronate apex, and a striated dorsal face. Velenovský[1] records this species from Lower Cretaceous rocks in Bohemia but without the essential evidence of cones: the same author describes a smaller type from Bohemia as *C. minimum*[2] (fig. 810, C). Schenk[3] figures some fragments from Lower Cretaceous rocks in the Tyrol as *Cyparissidium cretaceum*, but in the absence of cones there is no sufficient reason for their inclusion in *Cyparissidium*. Hollick's record of *C. gracile* from the Cretaceous of Block Island[4] is based on insufficient evidence.

BENSTEDTIA. Seward.

In 1862 Mackie[5] figured a fossil stem from Lower Greensand rocks in Kent which König afterwards named *Dracaena Benstedtii*. This name was adopted by Morris and Mantell for the supposed 'Dragon tree.' In 1868 Carruthers[6] expressed the opinion that the fossils are more likely to be Pandanaceous stems, while Gardner[7] spoke of a possible Cycadean affinity. An examination of Mackie's specimens and others in the British Museum led me to suggest a comparison with stems of recent species of Cycads, particularly stems of *Zamia* which do not retain the armour of leaf-bases (fig. 381 B, vol. III. p. 5) characteristic of most Cycadales, and to institute a new generic name *Benstedtia*[8] in preference to a designation implying an improbable relationship. The genus is defined as follows: stems characterised by irregular and interrupted grooves and broader ridges running transversely, with occasional small elliptical protuberances irregularly disposed on the surface. There are no distinct leaf-scars but branch-scars occasionally occur; the upward convergence of the transverse wrinklings indicates bifurcation in some specimens.

The English examples reach a length of over 40 cm. and a diameter of 15 cm. Smaller specimens with similar surface-characters are described by Fliche[9] from Lower Cretaceous beds in France as *Coniferocaulon colymbeaeforme* and compared by him to stems of *Araucaria imbricata*: other examples closely resembling

[1] Velenovský (85) B. p. 17, Pl. VIII.   [2] *Ibid.* Pls. IX., X.
[3] Schenk (76) B. p. 167, Pl. XXIX. figs. 10, 11.
[4] Hollick (06) p. 46, Pl. III. fig. 11.
[5] Mackie (62).   [6] Carruthers (68) p. 154 (footnote).
[7] Gardner (86²) p. 201.   [8] Seward (96²) p. 216.   [9] Fliche (00).

the French stems are recorded from the Uitenhage (Wealden)
series of South Africa and the Kimeridge rocks of Sutherland[1],
Scotland, in the former case as *Benstedtia* sp., and in the latter as
Coniferocaulon colymbeaeforme because of the presence of a small
pith more suggestive of a Conifer than of a Cycadean branch. A
large specimen from Jurassic rocks in India is described by Miss
Bancroft[2] as *Coniferocaulon* sp.; this agrees very closely in surface-
features with the casts from Kent, but an examination of transverse
and longitudinal sections demonstrated that the apparent surface
is not the actual surface, and such anatomical data as it was
possible to obtain clearly indicated the Coniferous nature of the
wood, conclusions in agreement with those reached by Dr Stopes
in the case of some English specimens. The reference of these
stems to Cycads or Conifers was based entirely on surface-characters
and it was recognised that no definite conclusion was possible
without anatomical confirmation. Dr Marie Stopes[3] succeeded in
obtaining preparations of tracheids from a Lower Greensand cast
showing uniseriate and separate bordered pits of the Abietineous
type thus disproving a Cycadean affinity. This discovery led to
the substitution of *Coniferocaulon* for *Benstedtia,* at least as regards
the specimens which afforded anatomical evidence; in the French
and South African examples no internal structure is preserved.
Dr Knowlton[4] pointed out that the generic name *Benstedtia* should
be preferred to *Coniferocaulon* on the ground of priority and he
named the English specimens *Benstedtia Benstedtii.* Dr Stopes[5]
replied to this criticism by asking why the Dragon tree, which is
merely a partially decorticated piece of badly preserved Coniferous
wood, should have a name. Specimens exhibiting distinctive
surface-features, whether complete or decorticated and even if
they are in some cases at least portions of Coniferous stems, are
none the less entitled to some recognition as a matter of convenience.
Some excellent illustrations of *Benstedtia* casts are given by Dr
Stopes in her recently published Catalogue of Lower Greensand
Plants[6].

[1] Seward (03) B. p. 34; (11²) p. 690.
[2] Bancroft (13) pp. 72, 85.   [3] Stopes (11).
[4] Knowlton (11).   [5] Stopes (11²).
[6] Stopes (15) p. 159, Pls. XIII., XIV.

CHAPTER LI.

PODOZAMITES AND NAGEIOPSIS;
GENERA INCERTAE SEDIS.

PODOZAMITES Braun, and **CYCADOCARPIDIUM** Nathorst.

The name *Podozamites*[1] was instituted for certain species previously included in *Zamites* characterised by the possession of distant alternate pinnae with a contracted base and veins slightly spreading in the proximal part of the lamina but for the most part approximately parallel. As defined by Braun *Podozamites* differs in no very important respect from *Zamites,* and the latter name is retained by Schenk for *Z. distans* Presl in preference to *Podozamites* applied to that species by Braun. By most authors *Podozamites* has been regarded as Cycadean, but Schenk's discovery of a specimen of *Podozamites*[2] in the Rhaetic beds of Franconia showing a cluster of small scale-leaves at the base of the axis led him to suggest a possible affinity to *Agathis* as an alternative to the generally accepted view that the appendages are leaflets or pinnae homologous with those of a pinnate Cycadean frond. In a later paper Schenk included in *Podozamites* some undoubted pinnate fronds on which Schimper founded the genus *Glossozamites*[3]. Schenk was, however, influenced in his preference for a Cycadean alliance by the structure of the epidermal cells (fig. 812, E) which have straight walls, and the same consideration weighed with Zeiller[4] who was strengthened in his view by the characters of the seed-bearing sporophylls described by Nathorst[5] and provisionally connected by him with *Podozamites*. Nathorst[6] described a specimen from the Rhaetic strata of Scania agreeing in the presence of basal scale-leaves with that figured by Schenk, and more recently

---

[1] Braun (43) p. 36.
[2] Schenk (67) A. Pl. xxxvi. fig. 3.
[3] *Ibid.* (71) Pl. ii.
[4] Zeiller (03) B. p. 159.
[5] Nathorst (86) p. 91, Pl. xxvi.
[6] *Ibid.* (86) Pl. xvi. fig. 10.

FIG. 811.   *Podozamites lanceolatus.*   (Nat. size; British Museum, 39,303.)

Schuster[1] has published photographs of some examples of *Podo-zamites distans* in which the base of the axis is invested by small imbricate scales and in connexion with it are two other clusters of similar scales, probably unexpanded buds. In 1900 I expressed the opinion that *Podozamites* is probably a Conifer[2], the supposed pinnate fronds (fig. 811) being foliage-shoots like those of recent species of *Agathis*. The most important recent contribution to our knowledge of *Podozamites* is due to Nathorst: in 1911 he published additional facts[3] with regard to some seed-bearing organs from the Rhaetic of Scania for which he proposed the generic name *Cycado-carpidium* in 1886 and in 1902[4] more fully described the type-species *C. Erdmanni*. Until the publication of Nathorst's more recent paper *Cycadocarpidium* was known only as detached sporophylls found in beds containing *Podozamites* leaves. The following de-scription is abridged from Nathorst's account.

Cycadocarpidium Erdmanni is represented by ovate sporophylls consisting of a sterile portion 9 mm. long and at most 6 mm. broad with 4—5 simple veins, tapering to a short and slender pedicel on each side of which is an oval seed (fig. 812, A—D) with an obliquely placed triangular lamina compared by Nathorst with a cupule and interpreted by Schuster as a leaflet. A specimen figured by Nathorst shows several sporophylls attached to a common axis, and supports his view that the seed-bearing organs were borne as imbricate carpellary scales. Fig. 812, A is drawn from Nathorst's restoration of a cone-like cluster of sporophylls. Another type of sporophyll, *Cycadocarpidium Swabii*, is distinguished by the larger dimensions of the lamina, 4·1 cm. long and 16 mm. broad, with 10 veins: in this type the two small seeds are apparently without any appen-dages (fig. 812, C, D). A third species, *C. redivivum*, is founded on small detached leaves and bud-like clusters previously as-signed to *Podozamites distans*. These are now recognised as small *Cycadocarpidium* sporophylls. There is a strong probability that *Cycadocarpidium* was borne on a *Podozamites* shoot; in form and venation the sterile lamina of the sporophylls agrees with the leaves of *Podozamites* and the two organs are constantly associated in the Scanian beds. Zeiller records *C. Erdmanni* in Rhaetic

[1] Schuster (11⁴).   [2] Seward (00) B. p. 242.
[3] Nathorst (11⁴).   [4] *Ibid.* (02) p. 8, Pl. I. figs. 5, 6.

S. IV   29

rocks in Tonkin[1] where *Podozamites* also occurs. Heer[2] in his account of some impressions of *Podozamites* from Spitzbergen figured a seed in close association with what he believed to be a carpellary leaf like that of a Cycad and suggested a connexion between the seed and *Podozamites*: this supposed connexion has, however, little to support it.

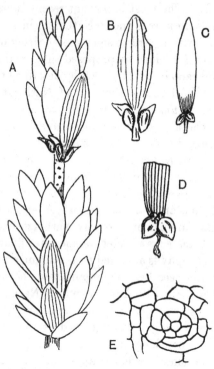

FIG. 812. A, B, *Cycadocarpidium Erdmanni*. A, Restoration of fertile shoot. B, Single sporophyll with seeds. C, D, *Cycadocarpidium Swabii*. (After Nathorst.) E, *Podozamites distans*. Epidermal cells. (After Schenk.)

Nathorst's view of the sporophylls is that each is a single fertile leaf bearing two ovules comparable morphologically with the megasporophylls of *Dioon*, but, as he points out, the terminal portion of the sporophyll of *Cycadocarpidium* is much more leaf-like than the sterile distal end of the megasporophylls of *Dioon*, *Encephalartos Zamia*. He considers that the evidence afforded

---

[1] Nathorst (114) p. 5.   [2] Heer (77) i. Pl. VIII. fig. 4.

by these seed-bearing organs favours a Cycadean alliance: on the other hand he has satisfied himself that some *Podozamites* specimens are shoots with spiral linear leaves like those of *Agathis*. It is in many cases very difficult to say whether the axis of a *Podozamites* bears the leaves in two ranks or spirally. Nathorst speaks of some examples in which the leaves are not spiral and suggests the existence of two kinds of branch some with spiral and some with two-ranked leaves as in certain Conifers. But if this is the case one can hardly imagine that the two-ranked arrangement is not due to the twisting of the leaves of shoots with spirally disposed foliage. In a recent contribution to the systematic position of *Podozamites* Schuster[1] speaks of *Cycadocarpidium Erdmanni* as differing from *C. Swabii* in the presence of two 'rudimentary leaflets' (the triangular lamina shown in fig. 812, B) which in the latter species are represented only by two small swellings at the upper ends of the seeds: he regards *C. Erdmanni* as the more primitive type. The lamina is homologised with the cover-scale or bract of the double cone-scale of the Abietineae; he compares the two leaflets of *C. Erdmanni* and the swellings in *C. Swabii* with abnormal seminiferous scales of an Abietineous cone. Schuster's view is that *Podozamites distaı s* is a primitive Conifer evolved from the base of a Cycadofilicinean line which gave rise to the Ginkgoales, a supposition based on a very slender foundation. Nathorst regards *Podozamites* as an intermediate type related both to Cycads and Conifers; he does not, however, overlook the fact that the sporophylls of *Cycadocarpidium* may be compared with those of some Conifers even though their resemblance to Cycadean sporophylls would seem to be closer. Additional data are needed before we can settle the position of *Podozamites*, but such information as we have may be said to point to the conclusion that it is nearer to the Conifers or the Ginkgoales than to any other group of Gymnosperms. Nathorst calls attention to a similarity between Yokoyama's *Ginkgodium Nathorsti*[2] and separate leaves of *Podozamites*; a similar comparison may be made between the latter genus and *Eretmophyllum*, a genus instituted by Thomas[3] and referred to the Ginkgoales (*cf.* figs. 658, 659, pp. 59, 62). There is indeed some resemblance between *Cycadocarpidium* sporophylls and abnormal

---

[1] Schuster (114).   [2] Yokoyama (89) B. Pls. VIII., IX.   [3] Thomas (13).

seed-bearing leaves of Ginkgo.   We have as yet but little to guide
us in our attempts to trace the ancestry of that remarkable survival
Ginkgo biloba, and it is highly probable that, if more satisfactory
records of older members of the Ginkgoales were available, we
should be able considerably to extend the range in morphological
characters which in the present representative of the group is
comparatively restricted.   The numerous leaf-bearing axes, many
of them branched, referred by Fontaine[1] to his genus *Nageiopsis*,
should not be overlooked from the point of view of their possible
relationship to *Podozamites*.   The branching habit of these Potomac
specimens is no bar to an affinity to *Podozamites* if examples of
the genus are no longer to be interpreted in terms of a Cycadean
frond.   Berry[2], in a recent revision of Fontaine's genus, refers
some species to *Podozamites* which he still regards as Cycadean.
It is interesting to find on a specimen of *Nageiopsis* figured by
Fontaine a zone of crowded scars[3] (fig. 816, *s*) such as may be seen
on an *Agathis* shoot.

Many of the leaves described as *Podozamites* are of little value
as evidence of the occurrence of the genus.   In the case of imperfect
specimens of detached leaves it is often impossible to distinguish
between *Podozamites*, *Phoenicopsis*, and the leaves of Araucarian
plants, or pinnae of some species of *Zamites*.   It is therefore not
possible to state with confidence the geological range of the genus.
Undoubted examples of *Podozamites* are essentially Rhaetic and
Jurassic fossils, and there can be no doubt as to the abundance
and wide geographical range of the genus in both these periods.
Such leaves as those recently figured by Hollick[4] from Cretaceous
beds of Long Island as *Podozamites lanceolatus* certainly agree
closely in form with that species, but they are all detached specimens: the fragmentary leaves from the Middle Cretaceous beds of
the Amboy clays described as *P. angustifolius* (Eich.) and *P. marginatus* Heer[5] afford no proof of the presence of *Podozamites*:
similarly Velenovský's species *P. miocenica* from Bohemia[6] might
equally well be referred to the genus *Dammarites*.   Well preserved specimens have been described by Zeiller[7] from the

[1] Fontaine (89) B. p. 195.   [2] Berry (10).
[3] Fontaine (89) B. Pl. LXXVI. fig. 5.   [4] Hollick (12) Pls. 162, 163.
[5] Newberry and Hollick (95) Pl. XIII. figs. 1—6.
[6] Velenovský (81) Pl. I. figs. 18—20.   [7] Zeiller (03) B.

Rhaetic flora of Tonkin[1]. There are few satisfactory records of
the genus from the southern hemisphere, and we have no actual
proof of its existence in India, though Feistmantel[2] refers to
Podozamites detached leaves, which, as an examination of the
original specimens shows, may have been borne on *Podozamites*
shoots, but they may also be examples of *Phoenicopsis*. One of
the leaves figured by Feistmantel from the
Jabalpur group as *Podozamites lanceolatus*
is reproduced in fig. 813 from a drawing
recently made from the actual fossil: the
lamina shows several fine parallel striations between the more clearly marked
veins. Miss Holden, who examined the
carbonised cuticles of some of the Indian
leaves lent to me by the Director of the
Indian Geological Survey, found that the
epidermal cells have straight walls and
the stomata, usually with six accessory
cells, occur in the intercostal regions on
both surfaces: the characters of the epidermis are favourable to a relationship
with the Coniferales and they are not

FIG. 813. A, Specimen figured
by Feistmantel as *Podo-
zamites lanceolatus*. B, Piece
of lamina enlarged. (Cal-
cutta Museum; Geol. Surv.,
India.)

inconsistent with the inclusion of the fossils in the genus *Phoenicopsis*. Halle[3] has recently described some imperfect leaves
from Patagonia as probably *Podozamites*, but as he pertinently
says the evidence is not enough to establish the correctness of the
determination. Some of the leaves from the Potomac beds included by Fontaine in *Podozamites* are of little value as authentic
records of the genus, but there is still considerable doubt as to
the relationship between this genus and *Nageiopsis* which was very
abundant in the Potomac flora. The leaves figured by Fontaine[4]
from the Jurassic of Oregon and from Alaska are also not above
suspicion as records of *Podozamites*, though there is no doubt that
the genus was represented in some of the North American floras.

---

[1] Walkom (17) p. 20.
[2] Feistmantel (82) p. 39, Pl. II. figs. 2—5.
[3] Halle (13).
[4] Fontaine in Ward (05) B. Pls. XXIV., XXV., XLIV. Knowlton (14) Pls. V., VI.

Podozamites distans (Presl).

This Rhaetic species[1] differs very slightly from the Jurassic type *Podozamites lanceolatus* and there has been much confusion on the part of authors between the two forms[2] which, indeed, cannot always be clearly distinguished. *P. distans* is often represented only by detached leaves but in some specimens the shoot reaches a length of 20 cm. The slender axis bears distant, lanceolate or ovate-lanceolate leaves, sometimes slightly falcate with a rounded or obtusely pointed apex and gradually contracted at the base which, as Zeiller says, may assume the form of a very short pedicel. The leaves may be 4—7 cm. long and 5—14 mm. broad; the veins, 0·4—0·7 mm. apart, are dichotomously branched in the proximal portion of the lamina but elsewhere parallel and simple, except that they slightly converge at the apex. The epidermal cells have straight walls and the stomata, which occur on the lower surface, either in rows or scattered, are surrounded by small subsidiary cells (fig. 812, E). The leaves are usually rather broader in proportion to their length than those of *P. lanceolatus* and the apex is less pointed. Braun[3] instituted two varieties, *longifolius* and *latifolius,* and to these Schenk[4] added others. The species is recorded from the Rhaetic of Scania[5], where it is abundant, from Persia[6], Tonkin, and many other regions: it occurs also in Jurassic strata[7], but on the whole *P. distans* is a characteristic member of Rhaetic floras.

The Rhaetic species *Podozamites Schenki* Heer[8] founded on Jurassic specimens from Siberia and described by Zeiller[9] and Nathorst from Tonkin, Persia, and Sweden is distinguished from *P. distans* by the smaller shoots and the more acuminate leaves.

Podozamites lanceolatus (Lindley and Hutton).

The type-specimen of *Zamia lanceolata* Lind. and Hutt.[10] in the Manchester Museum from the Middle Jurassic beds of Yorkshire consists of a slender axis bearing scattered and distant linear-

---

[1] Sternberg (38) A. Pl. xli. fig. 1.
[2] See Zeiller (03) B. p. 159 for examples of *P distans* referred to *P. lanceolatus.*
[3] Braun (47) p. 85.   [4] Schenk (67) A. Pls. xxxv., xxxvi.
[5] Nathorst (78) B. Pls. xiii., xv.
[6] Zeiller (05) p. 193.   [7] For references, see Zeiller (03) B. p. 159.
[8] Heer (77) ii. p. 45.   [9] Zeiller (03) B. Pl. xlii.
[10] Lindley and Hutton (36) A. Pl. xciv.

lanceolate leaves up to 7 cm. long and 7 mm. broad; the lamina
has a tapered acuminate apex and a less gradually though not
abruptly contracted base. The leaves differ from the leaflets of
Zamia which they superficially resemble in their less abruptly
contracted proximal end. In habit a shoot of *P. lanceolatus*
(fig. 811) very closely resembles the fronds of *Zamia media*. Some
of the leaves in the type-specimen appear to be laterally attached,
while others appear to be given off from the upper surface. The
leaves of this as of other species are frequently found detached.
The variability in the form of the leaves has led to the employment
of several varietal names, and if not used too freely the addition
of some descriptive term to the specific name may often serve a
useful purpose. Fig. 811 represents a good example of the species
from the Yorkshire coast. The method of attachment of the
leaves is not always clear, but their irregular distribution and the
slender axis are features more in accordance with a foliage-shoot
than a pinnate frond. *Podozamites lanceolatus* is a widely distributed Jurassic species[1] recorded from many European localities
extending to North Siberia and Spitzbergen as well as from North
America, Turkestan, Afghanistan, Japan, China, and elsewhere.
The specimens figured by Feistmantel from Upper Gondwana
rocks in India as *P. lanceolatus* (fig. 813) should, I am inclined to
think, be assigned to *Phoenicopsis*.

Podozamites Reinii Geyler; *Podozamites stonesfieldensis* Seward;
      Podozamites Griesbachi Seward.

These species from Jurassic strata serve as examples of a broader
type of the genus represented in the last two species by detached
leaves only. In these as in many other cases one cannot feel
absolute confidence as to the correctness of the determination. In
some of the Japanese examples of *P. Reinii*[2] (fig. 814) the broadly
oval leaves are attached to a slender axis. *P. stonesfieldensis*[3]
from the Great Oolite of Stonesfield is probably identical with the
leaves originally described by Buckman as *Naiadea ovata* and *Lilia
lanceolata*: the leaves are oblong-ovate, approximately 8 × 3·5 cm.;
the lamina is rather abruptly contracted at the base and more

---

[1] For references see Seward (00) B. p. 242; (07²); (11).
[2] Geyler (77) B. Pls. xxxiii., xxxiv.
[3] Seward (04) B. p. 121, Pl. iii. fig. 4; Pl. xi. figs. 1, 2.

gradually tapered towards the apex; the veins are slightly more
than 1 mm. apart and converge at each
end of the lamina. The species resembles *P. lanceolatus* var. *latifolius*
figured by Schenk[1] from China.

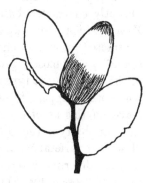

Podozamites Griesbachi[2] is a Jurassic
species from Afghanistan similar in the
shape and size of the leaves to the other
two species; it has a coarser venation
than *P. Reinii* but the venation is still
coarser in *P. stonesfieldensis*. A similar
form of leaf is figured by Velenovský[3]
from the Lower Cretaceous of Bohemia
as *P. striatus*.

FIG. 814.   *Podozamites Reinii*.
(After Geyler; ¾ nat. size.)

Podozamites Kidstoni Etheridge.

In this type[4], usually represented by detached leaves, the
lamina is smaller than in other species, short and broad with a
blunt apex and abruptly contracted at the base.   The species is
recorded from Afghanistan[5] and similar leaves are figured by
Chapman[6] from Jurassic rocks in Victoria, Australia.   Etheridge's
type-specimen is from the Burrum Coal Measures (? Triassic) of
Queensland.   This form of leaf agrees closely with some of
Fontaine's Potomac species of *Nageiopsis*, *e.g. N. obtusa* and *N.
heterophylla*[7], and a similar though not identical form is described
by Nathorst[8] from the Rhaetic of Sweden as *P. ovalis*, distinguished
by its broadly rounded and mucronate apex.

NAGEIOPSIS.   Fontaine.

This genus was established[9] for vegetative shoots abundantly
represented in the Potomac flora many of which closely resemble
Podozamites, though differing in their branched habit and in the
veins being less convergent in the apical region of the lamina.

[1] Schenk (83) A. Pl. XLIX. figs. 4 b, 5.
[2] Seward (12) p. 36, Pl. IV. fig. 58; Pl. VI. fig. 79.
[3] Velenovský (85) B. Pl. II. fig. 8.
[4] Jack and Etheridge (92) B. p. 317, Pl. XVIII. figs. 6, 7.
[5] Seward (12) Pl. IV. fig. 39.   [6] Chapman (09) Pl. XVIII.
[7] Fontaine (89) B. Pls. LXXXIV., LXXXV.
[8] Nathorst (78) B. Pl. XIII. fig. 5.   [9] Fontaine (89) B. p. 194.

Fontaine recognised the similarity between *Nageiopsis* and shoots of *Podocarpus* belonging to the section *Nageia*: this suggested the choice of the generic name. Berry[1] in his revision of *Nageiopsis* transfers some of Fontaine's species to *Podozamites*; he also reduces the number of the species retained in *Nageiopsis* on the ground that Fontaine attached too much importance to variations in the size and form of the leaves. I have elsewhere suggested[2] that some of the shoots referred to *Nageiopsis* may be Araucarian, as in habit they closely resemble *Araucaria Bidwilli* and *Agathis*. Until reproductive organs are discovered it is impossible to speak with confidence with regard to the position of the genus. It may be closely allied to *Podozamites* or, as Fontaine believed, it may be related to *Podocarpus*. It should be noted that some of the specimens included by Fontaine in *Nageiopsis* are hardly distinguishable from *Zamites Buchianus*[3].

Trees or shrubs characterised by irregularly branched foliage-shoots bearing leaves usually in two ranks but spirally attached; the leaves exhibit a wide range in size and shape, long and linear or lanceolate, acute or subacute, more or less abruptly contracted at the proximal end and attached by a very short stalk; there are several parallel veins dichotomously branched near the base of the lamina.

Nageiopsis anglica Seward.

This species[4] founded on the small specimen represented in fig. 815 has distichous leaves 1—1·5 cm. long with several parallel veins. The English Jurassic type agrees generally with *Nageiopsis microphylla* Font. and *N. descrescens* Font.: a similar form is recorded from

Fig. 815. *Nageiopsis anglica*. (From a specimen in the Whitby Museum; ⅔ nat. size.)

the Wealden beds of Sussex[5]. Though satisfactory evidence of affinity is lacking it is permissible to suggest an Araucarian affinity.

Nageiopsis longifolia Fontaine.

The linear-lanceolate leaves reach a length of 8—20 cm. and 5 mm. to 1·3 cm. in breadth; there are 9—12 veins unbranched

[1] Berry (10).　　　　[2] Seward (12) p. 33.　　　　[3] Berry (11) Pl. LXI.

[4] Seward (00) B. p. 288, fig. 51.　　　　[5] *Ibid.* (95) A. p. 211, Pl. XII. fig. 3.

except at the base (fig. 816); the lamina is abruptly narrowed and
attached by a short and slightly twisted stalk.   Though apparently
inserted laterally the leaves are in all probability spirally disposed.
In one of Fontaine's figures there is a group of small scars, fig. 816, *s*,
presumably of bud-scales, at one place on the axis.   This species

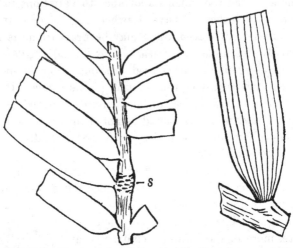

Fig. 816.   *Nageiopsis longifolia.*   (A, after Fontaine; nat. size.)

is very abundant in the Potomac beds of Virginia and Maryland[1]
and is recorded by Fontaine from several other localities though
for the most part on slender evidence.

Nageiopsis zamioides Fontaine.

In this species[2] the leaves are ovate-lanceolate and shorter than
in *N. longifolia*; they have a maximum breadth of 2 cm. and
reach a length of 7 cm.   The example described from English
Wealden beds as *Nageiopsis* sp. *cf. N. heterophylla* agrees closely
with Fontaine's Potomac species.

[1] Fontaine (89) B. p. 195, Pls. LXXV.—LXXIX.; LXXXV.; Berry (11) p. 384.
[2] *Ibid.* p. 196, Pls. LXXIX.—LXXXI.; Berry (11) p. 386, Pls. LXII., LXIII.

CHAPTER LII.

I. RECENT.

In this group of Gymnosperms are included three genera, *Ephedra*, *Gnetum*, and *Welwitschia*. They differ widely from one another in vegetative features, and *Ephedra*, the most primitive, is distinguished by certain important peculiarities of the reproductive organs.

EPHEDROIDEAE.   *Ephedra*.

GNETOIDEAE.   *Gnetum, Welwitschia*.

Having regard to our exceedingly meagre knowledge of fossil representatives, it is unnecessary to deal fully with the recent types[1], but the members of this aberrant section of seed-plants exhibit morphological characters of interest from the point of view of comparison with the Bennettitales and the Angiosperms. Though in external appearance the three genera are poles asunder, they have in common certain features both in the vegetative and reproductive organs which differentiate them from all other Gymnosperms and connect them more closely than the Cycads or Conifers with the Angiosperms. The leaves are opposite; the secondary xylem contains vessels in addition to tracheids; the male and female flowers are characterised by the possession of one or two envelopes in addition to the usual single integument; the inflorescences, occasionally though not as a rule bisporangiate[2], are distinguished by a dichasial system of branching, a character foreign to Gymnosperms as a whole though exhibited by the stem of *Wielandiella*, a member of the Bennettitales. There are good reasons for believing that pollination is effected by insects[3] in

---

[1] For a general account of the group, with illustrations, the student is referred to Wettstein (11) and Lotsy (11), or to Coulter and Chamberlain (10).

[2] Land (04); Berridge and Sanday (07) p. 127; Lotsy (11) p. 293.

[3] Pearson (06[2]) p. 274; (09) p. 343; Berridge and Sanday (07) p. 172; Karsten (92); Porsch (10).

Welwitschia, in some species of *Gnetum*, and occasionally in *Ephedra*. The seeds are albuminous and the embryos have two cotyledons. Archegonia are produced in the female prothallus of *Ephedra* while in *Gnetum* and *Welwitschia* these organs are represented by single cells as in the Angiosperms or by nuclei.

Ephedra[1] has a wide distribution in the warm temperate regions of the northern hemisphere: in America it occurs on both sides of the equator and from the Mediterranean region it reaches to Brittany in the west and North Africa in the south. *Gnetum* extends both east and west in the tropics: *Gnetum scandens* is a widely spread Asiatic species, and the genus occurs in Angola and in some other parts of Africa. *Welwitschia* is confined to a littoral strip of desert in extra-tropical South Africa from 14° S. to 23° S. and has not been found more than 50 miles from the coast.

Ephedra.

Shrubs, in some species with climbing branches, characterised by an *Equisetum*-like habit of the younger shoots which form long jointed and slightly fluted branches bearing whorls of two or sometimes three, scaly, concrescent leaves. In rare cases, *e.g.* *Ephedra altissima*, the leaves may reach 3 cm. in length and a breadth of 1—1·5 mm. Monoecious or dioecious; flowers unisexual; bisexual inflorescences are recorded in *E. campylopoda*[2]. The female flowers occur in strobili on a dichasially branched inflorescence; each strobilus consists of three pairs of bracts, in some species the bracts are more numerous. There is generally a single ovule in *E. altissima*, but in most species there are two or as many as six ovules in a single strobilus. The ovules are enclosed by two envelopes regarded by some authors as a perianth and an integument and by others as two integuments. In *E. distachya*, as described by Mrs Thoday and Miss Berridge[3], two vascular bundles supply the outer envelope (outer integument) one running up each angle of the flattened side of the flower. The thin inner integument becomes free from the nucellus at a distance of two-thirds its length and projects beyond the outer envelope as a long style-like micropylar tube. A ring of bundles runs a short distance up the inner integument but ends low down in a mass of transfusion-

---

[1] Stapf (89).   [2] Wettstein (11) p. 417.   [3] Thoday (Sykes) and Berridge (12).

tracheids. Attention is drawn to the resemblance of the outer integument to the integument of *Bennettites*, and the single ovule of *Ephedra* is considered to be the representative of the whole ovulate strobilus of *Bennettites* of which it is a much reduced derivative. There is a deep pollen-chamber at the apex of the nucellus[1] and there are 2—8 long-necked archegonia at the summit of the prothallus (endosperm). In its female prothallus and sexual apparatus *Ephedra* differs considerably from *Gnetum* and *Welwitschia*. The second envelope forms the hard shell of the seed which is enclosed by bracts either in the form of membranous wings (sect. Alatae) or as a red or yellow flesh (sect. Pseudobaccatae).

Ewart[2] found that the seeds of *Ephedra distachya* germinated after 93 days' immersion in sea-water.

The male flowers[3] occur also in strobili on dichasial inflorescences, a single flower occurs in the axil of each of the fertile bracts. A flower consists of a short axis bearing a pair of membranous appendages and the flower-axis is prolonged as a simple or bifid stalk bearing bilocular synangia, 2—6 according to the species. In some cases the central stalk or antherophore of the flower is flattened and laminar[4] instead of the usual cylindrical form: it has been interpreted both as an axial and a foliar structure, but the latter interpretation is probably correct. Arber and Parkin[5] regard the antherophore as having been formed from two fused members, and this view is adopted by Mrs Thoday and Miss Berridge. On the basis of this interpretation the microsporophylls of *Ephedra*, represented by the antherophore, are considered to be homologous with the disc of sporophylls of a Benettitean flower and with the stamens in a male flower of *Welwitschia*. Anatomically[6] *Ephedra* exhibits a closer agreement with the Conifers and in some respects with the Dicotyledons than with recent Cycads or the Bennettitales. The presence of vessels in the secondary xylem is an Angiospermous feature though in structure they differ from the Angiospermous type; the pitting of the tracheids is in the main Abietineous but the occurrence of compressed pits

[1] Land (04); for other references, see Lignier and Tison (12); also Sigrianski (13)
[2] Ewart (08).
[3] Thibout (96) gives a good account of the male flowers of the Gnetales.
[4] Thoday and Berridge (12) p. 970.   [5] Arber and Parkin (08).
[6] Thompson (12²). See also Jeffrey (17) p. 357.

furnishes a point of contact with the Araucarineae; rims of Sanio
occur and xylem-parenchyma is abundant; the medullary rays
are multiseriate as in Dicotyledons. The bast on the other hand
is essentially gymnospermous. The occasional occurrence of spiral
bands in the tracheids and the presence of lignified trabeculae in
the xylem-elements are other Coniferous traits. The leaf-trace is
double, a feature met with in *Agathis* as well as in recent Cycads
but not in the Bennettitales. The anatomy of seedlings affords
further indications of resemblance to *Araucaria* and the Podocarps[1].
It would seem, then, that the case for a relationship between the
Gnetales and the Bennettitales founded on the facts of floral
morphology does not derive support from the anatomical features
of the most primitive genus of the group.

Gnetum.

Small trees or climbers with long and slender stems; the internodes, sometimes reaching a length of 15 cm., bear pairs of ovate-
oblong or lanceolate-acuminate leaves 11—18 cm. long by 4—7 cm.
broad. The leaves[2] agree in form and venation with those of
many Dicotyledons and could not be distinguished from them in
a fossil state. The epidermal cells have undulate walls. The
flowers are in spikes; at each node two fused bracts form a cupular
structure in the axil of which the male or female flowers are borne
on an annular swelling. The male flowers[3] are in 3—5 whorls:
each consists of an envelope of two coherent leaves enclosing a
central column, as in *Ephedra,* which bears at the apex one, two,
or rarely four unilocular sporangia or reduced synangia. The
antherophore eventually elongates and pushes the anthers through
an aperture at the summit of the floral envelope[4]. In appearance
the antherophore of *Gnetum* approaches most closely to the stamen
of an Angiosperm. The female flowers[5] occur in a single series,
5—8 in a whorl; each consists of an ovule surrounded by three
envelopes; the outermost is coloured and succulent, the middle
envelope or outer integument is differentiated after fertilisation
into an inner sclerotesta and an outer sarcotesta, while the innermost covering is prolonged as a micropylar tube. There are no

[1] Hill and de Fraine (10) p. 329.   [2] Karsten (93); Lotsy (11) p. 347, fig. 209.
[3] Caporn (16).   [4] See also Pearson (15).
[5] Thoday (Sykes) (11); Lignier and Tison (13). See also Pearson (17).

archegonia: in some species the megaspore contains numerous free
nuclei all of which are potentially sexual; after fertilisation a
sterile nutritive tissue, or endosperm, is formed in the lower part
of the spore. In *Gnetum Gnemon* the endosperm is often formed
before fertilisation. In the great reduction of the female apparatus
and in the nature of the endosperm *Gnetum*[1] agrees much more
closely with *Welwitschia* than with *Ephedra*. Attention has been
called to certain resemblances between the seed of *Gnetum* and

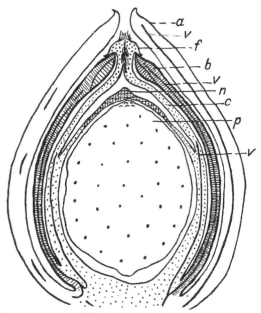

Fɪɢ. 817. Seed of *Gnetum africanum*. *a*, outermost envelope; *v*, vascular strands;
b, outer integument; *f*, flange of micropylar tube; *n*, nucellar cap; *c*, inner
integument; *p*, endosperm. (After Mrs Thoday.)

that of *Bennettites*[2]. For convenience of comparison a diagram-
matic section of a seed of *Gnetum africanum* is reproduced in
fig. 817. The outermost envelope, *a*, forms a green succulent
covering free to the apex; the middle envelope, *b* (outer integument),
is free from the other envelopes except at the apex where it is
locked to the inner integument by the downward growth of a
flange, *f*, from the apical region of the inner integument; the inner

---

[1] Pearson (09); see also Karsten (92); (93²); Lotsy (99).

[2] Berridge (11); (12); Thoday (Sykes) (11).

integument, *c*, is united to the nucellus for about two-thirds of the length of the seed-body. The nucellus consists of a few layers of cells and at the apex forms a nucellar cap, *n*, the cells of which are lignified; this cap is supported by a short tent-pole produced from the summit of the endosperm. A ring of vascular bundles enters the base of the seed and forms three series, the two outer pass up to the tip of both the two outer coverings, *v*, *v*, and the inner series extends up the inner integument as far as the level where nucellus and integument part company. In the seed shown in fig. 817 the micropyle is closed and the tissue in the closed region of the canal is lignified and dark. Among other features in which this seed agrees with that of *Bennettites Morierei* is the inner zone of the outer integument, composed of a palisade and a fibrous layer; the fibrous layer becomes five-angled in the upper part of the seed[1] and in transverse section presents a striking similarity to sections through the same region of a *Bennettites* seed (figs. 524, 527, Vol. III. pp. 397, 402). The oval fleshy seeds are able to germinate after lying some months in sea-water. There is a fairly close resemblance between *Gnetum* and *Ephedra* as regards anatomical characters, but some species of *Gnetum* (sect. *Thoa*) are character- ised by the formation of successive cambial cylinders as in *Cycas*. Strasburger[2] pointed out that in the vascular bundles of the leaves the parenchyma of the medullary rays forms continuous plates, a gymnospermous character.

Welwitschia.

This remarkable genus, discovered by Welwitsch in 1860 and described by Sir Joseph Hooker[3], presents striking peculiarities in the habit of the vegetative body. A *Welwitschia* plant has been aptly termed an adult seedling[4]; the large and squat tuberous stem, morphologically the swollen hypocotyl, may be as much as 4·5 m. in girth. The seedling has two cotyledons and an ex- ceptionally long radicle: at an early stage a pair of isobilateral leaves is produced at right-angles to the cotyledons and these

---

[1] Thoday (Sykes) (11) p. 1116, text-fig. 11.

[2] Strasburger (91) p. 148. For an account of the anatomy of *Gnetum*, see Duthie (12); La Rivière (16); Thomson, M. R. H. (16).

[3] Hooker, J. D. (63). For figures of *Welwitschia*, see also *Gard. Chron.* Jan. 22, p. 49, 1910.   [4] Sykes (10²) p. 333.

persist as the only leaves throughout the long life of the plant, attaining a length of 5 m. The tough lamina is torn into strips by the wind and 'the extraordinary appearance of the shapeless mass of coiled and twisted leaf-ribands standing out in bold relief from the sharp glistening dead landscape passes description[1].' The venation is parallel and there are numerous cross-connexions, some ending blindly in the mesophyll[2]. *Welwitschia* is dioecious and the flowers are borne in inflorescences with a dichasial branch-system produced from pits on the crown of the stem; the female inflorescences, which are larger than the male, reach a length of 30 cm. and bear cones about 7 cm. long. The female flowers occur singly in the axils of bracts which form four orthostichies giving a four-angled form to the cones. Each flower may produce two small leaf-rudiments[3], but the flower proper consists of an ovule with two envelopes; the outer, called by Hooker the perianth, is considerably extended tangentially and in the ripe seed forms a wing-like appendage producing an appearance almost identical with that of some *Samaropsis* seeds. The inner integument is prolonged upwards like a long and slender hollow bristle for a distance of 4—5 mm. beyond the upper edge of the subtending bract. The inner envelope has no vascular supply. The secretion of sugar in the micropylar tube attracts the pollinating insect *Odontopus sexpunctulatus*. The staminate cones are smaller and the subtending bracts connate. The outer envelope of the flower is formed of two membranous segments without vascular bundles which may be styled lateral prophylls of the axillary shoot: internal to these are two fused members forming a sac-like investment with free rounded lobes also without a vascular supply. Within these perianth-segments is the staminal tube bearing six free stamens each supplied with a vascular bundle and bearing a terminal trilocular synangium (fig. 818). The centre of the flower is occupied by a pyriform ovule surrounded by a thin integument continued as a slightly kinked stylar tube terminating in a flat stigmatic disc 1 mm. in diameter. There is no embryo-sac but the nucellus acts as a nectary, the drop-mechanism of the functional

---

[1] Pearson (06[2]) p. 270.

[2] de Bary (84) A. fig. 157; Sykes (10[2]); Takeda (13[2]).

[3] Lignier and Tison (12).

ovule in the female flower being retained in the sterile ovule of the male[1].

In contrast to the indefinite, spirally disposed, bracts or

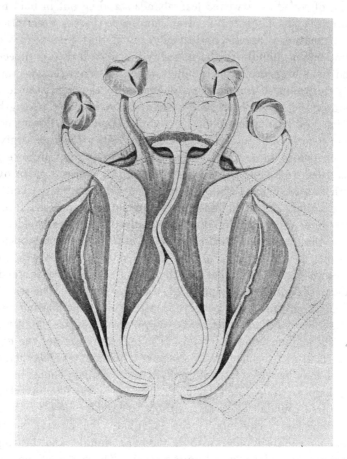

Fig. 818. *Welwitschia mirabilis*. Staminate flower (sectional elevation), subtending bract and the two missing stamens indicated by dotted lines. (From a drawing kindly supplied by Dr A. H. Church.)

perianth of *Cycadeoidea* flowers *Welwitschia* has only two connate segments, and the staminal disc of *Welwitschia* is considerably

[1] This account is based on the excellent description with illustrations by Dr Church (14).

reduced; the gynoecium consists of only one functional ovule instead of an indefinite number as in *Cycadeoidea*. Church regards the
resemblance between the flowers of these two genera as an instance
of parallel development, which does not imply relationship. He
thinks there is 'no indication whatever of any relation to the carpellary flowers of the Angiosperms.'

Hooker's account of the ovule has recently been considerably
extended by the important researches of Pearson[1]. The megaspore
consists of two regions, an upper fertile and a lower sterile portion;
each is composed of 'cells' with more than one nucleus; some of the
'cells' of the fertile region grow upwards as tubes into the nucellar
cone where pollen-tubes are encountered and fertilisation ensues.
The nuclei in each 'cell' of the sterile region fuse and uninucleate
cells are produced; this tissue now grows considerably in size and
cell-divisions occur resulting in the formation of an endosperm.
Pearson regards the free nuclei that are in the embryo-sac at the
time of septation into the multinucleate 'cells' as all alike, and all
potential gametes. It follows, therefore, that the endosperm
formed in the lower portion of the sac is the product of fusion of
sexual nuclei; it is not a gametophyte or a sporophyte and Pearson
proposes for it the new term trophophyte, 'a bye-product resulting
from the fusion of potentially sexual nuclei and functioning in the
same manner as the prothallus of the lower seed-plants.' More
recent work by this author confirms his opinion that the endosperm
of *Gnetum* is also a trophophyte.

For an account of the anatomy of *Welwitschia* and *Gnetum* the
student is referred to original sources. Miss Sykes[2] called attention to certain interesting characters,—the occurrence of reticulately
pitted protoxylem elements in the stem, the arrangement of
separate and not contiguous bordered pits in 1—2 rows on the
tracheids, and to the presence of concentric steles and inversely
orientated bands of vascular tissue in the stem and inflorescences
closely resembling Medullosean features.

In certain respects the Gnetales are closer than the Conifers or

[1] Pearson (06²); (09).

[2] Sykes (10); (10²); Worsdell (01²); Pearson (12); Bower (81); (82); Hill and de
Fraine (10); Boodle and Worsdell (94); Mary R. H. Thomson (16); Henriette
C. C. La Rivière (16).

the Cycads to the Angiosperms[1]. It has, for example, been suggested by Hallier[2] that they are reduced Dicotyledons comparable
with the Loranthaceae and Myxodendraceae; while Lignier and
Tison[3] regard them as a group of Angiosperms nearest to the
Amentales. The question of relationship between the Gnetales
and the Angiosperms, especially the difficult problems connected
with the endosperm, was fully considered by Pearson[4] in a paper
on the reproductive organs of *Gnetum Gnemon* published in 1915,
and in a later contribution[5], published after his death, the morphological problems are reviewed in the light of more recent work.
The same subject is dealt with by Prof. Thompson[6] in a recent
paper in which he calls attention to the form of the inflorescence,
the arrangement of the parts of the flowers, the presence of an ovary
with a style, the germination of the microspores at some distance
from the nucellus, as evidence of affinity to the Angiosperms, and
concludes that the ancestors of the Angiosperms were 'not far
removed from the genus *Gnetum*.' On the other hand some botanists
prefer to regard the Gnetales as a blindly-ending branch of Gymnosperms with no direct relationship to the Flowering plants. Difficult
as it is to believe that plants so different, when the sum of characters
is considered, as the Gnetales and the Bennettitales are off-shoots of
a common stock, it would be rash to assume that such resemblances
as have been emphasised by Miss Sykes and other authors have no
phylogenetic value.

At the time of his death (November, 1916) Professor Pearson
was engaged upon a volume on the Gnetales: in April, 1916, he
wrote, 'A large part of the book on the Gnetales is written, though
it will need some revision....As to the Gnetalean-Angiosperm
alliance, there must be one, I think, but at present I cannot bring
myself to believe that it is direct[7].' Had Pearson been able to
complete his work it is certain that a statement of his most recent
conclusions would have enabled botanists to obtain a clearer view

[1] Arber and Parkin (07); (08); references to other authors will be found in these
papers. See also Lignier and Tison (12); Lignier (03); Lignier and Tison (11).

[2] Hallier (05) p. 153.

[3] Lignier and Tison (11).

Pearson (15²): additional references to literature are given at the end of this
paper. See also Caporn (16).

[5] Pearson (17).   [6] Thompson, W. P. (16).   [7] Seward (17) p. ix.

of the true position of this puzzling group which, despite the lack of palaeobotanical evidence, is probably a survival from a remote past.

II. Gnetales (Fossil).

Arber and Parkin[1] and other authors have called attention to the lack of any trustworthy records of Gnetalean plants in the sedimentary strata of different periods. Several specimens have been described either as generically identical with *Ephedra* or as probable representatives of the two other members of the group, but while some are incorrectly determined others are too imperfect to be accepted as evidence. In view of the morphological features characteristic of the present members of the Gnetales and the geographical distribution of the species of *Ephedra*, *Gnetum*, and *Welwitschia*, it would seem safe to conclude that the absence of fossil forms is not explicable on the hypothesis of a recent origin of the group, but is rather the result of the imperfection of the geological record and of the difficulty of distinguishing between fragmentary remains of Gnetalean genera and vegetative or reproductive organs of similar external form belonging to other plants. Reference has already been made[2] to certain characters shared by the seeds of *Gnetum* and *Bennettites* and in spite of the great and obvious differences separating the Gnetales and Cycadales it would seem probable that the striking similarity between the seeds of *Gnetum* and those of the Bennettitales has some phylogenetic significance. But even granting a phylogenetic significance to the evidence brought forward by Mrs Thoday and other authors, we have still to admit that an indication of some former connexion between the Gnetales and the Bennettitalean line is rather the shadow of evidence with regard to the geological history of the Gnetales and not a substantial contribution to our knowledge of the antiquity of this section of the Gymnosperms.

The specimens described by Unger[3] from Eocene beds in Styria as *Ephedrites sotzkianus*, though very similar to those of *Ephedra fragilis* with which they are compared, are too fragmentary to be

---

[1] Arber and Parkin (08) p. 507. [2] See page 463.
[3] Unger (51) p. 159, Pl. xxvi.

accepted as trustworthy records. The pieces of vegetative branches and the paired nuts described by Heer[1] from Jurassic strata in Siberia as *Ephedrites antiquus* are of no botanical value. Portions of inflorescences preserved in amber from the Baltic coast and named by Goeppert and Berendt[2] *Ephedrites Johnianus* and similar specimens referred by Goeppert from the same Oligocene beds to *Ephedra Mengeana* have been identified by Conwentz[3] as fragments of flowering shoots of a Loranthaceous genus, *Patzea*. Engelhardt[4] refers some slender branches from Tertiary beds in Chile to *Ephedra* but they, like most of the specimens recorded as fossil representatives of the genus, are too incomplete to be accepted as evidence. In the absence of anatomical data or of well preserved flowers it would be exceedingly difficult to recognise impressions of vegetative shoots of *Ephedra* and to distinguish them from Dicotyledonous twigs of similar habit. Similarly the torn lamina of a *Welwitschia* leaf bears too close a resemblance to other linear parallel-veined leaves to be recognisable unless the preservation is such as to show traces of the characteristic venation mentioned in the account of the recent genus. Comparisons between some fossil seeds and the winged seeds of *Welwitschia*[5], though in some cases possibly justified by actual relationship, cannot be considered to have any importance unless supported by additional evidence. The seeds named by Renault *Gnetopsis* and subsequently investigated by Oliver and Salisbury[6] are now recognised as types closely allied to *Lagenostoma* and other Pteridosperm seeds from Carboniferous rocks.

In their monograph of the Pliocene Floras of the Dutch-Prussian Border Mr and Mrs Clement Reid figure under the name *Gnetum scandens* var. *robustum*[7] a piece of axis 8 mm. long and 4 mm. broad showing eight nodes bearing crowded scars of some deciduous appendages. The authors speak of the specimen as 'a portion of a male inflorescence of a *Gnetum*...so close to that of the living *G. scandens* that we cannot separate it.' If their

[1] Heer (77) ii. p. 82, Pls. xiv., xv.

[2] Goeppert and Berendt (45) A. Pls. iv., v.; Goeppert and Menge (83) A. Pl. xvi.

[3] Conwentz (86) pp. 136, 138, Pl. xiii. figs. 8—20.

[4] Engelhardt (91) p. 647.   [5] Seward (04) B. pp. 19, 20.

[6] Oliver and Salisbury (11) p. 34.

[7] Reid, C. and E. M. (15) p. 55, Pl. xx. fig. 27.

identification is correct—and though the evidence is hardly conclusive the resemblance between the fragment from Renver and an inflorescence axis of *Gnetum* is undoubtedly striking—it points to the occurrence in a Pliocene European flora of a genus that is now mainly tropical and which had not so far been recognised with any certainty in a fossil state.

The striking resemblance of *Gnetum* leaves to those of some Dicotyledons is an obvious difficulty in the way of the identification of impressions.

It is among the oldest examples of supposed Dicotyledons that search should be made for possible representatives of the genus *Gnetum*. Among the earliest records of Angiosperms are those described by Fontaine[1] from the Patuxent series of the Potomac formation which rests on Palaeozoic crystalline rocks and contains the remains of a flora that is clearly Jurassic or Wealden in its general facies; but with Jurassic Gymnosperms and Ferns are associated some Dicotyledon-like leaves of ovate and linear form for some of which Fontaine instituted the genera *Rogersia*, *Ficophyllum*, *Proteaephyllum* and referred others to *Ficus*, *Sapindopsis* etc. A revision of the Patuxent fossils by Berry[2] has led to a considerable simplification in nomenclature and to the conclusion that some at least of these Lower Potomac leaves are Gnetalean. A comparison of some of Fontaine's figures of *Ficus virginiensis*, species of *Ficophyllum*, *Proteaephyllum*, and *Rogersia* with a leaf of *Gnetum Gnemon* reveals a very close agreement, as regards form and venation, consistent with Berry's suggestion. It is by no means unlikely that these forerunners of the Dicotyledonous type that occur as foreign elements in a typical Jurassic flora, without an admixture of undoubted Angiosperms like those which occupy an important position in the upper beds of the Potomac formation, may belong to plants more closely allied to *Gnetum* than to any Angiosperm. Attention is especially called to the following species as revised by Berry and illustrated in Fontaine's monograph: *Ficophyllum oblongifolium* (Font.), *Rogersia longifolia* Font., *Proteaephyllum ovatum* Font.[3] It is possible that a careful study of

[1] Fontaine (89) B. pp. 281 *et seq.* See also Seward (14²).
[2] Berry (11) pp. 64, 148, 499, etc.
[3] Fontaine (89) B. Pls. 139, 141, 144, 145, etc.

the venation-characters of these and other fossil leaves may lead
to the discovery of criteria which may enable us to separate the
leaves of *Gnetum* from similar Dicotyledonous foliage.

It is with a keen sense of the incompleteness of my task that
Volume IV. is concluded without any attempt to deal with the
abundant if, in very many cases, undecipherable records of Angiosperms. The omission of this branch of Palaeobotany in what
purports to be a general text-book calls for a word of explanation.
A mere summary of conclusions so far published with regard to
the geological history of Flowering plants would not yield results
commensurate with the labour involved. What is needed is a
critical examination, as far as possible, of the actual specimens
and a careful scrutiny of the evidence on which determinations
are based. It is undoubtedly the fact that a large number of
leaf-impressions are practically valueless as trustworthy data, and
I venture to think that it is only with the cooperation of trained
systematists that any satisfactory estimate can be formed as to
the value of the fragmentary documents preserved in Cretaceous
and Tertiary strata. It is preferable to omit, at least for the present, this part of the subject than for the sake of completeness—
in a treatise that is very far from complete in its treatment of the
groups that have been considered—to essay a task for which the
author recognises that he is very inadequately equipped.

LIST OF WORKS REFERRED TO IN THE TEXT.

(Volumes III. and IV.)

[With a few exceptions this list does not include books and papers given in the Bibliographies in Volumes I. and II.]

The following are some of the Bibliographies which students will find useful for additional references:—Geological Literature added to the Geological Society's Library, published from time to time by the Society; Prof. Zeiller's lists in the 'Revue Générale de Botanique' (Paris); lists given by Arber in the 'Progressus Rei Botanicae' (Leiden), vol. I. Heft i. p. 218, 1907; Jongman's 'Die Palaeobotanische Literatur' (Jena), 1910–13; also the International Catalogue of Scientific Literature (Botany and Geology). For the Literature dealing with Cretaceous plants the student should refer to Dr Marie Stopes' 'Cretaceous Flora,' I. and II. (British Museum Catalogues, 1913, 1915).

The dates of books published in parts given in the footnotes to this volume are as a rule those of the concluding part. For the dates of separate parts of books relating to Palaeozoic floras the student is referred to Prof. Zeiller's valuable list at the end of the 'Flore Fossile du Bassin Houiller de Valenciennes.' Useful bibliographies of the writings of Saporta, Heer, and Ettingshausen have been compiled by Zeiller (96), Malloizel and Zeiller (N.D.), and Krasser (97).

Aase, Hannah C. (15) Vascular anatomy of the megasporophylls of Conifers. *Botanical Gazette*, vol. LX. p. 277.

Affourtit, M. F. A. and **H. C. C. La Rivière.** (15) On the ribbing of the seeds of *Ginkgo*. *Annals of Botany*, vol. XXIX. p. 591.

Andersson, J. G. (10) Die Veränderungen des Klimas seit dem Maximum der letzten Eiszeit. (Collection of papers published by the Int. Geol. Congress; edited by J. G. Andersson.) *Stockholm.*

Andrews, E. B. (75) Descriptions of Fossil Plants from the Coal Measures of Ohio. *Geol. Surv. Ohio.*

Antevs, E. (14) *Lepidopteris Ottonis* (Göpp.) Schimp. and *Antholithus Zeilleri* Nath. *K. Svensk. Vetenskapsakad. Hand.* Bd. LI. No. 7.

—— (14²) The Swedish Species of *Ptilozamites* Nath. *Ibid.* Bd. LI. No. 10.

—— (16) Das Fehlen resp. Vorkommen der Jahresringe in Paläo- und Mesozoischen Hölzern und das klimatische Zeugnis dieser Erscheinungen. *Geolog. Fören. Stockholm Förhand.* Bd. XXXVIII.

—— (17) Die Jahresringe der Holzgewächse und die Bedeutung derselben als klimatischer Indikator. *Progressus Rei Botanicae*, p. 285.

474 LIST OF WORKS

Arber, Agnes. (See also Robertson, A.) (10) On the structure of the Palaeozoic seed *Mitrospermum compressum* (Will.). *Ann. Bot.* vol. XXIV. p. 491.

—— (14) A note on *Trigonocarpus. Ibid.* vol. XXVIII. p. 195.

Arber, E. A. Newell. (02) Notes on the Binney collection of Coal-Measure Plants. Pt. iii. The type-specimens of *Lyginodendron oldhamium* (Binney). *Proc. Camb. Phil. Soc.* vol. XI. pt. iv. p. 281.

—— (03) On the roots of *Medullosa anglica. Ann. Bot.* vol. XVII. p. 425.

—— (03²) Discussion on Dr Kurtz's paper (1903). *Quart. Journ. Geol. Soc.* vol. LIX. p. 26.

—— (04) *Cupressinoxylon Hookeri* sp. nov. a large silicified tree from Tasmania. *Geol. Mag.* [v], vol. I. p. 7.

—— (05) On some new species of *Lagenostoma,* a type of Pteridospermous seed from the Coal Measures. *Proc. R. Soc.* vol. LXXI. B, p. 245.

—— (07) On Triassic species of the genera *Zamites* and *Pterophyllum,* types of fronds belonging to the Cycadophyta. *Trans. Linn. Soc.* vol. VII. pt. vii. p. 109.

—— (08) On a new Pteridosperm possessing the *Sphenopteris* type of foliage. *Ann. Bot.* vol. XXII. p. 57.

—— (09) On the Fossil Plants of the Waldershare and Fredville series of the Kent Coalfield. *Quart. Journ. Geol. Soc.* vol. LXV. p. 21.

—— (09²) On the affinities of the Triassic plant *Yuccites vogesiacus* Schimp. and Moug. *Geol. Mag.* [v], vol. VI. p. 11.

—— (12) On *Psygmophyllum majus* sp. nov. from the Lower Carboniferous rocks of Newfoundland, together with a Revision of the genus and Remarks on its affinities. *Trans. Linn. Soc.* vol. VII. p. 391.

—— (12²) Fossil Plants from the Kent Coalfield. *Geol. Mag.* [v], vol. IX. p. 97.

—— (13) A preliminary note on the Fossil Plants of the Mount Potts beds, New Zealand, collected by Mr D. G. Lillie, Biologist to Capt. Scott's Antarctic Expedition in the "Terra Nova." *Proc. R. Soc.* vol. LXXXVI. p. 344.

—— (13²) The structure of *Dadoxylon Kayi. Quart. Journ. Geol. Soc.* vol. LXIX. p. 454.

—— (14) A Revision of the Seed impressions of the British Coal Measures. *Ann. Bot.* vol. XXVIII. p. 81.

—— (14²) On the Fossil Flora of the Kent Coalfield. *Quart. Journ. Geol. Soc.* vol. LXX. p. 54.

Arber, E. A. Newell and J. Parkin. (07) On the origin of Angiosperms. *Journ. Linn. Soc.* vol. XXXVIII. p. 29.

—— (08) Studies on the Evolution of the Angiosperms. *Ann. Bot.* vol. XXII. p. 489.

Arnoldi, W. (01) Beiträge zur Morphologie einiger Gymnospermen. *Bull. Nat. Moscow,* No. 4, 1900.

Bailey, I. W. (09) The structure of the wood in the Pineae. *Bot. Gaz.* vol. XLVIII. p. 47.

——— (11) A Cretaceous *Pityoxylon* with marginal tracheids. *Ann. Bot.* vol. XXV. p. 315.

Baily, W. H. (69) Notice of Plant-remains from Beds interstratified with the Basalts in the county of Antrim. *Quart. Journ. Geol. Soc.* vol. XXV. p. 357.

Bain, F. and Sir W. Dawson. (85) Notes on the Geology and Fossil Flora of Prince Edward Island. *Canadian Rec. Sci.* vol. I. (1884–85) p. 154.

Baker, R. T. and H. C. Smith. (10) Research on the Pines of Australia. *Dpt. Public Instruction, Tech. Educ. Ser.* No. 16. *Sydney.*

Bancroft, Nellie. (13) On some Indian Jurassic Gymnosperms and *Rhexoxylon africanum*, a new Medullosean stem. *Trans. Linn. Soc.* vol. VIII. pt. ii. p. 69.

——— (14) Pteridosperm Anatomy and its relation to that of the Cycads. *New Phyt.* vol. XIII. p. 41.

Barber, C. A. (92) On the nature and development of the corky excrescences on stems of *Zanthoxylum*. *Ann. Bot.* vol. VI. p. 155.

——— (98) *Cupressinoxylon vectense*; a fossil Conifer from the Lower Greensand of Shanklin, in the Isle of Wight. *Ann. Bot.* vol. XII. p. 329

Bartholin, C. T. (94) Nogle i den bornholmske Juraformation forekommende Planteforsteninger. *Bot. Tidskrift* (Copenhagen), Bd. XIX. p. 87.

——— (10) Planteforsteninger fra Holsterhus paa Bornholm. *Danmarks Geol. Unders.* II. Raek. No. 24.

Bartlett, A. W. (13) Note on the occurrence of an abnormal bisporangiate strobilus of *Larix europaea* DC. *Ann. Bot.* vol. XXVII. p. 575.

Bassler, H. (16) A Cycadophyte from the North American Coal Measures. *Amer. Journ. Sci.* vol. XLII. p. 21.

Bayer, A. (08) Zur Deutung der weiblichen Blüten der Cupressineen nebst Bemerkungen über *Cryptomeria*. *Beiheft Bot. Cent.* Bd. XXIII. Abt. I. p. 27.

Beissner, L. (91) Handbuch der Nadelholzkunde. *Berlin.*

Bennett, J. J. and R. Brown. (52) Plantae Javanicae rariores. *London,* 1838–52.

Benson, Margaret. (08) On the contents of the pollen-chamber of a specimen of *Lagenostoma ovoides*. *Bot. Gaz.* vol. LV. p. 409.

——— (12) *Cordaites Felicis*, sp. nov., a Cordaitean leaf from the Coal Measures of England. *Ann. Bot.* vol. XXVI. p. 202.

——— (14) *Sphaerostoma ovale* (*Conostoma ovale* et *intermedianum*, Williamson), a Lower Carboniferous ovule from Pettycur, Fifeshire, Scotland. *Trans. R. Soc.* Edinburgh, vol. L. pt. i. No. i. p. 1.

Benson, M. and E. J. Welsford. (09) The morphology of the ovule and female flower of *Juglans regia* and of a few allied genera. *Ann. Bot.* vol. XXIII. p. 623.

Berger, R. (48) De fructibus et seminibus ex formatione lithanthracum. *Diss. Inaug. Vratislaviae.*

Bergeron, J. (84) Note sur les strobiles du *Walchia piniformis. Bull. soc. géol. France* [3], Tome XII. p. 583.

Berridge, E. M. (11) On some points of resemblance between Gnetalean and Bennettitean seeds. *New Phyt.* vol. X. p. 140.

—— (12) The structure of the female strobilus in *Gnetum Gnemon. Ann. Bot.* vol. XXVI. p. 987.

Berridge, Emily M. and Elizabeth Sanday. (07) Oogenesis and embryogeny in *Ephedra distachya. New Phyt.* vol. VI. p. 127.

Berry, E. W. (03) The Flora of the Matawan Formation. *Bull. New York Bot. Gard.* vol. III. No. 9, p. 45.

—— (05) Additions to the fossil Flora from Cliffwood. *Bull. Torrey Bot. Club,* vol. XXXII. p. 43.

—— (06) Contributions to the Mesozoic Flora of the Atlantic Coastal Plain. I. *Ibid.* vol. XXXIII. p. 33.

—— (07) Coastal Plain Amber. *Torreya,* vol. VII. p. 4.

—— (07²) Contributions to the Pleistocene Flora of North Carolina. *Journ. Geol.* vol. XV. No. 4, p. 338.

—— (08) Some Araucarian remains from the Atlantic Coastal Plain. *Bull. Torrey Bot. Club,* vol. XXXV. p. 249.

—— (08²) A Mid-Cretaceous species of *Torreya. Amer. Journ. Sci.* vol. XXV. p. 382.

—— (09) A Miocene Flora from the Virginian Coastal Plain. *Journ. Geol.* vol. XVII. p. 19.

—— (09²) Pleistocene Swamp deposits in Virginia. *Amer. Nat.* vol. XLIII. p. 432.

—— (10) A revision of the Fossil Plants of the genus *Nageiopsis* of Fontaine. *Proc. U. S. Nat. Mus.* vol. XXXVIII. p. 185.

—— (10²) The epidermal characters of *Frenelopsis ramosissimus. Bot. Gaz.* vol. L. p. 305.

—— (10³) Additions to the Pleistocene Flora of New Jersey. *Torreya,* vol. X. p. 261.

—— (10⁴) Contributions to the Mesozoic Flora of the Atlantic Coastal Plain. V. North Carolina. *Bull. Torrey Bot. Club,* vol. XXXVII. p. 181.

—— (11) The Lower Cretaceous deposits of Maryland. (Berry, Clark, and Bibbin.) *Maryland Geol. Surv.*

—— (11²) A Lower Cretaceous species of Schizaeaceae from Eastern North America. *Ann. Bot.* vol. XXV. p. 193.

—— (11³) The Flora of the Raritan Formation. *Geol. Surv. New Jersey, Bull.* 3.

—— (11⁴) A Revision of several genera of Gymnospermous plants from the Potomac group in Maryland and Virginia. *Proc. U. S. Nat. Mus.* vol. XL. p. 289.

—— (12) The age of the plant-bearing shales of the Richmond coalfield. *Amer. Journ. Sci.* vol. XXXIV. p. 224.

Berry, E. W. (12²) Notes on the genus *Widdringtonites*. *Bull. Torr. Bot. Club*, vol. XXXIX. p. 341.

—— (12³) Contributions to the Mesozoic Flora of the Atlantic Coastal Plain. VII. Texas. *Bull. Torrey Bot. Club*, vol. XXXIX. p. 387.

—— (12⁴) Pleistocene plants from the Blue Ridge in Virginia. *Amer. Journ. Sci.* vol. XXXIV. p. 218.

—— (14) The Upper Cretaceous and Eocene floras of South Carolina and Georgia. *U. S. Geol. Surv. Professional papers*, No. 84.

—— (15) The Mississippi River Bluffs at Columbus and Hickman, Kentucky, and their Fossil Flora. *Proc. U. S. Nat. Mus.* vol. XLVIII. p. 293.

—— (16) The Geological history of Gymnosperms. *The Plant World*, vol. XIX. p. 27.

—— (17) Contributions to the Mesozoic Flora of the Atlantic Coastal Plain. XII. Arkansas. *Bull. Torr. Bot. Club*, vol. XLIV. p. 167.

Bertrand, C. E. (74) Anatomie comparée des tiges et des feuilles chez les Gnétacées et les Conifères. *Ann. Sci. nat.* [v], vol. XX. p. 5.

—— (83) Note sur le genre *Vesquia*, Taxinée fossile du Terrain Aachénien de Tournai. *Bull. soc. bot. France* [3], Tome v. p. 293.

—— (89) Les Poroxylons végétaux fossiles de l'époque houillière. *Ann. soc. Belg. de Microscopie*, Tome XIII. Fasc. i.

—— (98) Remarques sur la structure des grains de pollen de *Cordaites*. *Assoc. Franç. pour l'avanc. des Sci. Nantes* (1898), p. 436.

—— (07) Les caractéristiques du genre *Rhabdocarpus* d'après les préparations de la collection B. Renault. *Bull. soc. bot.* [4], Tome VII, p. 654.

—— (07²) Les caractéristiques du genre *Diplotesta* de Brongniart. *Bull. soc. bot. France* [4], tome VII. p. 388.

—— (07³) Les caractéristiques du genre *Leptocaryon* de Brongniart. *Ibid.* p. 452.

—— (07⁴) Les caractéristiques du genre *Taxospermum* de Brongniart. *Ibid.* p. 213.

—— (07⁵) Remarques sur le *Taxospermum angulosum*. *Compt. rend. d'assoc. Franç. pour l'avanc. des sci.* (Reims, 1907), p. 410.

—— (08) Les caractéristiques du *Cycadinocarpus angustodunensis* de B. Renault. *Bull. soc. bot. France* [4], tome VIII. p. 326.

—— (08²) Les caractéristiques du genre *Cardiocarpus* d'après les graines silicifiées étudiées par Ad. Brongniart et B. Renault. *Ibid.* p. 391.

—— (08³) La spécification des *Cardiocarpus* de la collection B. Renault. *Ibid.* p. 454.

—— (09) Sur le genre *Compsotesta* de Ad. Brongniart. *Ann. Jard. Bot. Buit.* [2], suppl. 3.

—— (11) Le Bourgeon femelle des *Cordaites*. *Nancy*.

Bertrand, C. E. et B. Renault. (82) Recherches sur les Poroxylons. *Arch. bot. du Nord de la France*, vol. II. p. 243.

Bertrand, P. (08) Sur les stipes de *Clepsydropsis*. *Compt. Rend.* Nov. 16, 1908.

Bertrand, P. (11) Structure des stipes d'*Asterochlaena laxa* Sterzel. *Mém. soc. géol. Nord*, Tome VII. i.

—— (13) Les Fructifications de Neuroptéridées recueillies dans le terrain houiller du Nord de la France. *Ann. soc. géol. Nord*, Tome XLII. p. 113.

—— (14) État actuel de nos Connaissances sur les Genres 'Cladoxylon' et 'Steloxylon.' *Compt. Rend. de l'assoc. franç. pour l'Avancement des Sciences* (Havre, 1914), p. 446.

Beust, F. (85) Untersuchung über fossile Hölzer aus Grönland. *Neue Denksch. allgem. Schweiz. Ges. gesammt. Naturwiss.* Bd. XXIX.

Binney, E. W. (66) On Fossil wood in calcareous nodules found in the upper foot coal near Oldham. *Proc. Lit. Phil. Soc. Manchester*, vol. v. p. 113.

Bleicher and Fliche. (92) Contribution à l'étude des Terrains Tertiaires d'Alsace. *Bull. soc. géol. France* [3], Tome XX. p. 375.

Bodenbender, W. (96) Beobachtungen über Devon- und Gondwana Schichten in der Argentinischen Republik. *Zeitsch. Deutsch. geol. Ges.* Bd. XLVIII. p. 743.

—— (02) Contribucion al Conocimiento de la Precordillera de San Juan de Mendoza. *Bol. Acad. Nac. Cienc. Cordoba*, vol. XVII. p. 203.

Boodle, L. A. (15) Concrescent and solitary Foliage-leaves in *Pinus*. *New Phyt.* vol. XIV. p. 19.

Boodle, L. A. and W. C. Worsdell. (94) On the comparative anatomy of the Casuarineae, with special reference to the Gnetaceae and Cupuliferae. *Ann. Bot.* vol. VIII. p. 231.

Boulay. (79) Recherches de paléontologie végétale dans le terrain houiller du Nord de la France. *Ann. soc. scient. Bruxelles*, ann. iv. pt. 2, 1880.

—— (87) Notice sur la Flore tertiaire des environs de Privas (Ardiche) *Bull. soc. bot. France*, Tome XXXIV. p. 227.

—— (88) Notice sur les Plantes fossiles des grès tertiaires de Saint-Saturnin (Marne et Loire). *Journ. Bot. Ann.* 2, p. 921.

Bower, F. O. (81) On the germination and histology of the seedlings of *Welwitschia mirabilis. Quart. Journ. Micr. soc.* vol. XXI. pp. 15, 571.

—— (82) The germination and embryogeny of *Gnetum Gnemon. Ibid.* vol. XXII. [N.S.], p. 277.

—— (84) On the structure of *Rhynchopetalum montanum. Journ. Lin. Soc.* vol. XX. p. 440.

—— (12) Studies in the Phylogeny of the Filicales. II. *Lophosira*, and its relation to the Cyatheoideae and other Ferns. *Ann. Bot.* vol. XXVI. p. 269.

Bowerbank, J. S. (40) History of the Fossil fruits and seeds of the London Clay. *London*.

Braun, A. (75) Die Frage nach der Gymnospermie der Cycadeen. *Monatsber. K. Preuss. Akad. Wiss. Berlin*, p. 289.

—— (75²) Die Diagnosen dreier im Jahre 1873 von G. Wallis in Neu Granada entdeckter Cycadeen. *Ibid.* p. 376.

Braun, C. F. W. (43) Beiträge zur Petrefactenkunde Bayreuth (Graf zu Münster), Heft VI. *Bayreuth.*

—— (47) Die Fossile Gewächse aus den Grenzschichten zwischen dem Lias und Keuper des neu aufgefundenen Pflanzenlagers in dem Steinbrüche von Veitlahm bei Culmbach. *Flora*, p. 81.

—— (49) Beiträge zur Urgeschichte der Pflanzen. VI. *Weltrichia* eine neue Gattung fossiler Rhizantheen. *Progr. iii. Jahresber. K. Kreis-Landwirthsch. und Gewerbschule zu Bayreuth.*

Brauns, D. (66) Der Sandstein bei Seinstedt unweit des Fallsteins und die in ihm vorkommenden Pflanzenreste. *Paleont.* Bd. IX. p. 47.

Brenchley, Winifred E. (13) On Branching specimens of *Lyginodendron Oldhamium* Will. *Journ. Linn. Soc.* vol. XLI. p. 349.

Bristow, H. W. (62) The Geology of the Isle of Wight. *Mem. Geol. Surv. Great Britain.*

Brongniart, A. (25) Observations sur les végétaux fossiles renformés dans les grès de Hoer en Scanie. *Ann. Sci. nat.* vol. IV. p. 200.

—— (28) Essai d'une Flore du grès bigarré. *Ann. Sci. nat.* vol. V. p. 435.

—— (33) Notice sur une Conifère fossile du terrain d'eau douce de l'île d'Iliodroma. *Ann. Sci. nat.* tome XXX. p. 175.

—— (74) Études sur les graines fossiles trouvées à l'état silicifié dans le terrain Houiller de Saint Etienne. *Ann. Sci. nat.* tome XX. [5], p. 234.

—— (81) Recherches sur les graines fossiles silicifiées. *Paris.*

Bronn, H. G. (58) Beiträge zur triassischen Schiefer von Raibl. *Neues Jahrb. Min.* p. 129.

Brooks, F. T. and A. Sharples. (14) Pink disease. *Bull. No.* 21, *Depart. Agric. Fed. Malay States.*

Brooks, F. T. and W. Stiles. (10) The structure of *Podocarpus spinulosus* (Smith) R. Br. *Ann. Bot.* vol. XXIV. p. 305.

Buchman, J. (45) Outline of the Geology of the neighbourhood of Cheltenham (in collaboration with R. I. Murchison and H. E. Strickland). *London.*

Buckland, W. (28) On the Cycadeoideae, a Family of Fossil Plants found in the Oolite quarries of the Isle of Portland. *Trans. Geol. Soc.* [2]. vol. II. p. 395.

—— (37) Geology and Mineralogy considered with reference to Natural Theology. *London.*

Burckhardt, C. (11) Bemerkungen zu einigen Arbeiten von W. Gothan und A. G. Nathorst. *Cent. Min. Geol.; Paleont.* p. 442.

Burgestein, A. (06) Zur Holzanatomie der Tanne, Fichte und Lärche. *Ber. deutsch. Bot. Ges.* Bd. XXIV. Heft VI. p. 295.

—— (08) Vergleichende Anatomie des Holzes der Koniferen. *Wiesner-Festschrift, Wien.*

Burlingame, L. (08) The staminate cone and male gametophyte of *Podocarpus. Bot. Gaz.* vol. XLVI. p. 161.

Burlingame, L. (13) The Morphology of *Araucaria brasiliensis*. *Bot Gaz.* vol. LV. p. 97.

—— (15) The Morphology of *Araucaria brasiliensis*. *Ibid.* vol. LIX. p. 1.

—— (15²) The Origin and Relationships of the Araucarians. *Ibid.* vol. LX. p. 1.

Butterworth, J. (97) Some further investigations of Fossil seeds of the genus *Lagenostoma* (Williamson) from the Lower Coal Measures, Oldham. *Mem. Proc. Manchester Lit. and Phil. Soc.* vol. XLI. ix. p. 1.

Caldwell, O. W. (07) *Microcycas calocoma. Bot. Gaz.* vol. XLIV. p. 118.

Caldwell, O. W. and **C. F. Baker.** (07) The identity of *Microcycas calocoma. Ibid.* vol. XLIII. p. 130.

Cambier, R. et **A. Renier.** (10) *Psygmophyllum Delvali* n. sp. du Terrain houiller de Charleroi. *Ann. soc. géol. Belg.* Tome II. p. 23. (*Mém. in 4to.*)

Capellini, G. and **Conte E. Solms-Laubach.** (92) I Tronchi di Bennettitee dei Musei Italiani. *Mem. Reale Acad. Sci. Istit. Bologna* [5], tom. II. p. 161.

Caporn, A. St. C. (16) A note on the male inflorescence of a species of *Gnetum* from Singapore. *Ann. Bolus Herb.* vol. II. pt. i.

Carano, E (04) Contribuzione alla conoscenza della Morfologia e dello sviluppo del fascio vascolare delle foglie delle Cicadacee. *Ann. di Bot.* vol. I. p. 109 (*Rome*).

Carpentier, A. (11) Sur quelques fructifications et inflorescences du Westphalien du Nord de la France. *Rev. Gén. Bot.* tome XXIII. p. 1.

—— (13) Contribution à l'étude du Carbonifère du Nord de la France. *Mém. soc. géol. du Nord*, tome VII. ii. p. 1.

Carruthers, W. (66) On Araucarian cones from the Secondary beds of Britain. *Geol. Mag.* vol. III. p. 249.

—— (66²) On some fossil Coniferous fruits. *Ibid.* vol. III. p. 534.

—— (67) On *Cycadeoidea Yatesii* sp. nov. a fossil Cycadean stem from the Potton sands, Bedfordshire. *Ibid.* vol. IV. p. 199.

—— (67²) On Gymnospermous Fruits from the Secondary rocks of Britain. *Journ. Bot.* vol. V. p. 1.

—— (67³) On some Cycadean Fruits from the Secondary rocks of Britain. *Geol. Mag.* vol. IV. p. 101.

—— (68) British Fossil Pandanaceae. *Ibid.* vol. V. p. 153.

—— (69) On *Beania*, a new genus of Cycadean Fruit from the Yorkshire Oolite. *Ibid.* vol. VI. p. 1.

—— (69²) On some undescribed Coniferous fruits from the Secondary rocks of Britain. *Ibid.* vol. VI. p. 1.

—— (70) On Fossil Cycadean stems from the Secondary rocks of Britain. *Trans. Linn. Soc.* vol. XXVI. p. 675.

—— (71) On two undescribed Coniferous fruits from the Secondary rocks of Britain. *Geol. Mag.* vol. VIII. p. 1.

—— (77) Description of a new species of *Araucarites* from the Coralline Oolite of Malton. *Quart. Journ. Geol. Soc.* vol. XXXIII. p. 402.

Carruthers, W. (93) On *Cycas Taiwaniana,* sp. nov. and *C. Seemanni* R. Br. *Journ. Bot.* vol. xxxi. p. 1.

Carter, M. Geraldine. (11) A Reconsideration of the origin of Trans-fusion-tissue. *Ann. Bot.* vol. xxv. p. 975.

Caspary, R. and R. Triebel. (89) Einige fossile Hölzer Preussens. *K. Preuss. Geol. Landesanstalt,* Bd. ix. Heft iii. p. 113.

Čelakovský, L. (82) Zur Kritik der Ansichten von den Fruchtschuppe der Abietineen. *Abh. K. böhm. Ges. Wiss. Prag* [vi], Bd ii.

Chamberlain, C. J. (06) The ovule and female gametophyte of *Dioon. Bot. Gaz.* vol. xlii. p. 321.

—— (09) *Dioon spinulosum. Ibid.* vol. xlviii. p. 401.

—— (10) See Coulter and Chamberlain.

—— (10²) Fertilization and embryogeny in *Dioon edule. Bot. Gaz.* vol. l. p. 415.

—— (11) The adult Cycad Trunk. *Ibid.* vol. lii. p. 81.

—— (12) Morphology of *Ceratozamia. Ibid.* vol. liii. p. 1.

—— (12²) A round-the-world Botanical Excursion. *Pop. Sci. Monthly,* vol. lxxxi. p. 417.

—— (12³) Two species of *Bowenia. Bot. Gaz.* vol. liv. p. 419.

—— (13) *Macrozamia Moorei,* a connecting link between living and fossil Cycads. *Bot. Gaz.* vol. lv. p. 141.

Chapman, F. (09) Jurassic Plant-remains from Gippsland, Pt. ii. *Rec. Geol. Surv. Victoria,* vol. iii. pt. i. p. 103.

Chrysler, M. A. (15) The Medullary rays of *Cedrus. Bot. Gaz.* vol. lix. p. 387.

Church, A. H. (14) On the Floral Morphology of *Welwitschia mirabilis* (Hooker). *Phil. Trans. R. Soc.* vol. 205, p. 115.

Cockerell, T. D. A. (06) The Fossil Flora and Fauna of the Florissant (Colorado) shales. *Univ. Colorado Series,* vol. iii. No. 3.

—— (08) Description of Tertiary Plants. II. *Amer. Journ. Sci.* vol. xxvi [4], p. 537.

—— (08²). The Fossil Flora of Florissant, Colorado. *Bull. Amer. Mus. Nat. Hist.* vol. xxiv. p. 71.

—— (08³) Florissant; a Miocene Pompeii. *Pop. Sci. Monthly* (Aug. 1908, p. 112).

Coemans, E. (66) Description de la flore fossile du premier étage du terrain Crétacé du Hainaut. *Mém. Acad. R. Belg.,* tome xxxvi.

Coker, W. C. (03) On the gametophyte and embryo of *Taxodium. Bot. Gaz.* vol. xxxvi. p. 1.

—— (09) Vitality of Pine seeds and the delayed opening of cones. *Amer. Nat.* vol. xliii. p. 677.

Compter, G. (94) Die fossile Flora des untern Keupers von Ostthüringen. *Zeitsch. für Naturwiss. Leipzig,* Bd. lxvii. p. 205.

—— (03) Cycadeenfrüchte aus der Lettenkohle von Apolda. *Zeitsch. für Naturwiss. Stuttgart,* Bd. lxxv. p. 171.

Compton, R. H. (08) See South and Compton.

482 LIST OF WORKS

Conwentz, H. (78) Ueber ein tertiäres' Vorkommen Cypressenartiger Hölzer bei Calistoga in Californien. *Neues Jahrb. Min.* p. 800.

—— (82) Fossile Hölzer aus der Sammlung der König. geol. Landes-anstalt zu Berlin. *Jahrb. K. preuss. geol. Land. Bergakad. Berlin für das Jahr* 1881, p. 144.

—— (85) Sobre algunos arboles fosiles del Rio Negro. *Bol. Acad. Nac. Cienc. Cordoba*, tom. VII. p. 575.

—— (86) Die Angiospermen des Bernsteins. *Danzig.*

—— (89) Ueber Thyllen und Thyllen-ähnliche Bildungen, vornehmlich im Hölze der Bernsteinbäume. *Ber. Deutsch. Bot. Ges.* Bd. VII. p. (34).

—— (92) Untersuchungen über fossile Hölzer Schwedens. *K. Svensk. Vetenskapsakad. Hand.* Bd. XXIV. No. 13.

—— (01) The past history of the Yew in Great Britain and Ireland. *Rep. 71st Meeting Brit. Assoc. (Glasgow)*, p. 839.

Coulter, J. M. and C. J. Chamberlain. (03) The Embryogeny of *Zamia*. *Bot. Gaz.* vol. XXXV. p. 184.

—— (10) Morphology of Gymnosperms. *Chicago.*

Cramer, C. (68) Fossile Hölzer der Arctischen Zone. *Heer's Foss. Flor. Arct.* vol. I. p. 167.

Crié, L. (89) Beiträge zur Kenntniss der fossilen Flora einiger Inseln des Südpacifischen und Indischen Oceans. *Pal. Abhand. (Dames and Kayser)* (N.F.), Bd. I. Heft ii.

Daguillon, A. (90) Recherches morphologiques sur les feuilles des Conifères. *Rev. Gén. Bot.* tome 11, p. 154.

Dawson, J. W. (46) Notices of some Fossils found in the Coal Formation of Nova Scotia. *Quart. Journ. Geol. Soc.* vol. II. p. 132.

—— (62) On the Flora of the Devonian period in North-eastern America. *Ibid.* vol. XVIII. p. 296.

—— (63) Further Observations on the Devonian Plants of Maine, Gaspé, and New York. *Ibid.* vol. XIX. p. 458.

—— (63²) Synopsis of the Flora of the Carboniferous Period in Nova Scotia. *The Canadian Nat. and Geologist*, vol. VIII. p. 431.

—— (81) Notes on New Erian (Devonian) plants. *Quart. Journ. Geol. Soc.* vol. XXXVII. p. 299.

—— (85) On the Mesozoic Floras of the Rocky Mountains Region of Canada. *Trans. R. Soc. Canada*, sect. IV. p. 1.

—— (90) On new plants from the Erian and Carboniferous, and on the characters and affinities of Palaeozoic Gymnosperms. *Canadian Rec. Sci.* vol. IV. p. 1.

—— (93) On new species of Cretaceous plants from Vancouver Island. *Trans. R. Soc. Canada*, sect. iv. p. 53.

Dawson, Sir J. W. and D. P. Penhallow. (91) Note on the specimens of Fossil wood from the Erian (Devonian) of New York and Kentucky. *The Canadian Rec. Sci.* vol. IV. p. 242.

Depape, G. (13) Sur la présence du *Ginkgo biloba* L. dans le Pliocène nférieur de Saint-Marcel-d'Ardèche. *Compt. Rend.* vol. 157, p. 957.

Depape, G. and **A. Carpentier.** (13) Présence des genus *Gnetopsis* B. Ren. and R. Zeill. et *Urnatopteris* Kidst. dans le Westphalien du Nord de la France. *Ann. soc. géol. du Nord,* tome xlii. p. 294.

—— (15) Sur quelques graines et fructifications du Westphalien -du Nord de la France. *Rev. Gén. Bot.* vol. xxvii. p. 321.

Dorety, Helen A. (08) The embryo of *Ceratozamia,* a physiological study. *Bot. Gaz.* vol. xlv. p. 412.

—— (08²) The seedling of *Ceratozamia. Ibid.* vol. xlvi. p. 205.

—— (09) The extrafascicular cambium of *Ceratozamia. Ibid.* vol. xlvii. p. 149.

—— (09²) Vascular anatomy of the seedling of *Microcycas Calocoma. Ibid.* p. 139.

Dorrien-Smith, A. A. (11) A Botanizing expedition to West Australia. *Journ. R. Hort. Soc.* vol. xxxvi. p. 285.

Douvillé, H. et **R. Zeiller.** (08) Sur le terrain houiller du Sud Oranais. *Compt. Rend.* tome cxlvi. p. 732.

Drude, O. (90) Handbuch der Pflanzengeographie. *Stuttgart.*

Dümmer, R. (12) *Podocarpus formosensis. Gard. Chron.* Oct. 19, p. 295.

Dun, W. S. (10) Notes on some Fossil plants from the roof of the coal seam in the Sydney Harbour Colliery. *Journ. Proc. R. Soc. New South Wales,* vol. xliv. p. 615.

Duns, J. (72) On *Cardiocarpon. Proc. R. Soc. Edinburgh,* p. 692.

Dušánek, F. (13) Spaltöffnungen der Cycadaceen. (Abstract in the *Bot. Cent.* Bd. cxxv. p. 340.)

Dusén, P. (99) Über die Tertiäre Flora der Magellans-Länder. *Wiss. Ergeb. Schwed. Exped. nach den Magellansländern unter Leitung von O. Nordenskjöld,* Bd. i. No. iv. p. 87.

—— (08) Über die Tertiäre Flora der Seymour-Insel. *Wiss. Ergeb. Schwed. südpolar-Exped.* 1901–03, Bd. iii. Lief. iii. p. 1.

Duthie, Augusta V. (12) Anatomy of *Gnetum africanum. Ann. Bot.* vol. xxvi. p. 593.

Dutt, C. P. (16) *Pityostrobus macrocephalus,* L. and H. A Tertiary Cone showing Ovular structures. *Ann. Bot.* vol. xxx. p. 529.

Eames, A. J. (13) The Morphology of *Agathis australis. Ann. Bot.* vol. xxvii. p. 1.

Eberdt, O. (94) Die Braunkohlen Ablagerungen in der Gegend von Senftenberg. *Jahrb. K. Preuss. Geol. Land. Bergakad. Berlin,* Bd. xiv. p. 212.

Eichler, A. W. (81) Über die weiblichen Blüthen der Coniferen. *Monatsber. K. Akad. Wiss. Berlin,* p. 1020.

—— (82) Über Bildungabweichungen bei Fichtenzapfen. *Sitzber. K. Akad. Wiss. Berlin,* p. 40.

—— (89) Gymnospermae. *Engler and Prantl; Die Natürlichen Pflanzenfamilien,* Teil ii.

Eichwald, E. (53–68) Lethaea rossica. *Stuttgart.*

Elkins, Marion G. and **G. R. Wieland.** (14) Cordaitean wood from the Indiana Black Shale. *Amer. Journ. Sci.* vol. xxxviii. p. 65.

Elwes, H. J. (12) The Flora of Formosa. *Gard. Chron.* July 13, 1912, p. 25.

Elwes, H. J. and **A. Henry.** (06) The trees of Great Britain and Ireland, vol. I. *Edinburgh.*

Endlicher, S. (40) Genera plantarum secundum ordines naturales disposita. *Vindobonae,* 1836–40.

—— (47) Synopsis Coniferarum. *Sangalli.*

Engelhardt, H. (85) Die Tertiärflora des Jesuitengrabens bei Kundratitz in Nordböhmen. *Nov. Act. K. Leop.-Car. Deutsch. Akad. Natur-forsch.* Bd. XLVIII. No. 3, p. 299.

—— (91) Über Tertiärpflanzen von Chile. *Abh. Senckenberg. naturforsch. Ges.* p. 629.

—— (12) Weiterer Beitrag zur Kenntniss der fossilen Tertiärflora Bosniens. *Wiss. Mitt. aus Bosnien und der Herzegowina,* Bd. XII. p. 593.

Engelhardt, H. and **F. Kinkelin.** (08) Oberpliocene Flora und Fauna des Unter-Maintales. *Abh. Senck. Naturforsch. Ges.* Bd. XXIX. Heft iii. p. 151.

Engler, A. (89) Engler and Prantl; Die Natürlichen Pflanzenfamilien, Teil II. Abt. i.

—— (95) Die Pflanzenwelt Öst Afrikas, *etc.* Th. C. *Berlin.*

—— (97) Engler and Prantl; Die Natürlichen Pflanzenfamilien. Nachtrag zu Teilen II–IV.

Essner, B. (86) Ueber den diagnostischen Werth der Anzahl und Höhe der Markstrahlen bei den Coniferen. *Abh. naturforsch. Ges. Halle,* Bd. XVI. p. 1.

Etheridge, R. (93) On the occurrence of a plant allied to *Schizoneura* in the Hawkesbury Sandstone. *Rec. Geol. Surv. New South Wales,* vol. III. pt. iii, p. 74.

Ettingshausen, C. von. (51) Ueber einige neue und interessante *Taeniopteris* Arten aus den Sammlungen des Kais. Hof Mineralien Cabinetes und der K.k. geol Reichsanstalt. *Naturwiss. Abh. W. Haidinger,* vol. IV. p. 95.

—— (52) Beitrag zur näheren Kenntniss der Flora der Wealdenperiode. *Abh. K.k. geol. Reichs. Wien,* Bd. I. Abth. iii. No. 2, p. 1.

—— (52²) Die Steinkohlenflora von Stradonitz. *Ibid.* Bd. I. Abth. iii. No. 4.

—— (55) Die Tertiäre Flora von Häring in Tirol. *Ibid.* Bd. II. Abth. ii. No. 2.

—— (57) Die Fossile Flora von Köflach in Steiermark. *Jahrb. K.k. geol. Reichs.* Bd. VIII. p. 738.

—— (58) Beiträge zur Kenntniss der fossilen Flora von Sotzka in Untersteiermark. *Sitz. K. Akad. Wiss. Wien,* Bd. XXVIII. p. 471.

—— (67) Die Kreideflora von Niederschoena in Sachsen. *Ibid.* Bd. LV. Abth. i. p. 235.

—— (67²) Die fossile Flora des Tertiär-Bechens von Bilin. *Denksch. Wiss. Akad. Wien,* Bd. XXVI. p. 79.

Ettingshausen, C. von. (70) Beitrag zur Kenntniss der Tertiärflora Steiermarks. *Sitz. K. Akad. Wien,* Bd. LX. Abth. i. p. 17.

—— (72) Die Fossile Flora von Sagor in Krain. *Denksch. Wiss. Akad. Wien,* Bd. XXXII. p. 32.

—— (78) Beitrag zur Erforschung der Phylogenie der Pflanzenarten. *Ibid.* Bd. XXXVIII. p. 65.

—— (79) Report on Phyto-Palaeontological Investigations of the Fossil Flora of Sheppey. *Proc. R. Soc.* vol. XXIX. p. 388.

—— (80) Report on Phyto-Palaeontological Investigations of the Fossil Flora of Alum Bay. *Ibid.* vol. XXX. p. 228.

—— (85) Die Fossile Flora von Sagor in Krain. *Denksch. Wiss. Akad. Wien,* Bd. L. p. 1.

—— (86) Beiträge zur Kenntniss der Tertiärflora Australiens. *Ibid.* Bd. LIII. p. 81.

—— (87) Beiträge zur Kenntniss der Fossile Flora Neuseelands. *Denksch. K. Akad. Wiss. Wien,* Bd. LIII. p. 3.

—— (88) Contributions to the Tertiary Flora of Australia. *Mem. geol. Surv. N.S.W. Pal.* No. 2.

—— (88²) Die Fossile Flora von Leoben in Steiermark. *Denksch. K. Akad. Wiss. Wien,* Bd. LIV. Abth. I. p. 261.

—— (90) Die Fossile Flora von Schoenegg bei Wies in Steiermark. *Ibid.* Bd. LVII. p. 61.

Ewart, A. J. (08) On the longevity of seeds. *Proc. R. Soc. Victoria,* vol. XXI. [N.S.], pt. i. p. 1.

Feistmantel, O. (72) Beitrag zur Kenntniss der Ausdehnung des sogennanten Wyŕaner Gasschiefers und seiner Flora. *Jahrb. K.k. geol Reichs. Wien,* Bd. XXII. p. 289.

—— (76) Notes on the age of some Fossil Floras in India. *Rec. Geol. Surv. Ind.* vol. IX. pt. iv. p. 115.

—— (76²) Jurassic (Oolitic) Flora of Kach. *Fossil Flora of the Gondwana System,* pt. i. vol. II. 1880.

—— (77) Notes on Fossil Floras in India. *Rec. Geol. Surv. Ind.* vol. X. pt. iv. p. 196.

—— (77²) Jurassic (Liassic) Flora of the Rajmahal group in the Rajmahal hills. *Foss. Flor. Gond. Syst.* vol. I. pt. ii.

—— (77³) Jurassic (Liassic) Flora of the Rajmahal group from Golapili, near Ellore, S. Godaveri. *Ibid.* pt. iii.

—— (77⁴) Flora of the Jabalpur group (Upper Gondwanas) in the Son-Narbada region. *Ibid.* vol. II. pt. ii.

—— (77⁵) Ueber die Gattung *Williamsonia* Carr. in Indien. *Palaeontolog. Beit. Palaeontograph.* Suppl. III. Lief. iii.

—— (79) Upper Gondwana Flora of the outliers on the Madras coast. *Foss. Flor. Gond. Syst.* vol. I. pt. iv.

—— (79²) The Flora of the Talchir-Karharbari beds. *Ibid.* vol. III. pt. i.

—— (80) The Flora of the Damuda-Panchet Divisions. *Ibid.* vol. III. pt. ii.

Feistmantel, O. (80²) Note on the fossil genera *Noeggerathia* Sternberg, *Noeggerathiopsis* Feist. and *Rhiptozamites* Schmal. *Rec. Geol. Surv. India*, vol. XIII. pt. i. p. 61.

——. (80³) Further notes on the correlation of the Gondwana Flora with other Floras. *Rec. Geol. Surv. Ind.* vol. XIII. pt. iii. p. 190.

—— (81) Palaeontological notes from the Hazáribágh and Lohárdagga Districts. *Ibid.* vol. XIV. pt. iii. p. 241.

—— (81²) The Flora of the Talchir-Karharbari beds. *Foss. Flor. Gond. Syst.* vol. III. pt. i.

—— (81³) The Flora of the Damuda-Panchet Divisions. *Ibid.* vol. III. pt. iii.

—— (82) The Fossil Flora of the South Rewah Gondwana Basin. *Ibid.* vol. IV. pt. i.

—— (86) The Fossil Flora of some of the Coalfields in Western Bengal. *Ibid.* vol. IV. pt. ii.

—— (89) Übersichtliche Darstellung der geologisch-palaeontologischer Verhältnisse Süd-Afrikas. Th. i. *Abh. K. böhm. Ges. Wiss.* [vii], Bd. III.

Felix, J. (82) Beiträge zur Kenntniss fossiler Coniferen-Hölzer. *Engler's Jahrb.* Bd. III. p. 260.

—— (84) Die Holzopale Ungarns in Palaeophytologischen Hinsicht. *Mitt. Jahrb. K. Ung. geol. Anst.* Bd. VII.

—— (87) Untersuchungen über fossile Hölzer. *Zeitsch. Deutsch. geol. Ges.* p. 517.

—— (94) Untersuchungen über fossile Hölzer. *Ibid.* Heft I. p. 79.

—— (96) Untersuchungen über fossile Hölzer. *Ibid.* Heft II. p. 249.

Fiedler, H. (57) Die Fossile Früchte der Steinkohlen-formation. *Acad. Caes. Leop. Nov. Acta,* Bd. XXVI. p. 239.

Fliche, P. (96) Étude sur la flore fossile de l'Argonne (Albien-Cenomanien). *Bull. soc. sci. Nancy.*

—— (97) Note sur les nodules et bois minéralisés trouvés à St Parresles-vaudes (Aube) dans les grès verts infracrétacés. *Mem. soc. Acad. de l'Aube,* tome LX.

—— (99) Note sur quelques fossiles végétaux de l'Oligocène dans les Alpes Françaises. *Bull. soc. géol. France* [3], tome XXVII. p. 466.

—— (00) Contribution à la Flore fossile de la Haute-Marne (Infracrétacée). *Bull. soc. sci. Nancy.*

—— (00²) Note sur un bois fossile de Madagascar. *Bull. soc. géol. France* [3], tome XXVIII. p. 470.

—— (03) Note sur des bois silicifiés Permiens de la vallée de Celles (Vosges).

—— (05) Note sur des bois fossiles de Madagascar. *Bull. soc. géol. France* [4], tome V. p. 346.

—— (10) Flore Fossile du Trias en Lorraine et Franche-Comté avec des considérations finales par M. R. Zeiller. *Paris.*

Fliche, P. and Bleicher. (82) Étude sur la flore de l'oolithe inférieure aux environs de Nancy. *Bull. soc. sci. Nancy.*

Fliche, P. and R. Zeiller. (04) Note sur une florule Portlandienne des environs de Boulogne-sur-mer. *Bull. soc. géol. France* [4], tome IV. p. 787.

Fontaine, W. M. (93) Notes on some Fossil plants from the Trinity Division of the Comanche series of Texas. *Proc. U. S. Nat. Mus.* vol. XVI. p. 261.

Foxworthy, F. W. (11) Philippine Gymnosperms. *Philipp. Journ. Sci.* (C) *Botany*, vol. VI. No. 3. p. 149.

Fraine, E. de. (12) On the structure and affinities of *Sutcliffia*, in the light of a newly discovered specimen. *Ann. Bot.* vol. XXVI. p. 1031.

—— (14) On *Medullosa centrofilis*, a new species of *Medullosa* from the Lower Coal Measures. *Ibid.* vol. XXVIII. p. 251.

Fritel, P. H. and R. Viguier. (11) Étude anatomique de deux bois Éocènes. *Ann. sci. nat.* [9], tome XIV. p. 63.

Fujii, K. (96) On the different views hitherto proposed regarding the morphology of the flowers of *Ginkgo biloba. Bot. Mag. Tokyo,* vol. X. No. 109, p. 13.

—— (10) Some remarks on the Cretaceous Fossil Flora and the causes of extinction. *Bot. Mag. Tokyo,* vol. XXIV. No. 284, p. 197.

Fujioka, M. (13) Studien über den anatomischen Bau des Hölzes der japanischen Nadelbäume. *Journ. Coll. Agric. Imp. Univ. Tokyo,* vol. IV. No. 4, p. 201.

Gardner, J. S. (86) A monograph of the British Eocene Flora, vol. II. Gymnospermae. *Palaeont. Soc. London.*

—— (86²) Second Report of the Committee appointed for the purpose of reporting on the Fossil Plants of the Tertiary and Secondary beds of the United Kingdom. Rep. of the 56th Meeting Brit. Assoc.

Geinitz, H. B. (42) Charakteristik der Schichten und Petrefacten des Sächsischen Kreidegebirges. Heft III. *Dresden und Leipzig.*

—— (62) Dyas oder die Zechstein-formation und das Rothliegende, Heft II. *Leipzig.*

—— (63) Über zwei neue Dyadische Pflanzen. *Neues Jahrb. Min.* p. 525.

—— (71) Über fossile Pflanzen aus der Steinkohlen-Formation am Altai. *Leipzig.*

—— (73) Versteinerungen aus dem Brandschiefer der unteren Dyas von Weissig bei Pillnitz in Sächsen. *Neues Jahrb. Min.* p. 681.

—— (75) Über neue Aufschlüsse im Brandschiefer der unteren Dyas von Weissig bei Pillnitz in Sächsen. *Ibid.* p. 1

—— (80) Nachträge zur Dyas I. *Mittheil. aus dem K. Min.-geol. und Praehist. Mus. Dresden,* Heft 3.

Gerry, E. (10) The distribution of the Bars of Sanio in the Coniferales. *Ann. Bot.* vol. XXIV. p. 119.

Geyler, T. and F. Kinkelin. (90) Oberpliocän Flora aus den Baugraben des Klarbeckens bei Niederrad *etc. Abh. Senck. Naturforsch. Ges.* Bd. XV. p. 1.

Gibbs, L. S. (12) On the Development of the female strobilus in *Podocarpus. Ann. Bot.* vol. XXVI. p. 515.

Glück, H. (02) Eine fossile Fichte aus dem Neckertahl. *Mitt. Grossh. Bad. geolog. Landesanst.* Bd. IV. Heft iv. p. 399.

Goc, M. J. le. (14) Observations on the centripetal and centrifugal xylems in the petioles of Cycads. *Ann. Bot.* vol. XXVIII. p. 183.

Goebel, K. (05) Organography of Plants. Pt. ii. *Oxford.*

Goeppert, H. R. (40) Über die neulichst im Basalttuff des hohen Seelbachkopfes bei Siegen entdeckten bituminosen und versteinerten Hölzer, so wie über die der Braunkohlenformation überhaupt. *Arch. für Min., Geog., etc.* (*Karsten und von Decken*), Bd. XIV. p. 182.

—— (41) Über den Bernstein....*Uebersicht den Arbeiten und Veränderungen der Schles. Ges. für Vaterländ. Kultur. Breslau.*

—— (41²) *Taxites scalariformis,* eine neue Art fossilen Holzes. *Arch. für Min., Geog., etc.* Bd. XV. p. 727.

—— (44) Ueber die fossilen Cycadeen überhaupt, mit Rücksicht auf die in Schlesien vorkommenden Arten. *Uebersicht Arbeit. und Veränd. Schlesisch. Ges. vat. Kult.* 1843, p. 114. *Breslau.*

—— (45) F. Wimmer's Flora von Schlesien nebst einer Uebersicht der Fossilen Flora Schlesiens von H. R. Goeppert. *Breslau.*

—— (45²) Description des végétaux fossiles recueillis par M. P. de Tchihatcheff en Sibérie. *Voyage scientifique dans l'Altai oriental,* p. 379. *Paris.*

—— (46) Ueber die fossile Flora der mittleren Juraschichten in Oberschlesien. *Uebers. Arbeit. und Veränd. Schles. Ges. vat. Kultur im Jahre* 1845, Breslau, 1846, p. 139.

—— (47) Zur Flora des Quadersandsteins in Schlesien. *Nachtrag. Nov. Act. Ac. Caes. Leop.-Car.* vol. XXII. p. 355.

—— (50) Monographie der fossilen Coniferen. *Naturwerkundige Verhand. Holland. Maatschap. Wettenschappen Haarlem. Leiden.*

—— (52) Fossile Flora des Übergangsgebirges. *Nova Acta Caes. Leop. Carol. Nat. Cur.* Bd. XXII. (supplement).

—— (53) Ueber die gegenwärtigen Verhältnisse der Paläontologie in Schlesien, so wie über fossile Cycadeen. *Denksch. Schles. Ges. für Vaterländ. Kultur.*

—— (65) Die fossile Flora der Permischen Formation. *Palaeont.* Bd. II. p. 1.

—— (65²) Über die fossile Kreideflora und ihre Leitpflanzen. *Zeit. Deutsch. geol. Ges.* Bd. XVII. p. 638.

—— (66) Beiträge zur Kenntniss fossilen Cycadeen. *Neues Jahrb. Min.* p. 129.

—— (80) Beiträge zur Pathologie und Morphologie fossiler Stämme. *Palaeontol.* [N.F.], Bd. VIII. iii. p. 131.

—— (81) Revision meiner Arbeiten über die Stämme der fossilen Coniferen, insbesondere der Araucariten, und über die Descendenzlehre. *Bot. Cent.* Bd. V., VI. p. 378.

Goeppert, H. R. and **G. Stenzel.** (81) Die Medulloseae. *Palaeontol.* [N.F.], Bd. VIII. p. 113.

—— (88) Nachträge zur Kenntniss der Coniferenhölzer der Palaeozoischen Formationen. *Abh. K. Preuss. Akad. Wiss. Berlin.*

Gomes, B. A. (65) Vegetaes Fosseis. Flora fossil do Terrens Carbonifero das visinhanças dio Porto, serra do Bussaio, e moinho d'ordem proximo a alcacer do Sal. *Comm. geol. Portugal. Lisbon.*

Gordon, Marjorie. (12) Ray-tracheids in *Sequoia sempervirens. New Phyt.* vol. XI. p. 1.

Gordon, W. T. (10) On a new species of *Physostoma* from the Lower Carboniferous rocks of Pettycur (Fife). *Proc. Camb. Phil. Soc.* vol. XV. pt. v. p. 395.

—— (12) On *Rhetinangium Arberi,* a new genus of Cycadofilices from the Calciferous sandstone series. *Trans. R. Soc. Edinb.* vol. XLVIII. pt. iv. p. 813.

Gothan, W. (05) Zur Anatomie lebender und fossiler Gymnospermer-Hölzer. *Abh. K. Preuss. geol. Landes.* [N.F.], Heft XLIV. p. 1.

—— (06) Die fossilen Coniferenhölzer von Senftenberg. *Abh. K. Preuss. Geol. Landesanst. Bergakad.* [N.F.], Heft XLVI. p. 155.

—— (06²) Fossile Hölzer aus dem Bathonien von Russisch-Polen. *Verhand. K. Russ. Min. Ges. zu St Petersburg* [ii], Bd. XLIV. Lief. i. p. 435.

—— (06³) *Piceoxylon Pseudotsugae* als fossiles Holz. *Potonié's Abbild. und Beschreib. Foss. Pflanz.* Lief. IV. 80.

—— (07) Über die Wandlungen der Hoftüpfelung bei den Gymnospermen im Laufe der geologischen Epochen und ihre physiologische Bedeutung. *Sitz. Ges. Naturforsch. Freunde,* No. 2, p. 13.

—— (07²) Die Fossile Hölzer von Konig Karls Land. *K. Svensk. Vetenskapsakad. Hand.* Bd. XLII. No. 10, p. 1.

—— (08) Die Fossile Hölzer von der Seymour- und Snow Hill-Insel. *Wiss. Ergeb. Schwedis. Südpolar-Exped.* 1901–03, Bd. III. Lief. viii. *Stockholm.*

—— (08²) Die Frage der Klima-differenzirung im Jura und in der Kreideformation im Lichte paläobotanischen Tatsachen. *Jahrb. K. Preuss. Geol. Landes. für* 1908, Bd. XXIX. Th. ii. Heft 2, p. 220.

—— (09) Über Braunkohlenhölzer des rheinischen Tertiärs. *Jahrb. K. Preuss. geol. Land.* Bd. XXX. Teil i. Heft 3, p. 516.

—— (10) Die Fossile Holzreste von Spitzbergen. *K. Svensk. Vetenskapsakad. Hand.* Bd. XLV. No. viii.

—— (11) Über einige Permo-Carbonische Pflanzen von der unteren Tunguska (Sibirien). *Zeitsch. Deutsch. Geol. Ges.* Bd. LXIII. Heft 4. p. 418.

—— (13) Die oberschlesische Steinkohlenflora. Teil I. *K. Preuss. geol. Landes.* [N.F.], Heft LXXV.

Gourlie, W. (44) Notice of the Fossil Plants in the Glasgow Museum. *Proc. Phil. Soc. Glasgow,* vol. I. 1844. p. 105.

Grand'Eury, C. (00) Sur les tiges debout, les souches et racines de *Cordaites. Compt. Rend.* tome CXXX. (April 30).

—— (04) Sur les graines des Neuroptéridées. *Compt. Rend.* tome CXXXIX. p. 23.

—— (04²) Sur les graines des Neuroptéridées. *Ibid.* p. 782.

490   LIST OF WORKS

Grand'Eury, C. (05) Sur les *Rhabdocarpus*, les graines et l'évolution des Cordaitées. *Ibid.* tome CXL. p. 995.

—— (05²) Sur les graines de *Sphenopteris*, sur l'attribution des *Codonospermum* et sur l'extrême variété des 'graines de fougères.' *Ibid.* p. 812.

—— (13) Recherches géobotaniques sur les forêts et sols fossiles et sur la végétation et la flore houillères, en deux parties et dix livraisons. Pt. I. Livr. ii:, *Paris et Liège*.

Graner, F. (94) Die geographische Verbreitung der Holzarten. I. Die Coniferen. *Forstwissenschaft. Centralblatt, Berlin* (August).

Griffith, W. (59) Remarks on *Gnetum*. *Trans. Linn. Soc.* vol. XXII. pt. iv. p 299.

Groom, P. (10) Remarks on the Oecology of Coniferae. *Ann. Bot.* vol. XXIV. p. 241

—— (16) A Note on the Vegetative Anatomy of *Pterosphaera Fitzgeraldi* F. v. M. *Annals Bot.* vol. XXX. p. 311.

Groom, P. and W. Rushton. (13) The structure of the wood of East Indian species of *Pinus*. *Journ. Linn. Soc.* vol. XLI. p. 457.

Groppler, R. (94) Vergleichende Anatomie des Hölzes der Magnoliaceen. *Biblioth. Bot.* Bd. VI. Heft 31. *Stuttgart.*

Grossenbacher, J. G. (15) Medullary spots and their cause. *Bull. Torr. Bot. Club*, vol. XLII. p. 227.

Guppy, H. B. (06) Observations of a Naturalist in the Pacific between 1896 and 1899. *London.*

Gutbier, A. von. (49) Die Versteinerungen des Zechsteingebirges und Rothliegenden oder des Permischen Systems in Sächsen. *Dresden and Leipzig.*

Halle, T. G. (10) A Gymnosperm with Cordaitean-like leaves from the Rhaetic beds of Scania. *Arkiv för Bot. Upsala*, Bd. IX. No. 14.

—— (12) On the occurrence of *Dictyozamites* in South America. *Palaeobot. Zeitsch.* Bd. I. Heft i. p. 40.

—— (13) Some Mesozoic plant-bearing deposits in Patagonia and Tierra del Fuego and their Floras. *K. Svensk. Vetenskapsakad. Hand.* Bd. LI. No. 3.

—— (13²) The Mesozoic Flora of Graham Land. *Wiss. Ergeb. Schwed. südpolar Exped.* 1901–03, Bd. III. Lief. 14, p. 1.

—— See Möller, H. J. and T. G. Halle (13).

—— (15) Some xerophytic leaf-structures in Mesozoic plants. *Geol. Fören. Stockholm Förhand.* Bd. XXXVII. H. v. p. 493.

Hallier, H. (05) Provisional scheme of the Natural (Phylogenetic) system of Flowering Plants. *New Phyt.* vol. IV. p. 151.

Harker, A. (06) The Geological structure of the Sgurr of Eigg. *Quart. Journ. Geol. Soc.* vol. LXII. p. 40.

—— (08) The Geology of the small Isles of Inverness-shire. *Mem. Geol. Surv. Scotland.*

Harpe, P. de la. (62) See Bristow, H. W.

Harshberger, J. W. (98) Water-storage and conduction in *Senecio praecox* DC. from Mexico. *Contrib. Bot. Labt.* (*Univ. Pennsylvania*), vol. II. No. 1.

—— (11) Phytogeographic Survey of N. America. (*Die Veget. der Erde*; Engler and Drude, XIII. *Leipzig.*)

Hartz, N. (96) Planteforsteninger fra Cap Stewart i Østgrønland. *Meddel. om Grønland*, XIX. *Copenhagen.*

Hayata, A. (06) On *Taiwania*, a new genus of Coniferae from the Island of Formosa. *Journ. Linn. Soc.* vol. XXXVII. p. 330.

—— (07) On *Taiwania* and its affinity to other genera. *Bot. Mag.* (*Tokyo*), vol. XXI. p. 21.

—— (10) Botanical Survey by the Govt. of Formosa. *Congr. Int. Bot.* (*Bruxelles*), p. 59.

—— (17) Some Conifers from Tonkin and Yunnan. *Bot. Magazine*, vol. XXXI. .p. 113.

Heer, O. (62) On the Fossil Flora of Bovey Tracey. *Phil. Trans. R. Soc.* vol. CLII. p. 1039.

—— (68) i. Die in Nordgrönland, auf der Melville-Insel, im Banksland, an Mackenzie, im Island und in Spitzbergen entdeckten fossilen Pflanzen. *Flor. Foss. Arct.* vol. I. *Zürich.*

—— (69) Beiträge zur Kreide-Flora. I. Flora von Moletein in Mähren. *Neue Denksch. Allgem. Schweiz. Ges. gesammt. Naturwiss.* Bd. XXIII.

—— (71) iii. Die Miocene Flora und Fauna Spitzbergens. *Flor. Foss. Arct.* vol. II.

—— (71²) Beiträge zur Kreide-Flora. II. Kreide Flora von Quedlinburg. *Neue Denksch. Allgem. Schweiz. Ges. gesammt. Naturwiss.* Bd. XXIV.

—— (75) ii. Die Kreide-Flora der arctischen Zone. *Flor. Foss. Arct.* vol. III.

—— (75) iii. Nachträge zur Miocenen Flora Grönlands. *Ibid.*

—— (76) Flora Fossilis Helvetiae. *Zürich.*

—— (76²) Über Permische Pflanzen von Fünkkirchen in Ungarn. *Mitt. Jahrb. K. Ung. Geol. Anst.* Bd. v.

—— (77) i. Beiträge zur fossilen Flora Spitzbergens. *Flor. Foss. Arct.* vol. IV.

—— (77) ii. Beiträge zur Jura-Flora Ost Sibiriens und des Amurlandes. *Ibid.*

—— (78) i. Die Miocene Flora des Grinnell-Lands. *Ibid.* vol. v.

—— (78) ii Beiträge zur fossilen Flora Sibiriens und des Amurlandes. *Ibid.*

—— (78) v. Beiträge zur Miocenen Flora von Sachalin. *Ibid.*

—— (81) Contributions à la Flore du Portugal. *Sect. Trav. Geol. Port.* (*Lisbon*).

—— (81²) Zur Geschichte der Ginkgo-artigen Bäume. *Engler's Bot. Jahrb.* Bd. I. p. 1.

—— (82) i. Flora fossilis Grönlandica. *Flor. Foss. Arct.* vol. VI.

—— (83) Flora fossilis Grönlandica. *Ibid.* vol. VII.

Helmhacher, R. (71) *Sitzber. d. K. Böhm. Ges. Wiss.* p. 81.

Henry, A. (06) See Elwes and Henry.

Herzfeld, S. (10) Die Entwicklungsgeschichte der weiblichen Blüte von *Cryptomeria japonica* Don. Ein Beitrag zur Deutung der Fruchtschuppe der Coniferen. *Sitzber. Akad. Wiss. Wien*, Bd. cxix. Abt. i. p. 807.

Hick, T. (95) On *Kaloxylon Hookeri* Will. and *Lyginodendron oldhamium* Will. *Mem. Proc. Manchester Lit. Phil. Soc.* [4], vol. ix. p. 109.

Hilderbrand, F. (61) Die Verbreitung der Coniferen. *Rhein. und Westphal. Verhand.* Bd. xviii. p. 199.

Hill, T. G. and **E. de Fraine.** (10) On the seedling structure of Gymnosperms. IV *Ann. Bot.* vol. xxiv. p. 319.

Hirase, S. (98) Études sur la fécundation et l'embryogenie du *Ginkgo biloba. Journ. Coll. Sci. Imp. Univ. Tokyo*, vol. xii. p. 103.

Höhlke, F. (02) Ueber die Harzbehälter und die Harzbildung bei den Polypodiaceen und einigen Phanerogamen. *Beiheft Bot. Cent.* Bd. xi. p. 8.

Holden, H. S. (10) Note on a wounded *Myeloxylon. New Phyt.* vol. ix. p. 253.

Holden, Ruth. (13) Some fossil plants from Eastern Canada. *Ann. Bot.* vol. xxvii. p. 243.

—— (13²) Contributions to the anatomy of Mesozoic Conifers. No. 1. Jurassic Coniferous wood from Yorkshire. *Ann. Bot.* vol. xxvii. p. 533.

—— (13³) Cretaceous Pityoxyla from Cliffwood, New Jersey. *Proc. Amer. Acad. Arts and Sci.* vol. xvi. p. 609.

—— (14) Contributions to the anatomy of Mesozoic Conifers. II. Cretaceous Lignites from Cliffwood, N. Jersey. *Bot. Gaz.* vol. lviii. p. 168.

—— (14²) On the relation between *Cycadites* and *Pseudocycas. New Phyt.* vol. xiii. p. 334.

—— (15) A Jurassic wood from Scotland. *Ibid.* vol. xiv. p. 205.

—— (15²) On the cuticles of some Indian Conifers. *Bot. Gaz.* vol. lx. p. 215.

Hollick, A. (97) The Cretaceous clay marl exposed at Cliffwood, N. J. *Trans. N. Y. Acad. Sci.* vol. xvi. p. 124.

—— (04) Additions to the Palaeontology of the Cretaceous formation on Long Island. No. II. *Bull. N. Y. Bot. Gard.* vol. iii. No. 11, p. 403.

—— (06) The Cretaceous Flora of southern N. Y. and New England. *U. S. Geol. Surv. Mon.* vol. l.

—— (06²) Systematic Palaeontology of the Pleistocene deposits of Maryland. *Contributions from the New York Bot. Gard.* No. 85.

—— (12) Additions to the Palaeobotany of the Cretaceous formation on Long Island. *Bull. N. Y. Bot. Gard.* vol. viii. No. 28, p. 154.

Hollick, A. and **E. C. Jeffrey.** (06) Affinities of certain Cretaceous plant-remains commonly referred to the genera *Dammara* and *Brachyphyllum. Amer. Nat.* vol. xl. p. 189.

Holmes, W. H. (78) Fossil Forests of the Volcanic Tertiary formations of the Yellowstone National Park. *Ann. Rep. Geol. Geogr. Surv. U.S.A.* pt. II. p. 47.

Hooker, J. D. (52) *Dacrydium laxifolium. Icones Plant.* vol. v. pl. 815.

—— (60) Flora Tasmanica. *London.*

—— (62) On the Cedars of Lebanon, Taurus, Algeria, and India. *Nat. Hist. Rev.* p. 11.

—— (63) On *Welwitschia,* a new genus of Gnetaceae. *Trans. Linn. Soc.* vol. XXIV. p. 1.

Hooker, J. D. and E. W. Binney. (55) On the structure of certain limestone nodules enclosed in seams of bituminous coal, with a description of some Trigonocarpons contained in them. *Phil. Trans. R. Soc.* vol. CXLIX.

Hörich, O. (06) Potonié's Abbildungen und Beschreibungen fossilen Pflanzen-Reste. Lief. iv. 69, 70.

Howse, R. (88) A catalogue of Fossil Plants from the Hutton collection. *Nat. Hist. Trans. Northumberland, Durham, and Newcastle-upon-Tyne,* vol. X.

Hutchinson, A. H. (14) The male gametophyte of *Abies. Bot. Gaz.* vol. LVII. p. 148.

—— (15) On the male gametophyte of *Picea canadensis. Ibid.* vol. LIX. p. 287.

Jeffrey, E. C. (03) The comparative anatomy and phylogeny of the Conifers. I. The genus *Sequoia. Mem. Boston Soc. Nat. Hist.* vol. v. No. 10, p. 441.

—— (04) A fossil *Sequoia* from the Sierra Nevada. *Bot. Gaz.* vol. XXXVIII. p. 321.

—— (05) The comparative anatomy and phylogeny of the Conifers. The Abietineae. *Mem. Boston Soc. Nat. Hist.* vol. VI. No. 1.

—— (06) The wound Reactions of *Brachyphyllum. Ann. Bot.* vol. XX. p. 383.

—— (07) *Araucariopitys,* a new genus of Araucarians. *Bot. Gaz.* vol. XLIV. p. 435.

—— (08) Traumatic ray-tracheids in *Cunninghamia sinensis. Ann. Bot.* vol. XXII. p. 593.

—— (08²) On the structure of the leaf in Cretaceous Pines. *Ibid.* vol. XXII. p. 207.

—— (10) A new *Prepinus* from Martha's Vineyard. *Proc. Boston Soc. nat. Hist.* vol. XXXIV. No. 10, p. 333.

—— (10²) A new Araucarian genus from the Triassic. *Ibid.* vol. XXXIV. No. 9, p. 325.

—— (10³) On the affinities of *Yezonia. Ann. Bot.* vol. XXIV. p. 769.

—— (11) The affinities of *Geinitzia gracillima. Bot. Gaz.* vol. L. p. 21.

—— (12) The History, Comparative Anatomy and Evolution of the *Araucarioxylon* type. *Proc. Amer. Acad. Arts Sci.* vol. XLVIII. No. 13, p. 532.

Jeffrey, E. C. (14) Spore-conditions in hybrids and the mutation hypo-thesis of de Vries. *Bot. Gaz.* vol. LVIII. p. 322.

—— (17) The Anatomy of Woody Plants. *Chicago* 1917.

Jeffrey, E. C. and M. A. Chrysler. (06) On Cretaceous *Pityoxyla. Bot. Gaz.* vol. XLII. p. 1.

—— (06²) The Lignites of Brandon. *Contrib. from the Phanerogamic Labt. of Harvard Univ.* No. vi.

—— (07) The microgametophyte of the Podocarpineae. *Amer. Nat.* vol. XLI. No. 486, p. 355.

Jeffrey, E. C. and Ruth D. Cole. (16) Experimental Investigations on the genus *Drimys. Ann. Bot.* vol. XXX. p. 359.

Jeffrey, E. C. and R. E. Torrey. (16) *Ginkgo* and the microsporangial mechanisms of the seed plants. *Bot. Gaz.* vol. LXII. p. 281.

Johnson, T. (11) A seed-bearing Irish Pteridosperm, *Crossotheca Höninghausi* Kidst. *Sci. Proc. R. Dublin Soc.* vol. XIII. p. 1.

—— (12) *Heterangium hibernicum* sp. nov. a seed-bearing *Heterangium* from Co. Cork. *Ibid.* vol. XIII. No. 20.

—— (14) *Ginkgophyllum kiltorkense* sp. nov. *Ibid.* vol. XIV. p. 169.

—— (17) *Spermolithus devonicus*, Gen. et sp. nov., and other Pterido-sperms from the Upper Devonian beds at Kiltoscan, Co. Kilkenny. *Ibid.* vol. XV. p. 245.

Johnston, R. H. (86) Fresh contributions to our knowledge of the Plants of Mesozoic age in Tasmania. *Papers and Proc. R. Soc. Tasmania* for 1886, p. 160.

Johnstrup, M. F. (83) Recherches sur les fossiles appartenant aux formations Crétacée et Miocène, sur la côte occidentale du Grønland. *Medd. om Grønland*, vol. v.

Jones, W. S. (12) The structure of the Timbers of some common genera of Coniferous trees. *Quart. Journ. Forestry, April.*

—— (13) The minute structure of the wood of *Cupressus macrocarpa. Ibid.*

—— (13²) Ray-tracheids in *Sequoia sempervirens* and their pathological character. *Lampeter.*

Karsten, G. (92) Beitrag zur Entwickelungsgeschichte einiger Gnetum Arten. *Bot. Zeit.* p. 205.

—— (93) Untersuchungen über die Gattung *Gnetum.* I. *Ann. Jard. Bot. Buitenzorg*, tome XI. p. 195.

—— (93²) Zur Entwickelungsgeschichte der Gattung *Gnetum. Cohn's Beit. Biol. Pflanz.* VI. p. 337.

Kershaw, E. M. (09) The structure and development of the ovule of *Myrica Gale. Ann. Bot.* vol. XXIII. p. 353.

—— (12) Structure and development of the ovule of *Bowenia spec-tabilis. Ibid.* vol. XXVI. p. 625.

Kidston, R. (84) On a new species of *Schützia* from the Calciferous sandstone of Scotland. *Ann. Mag. Nat. Hist.* vol. XIII. p. 77.

—— (86) Notes on some fossil plants collected by Mr R. Dunlop, Airdrie, from the Lanarkshire coal-field. *Trans. Geol. Soc. Glasgow*, vol. VIII. p. 47.

Kidston, R. (90) The Yorkshire Carboniferous Flora. *Trans. Yorks. Nat. Union*, pt. XIV.

—— (92) Notes on some fossil plants from the Lancashire Coal Measures. *Trans. Manchester Geol. Soc.* pt. xiii. vol. XXI.

—— (04) On the Fructification of *Neuropteris heterophylla* Brongn. *Phil. Trans. R. Soc.* vol. CXCVII. p. 1.

—— (04²) On the Fructification of *Neuropteris heterophylla* Brongn. *Proc. R. Soc.* vol. LXXII. p. 487.

—— (04³) Some Fossil Plants collected by Mr A. Sinclair from the Ayrshire coalfield. *Kilmarnock Glenfield Ramblers Soc. Annals* (1901–04), No. iv. *Kilmarnock.*

—— (05) Preliminary Note on the occurrence of Microsporangia in organic connection with the Foliage of *Lyginodendron*. *Proc. R. Soc.* vol. LXXVI. p. 358.

—— (11) Les Végétaux houillers recueillis dans le Hainaut Belge. *Mém. Mus. Roy. d'hist. nat. Belg.* tome IV.

—— (14) On the Fossil Flora of the Staffordshire coalfields. Pt. iii. The Fossil Flora of the Westphalian series of the S. Staffs. coalfield. *Trans. R. Soc. Edinb.* vol. L. pt. i. p. 73.

Kidston, R. and **D. T. Gwynne-Vaughan.** (12) On the Carboniferous Flora of Berwickshire. Pt. i. *Stenomyelon tuedianum* Kidst. *Trans. R. Soc. Edinb.* vol. XLVIII. pt. ii. p. 263.

Kidston, R. and **W. J. Jongmans.** (11) Sur la Fructification de *Neuropteris obliqua* Brongn. *Arch. Neerl. sci. exact. nat.* [III. B], tome I. p. 25.

Kirby, J. W. (64) On some remains of Fishes and Plants from the 'Upper limestone' of the Permian series of Durham. *Quart. Journ. Geol. Soc.* vol. XX. p. 349.

Kirchner, O., E. T. Loew, and **C. Schröter.** (06) Die Coniferen und Gnetaceen Mitteleuropas. *Stuttgart.*

Kirk, T. (89) The Forest Flora of New Zealand. *Wellington.*

Kirsch, S. (11) The Origin and Development of resin-canals in the Coniferae, with special reference to the Development of Thyloses and their correlation with the Thylosed strands of the Pteridophytes. *Trans. R. Soc. Canada*, sect. iv. p. 43.

Kisch, Mabel H. (13) The Physiological Anatomy of the periderm of fossil Lycopodiales. *Ann. Bot.* vol. XXVII. p. 281.

Kleeberg, A. (85) Die Markstrahlen der Coniferen. (*Inaug. Diss.*) *Bot. Zeit.* Bd. XLIII.

Klein, L. (81) Bau und Verzweigung einiger dorsiventral gebaute Polypodiaceen. *Nov. Act. K. Leop. Car. Deutsch. Akad. Naturforsch.* Bd. XLII. No. 7, p. 335.

Knowlton, F. H. (89) Description of the fossil woods and lignites from Arkansas. *Ann. Rep. Geol. Surv. Arkansas*, vol. II. p. 249.

—— (89²) Fossil wood and Lignite of the Potomac formation. *Bull. U. S. Geol. Surv.* No. 56.

—— (90) A Revision of the genus *Araucarioxylon* of Kraus, with the compiled descriptions and partial synonymy of the species. *Proc. U. S. Mus.* vol. XII. p. 601.

Knowlton, F. H. (93) The Laramie and the overlying Livingstone formation in Montana. Report on the Flora. *Bull. U. S. Geol. Surv.* No. 105.

—— (99) Fossil Flora of the Yellowstone National Park. *Monographs.* XXXII. *U. S. Geol. Surv.* pt. ii. chap. XIV.

—— (00) Flora of the Montana formation. *Bull. U. S. Geol. Surv.* No. 163.

—— (05) The geology of the Perry basin in South-eastern Maine, with a chapter on the fossil plants. *U. S. Geol. Surv. Prof. Papers,* No. 35.

—— (11) The correct technical name for the Dragon tree of the Kentish Rag. *Geol. Mag.* [v], vol. VIII. p. 467.

—— (14) The Jurassic Flora of Cape Lisburne, Alaska. *U. S. Geol. Surv. Prof. Papers,* No. 85–D, p. 39.

Kny, L. (10) Über die Verteilung des Holzparenchyma bei *Abies pectinata* DC. *Ann. Jard. Bot. Buit.* [2], *Suppl.* III. p. 645.

Koettlitz, R. (98) Observations on the geology of Franz Josef Land. *Quart. Journ. Geol. Soc.* vol. LIV. p. 620.

Kosmovsky, C. (92) Quelques mots sur les couches à végétaux fossiles dans la Russie orientale et en Sibérie. *Bull. soc. Imp. Nat. Moscou* [N.S.], tome v. p. 170.

Kramer, A. (85) Beiträge zur Kenntniss der Entwickelungsgeschichte und des anatomischen Baues der Fruchtblätter der Cupressineen und der Placenten der Abietineen. *Flora,* XLIII. p. 519.

Krasser, F. (91) Über die fossile Flora der rhätischen Schichten Persiens. *Sitzber. K. Akad. Wiss. Wien,* Bd. C. Abth. i. p. 413.

—— (97) Constantine Freiherr von Ettingshausen. *Oesterr. bot. Zeitsch.* Nos. 9 and 10.

—— (03) Konstantin von Ettingshausen's Studien über die fossile Flora von Ouricanga in Brasilien. *Sitzber. K. Akad. Wiss. Wien,* Bd. CXII. Abt. i. p. 852.

—— (05) Fossile Pflanzen aus Transbaikalien der Mongolei und Mandschurei. *Denksch. K. Akad. Wiss. Wien,* Bd. LXXVIII. p. 589.

—— (09) Zur Kenntniss der fossilen Flora der Lunzer Schichten. *Jahrb. K.k. geol. Reichs. Wien,* Bd. LIX, Heft i. p. 101.

—— (12) *Williamsonia* in Sardinien. *Sitzber. K. Akad. Wien,* Bd. CXXI. Abt. i. p. 944.

—— (13) Die fossile Flora der Williamsonien bergenden Juraschichten von Sardinien. *Ibid.* Bd. CXXII.

Kraus, G. (64) Mikroskopische Untersuchungen über der Bau lebenden Nadelhölzer. *Würzb. Naturwiss. Zeitsch.* Bd. v. p. 144.

—— (66) Über den Bau der Cycadeenfiedern. *Prings. Jahrb.* Bd. IV. p. 305.

—— (83) Beiträge zur Kenntniss fossiler Hölzer. I. Hölzer aus den Schwefelgruben Siciliens. *Abh. Naturf. Ges. Halle,* Bd. XVI. p. 79.

—— (92) Beiträge zur Kenntniss fossiler Hölzer. *Abh. Naturforsch. Ges. Halle,* Bd. XVII. p. 67.

—— (96) Physiologisches aus den Tropen. *Ann. Jard. Bot. Buit.* vol. XIII. p. 217.

Kräusel, R. (13) Beiträge zur Kenntniss der Hölzer aus der Schlesischen Braunkohle. Teil ɪ. (*Inaug. Diss. Breslau.*) *Bot. Cent.* Bd. cxxɪɪɪ. p. 123.

Krystofovič, A. (10) Jurassic Plants from Ussuriland. *Mem. Com. Geol.* [N.S.], Livr. 56.

—— (15) Plant remains from Jurassic lake-deposits of Transbaikalia. *Mém. Soc. Imp. Russe Mineralog.* [2], ʟɪ.

Kubart, B. (08) Pflanzenversteinerungen enthaltende Knollen aus dem Ostrau-Karwiner Kohlenbecken. *Sitzber. K. Akad. Wiss. Wien,* Bd. cxvɪɪ. Abt. i. p. 573.

—— (11) Corda's sphaerosiderite aus dem Steinkohlenbecken Radnitz-Bŕaz in Böhmen nebst Bemerkungen über *Chorionopteris gleichenioides* Corda. *Ibid.* Bd. cxx. Abt. i. p. 1035.

—— (11²) *Podocarpoxylon Schwendae,* ein fossiles Holz von Altersee (Oberösterreich). *Österr. bot. Zeitsch.* No. 5, p. 161.

—— (14) Über die Cycadofilicineen *Heterangium* und *Lyginodendron* aus dem Ostrauer Kohlenbecken. *Österr. bot. Zeitsch.* No. ɪ. ii. p. 8.

Kurtz, F. (03) Remarks upon Mr E. A. Arber's communication on the Clarke collection of Fossil Plants from New South Wales. *Quart. Journ. Geol. Soc.* vol. ʟɪx. p. 25.

Kutorga, S. (42) Beitrag zur Palaeontologie Russlands. *Verhand. Russ.-Kais. Mineral. Ges. St Petersburg.*

—— (44) Zweiter Beitrag zur Palaeontologie Russlands. *Ibid.* p. 62.

Land, W. J. G. (04) Spermatogenesis and Oogenesis in *Ephedra trifurca. Bot. Gaz.* vol. xxxvɪɪɪ. p. 1.

Lang, W. H. (97) Studies in the Development and Morphology of Cycadean sporangia. I. The microsporangia of *Stangeria paradoxa. Ann. Bot.* vol. xɪ. p. 421.

—— (00) Studies in the Development and Morphology of Cycadean sporangia. II. The ovule of *Stangeria paradoxa. Ibid.* vol. xɪv. p. 281.

Lange, T. (90) Beiträge zur Kenntniss der Flora des Aachener Sandes. *Zeitsch. Deutsch. geol. Ges.* Bd. xʟɪɪ. p. 658.

Laurent, L. (12) Flore fossile des Schistes de Manat (Puy-de-Dôme). *Ann. Mus. d'hist. nat. Marseille (Geol.),* tome xɪv. p. 3.

Lawson, A. A. (04) The gametophytes, fertilization, and embryo of *Cryptomeria japonica. Ann. Bot.* vol. xvɪɪɪ. p. 417.

—— (09) The gametophytes and embryo of *Pseudotsuga Douglasii. Ibid.* vol. xxɪɪɪ. p. 163.

—— (10) The gametophytes and embryo of *Sciadopitys verticillata. Ibid.* vol. xxɪv. p. 403.

Lebour, G. A. (77) Illustrations of Fossil Plants; being an autotype reproduction of selected drawings. *London.*

Lesquereux, L. (74) Contributions to the Fossil Flora of the Western Territories. Pt. ɪ. The Cretaceous Flora. *Rep. U. S. Geol. Surv. Territ.* vol. vɪ.

Lesquereux, L. (78) On the Cordaites and their related generic divisions, in the Carboniferous formation of the United States. *Proc. Amer. Phil. Soc.* (*Philadelphia*), vol. XVII. p. 315.

—— (83) Contributions to the Fossil Flora of the Western Territories. Pt. iii. The Cretaceous and Tertiary Floras. *U. S. Geol. Surv. Territ.* vol. III.

—— (91) The Flora of the Dakota group. *Monographs U. S. Geol. Surv.* vol. XVII.

Leuthardt, F. (03) Die Keuper Flora von Neuewelt bei Basel. *Abh. Schweiz. palaeont. Ges.* Bd. XXX. p. 1.

Lignier, O. (92) La nervation taenioptéridée de folioles de *Cycas* et le tissu de transfusion. *Bull. soc. Linn. Normandie* [4], vol. VI. fasc. 1.

—— (94) La nervation des Cycadées est dichotomique. *Assoc. Franç. pour l'avancement de sci.* (*Caen*).

—— (94²) Végétaux Fossiles de Normandie. Structure et affinités du *Bennettites Morierei* S. and M. sp. *Mém. soc. Linn. Normand.* tome XVIII. p. 1.

—— (95) Végétaux fossiles de Normandie. II. Contributions à la flore liassique de Ste Honorine-la-Guillaume (Orne). *Ibid.* vol. XVIII.

—— (01) *Ibid.* III. Étude anatomique du *Cycadeoidea micromyela* Mor. *Ibid.* vol. XX. p. 331.

—— (03) Le fruit du *Williamsonia gigas* Carr. *Mém. soc. Linn. Normand.* vol. XXI. p. 19.

—— (03²) La Fleur des Gnétacées est-celle intermédiaire entre celle des Gymnospermes et celle des Angiospermes? *Bull. soc. Linn. Normand.* [5], vol. VII. p. 55.

—— (04) Notes complémentaires sur la structure du *Bennettites Morierei* S. and M. *Bull. soc. Linn. Normand.* [5], vol. VIII. p. 3.

—— (06) *Radiculites reticulatus*, radicelle fossile de Séquoïnée. *Bull. soc. bot. France*, tome VI. [iii.], p. 193.

—— (07) Sur un moule litigieux de *Williamsonia gigas* (L. and H.) Carr. *Ibid.* [6], vol. 1.

—— (07²) Végétaux fossiles de Normandie. IV. Bois divers (sér. 1). *Mém. soc. Linn. Normand.* vol. XXII. p. 239.

—— (09) Le *Bennettites Morierei* (S. and M.) Lignier ne serait-il pas d'origine infracrétacée? *Bull. soc. Linn. Normand.* [6], vol. II. p. 214.

—— (11) Le *Bennettites Morierei* (S. and M.) Lign. se reproduisait probablement par parthénogénèse. *Bull. soc. bot. France* [4], tome XI. p. 224.

—— (11²) Les "*Radiculites reticulatus*" Lign. soit probablement des radicelles de Cordaitales. *Assoc. Franç. Avanc. Sci.* XL. (*Dijon*), p. 509. [*See also* Lignier (06).]

—— (12) Stomates des écailles interséminales chez le *Bennettites Morierei* (S. and M.). *Ibid.* tome XII. p. 425.

—— (13) Végétaux fossiles de Normandie. VII. Contributions à la Flore Jurassique. *Mém. soc. Linn. Normand.* vol. XXIV. p. 69.

Lignier, O. (13²) Différenciation des tissus dans le Bourgeon végétatif du *Cordaites lingulatus* B. Ren. *Ann. sci. nat.* [7], vol. xvii. p. 233.

Lignier, O. et A. Tison. (11) Les Gnétales sont des Angiospermes apétales. *Compt. Rend.* Jan. 23.

—— (12) Les Gnétales, leurs fleurs et leur position systématique. *Ann. sci. nat.* [N.S.].

—— (13) L'ovule tritégumenté des *Gnetum* est probablement un axe d'inflorescence. *Bull. soc. bot. France* [4], tome xiii. p. 64.

—— (13²) Un nouveau Sporange Séminiforme, *Mittagia seminiformis*, gen. et sp. nov. *Mém. Soc. Linn. Normandie*, tome xxiv. p. 49.

Lima, W. de. (88) Flora Fossil de Portugal. Monographia do Genero *Dicranophyllum*. *Comm. dos Trab. geol. Portugal.*

Lindley, J. See Murchison and Lyell (29).

Lingelsheim, A. (08) Über die Braunkohlenhölzer von Saarau. *Jahres. Ber. Schles. Ges. Vaterland. Cultur.* Bd. lxxxv.

Lloyd, F. E. (02) Vivipary in *Podocarpus*. *Torreya*, ii. p. 113.

Lomax, J. (02) On some features in relation to *Lyginodendron oldhamium*. *Ann. Bot.* vol. xvi. p. 601.

Lotsy, J. (99) Contributions to the life-history of the genus *Gnetum*. *Ann. Jard. Bot. Buit.* [2], vol. i. p. 46.

—— (09) Vorträge über botanische Stammesgeschichte. Bd. ii. Jena.

—— (11) *Ibid.* Bd. iii.

Ludwig, R. (61) Fossile Pflanzen aus der ältesten Abtheilung der Rheinisch-Wetterauer Tertiär-Formation. *Palaeontograph.* Bd. viii. p. 39.

—— (69) Fossile Pflanzenreste aus den paläolithischen Formation der Umgegend von Dillenburg, Bidenhopf und Friedberg und aus den Saalfeldischen. *Ibid.* Bd. xvii. p. 105.

Lyon, H. L. (04) The Embryogeny. of *Ginkgo*. *Minnesota Bot. Stud.* xxiii. p. 275.

McBride, T. H. (93) A new Cycad. *Amer. Geologist*, vol. xii. p. 248.

Mackie, S. J. (62) The Dragon Tree of the Kentish Rag. *Geologist*, vol. v. p. 401.

McLean, R. C. (12) Two Fossil prothalli from the Lower Coal Measures. *New Phyt.* vol. xi. p. 305.

McNab, W. R. (70) On the structure of a Lignite from the Old Red Sandstone. *Trans. Bot. Soc. Edinburgh*, vol. x. p. 312.

Mahlert, A. (85) Beiträge zur Kenntniss der Anatomie der Laubblätter der Coniferen mit besonderer Berichsichtigung des Spaltöffnungs-Apparates. *Bot. Cent.* Bd. xxiv. p. 54.

Malloizel, G. and R. Zeiller. (N.D.) Bibliographie et tables iconographiques (O. Heer). *Stockholm.*

Mansell-Pleydell, J. C. (85) Notes on a cone from the Inferior Oolite Beds of Sherborne. *Proc. Dorset Nat. Hist. Antiq. Field Club*, vol. v. p. 141.

Mantell, G. (27) Illustrations of the Geology of Sussex. *London.*
—— (43) Description of some fossil plants from the chalk formation of the south-east of England. *Proc. Geol. Soc.* vol. IV. p. 34.
—— (46) Description of some Fossil Fruits from the Chalk formation of the south-east of England. *Quart. Journ. Geol. Soc.* vol. II. p. 51.

Marion, A. F. (84) Sur les caractères d'une Conifère tertiaire, voisine des Dammarées (*Doliostrobus Sternbergii*). *Compt. Rend.* vol. XCIV. p. 821.

Marsh, A. S. (14) Notes on the Anatomy of *Stangeria paradoxa.* *New Phyt.* vol. XIII. p. 18.

Marty, P. (08) Sur la Flore fossile de Lugarde (Cantal). *Compt. Rend.* vol. CXLVII. p. 395.

Maslen, A. J. (10) See Scott and Maslen.
—— (11) The structure of *Mesoxylon Sutcliffi* (Scott). *Ann. Bot.* vol. XXV. p. 381.

Massalongo, A. (59) Studii sulla Flora Fossile e geologia stratigrafica del Senigalliese. *Verona.*

Massalongo, A. and G. Scarabelli. (58) Studii sulla Flora Fossile e geologia stratigrafica del Senigalliese. *Verona.*

Masters, M. T. (91) Review of some points in the comparative morphology, anatomy, and life-history of the Coniferae. *Journ. Linn. Soc.* vol. XXVIII. p. 236.
—— (93) Notes on the genera of Taxaceae and Coniferae. *Ibid.* vol. XXX. p. 1.
—— (00) *Taxodium* and *Glyptstrobus. Journ. Bot.* (February, 1900).

Matte, H. (04) Recherches sur l'appareil libéro-ligneux des Cycadées. *Caen.*
—— (08) Sur le développement morphologique et anatomique des Cycadacées. *Mém. soc. Linn. Normand.* tome XXIII.

Matthew, G. F. (10) Revision of the Flora of the Little River group. *Trans. R. Soc. Canada,* sect. iv. vol. III. [3], p. 77.

Mercklin, C. E. von. (55) Palaeodendrologicon Rossicum. *St Petersburg.*

Miller, H. (58) The Cruise of the Betsey. *Edinburgh.*

Miquel, F. A. W. (42) Monographia Cycadearum.
—— (47) Collectanea nova ad Cycadearum cognitionem. *Linnaea,* Bd. XIX. p. 411.
—— (51) Over de Rangschikking der fossiele Cycadeae. *Tijdsch. Wiss. Nat. Wet.* vol. IV. p. 205.
—— (69) On the sexual organs of the Cycadaceae. *Journ. Bot.* vol. VII. p. 64.

Mirande, M. (05) Recherches sur le développement et l'anatomie de Cassythacées. *Ann. Sci. nat.* [ix], vol. I. p. 181.

Miyake, K. (06) Über die Spermatozoiden von *Cycas revoluta. Ber. Deutsch. bot. Ges.* Bd. XXIV. p. 78.
—— (10) The Development of the gametophytes and embryogeny in *Cunninghamia sinensis. Beiheft Bot. Cent.* Bd. XXXVII. Abt. i. Heft 1.

Mogan, L. (03) Untersuchungen über eine fossile Konifere. *Sitzber. K. Akad. Wiss. Wien,* Bd. CXII. Abt. i. p. 829.

REFERRED TO IN THE TEXT   501

Mohl, H. von. (62) Einige anatomische und physiologische Bemerkungen über das Holz der Baumwurzeln. *Bot. Zeit.* p. 225.

Möller, H. (03) Bidrag till Bornholms Fossila Flora (Rhät och Lias). Gymnospermer. *K. Svensk. Vetenskapsakad. Hand.* Bd. xxxvi. No. vi.

Möller, H. J. and **T. G. Halle.** (13) The Fossil Flora of the Coal-bearing deposits of south-eastern Scania. *Arkiv Bot. (Stockholm),* Bd. xiii. No. 7.

Morière, J. (69) Note sur deux végétaux trouvés dans le département du Calvados. *Mém. soc. Linn. Normand.* vol. xv.

Morris, J. (40) Memoir to illustrate a Geological map of Cutch (Grant, C. W.). *Trans. Geol. Soc.* [2], vol. v. pt. ii. p. 289.

—— (41) Remarks upon the Recent and Fossil Cycadeae. *Ann. Mag. Nat. Hist.* vol. vii. p. 110.

—— (54) A Catalogue of British Fossils. *London.*

Müller, C. (90) Ueber die Balken in den Holzelementen der Coniferen. *Ber. Deutsch. bot. Ges.* Bd. viii. p. 17.

Murchison, Sir R. and **R. Harkness.** (64) On the Permian rocks of the North-west of England, and their extension into Scotland. *Quart. Journ. Geol. Soc.* vol. xx. p. 144.

Murchison, Sir R. and **C. Lyell.** (29) On the Tertiary Freshwater formations of Aix, in Provence, including the coal-field of Fuveau. Description of plants by J. Lindley. *Edinburgh New Phil. Journ.* p. 287.

Nakamura, Y. (83) Ueber den anatomischen Bau des Holzes der wichtigsten Japanischen Coniferen. *Unters. aus dem Forstbot. Instit. zu München.* iii. Berlin.

Nathorst, A. G. (75) Om en Cycadékotte från den rätiska formationens lager vid Tinkarp i Skåne. *Öfver. K. Vetenskapsakad. Förh.* No. 10.

—— (78) Om *Ginkgo crenata* Brauns sp. från sandstenen vid Senstedt nära Braunschweg. *Ibid.* No. 3.

—— (80) Några anmärkningar om *Williamsonia,* Carruthers. *Ibid.* No. 9.

—— (81) Berättelse, afgifven till Kongl. Vetenskaps-Akad. *Öfver. K. Vetenskapsakad. Förhand.* No. 1.

—— (86) Om Floran i Skånes kolförande Bildningar. I. Floran vid Bjuf. *Sver. geol. Unters.* ser. C, Nos. 27, 33, 85 (1878–86).

—— (88) Nya anmärkningar om *Williamsonia. Öfver. K. Vetenskapsakad. Förh.* No. 6.

—— (89) Sur la présence du Genre *Dictyozamites,* Old. dans les Couches Jurassiques de Bornholm. *Övers. K. Dansk. Vidensk. Selsk. Förhandl.* p. 96.

—— (93) Beiträge zur Geologie und Palaeontologie der Republik Mexico (Felix und Link), Th. ii. Heft i. *Leipzig.*

—— (97) Zur Mesozoischen Flora Spitzbergens. *K. Svensk. Vetenskapsakad. Hand.* Bd. xxx. No. 1, p. 5.

—— (97²) Nachträgliche Bemerkungen über die Mesozoische Flora Spitzbergens. *Öfvers. K. Vet.-Akad. Förhand.* No. 8.

Nathorst, A. G. (99) The Norwegian North Polar Expedition 1893–96. Scientific Results, edit. F. Nansen. III. Fossil Plants from Franz Josef Land. *London and Christiania.*

—— (02) Beiträge zur Kenntniss einiger Mesozoischen Cycadophyten. *K. Svensk. Vetenskapsakad. Hand.* Bd. xxxvi. No. 4.

—— (06) Om några Ginkgovaxter från Kolgrufvorna vid Stabbarp i Skane. *Lunds Univ. Årsskrift* [N.F.], Afd. ii. Bd. ii. No. 8.

—— (07) Über Trias- und Jurapflanzen von der Insel Kotelny. *Mém. Acad. Imp. Sci. St Petersburg,* vol. xxi. No. 2.

—— (07²) Paläobotanische Mitteilungen. 2. Die Kutikula der Blätter von *Dictyozamites Johnstrupi* Nath. *K. Svensk. Vetenskapsakad. Hand.* Bd. xlii. No. 5.

—— (07³) *Ibid.* I. *Pseudocycas,* eine neue Cycadophytengattung aus den Cenomanen Kreideablagerungen Grönlands.

—— (08) Über die Untersuchungen kutinisierter fossiler Pflanzenteile. Paläobot. Mitt. 4–6. *K. Svensk. Vetenskapsakad. Hand.* Bd. xliii. No. 6, p. 3.

—— (08²) Paläobot. Mitt. 7. Über *Palissya, Stachyotaxus* and *Palaeotaxus Ibid.* Bd. xliii. No. 8.

—— (09) Paläobot. Mitt. 8. Über *Williamsonia, Wielandia, Cycadocephalus* und *Weltrichia. Ibid.* Bd. xlv. No. 4.

—— (09²) Über die Gattung *Nilssonia* Brongn. mit besonderer Berücksichtung Schwedischen Arten. *K. Svensk. Vetenskapsakad. Hand.* Bd. xliii. No. 12.

—— (10) Excursion C 3. Dépôts fossilifères (plantes) quaternaires de Skåne. *Compt. Rend. du XIe Congrès Géol. Internat.* p. 1353.

—— (11) Paläobot. Mitt. 9. Neue Beiträge zur Kenntniss der Williamsonia-Blüten. *K. Svensk. Vetenskapsakad. Hand.* Bd. xlvi. No. 4.

—— (11²) Bemerkungen über *Weltrichia* Fr. Braun. *Arkiv Bot.* (*K. Svensk. Vetenskapsakad. Stockholm*), Bd. ii. No. 7, p. 1.

—— (11³) Fossil floras of the Arctic Regions as evidence of geological climates. *Geol. Mag.* [v], vol. viii. p 217.

—— (11⁴) Paläobot. Mitt. 10. Über die Gattung *Cycadocarpidium* Nath. nebst einigen Bemerkungen über *Podozamites. K. Svensk. Vetenskapsakad. Hand.* Bd. xlvi. No. 8.

—— (12) Die Mikrosporophylle von *Wielandiella. Arkiv Bot. Stockholm,* Bd. xii. No. 6, p. 1.

—— (12²) Paläobot. Mitt. 11. Zur. Kenntniss der *Cycadocephalus* Blüte. *K. Svensk. Vetenskapsakad. Hand.* Bd. xlviii. No. 2.

—— (12³) Einige paläobotanische Untersuchungs-methoden. *Paläobot. Zeitsch.* Bd. i. Heft i. p. 26.

—— (13) How are the names *Williamsonia* and *Wielandiella* to be used? A question of nomenclature. *Geol. Fören. Stockholm Förh.* Bd. xxxv. H. vi. p. 361.

—— (13²) Die Pflanzenführenden Horizonte innerhalb der Grenzschichten des Jura und der Kreide Spitzbergens. *Geol. Fören. Stockholm Förh.* Bd. xxxv. H. iv. p. 273.

Nathorst, A. G. (14) Zur Fossilen Flora der Polarländer. Teil I. Lief. iv. Nachträge zur Pälaozoischen Flora Spitzbergens. *Stockholm.*

——— (15) Zur Devonflora des westlichen Norwegens. *Bergens Mus. Aarbog,* No. 7.

——— (15²) Tertiäre Pflanzenreste aus Ellesmere-Land. Rep. second Norwegian Arct Exped. in the "Fram" 1898–02, No. 35. *Kristiania.*

Negri, G. (14) Sopra alcuni legni fossili del Gebel Tripolitano. *Boll. Soc. geol. Ital.* vol. xxxiii. p. 321.

Nestler, A. (95) Ein Beitrag zur Anatomie der Cycadeenfiedern. *Pringsheim's Jahrb.* Bd. xxvii. p. 341.

Newberry, J. S. (54) New Fossil Plants from Ohio. *Annals of Science; including the Transactions of the American Association for the Advancement of Science,* vol. I. *Cleveland,* 1853–54, p. 116.

——— (73) Report of the Geological Survey of Ohio, vol. I.

——— (88) Rhaetic Plants from Honduras. *Amer. Journ. Sci.* vol. xxxvi. p. 342.

Newberry, J. S. and A. Hollick. (95) The Flora of the Amboy clays. *U. S. Geol. Surv. Monographs,* vol. xxvi.

Newton, E. T. and J. J. H. Teall. (97) Notes on a collection of rocks and fossils from Franz Josef Land, made by the Jackson-Harmsworth Exped. during 1894–96. *Quart. Journ. Geol. Soc.* vol. liii. p. 477.

——— (98) Additional notes on rocks and fossils from Franz Josef Land *Ibid.* vol. liv. p. 646.

Noack, F. (87) Der Einfluss des Klimas auf die Cuticularisation und Verholzung der Nadeln einiger Coniferen. *Pringsheim's Jahrb.* Bd. xviii. p. 519.

Noelle, W. (10) Studien zur vergleichenden Anatomie und Morphologie der Koniferen Wurzeln mit Rücksicht auf die Systematik. *Bot. Zeit.* p. 169.

Norén, C. O. (08) Zur Kenntniss der Entwickelung von *Saxegothaea conspicua* Lind. *Svensk. Bot. Tids.* Bd. ii. H. ii. p. 101.

Oliver, F. W. (02) On some points of apparent resemblance in certain Fossil and Recent Gymnosperm seeds. *New Phyt.* vol. i. p. 145.

——— (03) The ovules of the older Gymnosperms. *Ann. Bot.* vol. xvii. p. 451.

——— (04) Notes on *Trigonocarpus,* Brongn. and *Polylophospermum,* Brongn., two genera of Palaeozoic seeds. *New Phyt.* vol. iii. p. 96.

——— (05) Über die neuentdeckten Samen der Steinkohlenfarne. *Biolog. Centralblatt,* Bd. xxv. No. 12, p. 401.

——— (06) The seed, a chapter in Evolution. *Rep. 76th Meeting Brit. Assoc.* (*York*), p. 725.

——— (07) Note on the Palaeozoic seeds *Trigonocarpus* and *Polylophospermum.* *Ann. Bot.* vol. xxi. p. 303.

——— (09) On *Physostoma elegans* Will. an archaic type of seed from the Palaeozoic rocks. *Ibid.* vol. xxiii. p. 73.

——— (13) Makers of Modern Botany. A collection of Biographies by living Botanists, edited by F. W. Oliver. *Cambridge.*

Oliver, F. W. and E. J. Salisbury. (11) On the structure and affinities of the Conostoma group of Palaeozoic seeds. *Ann. Bot.* vol. xxv. p. 1.

Oliver, F. W. and D. H. Scott. (03) On *Lagenostoma Lomaxi*, the seed of *Lyginodendron*. *Proc. R. Soc.* vol. LXXI. p. 477.

—— (04) On the structure of the Palaeozoic seed *Lagenostoma Lomaxi*, with a statement of the evidence upon which it is referred to *Lyginodendron*. *Phil. Trans. R. Soc.* vol. CXCVII. p. 193.

Osborne, T. G. B. (09) The lateral roots of *Amyelon radicans* Will. and their *Mycorhiza*. *Ann. Bot.* vol. XXIII. p. 603.

Palibin, J. (01) Quelques données relatives aux débris végétaux contenus dans les sables blancs et les grès quartzeux de la Russie méridionale. *Bull. Com. Geol.* tome XX. p. 447.

—— (04) Pflanzenreste vom Sichota-Alin Gebirge. *Verhand. K. Russ. Mineral. Ges.* Bd. XLII. Lief. i. p. 31.

Patrick, J. S. (44) On the Fossil vegetables of the Sandstone of Ayrshire. *Ann. Mag. Nat. Hist.* vol. XIII. p. 283.

Pavolini, A. F. (09) La *Stangeria paradoxa* Th. Moore. *Nuov. Giorn. Bot. Italiano* [N.S.], vol. XVI. p. 335. *Firenze*.

Pax, F. (07) Beiträge zur fossilen Flora der Karpathen. *Engler's Bot. Jahrb.* Bd. XXXVIII. p. 272.

Pearson, H. H. W. (06) Notes on South African Cycads. I. *Trans. S. Afr. Phil. Soc.* vol. XVI. p. 341.

—— (06²) Some observations on *Welwitschia mirabilis*. *Phil. Trans. R. Soc.* vol. CXCVIII. p. 265.

—— (09) Further observations on *Welwitschia*. *Ibid.* vol. CC. p. 331.

—— (12) On the Microsporangium and Microspore of *Gnetum*, with some notes on the structure of the Inflorescence. *Ann. Bot.* vol. XXVI. p. 603.

—— (15) A note on the inflorescence and flower of *Gnetum*. *Ann. Bolus Herb.* vol. I. pt. iv. p. 152.

—— (15²) Notes on the Morphology of certain Structures concerned in Reproduction in the genus *Gnetum*. *Trans. Linn. Soc.* vol. VIII. pt. viii. p. 311.

—— (17) On the Morphology of the female flower of *Gnetum*. *Trans. R. Soc. S. Africa*, vol. VI. pt. i. p. 69.

Penhallow, D. P. (91) See Dawson and Penhallow.

—— (97) *Myelopteris topekensis* n. sp. a new Carboniferous Plant. *Bot. Gaz.* vol. XXIII. p. 15.

—— (00) Notes on the North American species of *Dadoxylon*. *Trans. R. Soc. Canada* [2], vol. VI. sect. iv. p. 51.

—— (02) Notes on Cretaceous and Tertiary Plants of Canada. *Trans. R. Soc. Canada* [2], vol. VIII. sect. iv. p. 31.

—— (03) Notes on Tertiary Plants. *Ibid.* vol. IX. sect. iv. p. 83.

—— (04) The anatomy of the North American Coniferales together with certain exotic species from Japan and Australia. *Amer. Nat.* vol. XXXVIII. pp. 243, 523.

Penhallow, D. P. (07) A Manual of the North American Gymnosperms. *Boston.*

—— (08) Report on a collection of Fossil woods from the Cretaceous of Alberta. *The Ottaua Naturalist,* vol. XXII. No. iv. p. 82.

Phillips, J. (71) The Geology of Oxford and the valley of the Thames. *Oxford.*

Pilger, R. (03) Taxaceae. *Das Pflanzenreich* (A. Engler), Heft XVIII (iv. 5). *Leipzig.*

Platen, P. (08) Untersuchungen fossiler Hölzer aus dem Westen der Vereinigten Staaten von Nordamerika. *Leipzig.*

Pomel, A. (49) Matériaux pour servir à la flore fossile des terrains jurassiques de la France. *Amt. Ber. Versam. Ges. deutsch. Naturforsch. und Ärzte. Aachen.*

Porsch, C. (05) Der Spaltöffnungsapparat im Lichte der Phylogenie. *Jena.*

—— (10) *Ephedra campylopoda* C. A. Mey. eine entomophile Gymnosperme. *Ber. Deutsch. Bot. Ges.* Bd. XXVIII. p. 404.

Potonié, H. (88) Die fossile Pflanzen-Gattung *Tylodendron. Jahrb. K. Preuss. geol. Landes.* p. 311.

—— (96) Ueber Autochthonie von Carbonkohlen-Flötze und der Senftenberger Braunkohlen-Flötze. *Ibid.* (1895), p. 1.

—— (02) Fossile Hölzer aus der oberen Kreide Deutsch-Östafrikas. Die Reisen des Bergassessors Dr Dantz in Deutsch-Östafrika in den Jahren 1898–00. *Mitt. aus den deutschen Schutzgebieten,* Bd. XV. Heft iv. p. 227.

—— (03) Pflanzenreste aus der Jura-formation. *Durch Asien.* Bd. III. Lief. i. *Berlin.*

—— (04) Abbildungen und Beschreibungen fossilen Pflanzen-Reste. Lief. II. No. 40. *K. Preuss. geol. Landes. Bergakad.*

Potonié, H. and C. Bernard. (04) Flore Dévonienne de l'étage H. de Barrande. *Leipzig.*

Prankerd, T. L. (12) On the structure of the Palaeozoic seed *Lagenostoma ovoides* Will. *Journ. Linn. Soc.* vol. XL. p. 461.

Prestwich, J. (54) On the structure of the strata between the London Clay and the Chalk in the London and Hampshire Tertiary systems. *Quart. Journ. Geol. Soc.* vol. X. p. 75.

Raciborski, M. (91) Flora Retycka Pólnocnego Stoku Gór Świętokrzyskich. *Rozprawy Wydzialu Akad. Umiej. Krakowie,* tom. XXIII.

—— (92) Przyczynek do Flory Retyckiej Polski. *Ibid.* tom. XXII. p. 1.

—— (92²) *Cycadeoidea Niedzwiedzkii* nov. sp. *Akad. wiss. Krak.* *Oktober* 1892.

Radais, M. (94) L'anatomie comparée du fruit des Conifères. *Ann. Sci. nat.* [7], vol. XIX. p. 165.

Ratte, F. (87) Note on two new fossil plants from the Wianametta shales. *Proc. Linn. Soc. N.S.W.* [2], vol. I. p. 1078.

—— (88) Additional evidence on Fossil Salisburiæ from Australia. *Proc. Linn. Soc. N.S.W.* [2], vol. II. p. 159

Rattray, C. (13) Notes on the pollination of some South African Cycads. *Trans. R. Soc. S. Africa*, vol. III. p. 259.

Reid, Clement and Eleanor M. Reid. (08) On the Preglacial Flora of Britain. *Journ. Linn. Soc.* vol. XXXVIII. p. 206.

—— (10) The Lignites of Bovey Tracey. *Phil. Trans. R. Soc.* vol. CCI. p. 161.

—— (15) The Pliocene Floras of the Dutch-Prussian border. *Meded. Rijksopsporing Delfstoffen*, No. 6. *The Hague.*

Renault, B. (79) Sur un nouveau groupe de tiges fossiles silicifiées de l'époque houillière. *Compt. Rend.* tome LXXXVIII. p. 35.

—— (80) Sur une nouvelle espèce de *Poroxylon. Compt. Rend.* tome XCI. p. 860.

—— (80²) Cours de Botanique fossile. Tome I. *Paris.*

—— (83) Cours de Botanique fossile. Tome III.

—— (85) Cours de Botanique fossile. Tome IV.

—— (87) Note sur le *Clathropodium Morieri. Bull. soc. Linn. Normand.* [4], vol. I. p. 3.

—— (88) Les Plantes fossiles. *Paris.*

—— (89) Sur un nouveau genre fossile de tige cycadéenne. *Compt. Rend.* vol. CIX. p. 1173.

—— (96) Note sur le genre Métacordaite. *Soc. d'hist. nat. d'Autun.*

Renault, B. and R. Zeiller. (85) Sur un nouveau type de Cordaitée. *Compt. Rend.* vol. C. p. 867.

—— (86) Sur quelques Cycadées houillières. *Ibid.* vol. CII. p. 325.

Rendle, A. B. (96) Gymnospermae. The Plants of Milanji, Nyasa-Land, collected by Mr A. Whyte. *Trans. Linn. Soc.* vol. IV. [2], p. 60.

—— (04) The Classification of Flowering plants. Vol. I. *Cambridge.*

Renier, A. (07) Trois espèces nouvelles. *Sphenopteris Dumonti, S. Corneti* et *Dicranophyllum Richiri* du Houiller sans Houille de Baudour, Hainaut. *Ann. soc. géol. Belg.* vol. XXXIV. Mém. p. 181.

—— (10) Documents pour l'étude de la paléontologie du terrain houiller. *Liège.*

—— (10²) Paléontologie du Terrain Houiller. *Liège.*

Renner, O. (04) Über zwitter Blüthen bei *Juniperus communis. Flora*, Bd. XCII. p. 92.

Richards, J. T. (84) On Scottish Fossil Cycadaceous leaves contained in the Hugh Miller Collection. *Proc. R. Phys. Soc. Edinburgh.*

Rivière, H. C. C. La. (16) Sur l'anatomie et l'épaississement des tiges du *Gnetum moluccense* Karst. *Ann. Gard. Bot. Buitenzorg*, vol. XXX. [2], p. 23.

Robertson, Agnes. (02) Notes on the anatomy of *Macrozamia heteromera* Moore. *Proc. Camb. Phil. Soc.* vol. XII. pt. i. p. 1.

—— (04) Studies in the Morphology of *Torreya californica* Torrey. *New Phyt.* vol. III. p. 205.

—— (06) Some points in the Morphology of *Phyllocladus alpinus* Hook. *Ann. Bot.* vol. XX. p. 259.

—— (07) The Taxoideae; a phylogenetic study. *New Phyt.* vol. VI. p. 92.

Rosen, F. (11) Die biologische Stellung der abessinischen Baumlobelie (*Lobelia Rhynchopetalum*). *Beit. Biol. Pflanzen. Cohn und Rosen*, Bd. x. Heft ii. p. 265.

Rothert, W. (99) Ueber parenchymatische Tracheiden und Harzgänge im Mark von *Cephalotaxus* Arten. *Ber. deutsch. bot. Ges.* Bd. xvii. p. 275.

Rushton, W. (15) Structure of the wood of Himalayan Junipers. *Journ. Linn. Soc.* vol. xliii. p. 1.

—— (16) The Development of ' Sanio's Bars ' in *Pinus Inops. Ann. Bot.* vol. xxx. p. 419.

Russow, E. (72) Vergleichende Untersuchungen der Leitbündel-Kryptogamen. *Mém. l'Acad. Imp. Sci. St Pétersbourg* [vii], tome xix. p. 1.

—— (83) Zur Kenntniss des Holzes, insonderheit des Coniferenholzes. *Bot. Cent.* Bd. xiii. p. 29.

Salisbury, E. J. (13) Methods of Palaeobotanical reconstruction. *Ann. Bot.* vol. xxvii. p. 273.

—— (14) On the structure and relationship of *Trigonocarpus shorensis. Ann. Bot.* vol. xxviii. p. 39.

Sanday, Elizabeth. (07) See Berridge and Sanday.

Sandberger, F. von. (64) Die Flora der oberen Steinkohlenformation im badischer Schwarzwalde. *Verh. Nat. Ver. Carlsruhe*, vol. i. p. 30.

—— (90) Ueber Steinkohlenformation und Rothliegendes im Schwarzwald und deren Flora. *Jahrb. K.k. geol. Reichs.* Bd. xl. Heft i. p. 77.

Sanio, K. (74) Anatomie der Gemeinen Kiefer (*Pinus silvestris* L.). *Pringsheim's Jahrb.* Bd. ix. p. 78.

Saporta, G. de. (62) Études sur la végétation du sud-est de la France à l'époque Tertiaire. *Ann. Sci. Nat.* tome xvi. [4], p. 309.

—— (62²) *Ibid.* tome xvii. p. 191.

—— (65) *Ibid.* tome iii. [5], p. 5.

—— (65²) *Ibid.* tome iv. p. 5.

—— (68) *Ibid.* tome ix. p. 5.

—— (68²) Note sur la flore fossile des régions arctiques. *Bull. soc. géol. France* [2], tome xxv. p. 315.

—— (73) Études sur la végétation du sud-est de la France à l'époque Tertiaire. *Ann. Sci. nat.* tome xvii. [5], p. 81.

—— (74) Sur la présence d'une Cycadée dans le dépôt Miocène de Koumi (Eubée). *Compt. Rend.* vol. lxxviii. p. 1318.

—— (75) Sur la découverte de deux types nouveaux de Conifères dans les schistes Permiens de Lodève (Hérault). *Ibid.* vol. lxxx. p. 1017.

—— (78) Observations sur la nature des végétaux réunis dans le groupe des *Noeggerathia*; généralités et type du *Noeggerathia foliosa* Sternb. *Ibid.* vol. lxxxvi. p. 746.

—— (78²) *Ibid.* Types du *Noeggerathia flabellata* L. .and H. et du *N. cyclopteroides* Goepp. *Ibid.* vol. lxxxvi. p. 801.

—— (78³) *Ibid.* Type des *Noeggerathia expansa* et *cuneifolia* de Brongniart. *Ibid.* vol. lxxxvi. p. 860.

508 LIST OF WORKS

Saporta, G. de. (78⁴) Sur le nouveau groupe Paléozoique des Dolero-phyllées. *Ibid.* vol. LXXXVII. p. 393.

—— (82) Sur quelques types de végétaux récemment observés à l'état fossile. *Ibid.* vol. XLIV. p. 922.

—— (84) Paléontologie Française. Plantes Jurassiques, tome III.

—— (91) *Ibid.* tome IV. Types Proangiospermiques.

—— (93) Revue des travaux de Paléontologie végétale. *Rev. Gén.* tome V. p. 1.

Saporta, G. de and A. F. Marion. (76) Recherches sur les végétaux fossiles de Meximieux. *Arch. Mus. d'hist. nat. de Lyon. Lyon.*

—— (78) Révision de la flore Heersienne de Gelinden. *Mém. cour. et Mém. sav. étrang.* tome XLI. (*Acad roy. sci. etc. Belg.*).

—— (85) L'Évolution du Règne végétal. Tome I. Les Phanérogames. *Paris.*

Saxton, W. T. (10) Contributions to the life-history of *Widdringtonia cupressoides. Bot. Gaz.* vol. L. p. 31.

—— (10²) Contributions to the life-history of *Callitris. Ann. Bot.* vol. XXIV. p. 557.

—— (10³) Notes on the anatomy of *Widdringtonia* and *Callitris. S. African Journ. Sci.* p. 282.

—— (10⁴) The development of the embryo of *Encephalartos Bot. Gaz.* vol. XLIX. p. 13.

—— (12) Note on an abnormal prothallus of *Pinus maritima* L. *Ann. Bot.* vol. XXVI. p. 943.

- —— (13) Contributions to the life-history of *Actinostrobus pyramidalis* Miq. *Ann. Bot.* vol. XXVII. p. 321.

—— (13²) The classification of Conifers. *New Phyt.* vol. XII. p. 242.

—— (13³) Contributions to the life-history of *Tetraclinis articulata,* Masters, with some notes on the Phylogeny of the Cupressoideae and the Callitroideae. *Ann. Bot.* vol. XXVII. p. 577.

Schauroth, C. von. (52) Herr von Schauroth an Herrn Beyrich. *Zeit. Deutsch. geol. Ges.* Bd. IV. p. 538.

Schenk, A. (67) Ueber die Flora der schwarzen Schiefer von Raibl. *Würzburg. Naturwiss. Zeitsch.* vol. VI. p. 10.

—— (68) Ueber die Pflanzenreste des Muschelkalkes von Recoaro. *Benecke's Geog.-Pal. Beit.* Bd. II. *München.*

—— (68²) Beiträge zur Flora der Vorwelt. *Palaeontol.* Bd. XVI. p. 218.

—— (71) Die Fossilen Pflanzen der Wernsdorfer Schichten in der Nordkarpathen. *Ibid.* Bd. XIX. p. 1.

—— (80) Ueber fossile Hölzer aus der *Libyschen Wüste. Bot. Zeit.* Bd. XXXVIII. p. 657.

—— (82) Ueber *Medullosa elegans. Engler's Jahrb.* Bd. III. p. 156.

—— (82²) Die von dem Gebrüdern Schagintweit in Indien gesammelten fossilen Hölzer. *Engler's Bot. Jahrb.* Bd. III. p. 353.

—— (83) Fossile Hölzer der libyschen Wüste (*Die Libysche Wüste,* Bd. III.).

Schenk, A. (89) Ueber *Medullosa* Cotta. *Abh. K. Sächs. Ges. Wiss.* Bd. xv. þ. 523.

Schlechtendal, D. von. (02) *Thuja occidentalis thuringiaca. Zeitsch. Naturwiss. Stuttgart.* Bd. LXXV. p. 33.

Schmalhausen, J. (83) Die Pflanzenreste der Steinkohlenformation am östlichen Abhange des Ural Gebirges. *Mém. Acad. S. Pétersbourg,* vol. XXXI. No. 13.

—— (83²) Beiträge zur Tertiär Flora süd-west Russlands. *Paleontol. Abh. (Dames and Kayser),* Bd. I. Heft iv. p. 285.

—— (87) Die Pflanzenreste der Artinskischen und Permischen Ablagerungen im Osten des Europäischen Russlands. *Mém. Com. géol. St Pétersbourg,* vol. II. No. iv. p. 1.

—— (90) Wissenschaftliche Resultate der von der Akad. der Wiss. zur Erforschung des Janalandes und der Neusibirischen Inseln in den Jahren 1885, 1886 ausgesandten Expedition. Abt. II. Tertiäre Pflanzen der Insel Neusibirien. *Mém. l'acad. Imp. Sci. St Pétersbourg* [vii], vol. XXXVII. No. 5.

Schneider, W. (13) Vergleichend-morphologische Untersuchungen über die Kurztriebe einiger Arten von *Pinus. Flora* [N.F.], Bd. v. p. 385.

Schroeter, C. (80) Untersuchung über fossile Hölzer aus der arctischen Zone. (Heer's *Flor. Foss. Arct.* vol. VI. 1882.)

—— (97) Ueber die Vielgestaltigkeit der Fichte (*Picea excelsa* Link.). *Vierteljahr. Naturforsch. Ges. Zürich.* Jahrg. XLII. p. 125.

Schuster, J. (11) Ueber Goeppert's *Raumeria* im Zwinger zu Dresden. *Sitzber. K. Bayer. Akad. Wiss.* p. 489.

—— (11²) *Weltrichia* und die Bennettitales. *K. Svensk. Vetenskaps- akad. Hand.* Bd. XLVI. No. 11.

—— (11³) *Pagiophyllum Weismanni* im unteren Hauptmuschelkalk von Würzburg. *Geog. Jahresheft,* Bd. XIII. p. 149.

—— (11⁴) Bemerkungen über *Podozamites. Ber. Deutsch. Bot. Ges.* Bd. XIX. Heft 7, p. 450.

Schütze, E. (01) Beiträge zur Kenntniss der Triassischen Koniferen- Gattungen *Pagiophyllum, Voltzia, Widdringtonites. Jahresheft Ver. Vat. Naturkunde, Stuttgart,* p. 256.

Scott, D. H. (97) The anatomical characters presented by the peduncle of Cycadaceae. *Ann. Bot.* vol. XI. p. 399.

—— (99) On the structure and affinities of Fossil Plants from the Palaeozoic rocks. III. On *Medullosa anglica,* a new representative of the Cycadofilices. *Phil. Trans. R. Soc.* vol. CXCI. p. 81.

—— (99²) On the primary wood of certain Araucarioxylons. *Ann. Bot.* vol. XIII. p. 615.

—— (02) On the Primary structure of certain Palaeozoic stems with the *Dadoxylon* type of wood. *Trans. R. Soc.* vol. XL. p. 331.

—— (03) The origin of seed-bearing plants. *R. Instit. Great Brit. Weekly Evening meeting,* Febry. 15.

—— (06) On *Sutcliffia insignis,* a new type of Medulloseae from the Lower Coal Measures. *Trans. Linn. Soc.* vol. VII. pt. iv. p. 45.

Scott, D. H. (09) The Palaeontological Record. II. Plants. *Darwin and Modern Science*, Art. XII. *Cambridge.*

—— (12) The structure of *Mesoxylon Lomaxi* and *M. poroxyloides. Ann. Bot.* vol. XXVI. p. 1011.

—— (11) The Evolution of Plants. *London.*

—— (14) On *Medullosa pusilla. Proc. R. Soc.* vol. LXXXVII. p. 221.

—— (15) The Heterangiums of the British Coal Measures. *British Assoc.* (*Manchester Meeting* 1915). (Abstract.)

Scott, D. H. and **E. C. Jeffrey.** (14) On Fossil Plants showing structure, from the base of the Waverley shale of Kentucky. *Phil. Trans. R. Soc.* vol. 205, p. 315.

Scott, D. H. and **A. J. Maslen.** (07) The structure of the Palaeozoic seeds *Trigonocarpus Parkinsoni* Brongn. and *T. Oliveri* sp. nov. *Ann. Bot.* vol. XXI. p. 89.

—— (10) On *Mesoxylon,* a new genus of Cordaitales. *Ibid.* vol. XXIV. p. 236.

Sellards, E. H. (03) *Codonotheca,* a new type of spore-bearing organ from the Coal Measures. *Amer. Journ. Sci.* vol. XVI. p. 87.

—— (07) Notes on the spore-bearing organ *Codonotheca* and its relationship with the Cycadofilices. *New Phyt.* vol. VI. p. 175.

Sernander, R. (93) Die Einwanderung der Fichte in Skandinavien. *Engler's Bot. Jahrb.* Bd. XV. p. 1.

Seward, A. C. (90) *Tylodendron* Weiss and *Voltzia heterophylla* Brong. *Geol. Mag.* vol. VII. [3], p. 218.

—— (93) On the genus *Myeloxylon. Ann. Bot.* vol. XXV. p. 1.

—— (94) On *Rachiopteris Williamsoni* sp. nov., a new Fern from the Coal-Measures. *Ibid.* vol. VIII. p. 207.

—— (96) A new species of Conifer, *Pinites Ruffordi,* from the English Wealden formation. *Journ. Linn. Soc.* vol. XXXII. p. 417.

—— (96[2]) Notes on the geological history of Monocotyledons. *Ann. Bot.* vol. X. p. 205.

—— (97) On *Encephalartos Ghellinckii* Lem., a rare Cycad. *Proc. Camb. Phil. Soc.* vol. IX. p. 340.

—— (97[2]) A contribution to our knowledge of *Lyginodendron. Ann. Bot.* vol. XI. p. 65.

—— (97[3]) On the association of *Sigillaria* and *Glossopteris* in S. Africa. *Quart. Journ. Geol. Soc.* vol. LIII. p. 315.

—— (97[4]) On the leaves of *Bennettites. Proc. Camb. Phil. Soc.* vol. IX. p. 273.

—— (00) Notes on some Jurassic plants in the Manchester Museum. *Mem. Proc. Manchester Lit. Phil. Soc.* vol. XLIV. pt. iii. No. 8.

—— (03) On the occurrence of *Dictyozamites* in England, with Remarks on European and Eastern Mesozoic Floras. *Quart. Journ. Geol. Soc.* vol. LIX. p. 217.

—— (06) Notes on Cycads. *Proc. Camb. Phil. Soc.* vol. XIII. pt. v. p. 299.

—— (07) Permo-Carboniferous plants from Kashmir. *Rec. Geol. Surv. India,* vol. XXXVI. pt. i. p. 57.

Seward, A. C. (07²) Jurassic plants from Caucasia and Turkestan. *Mém. com. géol. St Pétersbourg* [N.S.], Livr. 38.

—— (10) Article "Gymnosperms," *Encyclop. Brit.* edit. 11, vol. xii. *Cambridge*.

—— (11) Jurassic plants from Chinese Dzungaria. *Mém. com. geol. St Pétersbourg* [N.S.], Livr. 75.

—— (11²) The Jurassic Flora of Sutherland. *Trans. R. Soc. Edinb.* vol. xlvii. pt. iv. p. 643.

—— (11³) Links with the Past in the Plant world. *Cambridge*.

—— (11⁴) The Jurassic Flora of Yorkshire. The *Naturalist*, January, 1911.

—— (12) Mesozoic plants from Afghanistan and Afghan-Turkestan. *Mem. Geol. Surv. India, Pal. Ind.* [N.S.], vol. iv. mem. No. 4.

—— (12²) A petrified *Williamsonia* from Scotland. *Phil. Trans. R. Soc.* vol. cciii. p. 101.

—— (12³) Jurassic plants from Amurland. *Mém. com. geol. St Pétersbourg* [N.S.], Livr. 81.

—— (13) A contribution to our knowledge of Wealden Floras, with special reference to a collection of plants from Sussex. *Quart. Journ. Geol. Soc.* vol. lxix. p. 85.

—— (14) Antarctic Fossil plants. *British Antarctic (Terra Nova) Expedit.* 1910. *Nat. Hist. Report, Geology,* vol. i. No. 1. *London*.

—— (14²) Wealden Floras. *Hastings and East Sussex Naturalist,* vol. ii. No. 3, p. 126.

—— (17) H. H. W. Pearson, F.R.S., Sc.D. (Cambridge) *Ann. Bot.* vol. xxxi. p. 1.

Seward, A. C. and N. Bancroft. (13) Jurassic Plants from Cromarty and Sutherland, Scotland. *Trans. R. Soc. Edinb.* vol. xlviii. pt. iv. p. 867.

Shaw, F. J. F. (08) A contribution to the anatomy of *Ginkgo biloba*. *New Phyt.* vol. vii. p. 85.

—— (09) The seedling structure of *Araucaria Bidwillii*. *Ann. Bot.* vol. xxiii. p. 321.

Shaw, W. R. (96) Contribution to the life-history of *Sequoia sempervirens*. *Bot. Gaz.* vol. xxi. p. 332.

Shirley, J. (98) Additions to the Fossil Flora of Queensland *Geol. Surv. Bull.* No. 7. *Brisbane*.

Siebold, P. F. von. (70) Flora Japonica, 1842–70. *Leipzig*.

Sifton, H. B. (15) On the occurrence and significance of "bars" or "rims" of Sanio in the Cycads. *Bot. Gaz.* vol. lx. p. 400.

Sigrianski, A. (13) Quelques observations sur l'*Ephedra helvetica* Mey. Univ. Genève. *Faculté des sciences, Prof. Chodat* [8], Fasc. x.

Sinnott, E. W. (09) Paracedroxylon, a new type of Araucarian wood. *Rhodora*, vol. ii. No. 129, p. 165.

—— (11) Some features of the anatomy of the foliar bundle. *Bot. Gaz.* vol. li. p. 258.

—— (13) The morphology of the reproductive structures in the Podocarpineae. *Ann. Bot.* vol. xxvii. p. 39.

512 LIST OF WORKS

Smith, F. G. (07) Morphology of the trunk and development of the microsporangium of Cycads. *Bot. Gaz.* vol. XLIII. p. 187.

Smith, J. E. (1797) Characters of a new genus of plants named *Salisburia. Trans. Linn. Soc.* vol. III. p. 330.

Solereder, H. (99) Systematische Anatomie der Dicotyledonen. *Stuttgart.*
—— (08) *Ibid.* Ergänzungsband.

Solms-Laubach, Graf zu. (84) Die Coniferenformen des Deutschen Kupferschiefers und Zechsteins. *Pal. Abhand.* (*Dames and Kayser*), Bd. II. Heft ii. p. 81.
—— (90) Die Sprossfolge der *Stangeria* und der übrigen Cycadeen. *Bot. Zeit. Jahrg.* XLVIII.
—— (91) On the Fructification of *Bennettites Gibsonianus* Carr. *Ann. Bot.* vol. V. p. 419.
—— (92) See Capellini and Solms-Laubach.
—— (93) Ueber die in den Kalksteinen des Kulm von Glätzisch-Falkenberg in Schlesien enthaltenen Structur bietenden Pflanzenreste. *Bot. Zeit.* Jahrg. LI. p. 197.
——— (97) Ueber *Medullosa Leuckarti. Ibid.* Heft X. p. 175.
—— (99) Das Auftreten und die Flora der rhätischen Kohlen-schichten von La Ternera (Chili). *Neues Jahrb. Min. Beilage*, Bd. XII. p. 581.
—— (04) Die strukturbietenden Pflanzengesteine von Franz Josefs Land. *K. Svensk. Vetenskapsakad. Hand.* Bd. XXXVII. No. 7, p. 3.
—— (06) Die Bedeutung der Palaeophytologie für die systematische Botanik. *Mitt. Philomath. Ges. in Elsass-Lothringen*, Bd. III. p. 353.
—— (10) Über die in den Kalksteinen des Culm von Glätzisch-Falkenberg in Schlesien erhaltenen structurbietenden Pflanzenreste. IV. *Völkelia refracta, Steloxylon Ludwigii. Zeitsch. Bot.* Jahrg. II. Heft viii. p. 529.

South, F. W. and R. H. Compton. (08) Notes on the anatomy of *Dioon edule* Lind. *New Phyt.* vol. VII. p. 222.

Spiess, Karl von. (02) *Ginkgo, Cephalotaxus,* und die Taxaceen. Eine phylogenetische Studie. *Öster. Bot. Zeitsch.* Jahrg. LII. p. 432.
—— (03) *Ibid.* Jahrg. LIII. p. 1.

Sprague, T. A. (16) Dioncophyllum. *Kew Bulletin,* No. 4, p. 89.

Sprecher, A. (07) Le *Ginkgo biloba* L. *Genève.*

Squinabol, S. (92) Contribuzioni alla flora fossile dei terreni terziarii della Liguria. *Gênes,* 1889–92.

Stapf, O. (89) Die Arten der Gattung *Ephedra. Denksch. K. Akad. Wiss. Wien*, Bd. LVI.
—— (96) On the Flora of Mount Kinabalu in North Borneo. *Trans. Linn. Soc.* [2], vol. IV. p. 69.
—— (14) *Encephalartos Hildebrandtii. Bull. Miscell. Information, R. Bot. Gard. Kew,* No. 10, p. 386.

Starr, Anna M. (10) The microsporophylls of *Ginkgo. Bot. Gaz.* vol. XLIX. p. 51.

Staub, M. (85) *Pinus palaeostrobus* Etting. in der fossilen Flora Ungarns. *Természetrajzi Füzetek,* IX. p. 47.

Staub, M. (96) Die Fossilen Ctenis-Arten und *Ctenis hungarica* n. sp. *Földtani Közlöny,* vol. XXVI.

Stefani, C. de. (01) Flore Carbonifere e Permiane della Toscana. *R. Inst. Stud. sup. pratici e di perfezionamento in Firenze.*

Stenzel, G. (76) Beobachtungen an durchwachsenden Fichtenzapfen. *Nov. Act. Leop. Carol.* Bd. XXXVIII.

—— (88) See Goeppert and Stenzel.

Sterzel, J. T. (83) Ueber *Dicksonia Pluckeneti* (Schloth.). *Bot. Cent.* Bd. XIII. p. 282.

—— (00) Gruppe verkieselter Araucariten Stämme. *Bericht Naturwiss. Ges. Chemnitz,* Bd. XIV.

—— (03) Ein verkieselter Riesenbaum aus dem Rothliegenden von Chemnitz. *Ibid.* Bd. XV. p. 23.

—— (07) Die Karbon- und Rotliegendfloren im Grossherzogtum Baden. *Mitt. Badisch. geol. Landes.* Bd. V. Heft ii. p. 347.

—— (12) Der "verkieselte Wald" im Garten des König Albert Museums und des Orth-Denkmal in Chemnitz-Hilbersdorf. *Bericht Naturwiss. Ges. Chemnitz,* Bd. XVIII. p. 51.

Stiles, W. (08) The anatomy of *Saxegothaea conspicua* Lind. *New Phyt.* vol. VII. p. 209.

—— (12) The Podocarpeae. *Ann. Bot.* vol. XXVI. p. 443.

Stirling, J. (00) Notes on the Fossil Flora of South Gippsland Jurassic beds. *Rep. on the Vict. coal-fields,* No. 7. (*Dpt. Mines, Victoria.*)

Stokes and Webb. (24) Descriptions of some fossil vegetables of the Tilgate Forest in Sussex. *Trans. Geol. Soc.* [2], vol. II. p. 421.

Stopes, Marie C. (03) On the leaf of *Cordaites. New Phyt.* vol. II. p. 91.

—— (04) Beiträge zur Kenntniss der Fortpflanzungsorgane der Cycadeen. *Flora,* Bd. XCIII. Heft iv. p. 435.

—— (05) On the double nature of the Cycadean integument. *Ann. Bot.* vol. XIX. p. 561.

—— (07) The Flora of the Inferior Oolite of Brora (Sutherland). *Quart. Journ. Geol. Soc.* vol. LXIII. p. 375.

—— (10) Adventitious budding and branching in *Cycas. New Phyt.* vol. IX. p. 235.

—— (10²) The internal anatomy of *Nilssonia orientalis. Ann. Bot.* vol. XXIV. p. 389.

—— (11) The Dragon tree of the Kentish Rag, with remarks on the treatment of imperfectly petrified wood. *Geol. Mag.* [5], vol. VIII. p. 55.

—— (11²) The name of the Dragon tree. *Ibid.* p. 468.

—— (11³) A reply to Prof. Jeffrey's article on *Yezonia* and *Cryptomeriopsis. Ann. Bot.* vol. XXV. p. 269.

—— (11⁴) On the true nature of the Cretaceous plant *Ophioglossum granulatum* Hr. *Ann. Bot.* vol. XXV. p. 903.

—— (14) The "Fern Ledges" Carboniferous Flora of St John, New Brunswick. *Dpt. Mines, Geol. Surv., Canada, Mem.* 41, No. 38, *Geol. Ser. Ottawa.*

514   LIST OF WORKS

Stopes, Marie C. (14²) A new *Araucarioxylon* from New Zealand. *Ann. Bot.* vol. xxviii. p. 341.

—— (15) Catalogue of the Mesozoic Plants in the British Museum (Nat. Hist.). The Cretaceous Flora. Pt. ii. Lower Greensand (Aptian) plants of Britain. *London.*

—— (16) An early type of the Abietineae (?) from the Cretaceous of New Zealand. *Ann. Bot.* vol. xxx. p. 111.

Stopes, M. C. and K. Fujii. (10) Studies on the structure and affinities of Cretaceous Plants. *Phil. Trans. R. Soc.* vol. cci. p. 1.

Stopes, M. C. and E. M. Kershaw. (10) The anatomy of Cretaceous Pine leaves. *Ann. Bot.* vol. xxiv. p. 395.

Stopes, M. C. and D. M. S. Watson. (08) The present distribution and origin of the calcareous concretions in coal-seams known as "coal-balls." *Phil. Trans. R. Soc.* vol. cc. p. 167.

Strasburger, E. (66) Ein Beitrag zur Entwickelungsgeschichte der Spaltöffnungen. *Pringsheim's Jahrb.* Bd. v. p. 297.

—— (91) Ueber den Bau und die Verrichtungen der Leitungsbahnen in den Pflanzen. *Histol. Beit.* Heft iii. Jena.

Strübing, O. (88) Die Vertheilung der Spaltöffnungen bei den Coniferen. (*Inaug. Diss. Univ. Königsberg.*)

Stur, D. (68) Beiträge zur Kenntniss der geologischen Verhältnisse der Umgegend von Raibl und Kaltwasser. *Jahrb. K.k. geol. Reichs. Wien,* Bd. xviii. p. 71.

—— (77) Die Culm Flora. Heft ii. *Wien.*

Suzuki, Y. (10) On the structure and affinities of two new Conifers and a new fungus from the Upper Cretaceous of Hokkaido (Yezo). *Bot. Mag. Tokyo,* vol. xxiv. No. 284, p. 183.

Sykes, M. G. (10) (See also Thoday, M. G.) The anatomy of *Welwitschia mirabilis* Hook. f. in the seedling and adult stages. *Trans. Linn. Soc.* vol. vii. pt. ii. p. 327.

—— (10²) The anatomy and morphology of the leaves and inflorescences of *Welwitschia mirabilis. Phil. Trans. R. Soc.* vol. cci. p. 179.

Takeda, H. (13) A theory of transfusion-tissue. *Ann. Bot.* vol. xxvii. p. 359.

—— (13²) Some points in the anatomy of the leaf of *Welwitschia mirabilis. Ibid.* vol. xxvii. p. 347.

Tassi, F. (05) Ricerche comparate sul tessuto midollare delle Conifere.... *Bull. Lab. ed Ort. bot. Siena,* vii.

Tate, R. (67). On the secondary fossils from S. Africa. *Quart. Journ. Geol. Soc.* vol. xxiii. p. 130.

Tenison Woods, J. E. (83) On a species of *Brachyphyllum* from Mesozoic coal beds, Ipswich, Queensland. *Proc. Linn. Soc. N.S.W.* vol. vii. p. 659.

Thibout, E. (96) Recherches sur l'appareil mâle des Gymnospermes. *Lille.*

Thiselton-Dyer, W. T. (72) On some Coniferous remains from the Lithographic stone of Solenhofen. *Geol. Mag.* vol. ix. p. 1.

Thiselton-Dyer, W. T. (01) The carpophyll of *Encephalartos*. *Ann. Bot.* vol. xv. p. 548.

—— (01²) Persistence of the leaf-traces in *Araucaria*. *Ibid.* vol. xv. p. 547.

—— (02) Enumeration of the plants known from China proper, Formosa, etc. Cycadaceae. *Journ. Linn. Soc.* vol. xxvi. p. 559.

Thoday, Mary G. (Sykes, M. G.). (11) The female inflorescence and ovules of *Gnetum africanum*, with notes on *G. secundum*. *Ann. Bot.* vol. xxv. p. 1101.

Thoday, M. G. and Emily M. Berridge. (12) The anatomy and morphology of the Inflorescences and flowers of *Ephedra*. *Ann. Bot.* vol. xxvi. p. 953.

Thomas, F. (66) Zur vergleichenden Anatomie der Coniferen-Laubblätter. *Pringsheim's Jahrb.* Bd. iv. p. 23.

Thomas, H. Hamshaw. (11) The Jurassic Flora of Kamenka in the district of Isium. *Mém. com. géol. St. Pétersbourg* [N.S.], Livr. 71.

—— (12) Note on the occurrence of *Whittleseya elegans* Newb. in Britain. *Palaeobot. Zeitsch.* Bd. i. Heft i. p. 46.

—— (13) On some new and rare Jurassic plants from Yorkshire: *Eretmophyllum*, a new type of Ginkgoalean leaf. *Proc. Camb. Phil. Soc.* vol. xvii. pt. iii. p. 256.

—— (13²) The Fossil Flora of the Cleveland district of Yorkshire. I. The Flora of the Marske quarry. *Quart. Journ. Geol. Soc.* vol. lxix. p. 223.

—— (15) On some new and rare Jurassic plants from Yorkshire: The male flower of *Williamsonia gigas*. *Proc. Camb. Phil. Soc.* vol. xviii. pt. iii. p. 105.

—— (15²) On *Williamsoniella*, a New Type of Bennettitalean Flower. *Phil. Trans. R. Soc.* vol. 207, p. 113.

Thomas, H. H. and Nellie Bancroft. (13) On the cuticles of some recent and fossil Cycadean fronds. *Trans. Linn. Soc.* vol. viii. pt. v. p. 155.

Thompson, W. P. (10) The origin of ray-tracheids in the Coniferae. *Bot. Gaz.* vol. l. p. 101.

—— (12) Ray-tracheids in *Abies*. *Ibid.* vol. liii. p. 53.

—— (12²) The anatomy and relationship of the Gnetales. I. The genus *Ephedra*. *Ann. Bot.* vol. xxvi. p. 1077.

—— (12³) The structure of the stomata of certain Cretaceous Conifers. *Bot. Gaz.* vol. liv. p. 63.

—— (16) The Morphology and Affinities of Gnetum. *Amer. Journ. Bot.* vol. iii. p. 135.

Thomson, R. B. (07) The Araucarineae, a Protosiphonogamic method of Fertilisation. *Science* [N.S.], vol. xxv. p. 272.

—— (09) On the pollen of *Microcachrys tetragona*. *Bot. Gaz.* vol. xlvii. p. 26.

—— (13) On the comparative anatomy and affinities of the Araucarineae. *Phil. Trans. R. Soc.* vol. cciv. p. 1.

516   LIST OF WORKS

Thomson, R. B. (14) The spur-shoot of the Pines. *Bot. Gaz.* vol. LVII. p. 362.

Thomson, R. B. and A. E. Allin. (12) Do the Abietineae extend to the Carboniferous? *Bot. Gaz.* vol. LIII. p. 339.

Thomson, Mary R. H. (16) A note on the wood of *Gnetum Gnemon. Ann. Bolus Herb.* vol. II. pt. ii. p. 81.

Tison, A. (09) Sur le *Saxegothaea conspicua* Lind. *Mém. soc. Linn. Normand.* vol. XXIII. p. 139.

—— (12) Sur la persistance de la nervation dichotomique chez les Conifères. *Bull. soc. Linn. Normandie* [vi], vol. IV. p. 30.

—— (12²) See Lignier and Tison.

Tupper, W. W. (11) Notes on *Ginkgo biloba. Bot. Gaz.* vol. LI. p. 374.

Tuzson, J. (09) Monographie der fossilen Pflanzenreste der Balatonsee-gegend. *Result. der Wiss. Erforsch. des Balatonsees,* Bd. I. Teil i. *Budapest.*

—— (09²) Beiträge zur fossilen Flora Ungarns. *Mitt. Jahrb. K. Ung. Geol. Reichs.* Bd. XXI. Heft viii. p. 233.

—— (14) Beiträge zur fossilen Flora Ungarns. *Mitt. Jahrb. K. Ungarisch. Geol. Reichs.* Bd. XXI. Heft viii.

Unger, F. (45) Synopsis Plantarum Fossilium. *Leipzig.*

—— (47) Chloris Protogaea. *Leipzig.*

—— (49) Einige interessante Pflanzenabdrücke aus der K. Petre-factensammlung in München. *Bot. Zeit.* p. 345.

—— (51) Die Fossile Flora von Sotzka. *Denksch. K. Akad. Wiss. Wien,* Bd. II. Abt. ii. p. 131.

—— (52) Ueber einige fossile Pflanzen aus den lithographischen Schiefer von Solenhofen. *Palaeontograph.* Bd. II. p. 251.

—— (52²) Iconographia plantarum fossilium. Abbildungen und Beschrei-bungen fossiler Pflanzen. *Denksch. K. Akad. Wiss. Wien,* Bd. IV. p. 73.

—— (54) Zur Flora des Cypridinenschiefers. *Sitzber. K. Akad. Wiss. Wien,* Bd. XII. p. 595.

—— (54²) Jurassische Pflanzenreste. *Palaeontograph.* Bd. IV. p. 39.

—— (59) Der versteinerte Wald bei Cairo. *Sitzber. K. Akad. Wiss. Wien,* Bd. XXXIII. p. 209.

—— (67) Die fossile Flora von Kumi. *Denksch. K. Akad. Wiss. Wien,* Bd. XXVII. p. 27.

Veitchs Manual of the Conifers. (00) *London.*

Velenovský, J. (81) Die Flora aus den Ausgebrannten Tertiären Letten von Vršovic bei Laun. *Abh. böhm. Ges.* [vi], Bd. II.

—— (87) Neue Beiträge zur Kenntniss der Pflanzen des böhmischen Cenomans. *Sitzber. K. böhm. Ges. Wiss.* Jahrg. 1886, p. 633.

—— (88) Ueber einige neue Pflanzenformen der böhmischen Kreide-formation. *Ibid.* Jahrg. 1887.

—— (89) Květena Českého Cenomanu. *Abh. K. böhm. Ges. Wiss.* [vii], Bd. III.

—— (07) Vergleichende Morphologie der Pflanzen. Teil ii. *Prag.*

Vernon, R. D. (12) On the geology and palaeontology of the Warwick-shire coalfield. *Quart. Journ. Geol. Soc.* vol. LXVIII. p. 587.

Vetters, K. L. (84) Die Blattsteile der Cycadeen. (Inaug. Diss.) *Leipzig.*

Vierhapper, F. (10) Entwurf eines neuen Systems der Coniferen. *Jena.*

Walkom, A. B. (17) Mesozoic Floras of Queensland. *Queensland Geol. Surv. Publications,* No. 259.

Ward, L. F. (85) Synopsis of the Laramie group. *6th Ann. Rep. U. S. Geol. Surv.*

—— (87) Types of the Laramie Flora. *Bull. U. S. Geol. Surv.* XXXVII.

—— (88) The geographical distribution of Fossil Plants. *8th Ann. Rep. U. S. Geol. Surv.* (1887–88).

—— (94) Fossil Cycadean trunks of North America, with a revision of the genus *Cycadeoidea* Buckland. *Proc. Biol. Soc. Washington,* vol. IX. p. 75.

—— (94²) Recent discoveries of Cycadean trunks in the Potomac formation of Maryland. *Bull. Torrey Bot. Club,* vol. XXI. No. vii. p. 291.

—— (94³) The Cretaceous Rim of the Black Hills. *Journ. Geol.* vol. II. No. iii. p. 250.

—— (96) Some analogies in the Lower Cretaceous of Europe and America. *16th Ann. Rep. U. S. Geol. Surv.*

—— (98) Descriptions of the species of *Cycadeoidea* or fossil Cycadean trunks thus far determined from the Lower Cretaceous rim of the Black Hills. *Proc. U. S. Nat. Mus.* vol. XXI. p. 21.

—— (00) Elaboration of the Fossil Cycads in the Yale Museum. *Amer. Journ. Sci.* vol. X. p. 327.

—— (00²) Description of a new genus and 20 new species of fossil Cycadean trunks from the Jurassic of Wyoming. *Proc. Washington Acad. Sci.* vol. I. p. 253.

—— (00³) Report on the petrified forests of Arizona. · *Depart. of the Interior, Washington.*

—— (04) A famous fossil Cycad. *Amer. Journ. Sci.* vol. XVIII. p. 40.

Warming, E. (77) Recherches et remarques sur les Cycadées. *Oversigt K. D. Vidensk. Selsk. Forh. (Copenhagen),* p. 16.

Warren, E. (12) On some specimens of fossil woods in the Natal Museum. *Ann. Natal Mus.* vol. II. pt. iii. p. 345.

Webber, H. J. (01) Spermatogenesis and fecundation of *Zamia. U. S. Dpt. Agric. Bur. Plant Industry, Bull.* No. ii.

Weiss, C. E. (72) Fossile Flora der jüngsten Steinkohlen-formation und des Rothliegenden im Saar-Rhein-Gebiet. *Bonn,* 1869–72.

—— (74) Note in *Zeitsch. Deutsch. Geol. Ges.* Bd. XXVI. p. 616.

—— (79) Die Flora des Rothliegenden von Wünschendorf bei Lauban in Schlesien. *Abh. geol. specialkarte von Preussen,* Bd. III. Heft i. p. 1.

Weiss, F. E. (12) Report of the 80th meeting (Portsmouth) of the British Assoc. for the Advancement of Science, p. 550.

Weiss, F. E. (13) The root-apex and young root of *Lyginodendron*. *Mem. Proc. Manchester Lit. Phil. Soc.* vol. LVII. pt. iii.

—— (13²) A Tylodendron-like fossil. *Ibid.* vol. LVII. pt. iii.

Wernham, H. F. (11) Floral Evolution; with particular reference to the sympetalous Dicotyledons. *New Phyt.* vol. x. p. 73.

Wettstein, R. V. (90) Die Omorika-Fichte, *Picea omorica* (Panc.). *Sitzber. K. Akad. Wiss. Wien*, Bd. XCVIII. Abt. i. p. 503.

—— (11) Handbuch der systematischen Botanik. (Edit. ii.) *Leipzig.*

Wherry, E. T. (12) Silicified wood from the Triassic of Pennsylvania. *Proc. Acad. Sci. Philadelphia*, vol. LXIV. pt. ii. p. 366.

—— (16) Two new fossil plants from the Triassic of Pennsylvania. *Proc. U.S. Nat. Mus.* vol. LI. p. 327.

White, D. (90) On Cretaceous Plants from Martha's Vineyard. *Amer. Journ. Sci.* vol. XXXIX. p. 93.

—— (01) The Canadian species of the genus *Whittleseya* and the systematic relations. *The Ottawa Naturalist*, vol. xv. No. iv. p. 98.

White, D. and **C. Schuchert.** (98) Cretaceous series of the West coast of Greenland. *Bull. Geol. Soc. America*, vol. IX. p. 343.

Wieland, G. R. (99) A study of some American Fossil Cycads. Pt. ii. The Leaf structure of *Cycadeoidea*. *Amer. Journ. Sci.* vol. VII. p. 305.

—— (02) Notes on living Cycads. I. On the Zamias of Florida. *Ibid.* vol. XIII. p. 331.

—— (06) American Fossil Cycads. *Washington.*

—— (08) Historic Fossil Cycads. *Amer. Journ. Sci.* vol. XXV. p. 93.

—— (08²) Two new Araucarias from the Western Cretaceous. *Geol. Surv. South Dakota.*

—— (09) The Williamsonias of the Mixteca Alta. *Bot. Gaz.* vol. XLVIII. p. 427.

—— (11) On the Williamsonia Tribe. *Amer. Journ. Sci.* vol. XXXII. p. 433.

—— (11²) A study of some American Fossil Cycads. Pt. v. *Ibid.* vol. XXXII. p. 133.

—— (12) *Ibid.* pt. vi. On the smaller flower-buds of *Cycadeoidea*. *Ibid.* vol. XXXIII. p. 73.

—— (13) The Liassic Flora of the Mixteca Alta of Mexico, its composition, age, and source. *Ibid.* vol. XXXVI. p. 251.

—— (14) A study of some American Fossil Cycads. Pt. vii. Further notes on disk structure. *Ibid.* vol. XXXVIII. p. 117.

—— (16) American Fossil Cycads. Vol. II. Taxonomy. *Washington.*

Wiesner, J. (03) Die Rohstoffe des Pflanzenreiches. Bd. II. *Leipzig.*

Wild, G. (00) On new and interesting features in *Trigonocarpus olivaeformis*. *Trans. Manchester Geol. Soc.* vol. XXVI. pt. xv. p. 434.

Williamson, W. C. (40) On the Distribution of Fossil remains on the Yorkshire coast from the Lower Lias to the Bath Oolite inclusive. *Trans. Geol. Soc.* [2], vol. v. p. 223.

Williamson, W. C. (51) On the structure and affinities of the plants known as Sternbergiae. *Manchester Lit Phil. Soc.* [2], vol. IX. p. 340.

—— (69) On the structure and affinities of some exogens from the Coal Measures. *Monthly Micros. Journ.* vol. II. p. 66.

—— (70) Contributions towards the history of *Zamia gigas* L. and H. *Trans. Linn. Soc.* vol. XXVI. p. 663.

—— (72) Notice of further researches among the plants of the Coal Measures. *Proc. R. Soc.* vol. XX. p. 435.

—— (72²) On the structure of the Dicotyledons of the Coal Measures. *Rep. 41st meeting (Edinburgh) of the Brit. Assoc.* p. 111.

—— (74) On the Organisation of the Fossil Plants of the Coal Measures VI. *Phil. Trans. R. Soc.* vol. CLXII. p. 675.

—— (76) On some fossil seeds from the Lower Carboniferous beds of Lancashire. *Rep. 45th meeting (Bristol) of the Brit. Assoc.* p. 159.

—— (86) On the morphology of *Pinites oblongus*. *Mem. Proc. Manchester Lit. Phil. Soc.* vol. X. [3], p. 189.

—— (87) On the Organisation of the Fossil Plants of the Coal Measures. XIII. *Phil. Trans. R. Soc.* vol. 178.

—— (90) *Ibid.* XVII. *Ibid.* vol. CLXXXI. p. 89.

Wiliamson, W. C. and **D. H. Scott.** (94) The root of *Lyginodendron oldhamium* Will. *Proc. R. Soc.* vol. LVI. p. 128.

—— (95) Further observations on the Organisation of the Fossil Plants of the Coal Measures. III. *Lyginodendron and Heterangium*. *Phil. Trans. R. Soc.* vol. CLXXXVI. p. 703.

Wills, Lucy. (14) Plant cuticles from the Coal Measures. *Geol. Mag.* [6], vol. I. p. 385.

Wills, L. T. (10) The fossiliferous Lower Keuper rocks of Worcestershire. *Proc. Geol. Assoc.* vol. XXI. p. 249.

Winkler, C. (72) Zur Anatomie von *Araucaria brasiliensis*. *Bot. Zeit.* Jahrg. XXX. p. 581.

Witham, H. (30) On the vegetable fossils from Lennel Braes near Coldstream, upon the banks of the Tweed in Berwickshire. *Phil. Mag.* vol. VIII. p. 16.

—— (31) Observations on fossil vegetables accompanied by representations of their internal structure as seen through the microscope. *Edinb. Journ. Sci.* vol. V. p. 183.

Worsdell, W. C. (96) The anatomy of the stem of *Macrozamia* compared with that of other genera of Cycadeae. *Ann. Bot.* vol. X. p. 601.

—— (97) On transfusion-tissue, its origin and function in the leaves of Gymnospermous plants. *Trans. Linn. Soc.* vol. V. [2], p. 301.

—— (98) The vascular structure of the sporophylls of the Cycadaceae. *Ann. Bot.* vol. XII. p. 203.

—— (98²) The comparative anatomy of certain genera of the Cycadaceae. *Journ. Linn. Soc.* vol. XXXIII. p. 437.

—— (99) Observations on the vascular system of the female flowers of Conifers. *Ann. Bot.* vol. XIII. p. 527.

Worsdell, W. C. (00) The comparative anatomy of certain species of *Encephalartos*. *Trans. Linn. Soc.* vol. v. pt. xiv. p. 445.

—— (00²) The affinities of the Mesozoic fossil *Bennettites Gibsonianus* Carr. *Ann. Bot.* vol. xiv. p. 717.

—— (01) Contributions to the comparative anatomy of the Cycadaceae. *Trans. Linn. Soc.* vol. vi. pt. ii. p. 109.

—— (01²) The vascular structure of the flowers of the Gnetaceae. *Ann. Bot.* vol. xv p. 766.

—— (04) The structure and morphology of the ovule. *Ann. Bot.* vol. xviii. p. 57.

—— (05) Fasciation, its meaning and origin. *New Phyt.* vol. iv. p. 55.

—— (06) The structure and origin of the Cycadaceae. *Ann. Bot.* vol. xx. p. 129.

Wright, W. B. (14) The Quaternary Ice Age. *London.*

Yabe, H. (13) Mesozoische Pflanzen von Omoto. *Sci. Rep. of the Tokoku Imp. Univ.* (ser. ii. Geol.), Bd. i. Heft iv. p. 57.

Yasui, K. (17) A Fossil Wood of *Sequoia* from the Tertiary of Japan. *Ann. Bot.* vol. xxxi. p. 101.

Yates, J. (55) Notice of *Zamia gigas*. *Proc. Yorks. Phil. Soc.* vol. i. p. 37.

Yokoyama, M. (94) Mesozoic plants from Kozuke, Kii, Awa, and Tosa. *Journ. Coll. Sci. Imp. Univ. Japan*, vol. vii. pt iii. p. 201.

—— (05) Mesozoic plants from Nagato and Bitchu. *Ibid.* vol. xx.

Young, J. (69) *Trans. Nat. Hist. Soc. Glasgow*, vol. i. pl. iv. (No text.)

—— (76) Catalogue of the Western Scottish fossils. *Glasgow.*

Young, Mary S. (07) The male gametophyte of *Dacrydium*. *Bot. Gaz.* vol. xliv. p. 189.

—— (10) The morphology of the Podocarpineae. *Ibid.* vol. l. p. 81.

Zalessky, M. (05) Über Früchte aus den Unter Carbon-Ablagerungen des Mstabeckens in Nord Russland. *Bull. acad. Imp. sci. St. Pétersbourg*, tome xxii. No. iii. p. 113.

—— (05²) Notiz über die obercarbonische Flora des Steinkohlenreviers von Jantai in der südlichen Mandshurei. *Verhand. K. Russ. Min. Ges. St. Pétersbourg* [2], Bd. xlii. p. 385.

—— (09) Communication préliminaire sur un nouveau *Dadoxylon* provenant du Dévonien supérieur du bassin du Donetz. *Bull. acad. Imp. sci. St. Pétersbourg.*

—— (10) On the discovery of the calcareous concretions known as coal balls in one of the seams of the Carboniferous strata of the Donetz basin. *Ibid.* p. 477.

—— (11) Étude sur l'anatomie du *Dadoxylon Tchihatcheffi* Goepp. sp. *Mém. com. géol. St. Pétersb.* [N.S.], Livr. 68, p. 1.

—— (11²) Note préliminaire sur le *Coenoxylon Scotti*, nov. gen. et sp. *Études Paléobotaniques*, pt. i. *St. Petersburg.*

—— (12) Sur le *Cordaites aequalis* Goepp. sp. de Sibérie et sur son identité avec la *Noeggerathiopsis Hislopi* Bunb sp. de la Flore du Gondwana. *Mém. com. géol. St. Pétersb.* [N.S.] Livr. 86.

Zalessky, M. (12²) On the impressions of plants from the coal-bearing deposits of Sudženka, Siberia. *Bull. soc. Natural. Orel.* pt. IV.

—— (13) Flore Gondwanienne du Bassin de la Pétchora. I. Rivière Adzva. *Bull. soc. Oural. d'amis des Sci. Nat. à Ekatérinebourg,* vol. XXXIII.

Zang, W. (04) Die Anatomie der Kiefernadel und ihre Verwendung zur systematischen Gliederung der Gattung *Pinus.* (Diss. Inaug.) *Giessen.*

Zeiller, R. (78) Sur une nouvelle espèce de *Dicranophyllum. Bull. soc. géol. France* [3], vol. VI. p. 611.

—— (80) Note sur quelques plantes fossiles du terrain permien de la Corrèze. *Ibid.* vol. VIII. [3], p. 196.

—— (96) Le Marquis G. de Saporta, sa vie et ses travaux. *Bull. soc. géol. France* [3], vol. XXIV. p. 197.

—— (00) Sur les végétaux fossiles recueillis par M. Villiaume dans les gîtes charbonneux du Nord-ouest de Madagascar. *Compt. Rend.* Jun. 5.

—— (02) Nouvelles observations sur la flore fossile du bassin de Kousnetzk (Sibérie). *Ibid.* tome CXXXIV. p. 887.

—— (05) Sur les plantes Rhétiennes de la Perse recueillies par M. J. de Morgan. *Bull. soc. géol. France* [4], tome V. p. 190.

—— (05²) Sur quelques empreintes végétales de la formation charbonneuse supracrétacée des Balkans. *Ann. Mines.*

—— (08) See Douvillé and Zeiller.

—— (11) Note sur quelques végétaux infraliasiques des environs de Niort. *Bull. soc. géol. France* [4], tome XI. p. 321.

—— (11²) Sur une flore triasique découverte à Madagascar. *Compt. Rend.* vol. CLIII. p. 230.

—— (12) Sur quelques végétaux fossiles de la Grande Oolite de Marquise. *Bull. soc. acad. Boulogne-sur-mer,* tome IX. p. 5.

—— (N.D.) See Malloizel and Zeiller.

Zeiller, R. and P. Fliche. (04) Découverte de strobiles de *Sequoia* et de Pin dans le Portlandien des environs de Boulogne-sur-mer. *Compt. Rend.* tome CXXXVII. p. 1020.

Zeiller, R. and L. M. Vidal. (02) Sobre algunas impresiones vegetales del Kimeridgense de Santa María de Meyá. *Mem. Real Acad. Cienc. y Artes Barcelona,* vol. IV. No. 26.

Zigno, A. de. (53) Découverte d'une flore Jurassique analogue à celle de Scarborough dans les couches oolitiques des Alpes Vénitiennes. *Bull. soc. géol. France* [2], vol. X. p. 268.

—— (73–85) Flora fossilis formationis Oolithicae. Vol. II. *Padua,* 1873–85.

Zopf, W. (92) Ueber einige niedere Algenpilze (Phycomyceten) und eine neue Methode ihre Keime aus dem Wasser zu isoliren. *Abh. Naturforsch. Ges. Halle,* Bd. XVII. p. 79.

INDEX

(Volumes I.—IV.)

Fossil Genera

INDEX

(Volume IV.)

CAMBRIDGE: PRINTED BY J. B. PEACE, M.A., AT THE UNIVERSITY PRESS